PHYSICIANS AND SURGEONS IN GLASGOW

Physicians and Surgeons in Glasgow

*The History of the Royal College
of Physicians and Surgeons of Glasgow
1599–1858*

Johanna Geyer-Kordesch and Fiona Macdonald

THE HAMBLEDON PRESS
London and Rio Grande

Published by The Hambledon Press, 1999

102 Gloucester Avenue, London NW1 8HX (USA)
Po Box 162, Rio Grande, Ohio 45674 (USA)

ISBN 1 85285 186 4

© Johanna Geyer-Kordesch and Fiona Macdonald, 1999

A description of this book is available from the
British Library and from the Library of Congress

Typeset in Minion by Carnegie Publishing, Lancaster
Printed and bound in the UK on acid-free paper
by Cambridge University Press

Contents

Illustrations		vii
Introduction		ix
Acknowledgements		xiii
Glossary		xv
Abbreviations		xvii
1	Rights and Privileges	1
2	The Founders	37
3	Surgeons and Barbers	79
4	The Regulation of Practice	115
5	Enlightenment	153
6	Botany, Anatomy and Chemistry	193
7	Midwifery and General Practice	251
8	Corporate Medicine and the Hospitals	293
9	Time of Crisis	339
10	The Best Place to Study Medicine	397
Appendixes		417
Bibliography		433
Index		443

Illustrations

(Between Pages 174 and 175)

1. The City of Glasgow, *c.* 1820, by J. Clark
2. Broomielaw, Shipping, by J. Fleming
3. View of the Hunterian Museum, by J. Fleming
4. The Old Town's Hospital, *c.* 1849, by T. Fairbairn
5. The University of Glasgow, c. 1693. by John Slezer
6. The Second Faculty Hall (1791–1860), artist unknown
7. Cupping set, *c.* 1790
8. Amputation set, *c.* 1840–1850
9. William Cullen (1710–1790), Scottish School
10. William Hunter (1710–1790), by Allan Ramsay
11. Robert Watt (1774–1819), artist unknown
12. John Moore (1729–1799), copy by J. Barr
13. Joseph Black (1728–1799), by Sir Henry Raeburn
14. John Burns (1774–1850), by John Graham-Gilbert
15. Glasgow Royal Infirmary, by David Allen
16. Blythswood Place, St Vincent Street, by J. Fleming

Text Illustrations

1	Plan of the city of Glasgow in the Sixteenth Century	5
2	Peter Lowe (*c.* 1550–1610), artist unknown	36
3	Portrait of a cleft lip, from P. Lowe, *A Discourse of the Whole Art of Chyrurgerie* (London, 1612)	63
4	Portrait of a dry suture, from P. Lowe, *A Discourse of the Whole Art of Chyrurgerie* (London, 1612)	67
5	The bones of the human body, from Andreas Vesalius, *De humani corporis fabrica* (Basle, 1543)	73
6	View of patient's position for Lithotomy procedure	121
7	View of the Cathedral Infirmary from Five Mills, drawn and engraved by Joseph Swan	192
8	Plan of the city of Glasgow, 1775	250
9	Plan of the city of Glasgow, 1783	338

Introduction

In this volume we tell the story of one of the oldest medical incorporations in Britain. We also hope to set it in context, describing the distinctive pattern of medicine in the west of Scotland. This has meant departing from a narrow history and including a variety of other themes, as well as a chapter on scientific trends in the eighteenth century. Botany, anatomy, chemistry, gardens and museums, plus the significant involvement of medical men in the literature of the day, are touched upon. The Glasgow incorporation sometimes fades into the background because progress in medical knowledge and teaching at times only makes sense in the broader context of European medicine.

Glasgow grew from a small burgh with an ancient cathedral and university to a handsome city with wide streets, mansions and squares in the first decades of the nineteenth century. Civic and mercantile concerns intertwined with talent and hard-nosed competition in many areas, including that of the medical profession. By 1858, when the Medical Act irrevocably changed local jurisdiction, medicine in Glasgow had fought its way to educational and professional distinction. The city provided the definitive setting for the shaping of medicine, with local medical institutions sustaining the fabric of practice for over two and a half centuries. External influences also had a bearing. Political change and wars, travel abroad and the desire to adopt achievements seen elsewhere linked Glasgow with the scientific, intellectual, mercantile and ideological fortunes of other places.

In 1599 the Faculty of Physicians and Surgeons of Glasgow united in one body the two main branches of the medical profession. Collective power was typical of craft guilds, but the Faculty uniquely included physicians. Their stature as learned men who had graduated from universities normally separated them from other practitioners, but to this day the Glasgow Royal College retains this collegiality.

In its first century the Faculty primarily examined the skills of surgeons and barbers, excluding from practice, and intermittently prosecuting, those who did not obey their summons. The power of an incorporation lay in its ability to oversee healing practice and surgical techniques. Inevitably this created insiders and outsiders, corporate membership eventually coming to define standards. The rise of the surgeon soon dislodged the more mundane role of the barber in common curative practice. Separation then came relatively quickly, not least because of the higher social position of surgeons in Scotland (as opposed to England). This allowed the surgeons to court local physicians once more and secure their support in the eighteenth century.

The eighteenth century was a watershed. Remarkable for the renewal of university medical education, Glasgow fostered the early teaching of modern scientific subjects. The University of Glasgow inaugurated teaching in botany and anatomy in 1704 and 1714 respectively, both subjects being taught by Faculty surgeons. For surgeons to give lectures at a university on modern scientific subjects was an innovation that set the tone for the future. Shortly afterwards, in 1726, Edinburgh formed a similar, powerful, town-dominated medical faculty, transferring skills and research from continental sources to meld with Scottish medcine.

While the universities are usually credited with medical educational reform and the incorporations cast in the role of jealous rivals, the Faculty lent support to reform – indeed pointed the way – by jettisoning master-apprentice training in favour of lectures. The eighteenth century saw innovative medical teaching undertaken by surgeons as well as by the incumbents of medical chairs. The Faculty and the University both managed not only to teach the full medical curriculum by mid century, they also assembled a good library and specimen and teaching collections. A flourishing medical school was a civic achievement, its clientele recruited in the free trade of itinerant medical students and others in search of medical knowledge. The city now offered, in university and private lectures, the full medical curriculum, including botany, anatomy, surgery, midwifery and chemistry. Those attending were apprentices, students, midwives and irregulars, many moving to complete degrees or be licensed elsewhere. The oft-cited rivalry between the Faculty and the University did not materialise before 1800. The steep rise in educational demands by both created the healthy competition in teaching and the sciences advocated, among others, by Adam Smith.

Men like John Gordon, John Moore, John Burns and the Glasgow Hamiltons deserve more than the obscurity to which neglect has consigned them. Famous names, too, like William Smellie, William Cullen and William Hunter, need to be reassessed in the light of the formative years they spent in the west of Scotland. While Smellie and Hunter chose London as the place to test new approaches to medical learning and practice, their schooling and outlook was Scottish. Glasgow medical teaching shaped their work. Nor did they forget Glasgow in later years, helping a younger generation there to medical success. The Hamiltons, the brothers Robert and Thomas, and Thomas's son William, supplied the personal and professional ties that bound the Hunters, William Cullen, and the circle of University and Faculty medical men together. Cullen, another native son, was preeminent in his promotion of an internationally oriented medical and scientific outlook. He taught in a medical community that approved his and his colleagues' clinical teaching in the Town's Hospital as well as his advocacy of chemistry as a part of medicine.

The hospitals are part of the Faculty's history as is the rise of male midwifery. In both the incorporation influenced developments. The practice of consulting on staffing rotas and the medical management of hospitals began with the Town's Hospital and was a strong influence in later management battles at the Glasgow Royal Infirmary. Clinical teaching rose in this context. Male midwifery, on the other hand, was not hospital-based in Glasgow. Contrary to usage in other major cities, men learned an essentially female healing art through lectures and tuition on dummies, together with the home-visiting of poor 'deserving' women giving birth. The Faculty was one of the first bodies to regulate midwifery credentials (male and female). An awkward branch of medicine, midwifery ultimately helped to unite physicians and surgeons because of the overlap in training needed. In Glasgow, surgeons rather than physicians first taught midwifery.

The decade between 1790 and 1810 was one of great change. Regulation became a bogey as local regeneration clashed with the national movement to standardise qualifications. The Faculty raised its own standards successively, beginning in 1802. In 1813, however, it made a fatal misjudgement, prosecuting at law holders of Glasgow MD degrees. This changed its relations with the University. Much bad feeling and litigation ensued. This spate of infighting bred something else, however, that was more positive. From it emerged the first academic degree of *chirurgiae magister* (the CM). Moreover, Scottish practitioners soon saw the advantage in combining qualifications. In the 1830s and 1840s more and more of them were taking both a university MD and a surgeon's licence from the FPSG. Thus the general practitioner was esteemed as a well-qualified academic in Scotland. In England the divide between surgeon and physician remained marked, something that affected the politics of medical reform when it entered the arena of national debate.

A little over five years is a short time to research and write two books covering the 400 year history of the Faculty. This history of medicine in Glasgow, we hope, will motivate other historians to do further research. Fiona Macdonald wrote the first four chapters of this book and Johanna Geyer-Kordesch the remaining six. Dr Geyer-Kordesch was responsible for directing the whole research programme for this and its companion volume, *The Shaping of the Medical Profession*, written by Andrew Hull, which continues the history of the FPSG to the present day. The challenge to do justice to the history of Glasgow medicine has been a formidable one. In attempting to meet it, it has been our aim to work from the many valuable original sources, taking material from books only where necessary. For ease of comprehension among a medical readership, the Scots of the early modern records has been rendered in English in the first four chapters.

Johanna Geyer-Kordesch   Fiona Macdonald

Acknowledgements

We wish to thank the following people and institutions for their sustained support: the Wellcome Trust for major research grants and a Wellcome Trust Leave Fellowship for Dr Johanna Geyer-Kordesch in 1997–98; the Royal College of Physicians and Surgeons and three successive Presidents, Professor Sir Donald Campbell, Professor Norman Mackay, Mr Colin Mackay; the Principal of Glasgow University, Professor Sir Graeme Davies, for his clear understanding of how much research time books of this nature need; Professor Brian Whiting, Dean of the Faculty of Medicine, for keeping medical historians happy while deeply immersed in the concerns of contemporary medicine; the Thomas Reid Institute of Aberdeen University (now sadly disbanded) whose Director, Professor George Rousseau, and secretary, Alison Auld, enabled Dr Geyer-Kordesch to find the peace to write. Edna Robertson's fine eye for clear prose and short sentences lightened and sharpened the whole manuscript; her help with research and writing in the chapter on botany relating to the Sandyford gardens was greatly appreciated. Ann Cameron was the research assistant for work on the William Hamilton papers and her dedicated eye for detail immensely improved the midwifery chapter. Campbell Lloyd was unflagging in helping unravel the tangled threads of the chapter on medical reform and checking footnotes. Sibylle Naglis never lost hope throughout the arduous and painstaking stages of preparing the book for publication. Kirsten Paterson transcribed much needed notes at a critical time. Rae Meldrum, Johanna's secretary, was patience itself, calmly retyping and keeping order in all things. Many librarians and archivists were unflaggingly helpful: James Beaton, Carol Parry and Anna Forrest of the RCPSG Library; the staff of Glasgow University Library Special Collections and its Keeper, David Weston; the staff of Glasgow University Archives; the Librarians of the History of Medicine Library at the Wellcome Trust in London; the Librarians of the Royal College of Surgeons of Edinburgh; the Librarians of the Royal College of Physicians of Edinburgh, particularly Iain Milne; the Librarians of the Royal College of Surgeons of England, especially the late Iain Lyle; the Librarians of the Royal College of Physicians of England especially Geoffrey Davenport; the Archivists of PRONI in Belfast; Mike Barfoot, Lothian Health Service Archivist of the Edinburgh University

Library. To Martin Sheppard a special thank you for seeing this book through to publication.

Special and friendly help was also given by the core staff (Dr Marguerite Dupree and Dr Malcolm Nicolson) and members of the Wellcome Unit for the History of Medicine, Glasgow University, and the Unit's Administrator, Ann Mulholland. Professor Alan Smith of the Department of Modern History read the full manuscript close to our deadline and gave us the needed support to press onward: it was a crucial word of comfort during an exhausting time. Many thanks also to Helen Dingwall and Andrew Wear for their comments on chapters one to four and to Vivian Nutton for Chapter 4.

Most appropriately, however, we give heartfelt thanks to our families. James Paterson, Johanna's husband, was staunch in support and prepared to marry in the middle of all the pressures. Johanna's stepchildren, Judith, Jennifer, Kirsten and David, deserve love and praise for putting up with absences and with books and papers everywhere. Personal thanks also go to the Church of Scotland parish at Ardfern, Argyll, and the Rev. Michael Erskine, whose quiet support made all the difference.

Glossary

Act of Adjournal: A decision of court requiring one person to give satisfaction to another within a specified time.

Anderson's College: Founded in 1796 with subsequent changes of name – Anderson's Institute or Anderson's University. Anderson's College is used throughout this volume.

Bistoury: Any small knife for surgical purposes from the French *bistoire*.

Bond of Desistance: A bond signed by an unlicensed healer pledging to desist from irregular practice.

Cathartic medicines: Purgative drugs.

Clysters (clisters or glisters): Enemas, or the introduction of liquid anally into the intestines by means of a syringe.

Compear: To appear before a court or other authority with legal powers.

Couching for cataract: Pushing an opaque lens which was obscuring the vision into the lower part of the eye with a needle.

Court of Session: The supreme civil court (judicature) in Scotland.

Declarator: An action brought by an interested party to have some legal right or status declared but without claim on any person named as the defender.

Electuary: A powdered medicine mixed with jam, honey or syrup.

Emetic: A vomiter, or medicine that induces vomiting.

Empiric: A practitioner who administered treatment on the basis of what he found actually worked rather than according to received medical knowledge.

Fistula: An ulcerating, hollow canal in the body.

Gentlemanly sickness: Venereal disease.

Glasgow College: Glasgow University.

Irregular healers: A disparate group of healers (some with substantial empirical experience and training and some without) who practised without a licence.

Julep: Medicine administered in a sweet, liquid cordial.

Letters of Horning: A warrant, in Scots law, charging those named within either to act as instructed or to be 'put to the horn': proclaimed as an outlaw and banished.

Lithotomy: Cutting for the stone, or as it was known in Scotland, cutting for the gravel: a surgical operation to remove a bladder stone.

Lord Advocate: The principal law officer of the Crown in Scotland, and in the eighteenth century involved in framing legislation. In the eighteenth and early nineteenth centuries the Lord Advocate was a virtual government minister for Scotland.

Physic: Medicine.

Plaster: A poultice.

Procurator fiscal: The public prosecutor in a sheriff court.

Quicksilver: Mercury or hydrargyrus purification.

Sheriff: A legal officer who performed judicial duties and certain administrative duties.

The three venters: The three body cavities: skull, thorax and abdomen.

A Note on Dates and Money

Dates: Dates are given in New Style. Prior to 1600, the new year is taken to begin on 1 January (Scottish usage) rather than 25 March (English usage), and for these months the year is given in its modern form.

Money: Monetary values are generally given as either Sterling or Scots, but where unstipulated are Scots prior to 1707 and Sterling afterwards. The value of a merk was £13s. 4d. Scots; £12 Scots was equal to £1 Sterling.

Abbreviations

BL	British Library
Coutts, *A History of the University of Glasgow*	J. Coutts, *A History of the University of Glasgow from its Foundation in 1451–1909* (Glasgow, 1909)
Duncan, *Memorials*	A. Duncan, *Memorials of the Faculty of Physicians and Surgeons of Glasgow, 1599–1850* (Glasgow, 1896)
EUL Special Collections	Edinburgh University Library Special Collections
FPSG	Faculty of Physicians and Surgeons of Glasgow (title until 1909)
GCA	Glasgow City Archives
GLAHA	Glasgow Hunterian Art Gallery
GMC	General Medical Council
GRI	Glasgow Royal Infirmary
GUABRC	Glasgow University Archives and Business Record Centre
GUL Special Collections	Glasgow University Library Special Collections, GUL also contains GUL Eph (Ephemera)
GUL Pam	Glasgow University Library Pamphlets
GUL BPP	Glasgow University Library British Parliamentary Papers
HB	Health Board
MLRBM	Mitchell Library Rare Books and Manuscripts
NLS	National Library of Scotland
PMSA	Provincial Medical and Surgical Association
PRONI	Public Record Office of Northern Ireland
RCPLond	Royal College of Physicians of London

RCPSG	Royal College of Physicians and Surgeons of Glasgow (title after 1962). In footnotes, when accompanied by a reference number (e.g. in RCPSG, 1/1/1/2) this refers to the Royal College's Archive Collection
RCPEd	Royal College of Physicians of Edinburgh
RCSEd	Royal College of Surgeons of Edinburgh
RCSLond	Royal College of Surgeons of London (to 1843, thereafter the RCSEng)
RCSEng	Royal College of Surgeons of England
RCSI	Royal College of Surgeons of Ireland
RSML	Royal Society of Medicine Library
SRO	Scottish Record Office (from January 1999 National Archive of Scotland (NAS))
SUA	Strathclyde University Archives
Tennent, *Incorporation of Barbers*	J. B. Tennent, *Records of the Incorporation of Barbers of Glasgow, formerly the Incorporation of Chirurgeons and Barbers* (Glasgow, 1930)
UMM (Sen.)	University Meeting Minutes (Senate)

1

Rights and Privileges

The royal charter granted in 1599 to Mr Peter Lowe and Mr Robert Hamilton created, for the recipients and their successors, an autonomous jurisdiction in the examination and licensing of surgeons which the Faculty of Physicians and Surgeons of Glasgow spent most of the next 260 years (until registration was introduced by the 1858 Medical Act) defending. The charter established the Faculty in law as the monopoly regulator of medical qualifications and licensing in the west of Scotland; its position, in reality, however, was very different.[1]

While the authority vested in the Faculty established the body corporate as the regional touchstone of good and bad practice, the ways in which it was actually able to influence, especially to restrict, those who practised medicine in the west were more limited. It had no right, for instance, to examine physicians on their learning and practice but had only a right of veto over those who did not possess a medical degree. Its main powers were exercised over surgeons. A licence to practise surgery was given on the grounds of what a surgeon was capable of doing, not in what he was found wanting; poor and inadequate surgical practice was circumscribed by the issuing of licences on the basis of named skills only.[2]

The right to legitimise medical practitioners was useless without the power to take action against those deemed substandard, yet the nature of the seventeenth-century open medical market place meant that practitioners licensed by the Faculty competed with a large number of traditional and irregular healers.[3] Through its legal actions, the Faculty sought not only to define the regular medical community but also acted as a check against the traditional practitioners and irregulars. Since such action had to be supported

---

[1] A similar story of defence of its monopoly and privileges prior to the passing of the medical act of 1858 is told for the College of Physicians in London in G. Clark, *A History of the Royal College of Physicians of London*, 2 vols (Oxford, 1964, 1966).

[2] See below Chapter 4.

[3] See, for instance, the discussions in A. Digby, *Making a Medical Living: Doctors and Patients in the English Market for Medicine, 1720–1911* (Cambridge, 1994), pp. 24–30; Charles Webster, *The Great Instauration: Science, Medicine and Reform, 1626–1660* (London, 1975), p. 254.

by the local magistracy, until the late seventeenth century most prosecutions occurred within Glasgow rather than throughout the full extent of its bounds. As a consequence, the Faculty was constantly pushing for the reiteration and ratification of its powers in a variety of additional acts and decrees.

In its external relations, the Faculty's main problem during this period was the protection of its jurisdictional monopoly on licensing against encroachment from local vested interests. Focusing on internal issues, the most distinct feature of the first 120 years of the Faculty's history was the constant reshuffling and repositioning of the different divisions of healers constituted under the charter – physicians, surgeons, apothecaries and barbers – as each carved out its niche. The nature of an apprenticeship could vary greatly according to the status of the practitioner and his practice, and according to the apprentice's social background, and education. Because of this immense diversity, the different medical occupations during this period have been described not so much as nascent professions as 'a jumble of different types of practitioners who, for reasons of civic administration, fell into the same category'.[4]

Although the Faculty had royal sanction under the privy seal, this was not enough for it to operate efficiently within the routine hurly-burly of civic life. To be really effective in the region, the surgeons needed the cooperation of magistrates and burgesses in the west of Scotland's principal town; and to receive town privileges and protection, they had to pay burghal dues and buy in to the guild system. Eventually this led the surgeons to petition for simultaneous erection as a craft guild; admission, in a sense, of inadequate municipal protection. The creation of the Incorporation of Surgeons and Barbers as a town guild in 1656 led to an erosion of the status of the physicians within the body corporate. Surgeons' concerns undoubtedly dominated the Faculty simply because there were more of them, and from the beginning there was a particular drive to raise the standard of surgery above simple barbery. However, the surgeons came to regret their incorporation with the lowly barbers and, after the ratification of their charter in 1672, not only mended their relationship with the physicians but were ultimately to jettison the barbers.[5]

Nevertheless, for well over a century Glasgow was unique: in no other place were physicians, surgeons, apothecaries and barbers joined in corporate association.

---

[4] M. E. Fissell, *Patients, Power, and the Poor in Eighteenth-Century Bristol* (Cambridge, 1991), p. 51.

[5] For the split between the barbers and surgeons, see Chapter 3, below. The surgeons of Edinburgh had similarly sought to marginalise the barbers from the founding of the incorporation.

Glasgow's Charter and its Provisions

In his history of Scotland (1573), John Leslie, the Catholic Bishop of Ross, insisted, that Glasgow was a 'noble' market town:

> Surely ... the most renowned market in all the west, honorable and celebrated: Before the heresy [Reformation] there was an Academy not obscure neither infrequent [poorly attended] or of a small number, in respect both of Philosophy and Grammar and politic study. It is so frequent [drawing many people], and of such renown, that it sends to the East countries very fat cows, herring likewise and salmon, oxen-hides, whole and skins, butter likewise that none better, and cheese. But, contrary, to the west (where is a people very numerable in respect of the commodity of the sea coast), by other merchandise, all kind of corn to them sends.[6]

Little had changed by the turn of the century. Glasgow was a small, provincial market centre with a population of about 7000 souls; although it was, as Leslie had insisted, the main burgh in the west, and growing, it did not rank as one of the four major towns in Scotland until the mid seventeenth century, and was then only a third of the size of Edinburgh.[7] Glasgow differed from the capital city, Edinburgh, in the comparative proportion of craftsmen to merchants.[8] The mercantile base in Glasgow was relatively modest in comparison with the breadth of its manufactures, and the town had particular strengths in clothing, textiles and food processing.[9] Hence, in the late sixteenth century, Glasgow was more oriented to internal and coastal trade than the overseas commerce dominated by Edinburgh. But the importance of trade for the town was reflected in the 1609 Act of Parliament that the provost could only be elected from among the resident merchants. The 1603 union of the Scottish crown with England had led to an increase in overland trade in linen and yarn, and in 1611 the town was eventually elevated as a royal burgh, granting it valuable internal and overseas trading privileges. The boost which this gave to the town's commercial and trading activity led to its substantial expansion throughout the course of the seventeenth century, as it acquired a more stable economic base (though it did not gain a repu-

---

[6] John Leslie, *De origine, moribus et rebus Scotorum*, quoted in D. Daiches, *Glasgow* (London, 1977), pp. 17–18.

[7] M. Lynch (ed.), *The Early Modern Town in Scotland* (London, 1987), p. 4.

[8] A table of the numbers of craftsmen and merchants in Glasgow in 1605 is given in M. Lynch, 'The Social and Economic Structure of the Larger Towns, 1450–1600', in M. Lynch, M. Spearman and G. Steel (eds), *The Scottish Medieval Town* (Edinburgh, 1988), p. 274.

[9] M. Lynch, 'The Face of the Town', *Scottish Records Association*, Conference Report No. 8 (1987), p. 11.

tation as an expanding international trading centre until the final decades of the century).[10]

The town was ruled by the magistrates and town council, an oligarchy comprising some two dozen men and dominated by the merchants. Though the craft incorporations acquired some representation on the council in the 1570s and 1580s, wealthy craftsmen did not achieve parallel status to the leading merchants until the 1605 letter of guildry established an elite tier of guild brethren above the less wealthy burgesses. In the following year, a crown letter ordained that the council should consist of equal numbers of merchants and craftsmen. As the population began slowly to expand in the late sixteenth century, the civic administration was forced to confront the problems of badly built timber-framed houses, overcrowding, and streets littered with human and animal waste. Lepers were consigned to St Ninian's hospital, a lazaretto on the south side of the Clyde bridge and *cordons sanitaires* were established to try and isolate the burgh from the plague which spread through Scotland in 1574, 1584–85 and 1605–06. Bailies were appointed to monitor different wards of the burgh and minor public health officials were told to make tours of inspection and to quarantine the sick (the concept of quarantine having been introduced in Scotland in the fifteenth century). The town also retained a surgeon whose salary in 1581 was £20 Scots, a huge sum exceeded only by the salary payed to the provost.[11]

It was to such a town that Peter Lowe returned from France in 1598 to be contracted as town surgeon in March 1599 at eighty merks Scots per year.[12] Lowe, one of the founding members of the Faculty, took his first steps to protect the people of the market town of Glasgow against untrained medical practitioners through the civic authorities. A month later, the provost, bailies and town council of Glasgow took action against irregulars at the desire of the kirk session, which was concerned about practitioners who 'are not able to discharge their duty therein, in respect they have neither cunning nor skill to do the same', and wanted to enlist the help of 'cunning men of that art' in examining them. Three bailies were assigned to

---

[10] T. M. Devine, 'The Development of Glasgow to 1830: Medieval Burgh to Industrial City', in idem and G. Jackson (eds), *Glasgow*, i, *Beginnings to 1830* (Manchester, 1995), pp. 4, 6–7; J. McGrath, 'The Medieval and Early Modern Burgh', ibid., pp. 29–30, 45–46.

[11] McGrath, 'The Medieval and Early Modern Burgh', pp. 31–32; idem, 'The Administration of the Burgh of Glasgow, 1574–1586', 2 vols (unpublished Ph.D. thesis, University of Glasgow, 1986), i, p. 54; A. Keller, 'The Physical Nature of Man: Science, Medicine, Mathematics', in J. MacQueen, *Humanism in Renaissance Scotland* (Edinburgh, 1990), pp. 101–2.

[12] R. Renwick and J. D. Marwick (eds), *Extracts from the Records of the Burgh of Glasgow, 1573–1642*, i, Scottish Burgh Records Society (Glasgow, 1874), p. 191. For more on Peter Lowe, see Chapter 2.

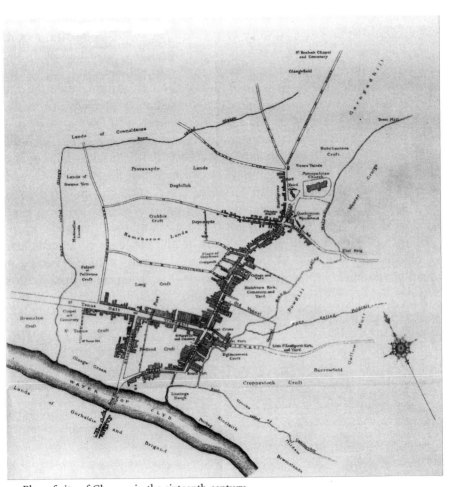

1. Plan of city of Glasgow in the sixteenth century.

a committee constituted to examine practitioners,[13] along with three members from the kirk session.[14] The committee was ordered to report back to the council, but there is no evidence that it took any further action.[15]

The episode was significant in that Peter Lowe was the likely originator of the initiative, and one of the 'cunning men of that art' chosen to advise the committee. That the call to set up a body to regulate medical practice came from a trained practitioner suggests a degree of competition from irregular healers. There is also little doubt that Lowe was shocked by the arbitrary nature of medical regulation in the burgh; he must have been impatient both of the town council and of the kirk session, resenting the interference of untrained amateurs in the regulation of medicine in the burgh. Having worked as an ambassador and surgeon in France, Lowe had cultivated influential connections and decided to approach James VI himself.[16] His readiness to do this not only reflects the high status held by the king's surgeon in Paris but the high regard in which Scots surgeons were held by royalty. The petition of Mr Peter Lowe, surgeon, and Mr Robert Hamilton, physician,[17] was accepted by the king who, on 29 November 1599, issued a royal charter or letter of gift to them, under the privy seal.

The charter's preamble drew attention to the existence of a regulatory vacuum in the west of Scotland in the late sixteenth century.[18] Particular mention was made of 'the great abuses which have been committed in time bygone, and yet daily continue, by ignorant, unskilled, and unlearned persons, who, under colour of Surgeons, abuse the people to their pleasure, passing away but trial or punishment, and thereby destroy an infinite number of Our subjects'.[19] The region henceforth subject to licensing controls

[13] James Forret, Alexander Baillie and Thomas Pettigrew.
[14] The principal, Blais Lowery and John Blackburn.
[15] Renwick and Marwick, *Records of the Burgh of Glasgow, 1573–1642*, p. 193.
[16] Tennent, *Incorporation of Barbers*, p. 8.
[17] Though a physician, Hamilton is often referred to as Mr. It may have been a title that stuck after he first took an MA.
[18] The original charter is no longer extant, having been lost in the long series of legal suits which the Faculty engaged in over the centuries. However, various notarial copies have survived. The charter is reproduced in Appendix 1, pp. 417–19 below.
[19] FPSG, *The Royal Charter and Laws of the Faculty of Physicians and Surgeons of Glasgow* (Glasgow, 1821), pp. 5–6. The general sentiment is similar to that expressed in the preamble of the 1512 English Act 3 Henry VIII, c. II: 'Forasmuch as the Science and Cunning of Physic and Surgery (to the perfect knowledge whereof be requisite both great Learning and ripe Experience) is daily within this realm exercised by a great multitude of ignorant persons, of whom the greater part have no manner of Insight in the same, nor in any other kind of Learning ... in the which they ... apply such Medicines unto the Disease as be very noious, and nothing meet therefore, to the high Displeasure of God, great infamy to the Faculty, and the grievous Hurt, Damage, and Destruction of many of the King's liege People, most especially

encompassed the burgh and barony of Glasgow, Renfrew, Dunbarton, and the sheriffdoms of Clydesdale, Renfrew, Lanark, Kyle, Carrick, Ayr and Cunningham.[20] This was a sizeable area, more or less coextensive with the diocese of Glasgow,[21] and far larger than the seven-mile radius within which both the London College of Physicians and the Dublin College of Physicians operated.[22]

In order to curtail abuse by the unskilled, the charter gave Peter Lowe, 'Our Surgeon, and chief Surgeon to Our dearest Son the Prince' with the assistance of Mr Robert Hamilton, 'Professor of Medicine' and their successors in Glasgow,[23] 'full power to call, summon and convene before them within the said Burgh of Glasgow, or in any other of our said Burghs or public places in the foresaid bounds, all persons professing or using the said art of Surgery and to examine them upon their literature, knowledge and practice'.[24] From the outset, therefore, it was an examining body.[25] Its putative power over everyone practising surgery in the region ensured that the Faculty was, throughout the seventeenth century, dominated by its surgeon members. Many historians have emphasised that the charter contains no specific reference to barbers.[26] This probably reflected a desire by a surgeon as eminent as Peter Lowe to keep surgery totally distinct from the lowly art of barber-surgery. After all, Lowe had been educated at the community of St-Côme in Paris, founded in the early fourteenth century, which was particularly distinguished as a company of 'pure surgeons' who did not undertake any barbers' work.[27] Lowe undoubtedly wished to raise the tone of the charter above any association with the simple art of barbery: an 'Act

of them that cannot discern the cunning from the uncunning'; and in the Latin text of the king's letters patent under the great seal which incorporated the College of the Faculty of Medicine of London, or the College of Physicians, on 23 September 1518. Clark, *Royal College of Physicians*, i, pp. 54, 59.

[20] FPSG, *The Royal Charter and Laws*, p. 6.

[21] R. M. Ross, 'Peter Lowe: Founder of the Faculty, Man of Mystery', *Dental Historian*, 28 (1995), p. 8.

[22] Clark, *Royal College of Physicians*, i, p. 54; J. F. Fleetwood, *The History of Medicine in Ireland* (2nd edn, Dublin, 1983), pp. 36–40.

[23] By contrast, six founder members were named in the charter of the College of Physicians of London, in addition to a seventh – Thomas Wolsey, cardinal archbishop of York and lord chancellor. Clark, *Royal College of Physicians*, i, p. 59.

[24] FPSG, *The Royal Charter and Laws*, p. 6.

[25] J. S. G. Blair, *History of Medicine in the University of St Andrews* (Edinburgh, 1987), pp. 12–13.

[26] For example, Tennent, *Incorporation of Barbers*; RCPSG 1/5/1, 'Weir's Faculty Memorandum' (MS volume of notes by William Weir on the history of the FPSG based on extracts from the minutes), 1869.

[27] D. de Moulin, *A History of Surgery* (Dordrecht, 1988), p. 106.

for Admission of Barbers', passed by the incorporation in June 1602, warned that barbers were 'not to meddle further in anything pertaining to surgery', under pain of £5 for each contravention.[28]

The charter dispensed power to license those found proficient 'according to their art and knowledge, that they shall be found worthy to exercise thereafter'. More important, licensees were to be discharged from practising 'any further than they have knowledge passing their capacity'. Power to limit practice was implemented from the outset.[29] In true Presbyterian style, everyone cited before the Faculty was to present testimonials of their conduct and behaviour from the ministers, elders or magistrates of the parish in which they lived. These were routinely required by the kirk in a pastoral context when anyone moved from one location to another.[30] There is little indication (from the Faculty minutes at least) that this specification of the charter was implemented,[31] though unsolicited verbal testimonials certainly led to citations for unlicensed practice.[32]

The charter gave Peter Lowe and Robert Hamilton, as Visitors, legal powers to execute letters of horning against the contumacious – those who were wilfully disobedient to the summons of a court. In Scots law, letters of horning constituted a warrant charging those within named to act as instructed or be 'put to the horn' – proclaimed as an outlaw and banished. Failure to comply with summons to a trial of skills was to be enforced, first, by poinding (legal seizure) of goods, and secondly, by imprisonment. The Visitors were obliged to 'visit every hurt, murdered, poisoned, or any other person taken away extraordinarily, and to report to the Magistrates the fact as it is'.[33] The practice of calling surgeons, particularly as expert witnesses in cases of suspicious or violent death, filtered through from the Romance countries where it was common as early as the thirteenth century. But the inclusion of forensic responsibilities in the 1599 charter probably resulted from Peter Lowe's familiarity with the French tradition, particularly with Ambroise Paré's *Traité des rapports* (Treatise on Reports) published in 1575

---

[28] RCPSG, 1/1/1/1b, Transcript Minutes of the FPSG, 1599–1688, pp. 15–16.

[29] For which, see Chapter 4, below.

[30] FPSG, *The Royal Charter and Laws*, p. 7. See, for instance, R. Houston, 'Geographical Mobility in Scotland, 1652–1811: The Evidence of Testimonials', *Journal of Historical Geography*, 11 (1985), pp. 379–94.

[31] One exception to this appears on 3 January 1677 when James Forrester in the parish of Kilmacolm compeared in obedience to a charge of horning for practising surgery within the Faculty's bounds and asked to be admitted to practise such parts of surgery as he was found qualified for upon trial. He produced a certificate in the hand of Mr Patrick Simpson, minister at his parish kirk 'of his life & conversation'. RCPSG, 1/1/1/1b, pp. 567–68.

[32] For examples of which, see Chapter 4, below.

[33] FPSG, *The Royal Charter and Laws*, pp. 7–8.

which offered blueprints for reports on sudden or accidental death.[34] They were further empowered to make statutes 'for the commonwealth [public good] of our subjects concerning the said arts'.[35] But public implementation of any of the Faculty's bylaws was dependent (as can be seen from the final clause below) on their legal execution by the civil magistracy.

The fourth clause specified that it would be unlawful to practise medicine 'without the testimonial of a famous University where Medicine is taught', or without the recommendation of the king's or queen's 'chief Physicians'. In this way, the charter gave the Visitors power only to inhibit the practice of physicians who did not present a university degree whereas they had the right to examine and license all those who wished to practise surgery. In so doing, the charter implicitly underwrote the elevated status of physicians and the ascendancy of academic, learned medicine.[36] For those who practised before submitting their qualifications, the Visitors had power 'to challenge, pursue and inhibit them from using and exercising the said arts of Medicine, under the pain of Forty pounds' – half the fine to accrue to the magistrates, the other half to be distributed for the benefit of the poor.[37] By giving powers to examine and refuse surgeon-apothecaries licence to practise, to inhibit physicians and to limit the nature of the practice of those whom it did license, the charter attempted to hold in check a well-established band of unqualified traditional practitioners and, in so doing, to define the regular medical community in the west of Scotland.[38]

In addition, the charter bestowed legal power to inspect drugs, insisting that no one sell drugs in Glasgow unless they had been examined by the Visitors or by William Spang, apothecary, under pain of confiscation of the drugs.[39] In Edinburgh, the appointment of visitors to apothecaries' shops in the early 1680s (after the founding of the College of Physicians in 1681) was responsible for a great deal of controversy. The College of Physicians in

---

[34] A. Wear, 'Medicine in Early Modern Europe, 1500–1700', in L. I. Conrad, M. Neve, V. Nutton, R. Porter and A. Wear, *The Western Medical Tradition, 800 BC to AD 1800* (Cambridge, 1995), p. 237.

[35] FPSG, *The Royal Charter and Laws*, p. 8.

[36] R. M. Stott, 'The Incorporation of Surgeons and Medical Education and Practice in Edinburgh 1696–1755' (unpublished Ph.D. thesis, University of Edinburgh, 1984), p. xxv.

[37] FPSG, *The Royal Charter and Laws*, p. 8. The charter of the College of Physicians of London also gave it power to prosecute unlicensed practice within the seven miles of its jurisdiction, under pain of paying £5 per month of their unlicensed practice – half to go to the king and half to the president and college. Clark, *Royal College of Physicians*, i, p. 60.

[38] See the point made in J. J. Keevil, *Medicine and the Navy*, 4 vols (Edinburgh and London, 1957), i, *1200–1900*, pp. 124–25.

[39] FPSG, *The Royal Charter and Laws*, p. 9.

London was later granted similar rights by an Act of 1740.[40] This was the nearest the charter came to legally sanctioning the position of apothecaries in the town. In particular, the charter restricted the right to sell rat poison (as arsenic or sublimate) to the apothecaries, the buyers of which had to stand surety for any costs or damage caused thereby under pain of 100 merks Scots. The seventh clause obliged the Visitors, their brethren and successors, to convene on the first Monday of every month, at some convenient location, 'to visit and give counsel to poor diseased folk, *gratis*'.[41] The provision for charitable consultation in the charter itself is highly significant. While the town had paid salaried surgeons to treat the poor since the sixteenth century, the charter reflected a particular interest in public health which must have been influenced by Lowe's experience as town surgeon. Glasgow was unique among British medical corporations in making such provision at that time. The London College of Physicians, for instance, did not make provision for free medical services to the metropolitan poor until the founding of its dispensary in 1696.[42]

Finally, valuable immunities and civic privileges were granted to the members of the new Faculty in recognition of their services in the public good, namely 'Immunity and Exemption from all Weapons-showing,[43] Raids, Hosts, Bearing of armour, Watching, Warding, Stenting,[44] Taxations, Passing on assize, Inquests, in Justice Courts, Sheriff or Burgh Courts, in Actions Criminal or Civil, notwithstanding of Our acts, laws and constitutions thereof, except in giving their Counsel pertaining to the said arts'.[45] Such civic privileges were commonly accorded by royal statute to European corporations: Mary, Queen of Scots, had granted similar exemptions to the Edinburgh surgeons in 1567; and members of the London College of Physicians received similar privileges, as did medical corporations in France.[46]

---

[40] H. M. Dingwall, *Physicians, Surgeons and Apothecaries: Medical Practice in Seventeenth-Century Edinburgh* (East Linton, 1995), p. 217; A. W. Sloan, *English Medicine in the Seventeenth Century* (Durham, 1996), pp. 92–93, quoting E. Kremers and G. Urdang, *History of Pharmacy* (4th edn, Philadelphia, revised by G. Sonnedecker, 1976), p. 101.

[41] FPSG, *The Royal Charter and Laws*, p. 9.

[42] Clark, *Royal College of Physicians*, ii, pp. 427–47.

[43] The weapon-showing or 'wappen-shaw' was the periodic muster of men of arms-bearing age in an administrative district or lordship.

[44] Liability for assessment of property and land for the purposes of taxation. The 'stent' was the amount thus legally fixed.

[45] FPSG, *The Royal Charter and Laws*, pp. 9–10. Members of the College of Physicians of London were granted similar (though not quite such extensive) privileges, exempting them for example from service on juries and inquisitions, not only in London but throughout the country. Clark, *Royal College of Physicians*, i, p. 60.

[46] See J. W. Willcock, *The Laws Relating to the Medical Profession* (London, 1830), pp. 137–41; L. Brockliss and C. Jones, *The Medical World of Early Modern France* (Oxford, 1997), p. 173.

The magistrates were charged with assisting the Faculty in executing these acts and in issuing charges within twenty-four hours.[47] This effectively emphasised a distinction between the Faculty convening as a court of judges and its prosecuting role. (The College of Physicians in London came to grief over such partiality in Bonham's Case in 1610, the two mutually exclusive roles apparently rendering void its statutory authority.)[48] But uncooperative magistrates prevented the Faculty from effectively cracking down on irregulars outside Glasgow until the late seventeenth century when, as Duncan puts it, 'a mania for prosecuting' seems to have seized it.[49]

Ratification

The charter of 1599 clearly placed the practice of surgery and medicine under the independent control of university-educated practitioners, thus distancing the founders from the craft-trained practitioners. The royal charter was subsequently ratified, three months later, on 9 February 1600, by the burgh of Glasgow which gave it official status and council support within the town. Having inspected 'our Sovereign Lord's letter of gift and faculty' granted to Peter Lowe, Robert Hamilton and William Spang, apothecary, 'professors of their arts', the council promised 'to hold, have, concur, fortify, and maintain them and their successors and liberties granted to them in the same in all points in time coming', provided that they did not act in a way prejudicial to the town's interests.[50]

It is significant that the town felt the need to include an apothecary in the ratification; technically granted to Lowe and Hamilton, the charter gave powers of drug inspection only to Spang. In this way, the town was recognising the apothecaries' need to organise themselves, as well as lending local legislative sanction to the amalgamation of the three groups of medical practitioners. But the willingness of physicians, surgeons and apothecaries in Glasgow to associate themselves together corporately is of greater significance. In most other places, they formed separate incorporations. Quite why all three sets of practitioners came together in one body has never been properly explained. Certainly, in London, it was considered that the union

---

[47] FPSG, *The Royal Charter and Laws*, p. 10.

[48] H. J. Cook, 'Against Common Right and Reason: The College of Physicians versus Dr Thomas Bonham', *American Journal of Legal History*, 29 (1985), p. 302.

[49] Interest in the area outside Glasgow was heightened with the readmission of the physicians to the corporation in the 1672, particularly after they were reinstated in their joint powers of visitation. Medical and surgical practice in areas to the west of Glasgow then came under their scrutiny. Duncan, *Memorials*, p. 74.

[50] Renwick and Marwick, *Records of the Burgh of Glasgow, 1573–1642*, p. 202.

and interaction of the barbers and surgeons would encourage learning and the dissemination of surgical knowledge.[51] And until 1617, when the first apothecaries' guild was formed in London, their interests were best protected by union with another trade. Presumably this would have been better achieved with all three branches of medicine in one association. It has also been suggested that paucity of numbers rendered it expedient in Glasgow. This is plausible; when the Faculty first met in June 1602 – some two and a half years after the granting of its charter – there were only seven members: Peter Lowe (surgeon), Robert Hamilton (physician), William Spang (apothecary), Adam Fleming, Robert Allason, Thomas Thomson (barber?),[52] and John Lowe. It was particularly unusual for physicians to incorporate with surgeons and apothecaries, but being so few in number in Glasgow – rarely more than two at this time – there were not enough to form a college. Therefore, in order to protect themselves from the growing number in Scotland who worked as general practitioners, especially those dispensing medicines (legally the physician's preserve), they decided to form an all-encompassing regulatory body with the surgeons. However, whether this union 'prevented repetition in Glasgow of some of the more heated and acrimonious conflicts which took place between the physicians and surgeons of the capital' is (as will be seen below) debatable.[53]

The word 'faculty' is first used in relation to the new incorporation in the council's ratification of the charter. The council refers to 'the privileges and statutes of our Sovereign Lord's letter of gift and faculty granted to master Peter Low, surgeon, master Robert Hamilton, William Spang and their successors'.[54] But the word is clearly made in reference to the charter – in the sense of power conferred under certain limitations – rather than as a title of the corporate body as applied later in the mid-seventeenth century. There can be little doubt that the use of the word Faculty in the corporation's title – meaning a legally constituted body with corporate rights – came from the Paris Faculté de Médecine with which Lowe was most familiar.

Just thirty-five years after the granting of the charter, and its ratification by the burgh in 1600, the Faculty felt the need for the reiteration of its original powers in general signet letters (writs issued under the king's signet

---

[51] R. Theodore Beck, *The Cutting Edge: Early History of the Surgeons of London* (London, 1974), p. 175.

[52] Though the terms under which he (and over half of the other early members) entered are not given, some nineteen days after the first meeting Thomson was ordained to take in his basins for contemptuous disobedience and failing to fulfil the duty he was charged with. RCPSG 1/1/1/1b, p. 12.

[53] Dingwall, *Physicians, Surgeons and Apothecaries*, pp. 36–37.

[54] Renwick and Marwick, *Records of the Burgh of Glasgow, 1573–1642*, p. 202.

enforcing a court decree) in favour of the surgeons of Glasgow against the magistrates and dated 14 August 1635.[55] This document recapitulated the principal tenets and powers given in the royal charter but its main import was that on 31 July 1635 James Hamilton obtained a decree from the Court of Session against burgh officials and local law officers. This charged them with taking action against all practitioners 'unless they be examined by the said Mr James Hamilton, present visitor ... and by his brethren of the said art and their successors'.[56]

The burgh authorities were evidently not upholding and defending the Faculty's position as established by the charter. Given the date of the signeting of this document, it was probably applied for in time to strengthen and mark the Faculty's position on the twenty-fifth anniversary of Peter Lowe's death (the Visitor in 1635 being the other founder, Mr James Hamilton, 'Professor of physic').[57] From this juncture, the Faculty was constantly defending the rights and privileges established in its charter. Indeed, the dominant theme of the period between 1599 and 1757 is the Faculty's attempt to establish itself as the monopoly regulator of medical practice in the west of Scotland by defending its jurisdictional rights, powers and privileges as granted in its original letter of gift.

In this, the FPSG undoubtedly suffered from a lack of official status with the town in the form of a seal of cause or letter of deaconry. So, when it made little ground in its attempts to enforce the magistrates' legal endorsement of its enactments against illegal practice in 1635, the Faculty's attitude changed to one of détente and reconciliation. At a meeting of 4 August 1656, the members 'all in one voice did condescend and agree That a seal of cause or letter of deaconry be purchased from the town council in favour of the faculty' without prejudicing the old gift given to them by James VI. At about the same time, the first distinct appearance of the name 'the faculty of surgeons and physicians of the burgh of Glasgow' occurs in the minutes on 26 May 1654.[58] The prospect of a local body of surgeons with two constitutions

---

[55] This is the earliest extant manuscript in the present College Archives. Though printed as an appendix in Alexander Duncan's *Memorials*, it is, nevertheless, not referred to in the text.

[56] RCPSG, 1/5/5(1), Miscellaneous Manuscript Material: Signet Letters in favour of Surgeons against the Magistrats, 1635.

[57] RCPSG, 1/1/1/1b, pp. 45–46.

[58] RCPSG, 1/1/1/1b, pp. 82, 119. It is apparent from a change in the extent and fullness of the entries that the first true minutes were taken in 1654. Accurate minuting probably began at this juncture because negotiations were already afoot which led to the granting of the letter of deaconry of 1656. But it does mean that the factual accuracy of the first half-century of the Faculty's history is compromised since these minutes were written retrospectively. Therefore, although the designation 'faculte' first appears in the minutes in 1627 (a correction of

called for different names for the body constituted under the charter and that constituted under the seal of cause. The latter was only to be in favour of the surgeons, apothecaries and barbers who had more need of civic accreditation than the university-trained physicians.

Were there any other reasons behind the application for incorporation with the town at this particular point? Possibly there was a certain economic imperative. Glasgow was continuing its steady but inexorable expansion in trade and commerce; its population had more than doubled in sixty years from 7000 in 1600 to 14,678 in 1660.[59] This revival of trade led, throughout Scotland, to an increase in the proportion of the population that was urbanised.[60] Thomas Tucker, registrar to the commissioners of customs and excise under Cromwell, was sent north in 1656 to give a report on the revenues of the customs and excise in Scotland. He stated that:

> The town, seated in a pleasant and fruitful soil, and consisting of four streets handsomely built in form of a cross, is one of the most considerable burghs of Scotland, as well for the structure as trade of it.[61] The inhabitants (all but the students of the college which is here) are traders and dealers: Some for Ireland with small smithy coals,[62] in open boats, from four to ten tons, from whence they bring hoops, rings, barrel-staves, meal, oats, and butter; some for France with pladding, coals, and herring (of which there is a great fishing yearly in the Western Sea), for which they return salt, paper, resin, and prunes; some to Norway for timber; and every one with their neighbours the Highlanders, who come hither from the Isles and Western parts ... Here have likewise been some who have adventured as far as Barbadoes; but the loss they have sustained by reason of their going out and coming home late every year has made them discontinue going thither any more.[63]

The easiest way for the Faculty to take a full share in the fruits of the

---

Duncan, *Memorials*, p. 57, which gives 1629), the use of the title is anachronistic and little credence can be given to its retroactive application. (There are similar problems with the first sixty years of the records of the London College of Physicians.) The Incorporation minutes, on the other hand, survive only from 1707, the year of the Union between Scotland and England, so little information can be gleaned from that source about the early history of surgeons in Glasgow.

[59] McGrath, 'The Medieval and Early Modern Burgh', p. 44.

[60] I. D. Whyte, *Scotland before the Industrial Revolution: An Economic and Social History, c. 1050–c. 1750* (London and New York, 1995), p. 170.

[61] As late as 1750, the town comprised only thirteen streets. Hugh Macintosh, *The Origin and History of Glasgow Streets* (Glasgow, 1902), p. vi.

[62] Small smokeless coals suitable for smith's work.

[63] T. Tucker, *Report upon the Settlement of the Revenues of Excise and Customs in Scotland, AD MDCIVI*, Bannatyne Club (Edinburgh, 1824 reprint), p. 38.

town's sustained commercial expansion was to be become fully incorporated with it and thus represented in its administration.

Various petitions by John Hall, deacon of the surgeons and barbers, stated that it was 'incumbent on us to have a letter of deaconry from your honours, as others of these incorporations have granted to them by your predecessors, for a joint and harmonious correspondence of brotherhood, as brother citizens willing to sympathise with the rest of the body of the city'. Lack of support by the town council – who were perhaps not prepared to accord privileges unless the surgeons paid into the joint funds and responsibilities of the trade incorporations – had probably resulted in an initial flouting of the Faculty's authority in Glasgow. While the charter made it incumbent on the town council legally to execute the Faculty's enactments against irregular practitioners, in reality the surgeons' lack of civic organisation made the magistrates less diligent in this respect. In petitioning for a seal of cause, the surgeons certainly made a point of stressing the similarity to the guilds in their organisational structure: 'we have been in use, yearly, to elect a deacon as visitor and overseer of the rest of the members of our calling as other callings have been in use, By virtue of any patent letter of deaconhead and seal of cause conferred upon them heretofore by any authority'.[64]

They asked that the letter of incorporation be granted 'in regard of our so long being a standing part of the crafts of this city', and because (like all the other trades) they gave an annual contribution to the poor of the Crafts' Hospital. So the body constituted under the 1599 charter manifestly regarded itself as a trade guild. A letter of deaconry – erecting the Incorporation of Surgeons and Barbers as a guild with representation in the Trades' House – was duly granted on 16 August 1656 to John Hall, 'present Deacon of the saids Surgeons and Barbers and whole present brethren of that art and craft, and to their successors'.[65] Granting a seal of cause was an act of recognition on behalf of the town council which gave a certain security to the association seeking it. Technically, a seal of cause permitted the election of a deacon, as well as endorsing the statutes or ordinances already made by the trade which were binding on its members.[66] Like the signet letters, it was passed within a day of the anniversary of Peter Lowe's death. Twenty-one years after its attempt to persuade the town to toe the line, the Faculty had come of age.

What precisely did this new erection mean for the Faculty? It is difficult to ascertain exactly how these two bodies interrelated: the Incorporation and the Faculty are not clearly distinguishable from this point as separate

[64] Tennent, *Incorporation of Barbers*, p. 17.
[65] Ibid., pp. 19, 20. The letter of deaconry or seal of cause is reproduced in Appendix 2, pp. 420–22 below.
[66] A. M. Smith, *The Nine Trades of Dundee*, Abertay Historical Society, 35 (Dundee, 1995), p. 28.

institutions; rather it seems that the physicians and surgeons had certain rights under their charter, while the privileges of the town in the 1656 letter of deaconry were given to the surgeons, being at the same time extended to the apothecaries and barbers.[67] One occupational association was simply in possession of two titles or constitutions governing different aspects of its members' professional interaction. The only clear difference between them is that the royal charter gave the Faculty powers and exclusive privileges and the right to license in four counties, while (despite the fact that the town council found its granting 'To tend to the good of the people, as well within as without the Burgh') the letter of deaconry was limited to the town of Glasgow.[68] The inclusion of apothecaries in a surgeon-dominated body is important because, traditionally, apothecaries had an occupational dependence on physicians, under whose authority they made up their medicines (though in nearby Edinburgh, the surgeons had long claimed the rights to practise pharmacy – a right strengthened by the erection of the Fraternity of Apothecaries and Surgeon-apothecaries by the town council in 1657, and confirmed in 1694 by the crown after a brief lapse).[69] A possible interpretation is that the more down-market practitioners were accredited by the letter of deaconry. Nonetheless, the lack of distinction persisting between the two bodies is implicit in the case of John Hall, barber from Edinburgh, who petitioned for admission to the *Faculty* in September 1671 'to exercise the office of a barber', when the obvious place for a barber was in the Incorporation of Surgeons and Barbers.

An earlier namesake, the John Hall, mentioned above, a prominent and distinguished surgeon, who entered the Faculty in 1647 as 'freeman with the calling as professor of Surgery', was Visitor in 1648, 1651–52 and 1654–55. When he was referred to in the 1656 negotiations as Deacon of the Surgeons and Barbers, Hall's leadership was clearly of the same body because barbers had been admitted into the Faculty's fold since 1602.[70] Therefore, this (in some respects academic) distinction between the body constituted under the charter and the guild Incorporation probably came about because the

---

[67] Though the term 'apothecary' is nowhere mentioned in the letter of deaconry, it is clear that 'the art of Surgeonry' therein mentioned, implies the art of the surgeon-apothecary. This is explicitly clarified much later in 1719, when the relationship between the barbers and surgeons was breaking down, and the town council found that 'the Barbers and their Sons, Sons in law and apprentices ought all Equally to be admitted to the practice of Surgery and pharmacy as well as Barberizing'. Tennent, *Incorporation of Barbers*, p. 41.

[68] H. Lumsden, 'Bibliography of the Guilds of Glasgow', *Records of the Glasgow Bibliographical Society*, 8 (1930), p. 9; Tennent, *Incorporation of Barbers*, p. 20.

[69] Dingwall, *Physicians, Surgeons and Apothecaries*, pp. 187–88, 222.

[70] RCPSG, 1/1/1/1b, pp. 48, 365. Individual admissions can be identified from 1636.

physicians found it easy to maintain a private practice while the more numerous surgeons and barbers needed the support and sanction of the town to work effectively.

A nephew of the Provost, John Hall had a high political profile in the town. When on 8 June 1657 the surgeons and apothecaries of Edinburgh informed the Faculty that the Protector, Oliver Cromwell, had granted letters patent 'for erecting a college of physicians there', the Faculty sent John Hall and Archibald Graham to Edinburgh 'to advocate and oppose the same, before the Council of State'.[71] After abortive attempts in 1621 and 1633, this was the third try at erecting a college of physicians in Edinburgh, which differed from the previous attempts in envisaging a college for Scotland and not simply Edinburgh. But the proposal was also opposed by the Surgeons of Edinburgh, the Edinburgh civic authorities, and the Scottish universities.[72] It was planned by Dr George Purves, who believed that a college of physicians was necessary to reduce the 'frequent murders committed universally in all parts ... by quacks, women, gardeners and others grossly ignorant'. The new foundation also aimed to circumscribe 'the unlimited and unaccountable practices of Surgeons Apothecaries and Empirics pretending to medicines', and place them firmly under the control of their socially elite brethren, the physicians.[73] In this, they were seeking to emulate the position of preeminence which London physicians enjoyed over apothecaries and barber-surgeons.[74] But surgeons, in particular, had always enjoyed a higher status in Scotland and were prepared to fight for it. According to the proposed patent, surgeons were to treat only external diseases, and if they recurred, a physician was to be consulted.[75] In addition, the physicians intended to invade the surgeons' traditional territory with the proposal that their new college have the right to bodies for dissection. Most significantly, the Edinburgh physicians sought not only greater control over local surgeons and apothecaries but power to license practice throughout Scotland (a theme which was to reappear at various junctures in the history of the Scottish Royal Colleges).[76]

[71] RCPSG, 1/1/1/1b, p. 132.

[72] G. McLachlan, *Medical Education and Medical Care: A Scottish-American Symposium* (Oxford, 1977), p. 30.

[73] D. Hamilton, *The Healers: A History of Medicine in Scotland* (Edinburgh, 1981), p. 68; A. C. Chitnis, 'Provost Drummond and the Origins of Edinburgh Medicine', in R. H. Campbell and A. S. Skinner (eds), *The Origins and Nature of the Scottish Enlightenment* (Edinburgh, 1982), p. 87.

[74] Dingwall, *Physicians, Surgeons and Apothecaries*, p. 110.

[75] W. S. Craig, *History of the Royal College of Physicians of Edinburgh* (Oxford, 1975), p. 51.

[76] This occurred, for instance, in 1848 and was successfully staved off by the passing, on 10 June 1850, of the 'Act for Better Regulating the Privileges of the Faculty of Physicians and Surgeons of Glasgow and Amending their Charter of Incorporation'.

The pursuit of such all-encompassing rights understandably produced strong reactions in the Scottish medical community. These came not only from the medical corporations (with predominantly surgical memberships), who feared encroachment of their rights and powers, but also from King's College, Aberdeen. The last stated that the physicians' proposals to license practice throughout Scotland were in direct contravention of their own charter which permitted Aberdeen graduates right of practice anywhere in the world. (This argument was revived, to effect, in the nineteenth-century dispute between the University of Glasgow and the FPSG over surgical licensing.) [77] But what action did the FPSG take on the physicians' proposal? Presenting copies of the petition submitted to the council of state and of a commission granted to William Brodie (or Brady) and Mr William Lightbody to act as agents in their absence, Hall and Graham reported back to the Faculty on 20 June 1657. Members were to discuss their objections to the physicians' patent before Hall and Graham were commissioned to return to Edinburgh. So important did the Faculty consider the case that by 6 August 1657 they had appointed no fewer than six members – John Hall, Dr Crichton, Messrs. James Hamilton, Archibald Graham, Thomas Lockhart, Robert Harris and Archibald Bogle – to 'Act & do what is necessary & Incumbent in reference to all transactions for the good of the faculty'.[78] At this juncture, considerable opposition was being mustered by Edinburgh town council (with which the Edinburgh surgeons had very close relations) against the Edinburgh physicians through the convention of royal burghs,[79] 'considering the dangerous consequences [which] may follow to the whole nation, and especially to their estate, if the patent granted for erecting a college of physicians in this burgh of Edinburgh be not escaped'.[80]

Having incorporated the Glasgow surgeons in the previous year, the town council also took keen interest in the Edinburgh negotiations. On 6 September 1657 it sent for the deacon of the surgeons, who 'was discharged publicly that neither he nor his brethren of calling should make any kind of agreement with the doctors of physic about the college of physicians craved to be erected

---

[77] Hamilton, *The Healers*, pp. 68–69.

[78] RCPSG, 1/1/1/1b, pp. 133–34.

[79] This was an independent body, beyond direct control of the crown, which met anything up to four times a year and to which each royal burgh sent representatives in order to regulate burgh affairs and to defend their privileges.

[80] For this, see J. D. Marwick (ed.), *Records of the Convention of the Royal Burghs of Scotland: With Extracts from Other Records Relating to the Affairs of the Burghs of Scotland, 1295–1738*, 5 vols (Edinburgh, 1866–90), iii, *1615–1676*, pp. 441. The debate continues on pp. 443, 448–51, 460, 462, 468–70.

by them, until he did first acquaint the council'.[81] Two days later, the town council asked the Visitor to convene a meeting of the Faculty to consider whether they wished to adhere to their old gift or join with the College of Physicians in Edinburgh. They chose, not surprisingly, to adhere to the original charter, and proposals for the new college were put aside at Cromwell's death. But the costs of Faculty negotiations and defence in Edinburgh were exorbitant; in all £117 14s. 0d. was expended on the two commissioners' trips to Edinburgh and on advocates and agents, with a further £12 to James Lockhart for his charges in attending the council of state and the surgeons of Edinburgh.[82] With Faculty membership in the seventeenth century routinely below fifty at this point, corporate funding cannot easily have covered such costs.

One of the unfortunate by-products of the Faculty's extended litigious action was the dispersal of their original documents; it led ultimately to the loss of their original charter. In the post-Restoration period, Daniel Brown was appointed on 9 May 1661 to receive from William Brodie, advocate in Edinburgh, a decree passed on the charter. The following September, Brown brought back the decree from Edinburgh, on which new letters of horning were raised in the name of Charles II, at a cost of £6 13s. 4d. The decree (which cost £5 16s. 0d. to retrieve) had been left with Brodie when the Faculty was at law with the College of Physicians. Left there by John Hall and Thomas Lockhart, it had lain for six years with the Edinburgh Incorporation of Surgeons.[83] Brown's brief, clarified in February 1662, was 'to cause denounce certain persons contraveners of their gift & letters thereon for their practice of physic and surgery who are neither warranted nor skilled'. Letters of caption (a warrant of arrest) were to be procured against them for the purpose.[84]

The members of the body incorporated in 1599 sought subsequently to broaden the provisions of the original charter to cover the additional personnel brought into their association under the 1656 letter of deaconry.[85] On 11 September 1672, the Faculty was granted a parliamentary ratification,

---

[81] Duncan, *Memorials*, pp. 69–70. Duncan states that this was 5 September, but since this was a Saturday and the Faculty met on a Monday, and the next meeting was 8 September, it would appear to have been the 6th.

[82] RCPSG, 1/1/1/1b, pp. 145, 151.

[83] The Edinburgh surgeons had bought Curryhill House in the south-east corner of Tron Kirk parish in 1657, but had to undertake extensive repair work on it which undoubtedly resulted in administrative disarray. Even then, it was not considered entirely appropriate, since they tried to raise money in 1669 for the building of a new hall. Dingwall, *Physicians, Surgeons and Apothecaries*, p. 59.

[84] RCPSG, 1/1/1/1b, pp. 188, 191, 196–97 (quotation), 205.

[85] FPSG, *The Royal Charter and Laws*, pp. 11, 13.

reiterating and upholding the provisions of its 1599 charter, and additionally extending it in favour of apothecaries and barbers. But the ratification served a far more serious purpose: it freed the Faculty from fear of unilateral crown forfeiture of their charter. Granted by the king, a royal charter could be similarly revoked; but if ratified by Parliament, rights and privileges could be annulled only by a subsequent Act. This gave the association a degree of security, because the Incorporation of Surgeons and Barbers was authorised only by local civic authority. (It was common for many corporations to try to procure another charter from the crown or Parliament which ensured its identity outside the town.)[86] Procured at his own expense, the ratification was the initiative of John Hall. It was read in the hearing of the entire Faculty, who approved it 'as being a deed done by him in all their favours and for their advantage'. Nonetheless, because his immediate repayment would weaken its stock and the poor fund, members decided that if Hall was willing to wait they would pay him when sufficient money was in hand to defray his costs. He was not reimbursed until 1681 when paid 500 merks for his trouble.[87]

In seeking ratification at this time, the Faculty was reacting to developments in Scottish burgh politics: 1672 was the year in which a parliamentary act removed anachronistic trading monopolies from the royal burghs (especially their monopoly on foreign trade), opening them to greater competition from the burghs of barony. This has generally been seen as a positive development which forced the burghs into reviewing 'the elaborate web of protectionism with which they had surrounded themselves'; it resulted in a relaxation of the tight controls on entry to burgess-ship and to merchant guilds, and the beginning of the breakdown of formal apprenticeship. Glasgow particularly flourished from this point because it was able to throw off the restrictions of the medieval urban economy, allowing traditional monopolies to go by the board in an era of intense commercial activity.[88] But initially, it led to attempts at consolidation. The 1605 letter of guildry which had established the Trades' House was ratified by an Act of Parliament on the same day[89]

---

[86] Smith, *Nine Trades of Dundee*, p. 27.

[87] RCPSG, 1/1/1/1b, pp. 403–4 (quotation), 680.

[88] M. Lynch, 'Continuity and Change in Urban Society, 1500–1700', in R. A. Houston and I. D. Whyte (eds), *Scottish Society, 1500–1800* (Cambridge, 1989), pp. 85–86. For aspects of these changes, see T. C. Smout, 'The Glasgow Merchant Community in the Seventeenth Century', *Scottish Historical Review*, 47 (1968), pp. 53–71; T. Devine, 'The Merchant Class of the Larger Scottish Towns in the Seventeenth and Early Eighteenth Centuries', in G. Gordon and B. Dicks (eds), *Scottish Urban History* (Aberdeen, 1983), pp. 92–111.

[89] R. Douie, and revised by F. Gibb Dougall, *Chronicles of the Maltmen Craft in Glasgow, 1605–1879* (Glasgow, 1895), pp. 6–7, 175.

in a flurry of ratification in which various institutions tried to preserve those features of privilege and restriction established by their charters to restrict outsiders from entering their ranks.

Surgeons and Physicians in Conflict, 1671–72

The question of the physicians' status in the FPSG was to the fore in the late seventeenth century. There may have been a knock-on effect from the Royal College of Physicians in London which – after trials and tribulations in the Interregnum – obtained, in 1663, a new charter that it hoped would strengthen its monopolistic regulation. But a disparate group of healers, including the Society of Apothecaries and the Barber-Surgeons' Company, as well as some of the leading practitioners in the field, managed to block the charter's reiteration in Parliament.[90] The FPSG was experiencing similar problems of professional alignment within its membership, and decided, in the following decade, to apply for ratification of its own charter.

The municipal alliance with the barbers discouraged the physicians from partaking fully in their Faculty membership for a number of years, unwilling though they were to merge with their Edinburgh counterparts. While Robert Mayne, Professor of Medicine at Glasgow University, had been prepared, when Visitor, to be the Faculty's delegate in the Trades' House,[91] after the erection of the Incorporation, the few practising physicians had little wish to associate themselves with mere craftsmen. But both physicians and surgeons experienced a loss in the separation. The former no longer reaped the benefits of professional association; the latter suffered loss of respect. The physicians having shown themselves unwilling members, the surgeons countered by raising a question on future admissions policy: did doctors of medicine living within the Faculty's jurisdiction have automatic right of admission?[92]

The point was soon tested. In 1671 three physicians, Dr John Colquhoun, Dr Mathew Brisbane and Dr Thomas Hamilton, sought admission to the Faculty. Meeting on 3 January 1671, the surgeons discussed their opinion of the articles drawn up by the calling to present 'to the physicians who are desiring to be incorporated with the faculty'. They decided, by a majority of votes, that the said articles were to stand as an agreement between them and the physicians, and were not to be altered. Chosen as representatives to meet

---

[90] Clark, *Royal College of Physicians*, i, pp. 304, 374–75; Cook, *Decline of the Old Medical Regime*, p. 24. They were challenged in addition by the Society of Chemical Physicians and by the growth in influence of the Royal Society.
[91] Mayne entered the FPSG in 1645.
[92] Duncan, *Memorials*, pp. 61–62.

with the physicians, the Visitor, John Hall, and Archibald Bogle, in presenting the articles, were to make it clear that the Faculty would agree to nothing further. They were to feed back the physicians' reactions. The question of whether physicians within the burgh of Glasgow had a natural right to the king's charter was put to the vote on 7 February. Six favoured the motion, but the majority ruled that it did not carry such a right. Nonetheless, on a second and contingent vote of 10 February, the members unanimously decided to admit the three physicians.[93] This decision is important because it establishes that, *de facto*, the Faculty (in the seventeenth century at least) was essentially an incorporation of surgeons. Since possession of a degree did not win them right of entry, the physicians accordingly chose to stay out – at this juncture.

Newly fortified by its parliamentary ratification in 1672, the Faculty took steps to repair the rift with the physicians who were seeking admission, as its institutional power-base would undoubtedly be strengthened if the physicians joined them. On this occasion, each group put forward the conditions on which they were prepared to reunite. The surgeons considered, on 14 November 1672, the ease with which agreement of certain physicians in their bounds might be had in sharing the legal exercise of their powers of visitation (and other duties and privileges) contained in the charter and its recent ratification. The Visitor, Archibald Bogle, and John Hall, late bailie, were therefore assigned to treat with Dr John Colquhoun and Dr Thomas Hamilton in that regard, and 'to deal with them For concurring with the said faculty about the regulation of medicine within their bounds and sitting with them upon honorable terms'.[94]

The Visitor and Bailie Hall reported on 16 December 1672 that they had made overtures to the physicians but that, before they could be fully accommodated, the Faculty had to pass an act in their favour. The following day, Dr Colquhoun was present when a draft invitation was read enjoining the physicians to unite with them. The surgeons expressed their opinion (with reference to the original charter) that cooperation with some physicians within their bounds 'is enjoined and necessary (if it can be had) for the right and legal exercise of the power of visitation and other duties and privileges contained in the said gift'. Furthermore, the growth of quackery and in irregular practitioners, who not only practised surgery but 'also take upon them in all internal diseases to prescribe and administer physic contrary to the tenor of the said gift', was held to be due to this defect in the Faculty's existing constitution. Rapprochement with the physicians was an attempt to

[93] RCPSG, 1/1/1/1b, pp. 356–57, 359 (quotation), 360.
[94] RCPSG, 1/1/1/1b, p. 405.

remedy this. The surgeons resolved that the time had come 'to procure (if it can be had) the concurrence and assistance of some physicians one or more within our bounds'.[95]

The surgeons conceived of a body where authority was jointly shared and exercised by representatives of the physicians and of the surgeons. To this end, Dr John Colquhoun was asked 'To take upon him and exercise the office of visitor conjunct with the present visitor of the Surgeons and pharmacists as fully, freely and honorably in all points as Mr Robert Hamilton did or might have done with Mr Peter Low according to the first intention and at the procuring of the said gift'. The Visitor and Bailie Hall continued in their role as negotiators; they were to satisfy Dr Colquhoun in any scruples he had in this regard, and to condescend to whatever proposals he might make on two conditions. First, any proposals were not to be 'destructive to the said Incorporation or to the Ratification of the said gift in our favour' – that is, the physicians were not to be permitted to undermine the newly-acquired status of the barbers. In second place, they were not to include other physicians within the Faculty boundaries or any adjoining town 'since letters of graduation or sojourning within the bounds give no right to physicians to claim an interest in the said gift But physicians of Glasgow who are called & acknowledged as such by the present faculty in whose favour the gift is now ratified in this current parliament'. In other words, the Faculty had not conceded their stance of 7 February 1671 that possession of a degree did not entitle a physician to admission to the Faculty. In this, they were not only contravening the spirit of the 1599 charter (if not going against its letter), but also applying restrictions on the geographical boundaries mentioned within it. Everything else which Dr Colquhoun might ask to secure his reputation from 'aspersions' (cast on him by the barbers), or for easing his burden in office, was to be conceded to him.[96]

Colquhoun put forward the physicians' case at a meeting of 17 December 1672. Expressing his desire to join the Faculty, he nonetheless wished its members to concede a number of points. The most important of these was: 'That mine and my brethren's share in the power of visitation be declared not to be precarious or dependant upon the bare [exclusive] call of the surgeons'; it was, instead, to be founded on the express tenor of the charter. Furthermore, Colquhoun asked for it to be minuted that the king's charter explicitly enjoined the concurrence of some physicians – a specification which went contrary to the surgeons' plea for secluding physicians until they had been called and acknowledged by the Incorporation. In so doing, the doctor

---

[95] RCPSG, 1/1/1/1b, pp. 428–29, 430–32 (quotations); Duncan, *Memorials*, p. 62.
[96] RCPSG, 1/1/1/1b, pp. 432–33; Duncan, *Memorials*, p. 63.

doubtless sought to accord a degree of dignity to the physicians' entry. Secondly, at least two physicians were to be admitted as Faculty members – one of whom was resident in the town (and could deputise for Colquhoun as Visitor, in his absence) and the other in the country, 'unto whom some points of the power of visitation may be committed for the exercise thereof'.[97]

The physicians insisted, thirdly, that all matters pertaining to the charter were to be determined by the Physician-Visitor and his brethren, with the Surgeon's Visitor and brethren *communi concilio*; by way of proviso, any act against which the physicians protested was prejudicial to their degree was to be declared void. Fourthly, the physicians did not intend to interest themselves in any courts or statutes enacted by the surgeons in relation to their craft business. But it was necessary, fifthly, 'That respect be had to the degree and dignity of physicians especially those incorporated, and to the physician's visitor'. The last was to be granted precedence by the others, and to insert his name first in all acts and letters. The Surgeon-Visitor was to take an oath, in addition, in the name of his brethren, 'to seek the honour and advantage of the physicians especially those incorporated'. Similarly, the Physicians' Visitor had to swear to uphold the welfare of the incorporated surgeons and pharmacists and to defend their privileges. In the sixth instance, a select number of surgeons and apothecaries – chosen by their Visitor – were to vote and hold court with the two Visitors. Finally, both Visitors were to have power to convene the Faculty when matters arose of concern to them.[98]

The physicians' articles were passed by the Faculty under the following provisos: 'That none of the physicians' servants shall be privileged or have any power To meddle with Surgery or pharmacy within [the] burgh except in cases of necessity for want of Surgeons and pharmacists'; the nomination of any assessor (to take on the office thereafter) was to be carried by plurality of votes, the assessor always being a doctor. Colquhoun accepted these conditions, taking up the office of Physician-Visitor conjointly with the Surgeon-Visitor. Both sides then exchanged oaths: 'The said doctor to maintain the just rights and privileges of the said incorporation with their welfare and the said members to maintain the honour and advantage of the said physician visitor and did take [each] other by the hand in further testimony of their unanimous assent to the premise'. Colquhoun proceeded to nominate Dr Thomas Hamilton as the fittest assessor for the town and Dr Michael Wallace in Ayr for the country. The Faculty passed the nominations. Colquhoun, the Surgeon-Visitor and his brethren chose eighteen members[99] – any nine of them convening as

---

[97] RCPSG, 1/1/1/1b, pp. 434–36; Duncan, *Memorials*, p. 63.
[98] RCPSG, 1/1/1/1b, pp. 436–38.
[99] John Hall, late bailie, Archibald Bogle, Mr Archibald Graham, Messrs David Sharp and

a quorum – to have power to sit and vote in court with the Visitors 'as capable to exercise several articles of visitation contained in the king's gift'.[100]

What was the ultimate outcome of these negotiations? Though the surgeons presented a united front, almost everything the physicians demanded was in effect conceded to them. They even secured a veto on the Faculty's resolutions. They dissociated themselves from those members operating under the letter of deaconry and were careful to associate themselves with those organised under the charter. With the formal acceptance of the physicians, the association was restored to some kind of professional balance but it resulted, to all intents and purposes, in one body with a dual constitution. One part was the medical and surgical members – admitted on different conditions and with their own leaders; the other was the craft element. Surgeons faced in two directions, interacting with the physicians at the top and with the barbers at the bottom of the occupational totem.[101]

The FPSG versus the Magistrates of Glasgow

The attempted inclusion of four grades of practitioner under one corporate umbrella resulted, as might be expected, in a plethora of problems. Discontent had reared its head as early as 1657: when an apothecary was appointed deacon of the Incorporation, his election had to be declared void because, on appeal by the surgeons, it was found that the letter of deaconry specified that the deacon had to be a surgeon.[102] A hierarchy had therefore been established by the letter of deaconry. Understandably, because it accorded them greater autonomy, the Faculty were unwilling to relinquish the rights and privileges granted by the charter and to operate solely under the constraints of the town council and Trades' House. The FPSG's autonomous rights were tested in 1679 (five years after the parliamentary ratification), when one Harry Marshall complained to the Glasgow magistrates and town council of the surgeons' and pharmacists' adamant refusal to admit him as a freeman with them. The council ordered them to specify why Marshall should not be admitted by their next meeting, failing which 'the town will grant him licence to set up'.[103]

Charles Mowat, James Thomson, Andrew Elphingstoun, John Hall, son of the above John Hall, Daniel Brown, John Robison, Adam Gray, Robert Houston, George Lockhart, Andrew Ralstoun, Evir McNeill, James Weir, John Liddell and John Fleming.

[100] RCPSG 1/1/1/1b, pp. 439–41.
[101] Duncan, *Memorials*, pp. 64–65.
[102] Tennent, *Incorporation of Barbers*, p. 24.
[103] J. D. Marwick (ed.), *Extracts from the Records of the Burgh of Glasgow, AD 1663–1690*, iii, Scottish Burgh Records Society (Glasgow, 1905), p. 271.

Henry Marshall's is a curious case. Recently translated from Kilsyth – where his father Patrick Marshall (1631–1697) practised as a surgeon [104] – he had petitioned the Deacon of the Surgeons, on numerous occasions, to be admitted a freeman since he was already a burgess and guildbrother professing the arts of surgery and pharmacy. Yet (according to his petition), though they had already admitted both burgesses and strangers, the surgeons refused outright to admit him even were he found qualified on trial and paid the dues.[105] The Incorporation refused even to try him! This, of course, made a mockery of burgh privileges. The situation had arisen in the first place because – using their power of making acts for the benefit of practitioners under the 1599 charter – the Faculty, 'taking to their serious consideration the prejudice that may arise through their promiscuous admission of strangers to practise Surgery and pharmacy within the city of Glasgow', had passed on 25 March 1679 an act restricting the admission of strangers. This ordained: 'That no person or persons whatsoever shall in any time coming Be admitted to practise either of the saids arts of surgery and pharmacy within the city of Glasgow But such as either have served their apprenticeship with a freeman or member of the faculty for the time for the space of Five years ... or otherwise be a freeman's son or married to a freeman's daughter'. Admission was to be refused to outsiders. One exception was granted to the magistrates, who were to be permitted, if there was a shortage of qualified surgeons, to introduce one or two experienced surgeons and pharmacists to the city, subject to trial by the Faculty and payment of freedom fines to the poor.[106] Admittance of pharmacists caused more problems than that of surgeons. Indeed, it is more than likely that Henry Marshall was deliberately introduced by an irate council to test the Faculty's new act of exclusion.

Marshall supplicated that the town council grant him 'such licence to set up as if he were really admitted with the said calling'. When the Visitor of

---

[104] Duncan, *Memorials*, p. 246. Through the marriage of Henry's daughter Lillian to Alexander Horsburgh, surgeon, and of his granddaughter to the merchant Robert Cowan, the Marshalls fed into the renowned medical dynasty of Cowans which produced Robert Cowan MD (Professor of Medical Jurisprudence, University of Glasgow, 1839–1841) and John Black Cowan MD, joint editor of the *Glasgow Medical Journal* (Lecturer in Medical Jurisprudence at the Andersonian University, 1856–1863; Professor of Materia Medica, University of Glasgow, 1865–1880). Ibid, pp. 186, 246; GUL Special Collections, Mu23-a. 3, Andersonian Institution, vol. 3, testimonials in favour of John B. Cowan, MD, Fellow of the FPSG, and late civil surgeon attached to the army in the Crimea; candidate for the Lectureship of Medical Jurisprudence in Anderson's University, Glasgow, William MacKenzie, 1856; J. D. Comrie, *History of Scottish Medicine*, 2 vols (London, 1932), ii, p. 662.

[105] Marwick, *Records of the Burgh of Glasgow, 1663–1690*, p. 273; Duncan, *Memorials*, p. 79.

[106] RCPSG 1/1/1/1b, pp. 629–31.

the Surgeons did not compear to counter Henry Marshall, the council duly colluded in granting him 'full power, licence and liberty to set up and exercise his calling as surgeon and apothecary and pharmacist with this burgh, as amply in all respects as if he were admitted freeman with the said calling of surgeons'.[107] This bold step, in overriding of the surgeons, caused tremendous ramifications in local politics. The case against Marshall was the first major jurisdictional battle in the history of the Faculty which set the standard for future cases. By 31 October 1679, the FPSG was preparing the way: 'The members of the faculty Finding it necessary and convenient that they consult their Rights and privileges For preventing any debates [that] can be raised thereabout', ordained Bailie Hall to go to Edinburgh for that purpose. Taking various Faculty writs to consult with learned counsel in the capital, he also submitted a query which was formally answered by Sir John Cunningham and Mr William Hamilton, advocates 'with some private instructions to the faculty for exercising their privileges with an unsubscribed double [copy] of the act of the town council in favour of Mr Henry Marshall'. He was reimbursed by the sum of £68 7s. 8d. Scots.[108] The Faculty ultimately raised an action of declarator [109] against the magistrates and town council of Glasgow in the Court of Session and the case dragged on for years. High-ranking counsel were employed on both sides.[110]

The magistrates' main line of defence was that the Faculty were merely seeking to create a monopoly for their own services (which they clearly were, since Marshall was qualified), but could injure the city by reducing the numbers eligible to be admitted as freemen surgeons. In other words, the FPSG was using the privileges of its charter contrary to the benefit of the town. The Faculty counterclaimed that power to make such laws was never granted to those who would make them to the detriment of the public good; the vetoing of its decision would result only in the subordination of royal and parliamentary authority to municipal authority.[111] Andrew Wear writes of early modern Europe (1500 to 1700) that: 'City colleges of physicians and guilds of apothecaries and surgeons were common in France, Germany, Italy, the Netherlands and England. Their attempts at regulation were supported

---

[107] Marwick, *Records of the Burgh of Glasgow, 1663–1690*, pp. 273–74.
[108] RCPSG, 1/1/1/1b, pp. 637, 654–55; Duncan, *Memorials*, p. 80.
[109] In Scots law, an action of declarator was an action brought by a party with a vested interest to have some legal right or status officially declared, but without claim on the defender to do anything.
[110] Sir Hugh Dalrymple, later Lord President, was engaged for the Faculty, and James Stewart, senior, later Sir James Stewart of Goodtrees, for the magistrates.
[111] Duncan, *Memorials*, p. 80.

by governments and city councils, but also at times frustrated by them'.[112] Of nowhere was this more true than Glasgow.

The history of the FPSG in the seventeenth century can easily be summed up as a power-game between the surgeons and the local oligarchy: the Faculty versus the magistrates of Glasgow. According to a town council minute of 13 March 1683, the town treasurer gave £5 sterling to one John Maxwell, wright, to pay to a mountebank for cutting off Archibald Bishop's leg.[113] It is hard to imagine that there were no indigenous surgeons capable of doing this. But, significantly, in the following year the council decided that it no longer had money to maintain a town's physician or surgeon. Was this really the case, or was it simply removing its patronage from the Incorporation, whose surgeon members had held the post since Peter Lowe was first appointed? A town's surgeon was not reappointed until November 1686.[114]

The Faculty continued to put pressure on the magistrates at local level. In May 1691, the surgeons petitioned the town council asking it to approve the act of exclusion made by the Incorporation on 25 March 1679 and to rescind its act in Marshall's favour. The FPSG stated in its petition that the magistrates:

> did maliciously adhere to a supplication put in by Mr Henry Marshall, whereby he desired to be privileged to set up and practise their arts over their bellies and without their consent ... The granting whereof has ruined the whole incorporation of surgeons to this day, and whereby they were necessitate ever since, to separate themselves from the rest of the incorporations of trades, so that the said whole incorporation of trades has found the prejudice and loss of having a particular calling disjoined from them.

The Incorporation of Surgeons and Barbers boycotted its automatic representation in the Trades' House; Marshall's admission 'worried the whole incorporation of Surgeons and made them separate from the rest of the incorporations of trade so that they all had sustained a considerable loss in the maintenance of their whole poor'.[115] On 9 May 1691 council capitulated, rescinding its act in favour of Henry Marshall,[116] and recognising that by virtue of their gift, ratification and possession, 'The said Visitors and their successors have The undoubted right of licensing and authorizing fit and

---

[112] A. Wear, 'Medicine in Early Modern Europe, 1500–1700', p. 235.
[113] Marwick, *Records of the Burgh of Glasgow, 1663–1690*, p. 330.
[114] Ibid., pp. 368, 392.
[115] RCPSG, Decree of Declarator, the Surgeons of Glasgow against the Magistrates, 28 July 1691, fol. 3.
[116] J. D. Marwick (ed.), *Extracts from the Records of the Burgh of Glasgow, 1691–1717*, Scottish Burgh Records Society, 4 (Glasgow, 1908), pp. 16–19.

skilled persons in the said art and trade and debarring all such as are not duly author[iz]ed From all practise and exercise of the said trade'. It also conceded 'that the Magistrates and Council of Glasgow have no right nor power to warrant or authorize any persons to exercise surgery or pharmacy within the city of Glasgow except such as are duly approven by the visitors conform to the rules and statutes made about the admission of fit persons for that effect'. Any warrant granted by the late magistrates in favour of persons not duly admitted to set up shops and to practise within the town was declared void.

On 9 July 1691, the Court of Session also decided in the Faculty's favour. The Lords of Session declared in favour of the Surgeons of Glasgow 'That the Magistrates and Council of Glasgow have no right nor power To warrant or authorize any person To exercise Surgery or pharmacy within the city of Glasgow Except such as are duly Approven of By the visitors'. In effect, then, the town council had been usurping the right to license medical practitioners in Glasgow. Since this right had been taken away, the town council's advocates insisted that the court also declare that the incorporation should be obliged to supply sufficient number of qualified persons for the city's needs.[117] This decision also officially established the preeminence of the Faculty under the charter (and not under the letter of deaconry). The Deacon of the Surgeons and Barbers did not return to the Trades' House for almost two decades (1709) – and by that time there were serious problems between the surgeons and the barbers – so the dispute had important consequences for the surgeons. Marshall, on the other hand, happily proceeded to the office of Visitor in 1703, 1704 and 1705.[118]

This was the end neither of the story of the act of exclusion nor the association of the Marshalls with it. In 1698 another action was heard by the Court of Session between Andrew Reid, surgeon in Glasgow (former apprentice of John Marshall, also surgeon there), and the FPSG. The trouble arose because Reid wished to practise in Glasgow but his master was not a member of the Faculty.[119] Therefore, although Reid had paid his fines, had been made a burgess of Glasgow on 15 April 1697,[120] and had subsequently been practising and paying his stent (a local land tax) for the practice since that time, he was denied admission to the Faculty. In effect, the FPSG were

[117] RCPSG, Decree of Declarator, the Surgeons of Glasgow against the Magistrates, 28 July 1691, fos 5 (quotation) and 6. This also appears in the conclusions on fol. 7.
[118] Duncan, *Memorials*, p. 84.
[119] John Marshall might be the brother of the aforementioned Henry Marshall. Patrick Marshall, surgeon in Kilsyth had another son, John, who died in 1719. Duncan, *Memorials*, p. 246.
[120] SRO, CS232 G/1/13, box 170, Court of Session Processes, Andrew Reid his tickit 1697.

re-establishing the boundaries of their act of exclusion but Reid contended that this was 'an unwarrantable act made by the surgeons themselves, discharging all people without respect to their qualifications with the burgh of Glasgow, except if they serve five years as apprentice to one of their Incorporation or marry one of their daughters'. Reid was ordered to pay £40 Scots for each of three transgressions 'in practising of the said Art', and told to desist from practice until found qualified and admitted to the Faculty. Reid argued that their charter gave the Surgeons no right to do this without recourse to the magistrates (who obviously supported him). The Surgeons' action had been pursued unilaterally, in closed court, without allowing anyone to plead against the corporation. Reid's defence was that although the town council had ratified the corporation's act of 1691, it only had power to ratify acts of its own incorporations whereas the Faculty was constituted under a royal charter.[121]

According to the terms of the charter, Reid had offered himself ready to take a trial on 3 June 169[7],[122] and had the sense to have this recorded by a notary public. Declaring that 'this matter is altogether invidious and patched up & set on foot by a set of people of the said Incorporation who endeavour to enhance the whole employment in that country and to seclude persons as well if not better qualified than themselves',[123] he produced his indentures of apprenticeship before James Weir, then Visitor of the Faculty,[124] dated 15 December 1691.[125] In view of the subsequent difficulty which Reid had in gaining entry to the profession, the most pertinent clause in the agreement is that 'the said John Marshall shall be hereby obliged not only to procure the said Andrew Reid his freedom with the town of Glasgow, but also with the Craft & vocation Conform as he either has already or shall hereafter procure the same himself'.[126]

---

[121] CS232, box 170, Court of Session Processes, 29 February 1700, summonds James Weir and John Boyd.

[122] The final year is blank in the original, but Reid made his original application for a trial in 1697.

[123] CS232, box 170, Court of Session Processes, 29 February 1700, summonds James Weir and John Boyd.

[124] SRO, CS232, box 170, Court of Session Processes, instrument Andrew Reid Contra Visitor of the Chirurgeons, 1698.

[125] These were between him, lawful son of the late Mr William Reid of Dalldilling, and John Marshall, surgeon-apothecary. Note that the last figure of the date is in the fold of the manuscript and is unclear.

[126] SRO, CS232, box 170, Court of Session Processes, indentures betuixt John Marshall and Andrew Reid 1691. The booking of Reid's apprenticeship with Marshall was noted in the records of the craft's rank of the city by the clerk of the incorporation on 24 February 1694.

Submitting a discharge from Marshall, dated 1 March 1697, that he had faithfully fulfilled the terms of his indenture, Reid asked Weir for a trial in surgery and pharmacy in front of the Faculty so that he could be found qualified and licensed to practice. Weir responded that the Faculty was willing only to admit him to practise outside Glasgow, because 'John Marshall with whom He served his apprenticeship was not free with the Faculty'. Reid replied that since his master had practised in Glasgow, and kept a public shop in the most eminent part of the town, that the Faculty's private connivance must be deemed sufficient to admit him. He also protested that the 1599 charter did not restrict residence if an entrant was found qualified (which was true); therefore the Faculty might be liable for all costs, damages and expenses sustained by him through their prohibition and denial of his admission.[127] The case went to the Court of Session in 1698. On 15 January 1700,[128] Reid proved that he had paid the appropriate dues for his trade, namely £15 'as his stent for Trade for Ten months' from Martinmas 1696 to 1697, and obtained a bill of suspension against the Faculty.[129] The amount spent by Andrew Reid between 1697 and 1701 (a grand sum of £607 19s. 10d. Sterling), 'in the tedious process pursued by the Surgeons of Glasgow against him',[130] tends to indicate that the occupation of surgeon-apothecary in Glasgow in the late seventeenth century was relatively lucrative.[131] Against this, Reid claimed for 'Loss of Employment and patients and my customers which has tended to the utter ruin of my family and Employment which every session could not be under ten pound[s] sterling'.[132]

In a printed petition to the Lords of Council and Session against the Faculty, dated 7 January 1702, Reid stated that before his return to Glasgow in 1697 he had spent time in Flanders, in the king's service, 'whither he went

---

[127] SRO, CS232, box 170, instrument Andrew Reid Contra Visitor of the Chirurgeons, 1698.
[128] SRO, CS232, box 170, instrument and protest Andrew Reid contra Weir, Boyd and others, 1700.
[129] SRO, CS232, box 170. Two receipts for Reid's payment of stent for trade (as stented in the stent roll) appear in box 170; one is signed by George Lander and dated 22 January 1698; the other is signed by John French and dated 15 April 1698.
[130] In Scots law, the term process refers to the step-by-step action and procedure in a judicial case, comprehending all those writs, forms and pleadings brought under judicial cognisance.
[131] Reid's costs included £48 for incident expenses for five weeks in Edinburgh in 1699, £5 10s. 0d. for the hire of a horse from Glasgow to Edinburgh in 1700, £66 13s. 0d. for constant attendance in the session in 1701.
[132] SRO, CS232, box 172, Court of Session Processes, accompt of expenses for Andrew Reid in the actione att the Instance of the Chirurgeones of Glasgow against him 1702. Note that one of the amounts on p. 2 is overwritten and therefore unclear, but has been read as 05 12 00. In any case, the total still exceeds £600.

for accomplishing himself in his said Art and Calling'.[133] Legal cases later in the eighteenth century show that the Faculty often had problems with surgeons returning from military or naval service who considered that they had been rigorously enough tried by the examination boards of these bodies.[134] But the most significant aspect of this case is surely that, a century after its inception, surgeon-apothecaries were still practising in Glasgow without being under any compulsion to join the Faculty, which was trying its best to exclude them but without the juridical teeth to ban them completely from practice.[135]

Losing on the medical licensing front, the magistrates vented their spleen on the surgeons over aspects of their special exemptions and immunities from taxes and public duties which they viewed as anachronistic. In spite of the special privileges laid down in the charter, the magistrates sought to tax the Faculty in a similar fashion to other people, so that they were manoeuvred into paying cess, stent and poor rates as well as having soldiers billeted on them. On 17 July 1694 the magistrates obtained a decree of the Court of Session against eleven defenders – 'all Surgeons and Pharmacists in Glasgow' – from whom they had tried unsuccessfully to levy stent. Finding the council's charges in order, the Lords ordained the members of Faculty to make payment.[136] In relation to local taxation at least, a legal precedent was established in favour of the council, but this did not stop the Faculty from attempting to uphold their original gift. On 28 October 1704, for instance, the council met to discuss a petition given in earlier in the month by the Visitor, John Boyd, and members of 'the faculty of surgeons and pharmacists

---

[133] SRO, CS 232, box 170, Court of Session Processes, headed: unto the Right Honourable the Lords of Council and Session, the Petition of Andrew Reid Chyrurgeon in Glasgow, against the Chyrurgeons of Glasgow, 7 January 1702. John Marshall, brother of Henry Marshall is known to have studied in Paris in the late 1670s. See Duncan, *Memorials*, p. 246.

[134] See, for example, the case of Alexander Dunlop.

[135] Since these processes are unextracted and no date survives for the verdict, it has proved impossible to obtain the final outcome in Reid's case. In the Court of Session (and also in inferior courts), after the pronouncement of any judgment, a certain time – twenty days in the Court of Session – was allowed for submitting it to review; if such judgment was not reviewed within the allotted time, it became the final judgment in the cause (subject to no review except that of the House of Lords). Where it was necessary to enforce implementation of a decree, legal execution could be obtained only under an extract of the decree, authenticated according to the forms of the court in which it was pronounced. Robert Bell, *A Dictionary of the Law of Scotland*, 2 vols (3rd edn, Edinburgh, 1826), i, pp. 379–80. This, of course, was expensive. After what had already been expended on the process itself, where no further action was intended on the decree, it was often left unextracted and the verdict lost.

[136] GCA, A2:25, decreet absolvitor in part and for expences the Magistrates and Town Council of Glasgow against the Faculty of Physicians and Surgeons of Glasgow, 1794, p. 24. The decree begins with the history of earlier legal cases.

of this burgh', showing that though they had been warned by the town officers to weaponshowing, watching and warding with the other inhabitants, they were not liable

> because the same is inconsistent with their employment of surgery and pharmacy, being operations on the bodies of men and dispensing of medicines to them, which in a great many particulars are ordinary daily occurrences can admit of no delay without hazard of life or danger of the health of the patient, which the petitioners conceive is (without any order) sufficient ground for exemption, especially considering the paucity of their number in such a populous place as this.

In Edinburgh, Linlithgow, Perth and most other burghs, medical practitioners were apparently so exempted. But the town council found the surgeons and apothecaries liable. The argument made for their liability was that, on joint application, the surgeons and apothecaries of the burgh had been 'incorporated with the barbers in a deaconry' by the town council in 1656. In consequence, they were bound by their burgess oath to all taxations, watchings and wardings laid on the burgh; indeed, they were under command of the magistrates to participate and not to purchase or use exemptions. They were bound further 'in respect that they keep shops and open traffic of trade and take apprentices who by their apprenticeship have the privilege of burgess and guildbrother as well as any other apprentices of merchants or trade'.[137]

Nonetheless, on almost each occasion that the question arose during the next century, the Faculty continued stubbornly to pursue the same argument, in an attempt to retain their privileged immunity from local taxation, and in the forlorn hope that the decision might eventually be overturned. In 1709 the Faculty won a small victory when its position was upheld in an act of adjournal of the circuit court of justiciary declaring that members of Faculty were, by their charter, exempted from serving as jurors.[138] Almost fifty years later they were still fighting the same corner. When the magistrates quartered soldiers on Faculty members in 1757, they refused to accept it, 'conceiving the same disagreeable to the view and intention of the law in general; and an innovation or encroachment upon their privileges in particular'. On taking advice from learned counsel, the Faculty was advised that it did not have a good case. It took no heed, standing firm on its rights of corporate privilege.[139]

---

[137] Marwick, *Records of the Burgh of Glasgow, AD 1691–1717*, pp. 388–89.

[138] SRO, CS 233/97/10, appendix to Case for the Faculty of Physicians and Surgeons of Glasgow, against Thomas Menzies and Others, 1827, pp. 23–25.

[139] RCPSG, 1/5/4a, bundle 11, Memorial and Queries for the Faculty anent Quartering of soldiers with the answers of Ferguson and Pringle advocates 1757, pp. 2, 4–5.

The magistrates' antagonistic stance towards the Faculty on the question of licensing led to a broadside against the FPSG. From the mid eighteenth century, a number of individuals, after Marshall, brought cases against the corporation. This was compounded by the fact that many Scots then received training as naval or army surgeons, and the medical school had also begun to flourish in Glasgow, so that the Faculty was no longer the only medical institution capable of maintaining standards in medical practice in the west of Scotland. James Calder, a gardener who set himself up as a quack, was summoned for trial by the Faculty in August 1757 'for Letting of Blood and Dispensing Medicine'. Fined £20 Scots, he was prohibited from doing either until found qualified and licensed. The Faculty took legal action when he refused to pay the fine. The case finally went to the Court of Session which decided against Calder in 1763. The dispute was of lasting significance. Calder was the first person to challenge the Faculty's existence as a corporate body, arguing that the charter had been granted to Lowe and Hamilton as individuals, in their capacities as king's surgeon and professor of medicine, respectively, and not to any corporate successors. It was a powerful argument – even though basically unfounded – and provided a legal stick with which to beat the Faculty for at least eighty years to come in various other suits.[140] The magistrates had consistently challenged the provisions of the 1599 charter, but Calder's argument – that the Faculty was not legally constituted as a corporation – formed not only the basis of defence for further actions by individuals, but was used to great effect by the University of Glasgow in the FPSG's extensive litigation against it in the nineteenth century.

Wresting it from the hands of kirk officials, the royal charter of 1599 assigned the regulation of medical practice in the west of Scotland to the hands of trained medical practitioners for the first time. In granting significant powers in the examination and licensing of surgeons, in inhibiting inadequately qualified physicians from practice, and investing legal power in the Visitors to proceed against the unlicensed, the charter aimed not only at containing a vast underclass of traditional healers but at delineating the regular medical community. The charter was granted to a surgeon and a physician, and gave discretionary powers of drug regulation to an apothecary, thus associating all three under one corporate umbrella, and in so doing, underwrote the established paradigm of regular medical organisation. This occupational paradigm – from physician at the top, to surgeon (or surgeon-apothecary)

[140] RCPSG, 1/1/1/2, Minutes of the FPSG, 1733 to 1757, fol. 182v; RCPSG, 1/1/1/3, Minutes of the FPSG, 1757 to 1785, pp. 46, 73; Duncan, *Memorials*, pp. 102–3; see Signet Library, vols 24, 26; vol. 587, 1; vol. F20, 42, Glasgow Faculty (Wallace) v. Calder, 1761.

and apothecary at the bottom – may have been typical of medical practice in Britain at this time, but the organisation of healers in one professional association, as in Glasgow, was decidedly not.

However, the coexistence of this occupational *ménage-à-trois* was not without its difficulties. All the significant achievements of this period, from the major role played by Peter Lowe in the negotiations for the 1599 charter, through the 1656 erection of the Incorporation of Surgeons and Barbers to the eventual sloughing off the barbers in 1722, repeatedly emphasise the primacy of the surgeon members. In the first half-century, there appears to have been a relatively unproblematic coexistence between the different divisions of healers but, after the civic rapprochement in 1656, the physician members drifted away only to be readmitted under special dispensation from the surgeons in 1672. Nor did the surgeons ever fully accept their civic association with the barbers.

What did the Faculty achieve between its erection in 1599 and 1757? It managed to implement its regulatory powers (within certain limits) in the geographical boundaries of its given jurisdiction, though in the first half of the seventeenth century, it was not particularly effective outside the town of Glasgow. After seeing wisdom in having the surgeon members organised within a town guild, and having established their civic status through a letter of deaconry, the prosecution of irregulars began in earnest in the late seventeenth century. This can only indicate a degree of contentment by the Faculty with its own position.

Yet, in spite of sterling efforts, there was a number of reasons why the FPSG was unable to regulate medical practice effectively during this period. First and foremost, an extensive clientele existed for irregular practitioners. As long as a market for their services persisted, healers were unlikely to renounce either their livelihood or their lay medical authority. In the second place, as already indicated, the membership of both the Faculty and the Incorporation of Surgeons and Barbers consisted mainly of surgeons, and apprenticeship trained barber-surgeons did not wield much power with either local or central government. Thirdly, while the Faculty undoubtedly thought it was strengthening its civic position in seeking erection as a trade guild, of necessity it subordinated itself to municipal authority in so doing. It is doubtful, otherwise, whether the town council would have dared to license Henry Marshall over the head of the Deacon of Surgeons. In all these ways, the Faculty's technical monopoly on licensing in the region was challenged and undermined.

Peter Lowe (*c.* 1550–1610), artist unknown.

2

The Founders

Of the two founding members of the Faculty of Physicians and Surgeons of Glasgow – Maister Peter Lowe and Dr Robert Hamilton – Peter Lowe has come down to posterity as the motivating figure behind the Glasgow college, mainly because his extant surgical corpus has created for him a sharp historical profile. His published work is, in itself, an indication not only of his energy and enthusiasm but also of the achievement, through study on the Continent, of a level of surgical learning almost unrivalled among his countrymen. Looking beyond the institutional interactions of the Faculty, crown and Glasgow town council, however, it is very difficult to reconstruct an accurate picture of medical practice at this time because in the west of Scotland few sources survive which give much information about it.[1] In this respect, the surgical writings of this founder not only give an idea of how contemporary surgical knowledge and practice was envisaged and represented straight from Lowe's own pen,[2] but also, as an institution largely founded out of Lowe's experience, give an indication of what it aspired to.[3]

Robert Hamilton, physician, remains a more elusive figure. Possibly an arts graduate of Glasgow University,[4] Hamilton was the second recipient of

---

[1] Far more survives for Edinburgh, in this respect, undoubtedly assisted by the fact that the collections of the Advocates' Library in Edinburgh were incorporated into the holdings of the National Library of Scotland.

[2] In a similar way, Lucinda McCray Beier referred to the surgical literature of Richard Wiseman, surgeon to Charles II, to elucidate the differences between theoretical practice and the daily practice of the seventeenth-century surgeon Joseph Binns. L. McCray Beier, 'Seventeenth-Century English Surgery: The Casebook of Joseph Binns', in C. Lawrence (ed.), *Medical Theory, Surgical Practice: Studies in the History of Surgery* (London, 1992), pp. 48–84. Like Lowe, Binns was a member of the London Barber-Surgeons' Company, practised full-time and engaged in no barber's work. He often worked with the physician Dr John Bathurst as Lowe did with Dr Robert Hamilton.

[3] Similarly, Sir George Clark gave an account of the life and published work of Thomas Linacre, the founder of the London College of Physicians, concluding that 'Like all great corporate institutions it has lived its own life, and its founder's greatness is measured by its achievement'. G. Clark, *A History of the Royal College of Physicians of London*, 2 vols (Oxford, 1964, 1966), i, p. 53.

[4] Listed among the students who received the degree of Master of Arts in the University

the charter. He may simply have practised as a physician without ever having gained an MD, but it was the custom for Scottish students to take an MA and then finish their education abroad, a pattern which distinguishes the intellectual history of Scotland from that of sixteenth-century England and other European countries at this time. There was no specialisation in medicine in Scottish universities (though it was regarded as part of the general education taught there and Aberdeen University, for instance, retained a 'mediciner'), and those who wished to obtain a medical degree had to go to the continental medical schools. But assuming he had some formal education in medicine, perhaps from a continental university, he would have studied the ancient medical texts in Latin and possibly in Greek, and learned the dialectical skills basic to medieval medical education. In addition, he was probably introduced to the humanists' intellectual investigations of the ancient texts.[5] Robert Hamilton was bequeathed lands in the Protestant plantation of Ulster by his kinsman, Sir Claud Hamilton of Shawfield.[6] He was related to the Hamiltons, Earls of Abercorn. Peter Lowe dedicated the second edition of his *Chirurgerie* to James Hamilton, Earl of Abercorn, who was a planter in Ulster.[7] These connections were probably not coincidental. On the only occasion that Robert Hamilton appears in the privy council records – as a physician validating the absence of the Countess of Abercorn before the Lords of Council, it is in connection with the Hamiltons of Abercorn (who were among the few Catholic settlers in the Plantation).[8]

---

of Glasgow in 1584 is *Robertus Hammilton*, and among those who received the same in 1604 is *Robert Hammiltoun*. C. Innes (ed.), *Munimenta Alme Universitatis Glasguensis: Records of the University of Glasgow from its Foundation till 1727*, 4 vols Maitland Club, 72 (Glasgow, 1854), iii, pp. 4, 9, noted from Duncan, *Memorials*, p. 233.

[5] A. Keller, 'The Physical Nature of Man: Science, Medicine, Mathematics', in J. MacQueen, *Humanism in Renaissance Scotland* (Edinburgh, 1990), pp. 97, 99; H. J. Cook, *The Decline of the Old Medical Regime in Stuart London* (Ithaca and London, 1986), p. 50. For medical humanism in the English context, see R. J. Durling, 'Linacre and Medical Humanism', in F. Maddison, M. Pelling and C. Webster (eds), *Essays on the Life and Work of Thomas Linacre, c. 1460–1524* (Oxford, 1977), pp. 76–106.

[6] T. Gibson, *The Royal College of Physicians and Surgeons of Glasgow* (Loanhead, 1983), p. 29. But note that the 'Doctor Hamilton' mentioned on 21 July 1594 as landing in a Flemish ship at Aberdeen with various Catholic exiles is not Dr Robert Hamilton, physician, but the Jesuit, Mr John Hamilton, Doctor of Sacred Theology. A. I. Cameron, *Calendar of the State Papers Relating to Scotland and Mary Queen of Scots, 1547–1603*, xi, AD 1593–95 (Edinburgh, 1936), p. 378.

[7] For an account of a slightly later medical contemporary with significant roots in Ulster, see J. F. McHarg, *In Search of Dr John MakLuire: Pioneer Edinburgh Physician Forgotten for over Three Hundred Years* (Glasgow, 1997).

[8] A Scots portion of the Protestant plantation of Ulster was retained, in the barony of Strabane, for the Roman Catholic religion through the agency of the planter, Sir George

Following a complaint by the presbytery of Paisley against Marion Boyd, Countess of Abercorn, and Thomas Algeo, her servant, who were charged on 26 June 1628 with being papists and defying the censure of the kirk, John Hay, minister, appeared for the presbytery while the countess was represented by her son, William Hamilton. The latter produced a testimonial signed by Mr Andrew Hamilton, minister at Kilbarchan, the vicar and two elders of the parish, as well as by Robert Hamilton, Doctor of Medicine, that the countess was unable to travel because of weakness and infirmity.[9] This indicates that Robert Hamilton was probably *Magister Artium* (Master of Arts) and that he allied himself unequivocally with the underground Catholic faction in Scotland at a time when it was safer to be pragmatic in religious affairs. Lowe, too, is likely to have been a Catholic – another bond between them – since, as the college's nineteenth-century historian, Alexander Duncan, puts it: 'Professional offices in those days of embittered religious strife would hardly be bestowed or received independent of creed'.[10] The kirk session had once disciplined him for an unrecorded offence (perhaps for practising according to Catholic rites?).[11] Robert Hamilton, 'doctor of physic', died in the following year in October 1629. His estate was nowhere near the size of Lowe's. He left just £84 Scots or about £7 Sterling.[12]

Although he was not one of the official recipients of the charter, the apothecary William Spang was mentioned by name (as an inspector of drugs) in its fifth clause: 'That no manner of person sell any Drugs in the City of Glasgow except the same be sighted by the said Visitors, and by William Spang, Apothecary, under pain of confiscation of the Drugs'.[13] He was also mentioned in the ratification of the charter by the town council on 9 February

Hamilton of Greenlaw. When his brothers died leaving minors as issue, Sir George ensured, through his influence, that the children of both the Earl of Abercorn and Sir Claud Hamilton of Schawfield were converted to Catholicism. M. Perceval-Maxwell, *The Scottish Migration to Ulster in the Reign of James I* (London, 1973), p. 272.

[9] H. Brown (ed.), *The Register of the Privy Council of Scotland*, 2nd series, ii, AD 1627–1628 (Edinburgh, 1900), pp. 343–44.

[10] Duncan, *Memorials*, p. 22.

[11] On 8 August 1598, 'The presbytery ordains Mr Peter Lowe Doctor of surgery to be convened before the session there to answer for his entry on the pillar not having satisfied the treasurer of the kirk, and without his Injunctions and not behaving him on the pillar as become[s]. And further to make as yet two Sundays his repentance on the pillar, and first to satisfy the said treasurer as the said session has ordained him to do'. GCA, CH2/171/32, transcript minutes of the presbytery of Glasgow, i, pt 2, p. 274. Standing at the pillar, throughout the duration of Sunday service, was a form of ritual humiliation.

[12] RCPSG, Archibald Goodall papers, Commissariot of Glasgow, Testaments, 21, Copy Testament of Mr Robert Hammiltoun, Doctour of Phisick, Glasgow. Confirmed 15 June 1630.

[13] FPSG, *The Royal Charter and Laws of the Faculty of Physicians and Surgeons of Glasgow* (Glasgow, 1821), p. 9.

1600.[14] For these reasons, he is often accorded something approaching the status of a founder. Mentioned in practice in Glasgow in 1574, he was made a burgess of the town in August 1576, 'to keep his booth in medicine drugs for serving of the town'.[15] This is an indication of Spang's status as a senior apothecary there. One of the original seven members present at the first meeting of the Faculty in June 1602, Spang's appointment as Visitor in September 1606 is a further indication of his seniority among the early members of the incorporation. His son, William Spang, the younger, was also admitted as a freeman of the Faculty in September 1602, after trial and examination. He became quartermaster (an examiner and executive of the Faculty) from September 1607 to 1609.[16]

The craft of surgery slowly gained in status in the sixteenth century. With the work of the great anatomists like Andreas Vesalius, surgery and anatomy began to grow in academic importance.[17] Early modern surgical practice in Britain, however, has attracted scant attention from historians of medicine: the few specific case studies mostly refer to England,[18] though Helen Dingwall has recently begun to bridge the gap in Scotland with her study of medical practitioners in seventeenth-century Edinburgh.[19] All the evidence suggests that Scottish surgeons were accorded a higher status than English ones. One possible reason was that the military and mercenary traditions were important in Scottish culture (surgeons in the army had the rank of lieutenant). James IV, 1488–1513, described as 'well learned in the art of medicine and also a cunning Surgeon', took a particular interest in medicine.[20] Scotland may also have been influenced by its Auld Alliance with France where the craft of surgery was more highly esteemed.[21] Indeed, Peter Lowe trained there. But where exactly did he fit into this tradition?

---

[14] R. Renwick and J. D. Marwick (eds), *Extracts from the Records of the Burgh of Glasgow, 1573–1642*, i, Scottish Burgh Records Society (Glasgow, 1874), p. 202.

[15] J. R. Anderson (ed.), *The Burgesses and Guild Brethren of Glasgow, 1573–1750*, Scottish Record Society (Edinburgh, 1925), p. 4.

[16] RCPSG, 1/1/1/1b, Transcript Minutes of the FPSG, 1599 to 1688, pp. 17, 20.

[17] V. Nutton, 'Humanist Surgery', in A. Wear, R. K. French, and I. M. Lonie (eds), *The Medical Renaissance of the Sixteenth Century* (Cambridge, 1985), p. 81.

[18] See for instance, Beier, 'Seventeenth-Century English Surgery', pp. 48–84.

[19] For day-to-day surgical practice, see particularly 'Seeing the Doctor: General Medical and Surgical Practice', chapter 4 in H. M. Dingwall, *Physicians, Surgeons and Apothecaries: Medicine in Seventeenth-Century Edinburgh* (East Linton, 1995), pp. 147–64.

[20] Dingwall, *Physicians, Surgeons and Apothecaries*, p. 34, quoting D. Guthrie, 'The Medical and Scientific Exploits of King James IV of Scotland', *British Medical Journal* (1953), i, p. 1191.

[21] D. Hamilton, *The Healers: A History of Medicine in Scotland* (Edinburgh, 1981), pp. 58–59, 61.

The Life of Peter Lowe

Much of what is known about Peter Lowe is derived from his book *The Whole Course of Chirurgerie* (1597). In the preface to the second edition of his *Chirurgerie*, he gave details of his thirty years abroad between 1566 and 1596, stating that he had been twenty-two years in France and Flanders; two years after that as surgeon major to the Spanish Regiments helping the Catholic League at the Siege of Paris; and, finally, military surgeon to the French king, Henry IV of Navarre, being 'in the wars six years, where I took advantage to practise all points and operations of Surgery'.[22] John Mason Good in his *The History of Medicine, so far as it Relates to the Profession of the Apothecary* (1795), inaccurately stated: 'In the fifteenth century, the ENGLISH surgeons, however, must have been regarded in a very respectable point of view; for PETER LOWE, who flourished about the middle of this century, and wrote a volume on this subject, was appointed Surgeon to the King of FRANCE and NAVARRE'.[23] This not only placed him a century before his time but (more heinously) attributed English nationality to him.[24] Lowe, whose own account places his birth around the mid sixteenth century,[25] described himself as 'Peter Lowe, Scotchman, Arellien'.[26] There has been conjecture about the term Arrellien: was it a Latinised form of his birthplace in Scotland – possibly Errol, in Perthshire[27] Airlie, or Ayr;[28] or of *Orléans* in France, the Latin for which was Aurelianus?[29] The term was

---

[22] G. H. Edington, 'The "Discourse" of Maister Peter Lowe: Extracts and Comments', *Glasgow Medical Journal*, 6th series, 98 (1922), p. 44.

[23] J. Mason Good, *The History of Medicine, so far as it Relates to the Profession of the Apothecary, from the Earliest Accounts to the Present Period: The Origin of Druggists, their Gradual Encroachment on Compound Pharmacy, and the Evils to which the Public are from thence Exposed; As also from the Unskilful Practice of Ignorant Medicasters, and the Means which have Lately been Devised to Remedy these Growing Abuses*. Published at the Request of the Committee of the General Pharmaceutic Association of Great Britain (London, 1795), p. 97.

[24] It is doubly hard to imagine how Good managed this when the edition of Lowe's *Discourse* which he allegedly referred to in the library of the Medical Society of London was the 1612. This states on the title page that Lowe was Scottish. Good, *History of Medicine*, p. 98n.

[25] University study was commonly embarked upon at the age of fifteen at this time.

[26] P. Lowe, *The Whole Course of Chirurgerie*, Classics of Medicine Library (Birmingham, 1981). The first edition had no pageination, hence section headings have been kept in the original Scots as essential references.

[27] On this, see for instance, C. Irvin, *Historiae Scoticae Nomenclatura Latino-Vernacula; or Latino-Vernacular Nomenclature of Scotish [sic] History. Enriched with Many Select Phrases from the Ancient Monuments of the Scots, and the Aboriginal Language of the Gael* (Glasgow, 1819), p. 21, which reads '*Arellius*: earl of Errol, high constable of Scotland'.

[28] Duncan, *Memorials*, p. 23, quoting the physician and writer Jean Astruc, *De morbis veneris* (2nd edn, Venice, 1748), ii, p. 283.

apparently more closely connected with his career on the Continent for it was dropped from the second edition of his book, in which he was designated 'Peter Lowe Scottishman'.[30] But, all that is known about his birth family is that he had a brother called John.[31]

That Lowe received his medical education in France is also undisputed. He probably graduated master of arts, since he is always referred as Maister (or Master) Peter Lowe, but from which university is not known.[32] Most of his medical links were with Paris where he pursued his postgraduate education. He studied medicine there because there was then no systematic teaching of the subject in the Scottish universities, and Paris was at that time a centre of excellence for medical education. Its doctors were learned, its professors were eloquent orators, and the Faculty of Medicine was the largest in Europe.[33]

Lowe was educated at the community of St-Côme in Paris, a surgical company founded in the early thirteenth century[34] which prided itself on consisting of 'pure surgeons' who did not undertake any barbers' work.[35] One of the community's most famous sons, Lanfranc of Milan – generally regarded as the founder of French surgery – established the importance of the interrelation of the different branches of medicine: 'I say that no man

---

[29] A. L. Goodall, 'The Royal Faculty of Physicians and Surgeons', *Journal of the History of Medicine and Allied Sciences*, 10 (1955), pp. 208–9; RCPSG, Archibald Goodall Papers, Letter, Archives du Loiret, 30 March 1948. The register of Orléans MDs was destroyed in bombing in 1940. However, one Dr Dureau, writing to Dr James Finlayson on 8 March 1877, said that a medical graduate from Orléans was, in 1596, designated *medicus Aurelianus*. On Orléans, Irvin, *Historiae*, p. 19, states '*Aurelia*: Orléans in France; it is a delicate town, situate upon the river Loire, thirty-four leagues above Paris'.

[30] *A Discourse of the Whole Art of Chyrurgerie: Wherein is Exactly Set Downe the Definition, Causes, Accidents, Prognostications and Cures of all Sorts of Diseases, Both in Generall and Particular, Which at Any Time Heretofore have been Practized by any Chirurgeon. According to the Opinion of All the Ancient Professors of that Science. Which is not Onely Profitable for Chyrurgions; but Also for All Sorts of People: Both for Preventing of Sicknesse; and Recoverie of Health* (London, 1612), title page.

[31] RCPSG, Archibald Goodall Papers, Letter, Town Clerk Deput, the Corporation of Glasgow, 13 June 1949, quoting *Glasgow Protocols, 1573–1600*, xi, no. 3556.

[32] Goodall, 'The Royal Faculty', p. 210; A. Duncan, *Memorials*, p. 21.

[33] Nutton, 'Humanist Surgery', pp. 82–83; I. M. Lonie, 'The 'Paris Hippocratics': Teaching and Research in Paris in the Second Half of the Sixteenth Century', ibid., p. 155.

[34] For a brief account of the community of St-Côme see T. Gelfand, *Professionalizing Modern Medicine: Paris Surgeons and Medical Science and Institutions in the Eighteenth Century* (Westport, Connecticut, and London, 1980), pp. 21–27, and for its seventeenth-century consolidation, pp. 28–44.

[35] D. de Moulin, *A History of Surgery with Emphasis on the Netherlands* (Dordrecht, 1988), p. 106.

can be a good physician who has no knowledge of operative surgery, a knowledge of both is essential'.[36] However, medical training and practice were hampered, during this period, by rigid social prejudices which distinguished between artisans or craftsmen, among whom the surgeons were numbered, and their middle-class counterparts, the university-trained intellectuals, including the physicians. In the pre-antiseptic era, the surgeon's art had no truck with major operative surgery. The meaning of 'surgery' was literally 'hand work'; it meant the cure of wounds, inflammations, ulcers, dislocations and fractures, removal of foreign bodies and the treatment of skin and venereal disease.[37] There was some invasive surgery, but operations in the depths of the abdominal cavity were avoided mainly because they tended to kill the patient. Physicians, on the other hand, dealt with internal ailments. There was an obvious separation between a methodical, rational, intellectual training, and the more mechanical but equally essential skills of dissection, experimentation and quantitative methodology.[38] John Mason Good remarked that it was obvious from the extended title of Lowe's 1612 edition – 'which is not only profitable for Surgeons, but also for all Sorts of People, both for *preventing* of *Sickness*, and *Recovery of Health*'- and more especially from the book itself that 'the term "Surgeon" was often used for *pharmaceutist* or *apothecary*; or, at least, that the two branches of the profession were frequently united'.[39] This may simply be an anachronistic observation from an eighteenth-century perspective, but in uniting physicians and surgeons within the Faculty of Physicians and Surgeons of Glasgow, Peter Lowe did much to break down these social barriers, and fuse the two (if not three) branches of medical learning in the town.

While in France, Lowe probably operated – at least in some capacity – as a spy and ambassador for Scotland. The Scottish State Papers note for 23 May 1595 that 'a new admitted surgeon' to the king of France had prepared a casket containing some secret plans of England, domestic and foreign, with a description of all ports and havens there.[40] However, the association mooted by earlier commentators between Peter Lowe and a 'Scotchman and pirate' who is mentioned around this time in the *Calendar of Scottish Papers*, is absolutely spurious.[41] It would be a wonderfully romantic story, but the

[36] R. Theodore Beck, *The Cutting Edge: Early History of the Surgeons of London* (London, 1974), p. 24.
[37] O. Temkin, *The Double Face of Janus* (Baltimore and London, 1977), p. 490.
[38] E. Zilsel, 'The Sociological Roots of Science', *American Journal of Sociology*, 47 (1942), p. 544.
[39] Good, *History of Medicine*, pp. 97–98n.
[40] *Calendar of Scottish Papers*, xi, quoted in a letter, by John Durkan, 11 December 1949.
[41] L. Scott, 'A Note on the Life of Peter Lowe, Scotchman', Classics of Medicine edition, p. 8.

pirate's name was Peter Love (the association occurring simply because of the lack of standard orthography of the time). Nevertheless, the granting of a charter by James VI has been seen as recognition for services rendered to the king on the Continent. Lowe had the backing of Robert Hamilton, a Glasgow physician, in his venture. Their petition for the establishment of a regulatory body to control the practice of medicine in the west of Scotland was duly accepted by the king who, on 29 November 1599, issued a Royal Charter to them under the Privy Seal. In a salute to his peers at the front of the 1612 edition of his *Chirurgerie*, Lowe outlined, in his own words, the grounds of his petition to James VI:

> It pleased his Sacred Majesty to hear my complaint about some fourteen years ago, upon certain abusers of our Art, of diverse sorts and ranks of people, whereof we have good store, and all things failing, prodigals, and Idle people do commonly meddle themselves with our Art, who ordinarily do pass without either trial or punishment. The matter being considered, and the abuse weighed by his Majesty and Honorable council, thought not to be tolerated, for the which I got a privilege under his highness' privy seal, to try and examine all men upon the Art of Surgery, to discharge, & allow in the West parts of Scotland, who were worthy, or unworthy, to profess the same.[42]

Particularly ironic in view of the Faculty's subsequent battles over regulation, were his words 'try and examine'.[43] Indeed, in Lowe's opinion, such was the iniquity of the time that 'abusers are commonly overseen by such as ought to punish them: in such sort that one blind[man] guides another, and most commonly fall both into the ditch'.

Published Works

Peter Lowe's publications, on the other hand, give some idea of what well-trained apprentices were taught at the tail-end of the sixteenth century. Though he states in the introduction to his *Chirurgerie* that he began collecting practical wisdom 'at vacant hours, into this Book' (which was based on his surgical practice on the Continent),[44] Lowe did not begin to publish until he returned to Britain. Like the most eminent physicians, he

---

[42] P. Lowe, *A Discourse of the Whole Art of Chyrurgerie* (London, 1612), 'To My very Worshipfull, learned, and well experimented good friends, Gilbert Primrose Sergeant Chirurgian to the King's Majestie; James Harvie cheife Chirurgian to the Queenes Majestie; those of the Worshipfull companie of Chirurgians in London, and Edenborough, and all such well experimented men in this Kingdome who are licensed to professe the Divine art of Chirurgerie'.
[43] For which, see Chapter 1, above, and Chapter 4, below.
[44] Introduction to the second edition of the *Chirurgerie*.

undoubtedly wrote to enhance his professional stature as a surgeon. In his essay on humanist surgery, Vivian Nutton contends that in the sixteenth century 'the gulf between surgeon and physician was not as wide as has been thought; that particularly when contrasted with the barbers and barber-surgeons, the two groups had much in common ...'[45] Repeated allusion to internal physic, as well as frequent denigration of barber-surgeons in Lowe's *Chirurgerie*, shows that he identified himself far more with physicians than barber-surgeons. The 'Common Enemy' of both was the irregular practitioner.[46]

Though trained as a surgeon, Lowe considered himself well qualified to deal with internal medicine. 'There are many other things, which might be spoken of this matter, which I leave to the physicians, being more medical, than Surgical, but by reason that sometime it falls under the Surgeon's hands, I thought good thus much to treat of it', he commented on the treatment of hydropsy (water retention in the belly and legs). However, he was unable to refrain from insisting to his Scottish readership that he 'knew a man that was cured by abstaining from drink half a year'.[47] In the 1612 edition of the *Chirurgerie* he also confessed, having been asked by sickly, aged friends to prescribe some form of regimen for them, that 'I am willing to do, although it be more medical than Surgical'.[48] Nevertheless, he was always mindful of the contemporary pecking order, suggesting for the ameloriation of scrotal hernia that the patient first be purged, then treated with juleps, pills, clysters, and such like 'as shall be thought expedient by the skilful Physician'.[49]

Lowe's first work, published in English in 1596, *An Easy, Certaine and*

[45] Nutton, 'Humanist Surgery', p. 75.

[46] D. Turner, *Apologica Chyrurgica: A Vindication of the Noble Art of Chyrurgery from the Gross Abuses Offer'd Thereunto by Montebanks, Quacks, Barbers, Pretending Bone-Setters* (London, 1695), p. 25, quoted in Lawrence (ed.), *Medical Theory, Surgical Practice*, p. 4.

[47] Lowe, *The Whole Course of Chirurgerie*, fifth treatise, chapter 18, 'of the Tumour in the belly, called *Hidrupsie*'. The surgeon's only involvement in its treatment was in making an incision and placing in it a hollow tent of silver or lead (hollow tube for dilating an opening) through which the fluid could drain. S. Cooper, *A Dictionary of Practical Surgery: Containing a Complete Exhibition of the Present State of the Principles and Practice of Surgery, Collected from the Best and Most Original Sources of Information and Illustrated with Critical Remarks* (London, 1809), p. 422.

[48] Lowe, *A Discourse of the Whole Art of Chirurgerie*, p. 32. His regimen, in eight points, pertained entirely to their method of eating which was always to be moderate. He recommended that they not overload their stomach; that they eat foods with a laxative effect – 'those meats that digest easiest, should first enter into the stomach, like thick soup, prunes, and such other[s], as have the virtue to loose[n] the belly'; and that 'they be nourished often and little at one time, which is meet to be observed both by old and young, as is at large set down by Galen'. Ibid., pp. 33–34.

[49] Lowe, *A Discourse of the Whole Art of Chirurgerie*, p. 32.

Perfect Method, to Cure and Prevent the Spanish Sickness[50] was a treatise on the subject of syphilis, whose spread he saw little hope of stopping. However, he thought prostitutes less likely to pass on the disease, according to the humoral theory of the day, because they were not emotionally involved in their work and were therefore less 'hot and do not pass on the venom'.[51] Unknown by either Hippocrates or Galen, sixteenth-century practitioners experimented with infusions of guiacum and with mercury.[52] Lowe's preferred treatment was administering mercury pills for thirty or forty days, implemented with a combination of enemas, purging and sweating – a harsh cure which diminished bodily strength and inevitably led to mercury poisoning.[53]

In the following year, Lowe published the book for which he is best known, *A Course of the Whole Art of Chirurgerie*, dedicated to James VI. This dedication, in itself, probably eased the passage of the charter granted to Lowe and Hamilton in 1599. Lowe's book was the first surgical textbook written in English. The few previous publications along these lines had been designed for surgeons already in practice and not for students. Glasgow, indeed, had a link with medical books in the vernacular. In the early sixteenth century, Andrew Boorde, 'an ill-adjusted Carthusian monk',[54] and sometime physician to Henry VIII, who defected to physic prior to the worst excesses of the dissolution of the monasteries, came and spent a year in Scotland. He lived at 'a little university named Glasgow, where I study and practise physic for the sustentation of my living'. A man of dark humour,[55] he referred to himself as Andreas Perforatus, or Andrew 'bored' (as in bored to death), and had no great affection for the Scots: 'Trust no Scot, they use flattering words and all is falsehood'. His *Breviarie of Health* (1547), a brief, household medical dictionary,[56] is significant as the first medical book by a medical

---

[50] It was so called because of the Spanish mercenaries in the army of Charles VIII of France which is held responsible for the spread of the disease in late fifteenth- and early sixteenth-century Europe.

[51] Quoted in Keller, 'The Physical Nature of Man', p. 103.

[52] A. W. Sloan, *English Medicine in the Seventeenth Century* (Durham, 1996), p. 43.

[53] P. Lowe, *An Easy, Certaine and Perfect Method to Cure and Prevent the Spanish Sickness: Wherby the Learned and Skilfull Chirurgian May Heale a Great Many Other Diseases* (London, 1611), 'The 18. Chapter, to heal this disease by Pills of Mercure, which is the first way that I have used'.

[54] Clark, *Royal College of Physicians*, i, p. 68.

[55] The clown which accompanied travelling mountebanks and performed on a portable stage to soften up the crowd, was named Merry Andrew after the witty Andrew Boorde. Sloan, *English Medicine in the Seventeenth Century*, pp. 140–41.

[56] D. Guthrie, *A History of Medicine* (London and Edinburgh, 1947), pp. 172–73.

practitioner to be both written and printed in English. A pioneering work in family medicine, Boorde's *Breviarie* went to six editions.[57]

Lowe's *Chirurgerie* was also important because the first English translation of Hippocrates' *Prognostics* (probably from a French translation),[58] was published as an appendix to the first edition describing what the practitioner could deduce about illness from signs in various parts of the body.[59] This allied Lowe with a group of leading surgeons, championed by the London College of Surgeons, who were actively engaged in translating humanist surgery (particularly the writings of Galen, who had dissected apes and other animals, and was still the main source for anatomical knowledge) into English. Many of them, including George Baker, Thomas Gale,[60] and Roger Coplande, were among Lowe's close friends.[61]

Medical humanism had its origins in the Renaissance. Driven by a small band of academics led by Nicolao Leoniceno at the University of Ferrara, it had spread through all the Italian medical universities and into other European countries by the early sixteenth century; for instance the Belgian anatomist, Andreas Vesalius, who studied in Paris and taught at Padua, was a humanist.[62] Reaching its peak at the University of Padua, medical humanism has been characterised by Jerome Bylebyl as dichotomous, simultaneously conservative and reforming: 'conservative because its guiding assumption was that medical theory and practice had reached unparalleled heights among the ancient Greeks, especially through the work of Hippocrates and Galen, and reforming because its proponents took a careful look at medicine as taught and practised by contemporary physicians and found it to be distressingly corrupt and incomplete in relation to the surviving monuments of ancient Greek medicine'.[63] The reform programme was to be based on translation, editing and study of the ancient Greek physicians'

[57] J. L. Thornton, *Medical Books, Libraries and Collectors: A Study of Bibliography and the Book Trade in Relation to the Medical Sciences* (2nd edn, London, 1966), p. 57; further editions: 1552, 1557, 1575, 1587, 1598. Copies survive in the Royal College Rare Books Collection and in the Hunterian Collection in Glasgow University Library.

[58] Duncan, *Memorials*, p. 21.

[59] *The Presages of Divine Hippocrates: Divided into Three Parts. With the Protestation or Oath which Hippocrates Caused his Schollers to Make at their Entrie with him to their Studies*. Printed as an appendix to the 1612 edition of Lowe's *Chirurgerie*.

[60] Thomas Gale was the author of two surgical works in the vernacular: *An Excellent Treatise on Wounds Made with Gunneshot* (London, 1563), which includes the first mention of syphilis in English; and *Certaine Workes of Chirurgie, Newlie Compiled* (London, 1586), a copy of which survives in the College Library.

[61] Nutton, 'Humanist Surgery', pp. 80, 82.

[62] C. D. O'Malley, *Andreas Vesalius of Brussels, 1514–1564* (Berkeley and Los Angeles, 1964), pp. 64–69.

[63] J. J. Bylebyl, 'The School of Padua: Humanistic Medicine in the Sixteenth Century', in

texts. Galen, in particular, strode as a colossus over the European medical tradition of the sixteenth century and his works were widely read.[64] Humanism's main legacy was the importance it placed on learning. Its critical approach and rationalist attitudes inspired changes, and in Scotland especially, where humanist ideas arrived relatively late, there was a marked inclination to adopt the novelties of the Continent in reassessing the ancient texts. The rediscovery of Galen's surgical tracts provided the surgeon with practical new knowledge on surgical technique, a model for the unification of medicine and a new sense of identity.[65] The status of surgery was thereby enhanced.

Many humanist scholars of this period held the liberal view that the gulf between physician and surgeon could be bridged. This was to be achieved by educating surgeons in the theory of what they were doing, while encouraging physicians to come more into contact with the body. Peter Lowe achieved this practically: through his interest in prognosis, he showed that he believed surgery was not just a craft skill but one which entailed acute intellectual discrimination. Like his friends among the eminent English surgeons, Lowe concerned himself not only with the containment of quackery, but with raising the status of the surgeon through proper education and training and with raising the abysmal standards of barber-surgery.[66] The admiration expressed in his main text for the published work in anatomy of Robert Auchmowtie, surgeon in Edinburgh, 'sometime Surgeon in the great Hospital of Paris', had disappeared by the second edition. There was a very good reason for this: in April 1600, Auchmowtie, contemptuously referred to as 'barber' in a later history, was convicted in Edinburgh of murder.[67] Lowe insisted, with reference to Plato, that there was 'nothing harder than Surgery' when practised as a learned science and art, but that it attracted its fair share of deceivers 'with their charms, hearts full of poison and false promises'.[68]

C. Webster, *Health, Medicine and Mortality in the Sixteenth Century* (Cambridge, 1979), pp. 339–40, 342.

[64] For an estimation of the scale of Galen publications, see the introduction to R. J. Durling, 'A Chronological Census of Renaissance Editions and Translations of Galen', *Journal of the Warburg and Courtauld Institutes*, 24 (1961), pp. 230–305. The RCPSG, holds, e.g., *Libri V iam primum conversi. De causis, utilitate, difficultate respirationis; Libri IV. De uteri dissectione. De foetus formatione. De semine.* Tr. J. C. In Marcellus, Empircus, *De medicamentis empiricis*, folio (Basle, H. Froben and N. Episcopius, 1536). Ibid., p. 259.

[65] Nutton, 'Humanist Surgery', p. 79; Keller, 'The Physical Nature of Man', p. 98.

[66] Nutton, 'Humanist Surgery', pp. 80, 82–88, 99.

[67] For further details, see Dingwall, *Physicians, Surgeons and Apothecaries*, p. 47.

[68] Quoted in Lawrence, *Medical Theory, Surgical Practice*, p. 3. See the introduction for an informative survey of the history and historiography of surgery.

Lowe's main purpose, therefore, in writing his *Chirurgerie*, was to provide a comprehensive reference work for the conscientious trainee surgeon. In an address 'To the Friendly Reader' (1612 edition), he stated that his aim in writing the book was:

> to advance your study, and to instruct you in the whole course of Surgery, with the manner to cure the most part of all diseases usually practised by Surgeons, as well by ancient as late writers, of diverse and sundry practises, & hidden secrets, not practised heretofore by any Surgeon.

In order to make it as accessible to the widest range of reader, he planned to employ 'such plain terms as I could, for the use of the common sort'. And, in excusing any possible errors, Lowe declared his intention that 'in good will and love I have done this for the advancement of all men, specially young Surgeons'. Lowe had obviously benefited from both the intellectual distillation which had occurred while teaching apprentices in Glasgow, as well as the extensive reading which – released from his busy military practice on the Continent – he subsequently had the time to do there.

Lowe's *Chirurgerie* eventually achieved four editions, published in 1597, 1612, 1634 and 1654.[69] The second edition, its title amended from the original *The Whole Course of Chirurgerie* to *A Discourse of the Whole Art of Chyrurgerie*, was revised and extended. The bibliography more than doubled and there was a nine-page index. There was also a new address to Lowe's medical peers – directed especially to his 'learned and well experimented good friends, Gilbert Primrose Sergeant Surgeon to the King's Majesty; James Harvie chief Surgeon to the Queen's Majesty' – outlining the abuse of the art of surgery by nine different groups of irregulars: 'like cosoners [cheats], quacksalvers, charlatans, witches, Charmers, & diverse other sorts of abusers'.

Lowe planned at least two other books which were never finished: one on the diseases of women and young children and the other a domestic, self-help guide.[70] The latter, *The Poore Man's Guide*, was in a more advanced state of preparation. It seems to have been a collection of remedies and recipes for popular use with hints for their application, and suggests that Lowe considered himself able to make up and administer them. Sometimes Lowe noted the provenance of his remedies. Of hydropsy, he claimed 'an excellent

---

[69] A. Besson (ed.), *Thornton's Medical Books, Libraries and Collectors* (3rd edn, Aldershot and Vermont, 1990), p. 66.

[70] In the dedication of his *Spanish Sickness*, Lowe intimated that he intended 'hereafter to publish diverse other Books of Surgery'. He referred, in its preface, to his book *De partu mulierum*, and in chapter 21 to his treatise *The Infantment*, later called *The Sicknes of Women*. From the more standardised citation of its title (to which Lowe referred in the *Spanish Sickness* and the first edition of his *Chirurgerie*), his other book seems to have been almost finished.

remedy in the poor man's Guide for the cure of this disease, if the fault be not in the liver, whereby I healed many in Paris during the time of the siege'.[71] In the second edition, he explained that this was 'a remedy which I brought from a Turk, who was bond-slave to *Dondego de varro Viador*, General of the Spanish regiment there'.[72] Lowe also claimed to have set out in *The Poore Man's Guide* a method for embalming which preserved a corpse for up to eighty years. This was an aspect of surgical practice which he continued to ply in Glasgow: just before his death, he was paid, in May 1610, the sum of £40 Scots 'partly for his fee and partly for the expenses made by him in bowelling of the laird of Houston, late provost'.[73]

Lowe's *Chirurgerie* consisted of ten treatises or (in the language of the second edition) 'Books', and proved popular as a text book.[74] The First Treatise begins with an historical appraisal of the ancient art of surgery and includes a catechism between John Cointret, Dean of the Faculty of Surgery in Paris, and Peter Lowe 'his Scholar',[75] which identifies at least one of his teachers there. But reflecting Lowe's increasing professional status, by the second edition, the participants had changed to Lowe 'Doctor of Surgery', and his son, John Lowe. He also denounced 'land loupers',[76] and simple Barbers who promised 'for lucre to heal infirmities', worthless people who usurped the name of surgeon but 'have scarce the skill to cut a beard which properly pertains to their trade'. In contrast, he expounded upon the essential qualities of an ideal surgeon:

> that he be learned chiefly in those things that pertain to his art, that he be of a reasonable age, that he have a good hand, as perfect in the left as the right, that he be ingenious, subtle, wise, that he tremble not in doing his operations, that he have a good eye, that he have good experience in his art, before he begin to practise the same. Also that he have seen and observed for a long time, of learned Surgeons, that he be well mannered, affable, hardy in things certain, fearful in things doubtful and dangerous, discreet in judging of sicknesses, chaste, sober,

---

[71] Lowe, *The Whole Course of Chirurgerie*, fifth treatise, chapter 18.

[72] Lowe, *A Discourse of the Whole Art of Chyrurgerie*, pp. 225–26. For these, see J. Finlayson, *Account of the Life and Works of Maister Peter Lowe, the Founder of the Faculty of Physicians and Surgeons of Glasgow* (Glasgow, 1889), pp. 40–48, 71, 73.

[73] Renwick and Marwick, *Records of the Burgh of Glasgow, AD 1573–1642*, p. 314. For more on Lowe's method of embalming, see discussion of the eighth treatise below.

[74] Note that in the 1597 edition, the seventh treatise is twice used as a heading, by mistake, a misprint which, in effect, makes the ninth treatise into the tenth.

[75] This is the terminology used in Besson, *Thornton's Medical Books*, p. 66; Lowe, *The Whole Course of Chirurgerie*, chapter 2. This form is also employed in the first chapters of treatise six, seven and eight.

[76] Land loupers means itinerants.

pitiful, that he take his reward according to his cure and ability with the sick, not regarding avarice.[77]

Most of the remaining chapters in the First Treatise were a detailed exposition of the six naturals constituting the body – the elements; the temperament or complexion; the humours; the members; the virtues or faculties; the effects of virtues; and the spirits.[78] In the early modern period, disease was regarded as an imbalance of naturals. This Aristotelian-Galenic philosophy of healing and disease causation centred on the state of balance and imbalance of body fluids,[79] and the associated four temperaments (hot, cold, dry and moist), four members (brain, heart, liver and testicles) and five ages (infancy, adolescence, young age, man's estate and old age).[80] Humoral theory was accepted relatively uncritically until the late seventeenth century.

The Second Treatise dealt with the non-naturals – those things which conserve the body if used correctly – air; food and drink; the motion and rest of the body; sleeping and waking; repletion and evacuation; and things which disturb the mind.[81] Lowe maintained that sicknesses considered cold and humid, like putrified fevers, catarrhs and hydropsies, and tumours of the pituitary, were best treated in houses where the air was hot and dry.[82] He believed that 'That which nourishes well, engenders good juice, of the which Galen his writing in his books of the Faculty of Aliments, and in the book of Conservation of health, as also Hippocrates in diverse places'. Food which engendered 'good juice' was lightly digested, made little excrement and good blood; meat or fish flesh, good wheaten bread, and good ale were in this category. Water was not then as uniformly pure as now, but the purifying effect of the yeast in beer rendered it less pernicious to health.[83] The innate characteristics of different kinds of food would have been considered in Lowe's *The Poore Man's Guide*, had it appeared.[84] However, the

---

[77] Lowe, *The Whole Course of Chirurgerie*, first treatise, chapter 2, 'Conditions of a Chirurgian'.

[78] Lowe, *The Whole Course of Chirurgerie*, first treatise, chapter 3, 'Of Naturall Thinges in Generall'.

[79] See A. Wear, 'Medicine in Early Modern Europe, 1500–1700', in L. I. Conrad, M. Neve, V. Nutton, R. Porter and A. Wear, *The Western Medical Tradition, 800 BC to AD 1800* (Cambridge, 1995), pp. 255–64.

[80] This is in the second edition only, Lowe, *A Discourse of the Whole Art of Chyrurgerie*, pp. 27–32.

[81] Lowe, *The Whole Course of Chirurgerie*, introduction to second treatise, 'of unnatural thinges, the consideration whereof is most needfull for the preparation of health, and containeth six Chapters'.

[82] Ibid., chapter 1, 'Of the Aire'.

[83] R. Theodore Beck, *The Cutting Edge: Early History of the Surgeons of London* (London, 1974), p. 178.

[84] Lowe, *The Whole Course of Chirurgerie*, second treatise, chapter 2, 'Of Meate and Drinke'.

best drink was clean wine of good colour and taste, taken in moderation, 'for it has the virtue to nourish, strengthen, corroborate the natural heat, tempers the humours, and purges both by sweat and wine, gives appetite, helps the faculty of concoction, and makes men joyful'. Few would argue with the last.[85]

In his pronouncements on 'Moving and Exercise', Lowe fixed women firmly in the realm of sport: 'Hippocrates says, that labour, meat, drink, sleeping, playing, and women, ought to be moderately used, like all other exercises'.[86] Humours which diminished and molested the body were evacuated in one of two ways: either artificially with the use of purges, cathartic medicines, diuretics, vomiters, or phlebotomy (universal evacuations); or naturally through exercise, friction, bathing, menstruation, or copulation (particular evacuations). But, he cautioned, desire for the 'act venereal' was to be indulged only occasionally:

> The body is evacuated by the immoderate act of Venus, like as diverse other mischiefs ensue thereupon: and first of all, it is hurtful to the eyes, and all the organs sensitive, to the nerves, the thorax, the kidneys, and diverse other parts of our bodies, and makes men forgetful, provokes the gout,[87] & dolours nephritic, and diverse diseases of the bladder, brings soon old age, consequently death, it does hurt, being immoderately used, not only man, but all animals.[88]

Lowe, however, was not prepared completely to dismiss sexual intercourse since it increased the opportunities of his trade 'because such as delight in this pastime will formalize,[89] as also because the usage hereof is sometime profitable to the Surgeon, I will not altogether condemn it'. Particular evacuations were thought to occur when the brains were discharged (in the form of mucous) through the roof of the mouth and nose, but the belief that catarrh could drain from the brain to the nostrils was eventually disproved in the 1660s by Richard Lower, a distinguished Oxford anatomist, through detailed study of the anatomy of the brain and skull.[90] Particular evacuations were also discharged obscurely through the eyes and ears, the lungs through

---

[85] Lowe, *A Discourse of the Whole Art of Chirurgerie*, pp. 44–46.

[86] Lowe, *The Whole Course of Chirurgerie*, second treatise, chapter 3, 'Of Moving and Exercise'.

[87] Gout was, indeed, a disease of sexually-active males. For a medical and cultural history of the disease, see R. Porter and G. S. Rousseau, *Gout: The Patrician Malady* (New Haven and London, 1998).

[88] Lowe, *The Whole Course of Chirurgerie*, second treatise, chapter 5, 'Of Repletion and Evacuation'.

[89] Presumably marry.

[90] Sloan, *English Medicine in the Seventeenth Century*, p. 77, with reference to R. Lower, *De catarrhis* (1672), translated by R. Hunter and I. Macalpine (London, 1963).

the trachea, the stomach by vomiting, the intestines through the buttocks, the liver, spleen, kidneys and bladder through the urine, and women's privy parts in the course of menstruation (on which Lowe referred the reader to his *Booke of Women's Diseases*).[91]

The Third Treatise deals with 'things altogether contrary to our nature' (a phrase borrowed from Galen's *De temperamentis et libris de symptomatum causis*), that is, malady, cause of malady, and accidents of malady, with emphasis on the theoretical definition of each.[92]

Operative Surgery

The real surgical context of Peter Lowe's book begins at the Fourth Treatise; in the second edition, Lowe informs his son: 'Now it shall be necessary that I instruct you in the exercise of Surgery'. The author deals with 'tumours or apostemes against nature in general' at a time when the word 'tumour' was applied to a swelling rather than specifically to carcinoma (which won a chapter of its own only in the second edition).[93] Its all-inclusive nature may best be gauged by the appearance in this section of Herpes as a 'little ulcer accompanied with tumour',[94] and of Oedema.[95] In the 1612 edition of the *Chirurgerie*, the Fourth Treatise contained three extra chapters – one dealing with various different types of pustule, tubercle and bubo which Lowe included 'Because that the most part either of old or new writers, have made small mention of those tumors. So that the young Surgeon may be

[91] Lowe, *The Whole Course of Chirurgerie*, second treatise, chapter 5, 'Of Repletion and Evacuation'.

[92] Lowe, *The Whole Course of Chirurgerie*, third treatise, 'Of Things Altogether Contrary to our Nature, which Containeth three Chapters'.

[93] Edington, ' "Discourse" of Maister Peter Lowe', p. 46.

[94] Lowe, *The Whole Course of Chirurgerie*, fourth treatise, 'Of Tumors or Aposthumes against Nature in Generall which Contaynes XIII Chapters: Written by Peter Low Arellian, Doctor in Chirurgerie, and Chirurgian ordinarie to the King of France and Navair', chapter eight, 'Of Herpes'. To dry up the herpetic ulcers, in the first instance, Lowe suggested the application of fomentations of roses and plantain, sodden in wine or water. Applying alcohol would certainly have been effective.

[95] Lowe, *The Whole Course of Chirurgerie*, fourth treatise, chapter 9, 'Of Tumors Which Proceede of the Pituite, and First of Edema'. Oedema was to be treated by taking a remedy such as Diacatholicon to divert the fluxion of humour, 'the administration of which, you shall use the counsel of the learned Physician', by eating roasted meats (rather than sodden), and abstaining from foods which produced phlegm such as fruit, pottage, cheese, fish, herbs, water, sadness, too much sleep and also from women, 'especially if the sick be weak'. However, note that by the second edition of the *Chirurgerie*, Lowe was confident enough to substitute 'the counsel of the expert Physician or Surgeon'. Lowe, *A Discourse of the Whole Art of Chirurgerie*, p. 105.

the better instructed, I thought good to speak of them in this place';[96] another on the extirpation of fingers and toes; and one on the treatment of erysipelas.

The formation of apostemes – caused either by fluxion (a discharge of fluid) or congestion – was explained on a humoral basis. According to whether the swelling was in the making or already made, cure was achieved by employing one of two methods: 'the one to stay the fluxion of the humour to the place, the other [to] evacuate the humour gathered in the place'.[97] As for gangrene and sphacelus (the complete mortification of the post-gangrenous phase), the latter was treatable only by amputation which Lowe, like all sixteenth-century surgeons, believed was a last resort. Serious injury to the extremities was also an indication.[98]

> The cure, in so much as may be, consists only in amputation of the member, which shall be done in this manner, for the friends must first be advertised of the danger, because often death ensues, as you have heard, either for apprehension, weakness, or flux of blood. For this cause the learned Celsus calls it a miserable remedy, yet we use it, by reason in so doing, there is some hope, and in not using of it, there is none, but sudden death, for better it is to lose one member, than the whole body.

The limb was seldom amputated above the knee prior to the sixteenth century, and Lowe seems to have preferred not to go above the knee-joint. He advised surgeons 'to cut four inches from the joint in all amputations, saving only if the mortification or pulling apart of the bone end in the joint, then it may be cut in the joint, chiefly in the knees'. He set down the specifics of the procedure, stressing his use of the ligature. Two men were to hold the patient and the surgeon ordered him to bend and stretch out the limb so that the veins would become more prominent and make cauterisation or knitting easier. Then:

> the Surgeon shall pull up the skin & muscles, as much as he can, afterwards he shall take a strong ribbon, and bind the member fast, above the place two inches, where the amputation shall be ... The bandage thus made, we cut the flesh with a razor or knife, that is somewhat crooked like a hook, the flesh being cut to the bone, it must be scraped with the back of the said knife made purposely for that effect, to the end the periosteum, that covereth the bone, be not painful in cutting the bone, otherwise it tears with the saw, and causes great dolour, and also

---

[96] Ibid., p. 80.

[97] Lowe, *The Whole Course of Chirurgerie*, fourth treatise, 'Of Tumors or Aposthumes Against Nature in Generall'.

[98] G. Lawrence, 'Surgery (Traditional)', in W. F. Bynum and R. Porter (eds), *Companion Encyclopedia of the History of Medicine*, 2 vols (London and New York, 1993), ii, p. 978.

hinders the cutting. This done, saw the bone, & being cut, we loose the ligature, and draw down the skin to cover the bone in all parts.[99]

Pulling the skin and muscles as high as possible above the site of amputation, where they were secured with a tight ligature before severing the soft tissues at one blow, was the technique of the Frenchman, Ambroise Paré (1510–90), possibly the most famous surgeon of his century, and like Lowe, an army surgeon. Paré introduced a new technique for controlling haemorrhage after amputation, substituting the red-hot iron for vascular ligature. His technique was used throughout the seventeenth century.[100] Jacques Guillemeau (1550–1613), the French royal surgeon, to whose work Lowe had been introduced in Paris,[101] also adopted this method. Peter Lowe was familiar with it too. Later in his treatise on wounds, Lowe describes the ligature as:

> a piece of cloth made long two or three ells, and in breadth three or four inches, according to the member and hurt; the cloth must be soft, clean without hem or seam, and more slack in wounds than in fractures, and of it there are diverse sorts, for some are to contain, as in simple wounds, some are to expel matter, as we see in [con]cave wounds, some are defensives to stay fluxion, some to retain the medicaments on the part, as in the throat and belly, some are mortificative, which we use in legs or arms gangrened to cut them off.[102]

In general, Lowe preferred to cauterise because it entailed less blood loss than knitting (which sometimes came loose). He also recommended the actual cautery where there was putrefaction but would employ the ligature where the wound was clean. The actual cautery, an instrument of gold, silver, brass, iron or lead, was so called because it burned by the direct application of heat. It was used for staunching blood in veins and arteries, where there was gangrene, on rotten bones, and for ulcers and venomous bites. The potential cautery burned with acid or similar applications such as tartar, arsenic, sublimate, oil of vitriol, brimstone, ashes of the oak and fig tree, quick lime, salt nitre 'all of which you may use together, or some of them in water', and was applied as a plaster.[103] Lowe explained: 'We must have three or four little instruments of iron, crooked at the end, the point in form of a button, made red hot, which we take, and apply on the veins one after another, holding them a certain space, till the scale be made, yet not burning

---

[99] Lowe, *The Whole Course of Chirurgerie*, fifth treatise, chapter 6, 'Of Sphasell'.
[100] Lawrence, 'Surgery (Traditional)', p. 978.
[101] In the fifth treatise, chapter 11, on tumours (swellings), Lowe refers to *rannuculus* or a tumour of the veins of the tongue: 'In opening of it the sick sometime[s] becomes mad as I did once see: Jaques Guillemeau Surgeon reports to have seen it four times'.
[102] Lowe, *The Whole Course of Chirurgerie*, sixth treatise, chapter 1.
[103] Ibid., ninth treatise (recte), chapter 5, 'Of Cauters Actuall and Potentiall'.

much of the vein'.[104] Use of the actual cautery to staunch haemorrhage, or of styptics (drugs to contract the blood vessels and tissues) and compression bandaging was more general in Europe until the eighteenth century than Paré's method of ligaturing of arteries. Applying tight bandaging or adhesive strapping was also more popular than Paré's suturing.[105]

After staunching the flow of blood, Lowe would apply an astringent powder to the stump on dry flax; thereafter, more astringent powders, mixed with whites of eggs and oil of roses, were applied, chiefly on the blood vessels, and the stump covered in bandages wet in oxycrate (astringent water). Bandaging was removed within three days.[106] Redressing proceeded in the same manner until eight to ten days had elapsed. The rest of the wound was to be covered with a mixture of 4 oz. of turpentine well washed in plantain water (renowned for stopping blood flow); 1 oz. of honey of roses; and ½ oz. of barley flour, all bound with two egg yolks, until it had finished suppurating. Any excrescence of the flesh was to be corrected with a small quantity of powder of alum, savine,[107] ochre, or powder of mercury, either applied separately or mixed with an ointment. Lowe particularly recommended the text on amputation 'most learnedly set down by my good friend M. William Clowes one of her Majesty's Surgeons'.[108] Like Lowe, a surgeon in the learned tradition who drew on military and naval experience, Clowes (1544–1604) was one of a group, in late sixteenth-century England, promoting use of the vernacular. He also fiercely opposed unlicensed practitioners.[109] His 1579 treatise on syphilis is one of the earliest in English on a specific disease.[110]

---

[104] Ibid., fifth treatise, chapter 6, 'Of Sphasell'. An illustration of a cauterising iron, as well as two kinds of wooden prosthesis are given in *A Discourse of the Whole Art of Chirurgerie*. See pp. 92, 96, 'Instruments and Cauters Actuals, for Extirpation; Legges of Wood'. Prostheses were devised by Ambroise Paré and other surgeons.

[105] Lawrence, 'Surgery (Traditional)', p. 978.

[106] If, perchance, a vein or artery had reopened, blood could be stayed by dissolving a little vitriol (sulphuric acid) in vinegar and laying the vein on a breadth of flax. Where two or more blood vessels reopened, a servant was to apply pressure with his fingers and knit them again.

[107] Oil from the shoots or leaves of the Savine (a small juniper bush) was used as a medicine in the treatment of rheumatism.

[108] Lowe, *The Whole Course of Chirurgerie*, fourth treatise, chapter 6. William Clowe's *A Profitable and Necessarie Booke of Observations for All Those that are Burned with the Flame of Gun Powder* (London, 1596), survives in the College's collection of rare books.

[109] A. L. Wyman, 'The Surgeoness: The Female Practitioner of Surgery, 1400–1800', *Medical History*, 28 (1984), p. 35.

[110] Lawrence, 'Surgery (Traditional)', p. 970. Acknowledging his services to surgery, Clowes commented in 1585 on the achievements of Paré that though he 'has small understanding in the Latin tongue, howsoever it is known, that he is not unskilful in any part of this art of Surgery'. Wear, 'Medicine in Early Modern Europe', pp. 297–98. By the 1612 edition, Lowe

Peter Lowe's book, too, was particularly useful for naval surgeons because it covered not only surgery but fevers and diseases of specific organs of the body as well.[111]

Tumours could be caused by imbalance of any one of the four humours – blood, phlegm, choler (yellow bile), and melancholy (black bile). Carcinoma was of two types, ulcerated and unulcerated: 'Cancer occult' or 'hidden Cancer' (as the second edition has it). Cancer was thought to be caused by a dry, melancholic humour, and also by 'evil diet': eating things that bred thick, corrupted blood. The prognosis for tumours in the stomach, head, shoulders, neck and under the arms was bad, and they were best left 'because these places cannot be cut, for the great flux of blood, which may happen in them'. Treatment was directed at stopping the melancholic humour from settling on site by administering remedies which purged it (the commonest way of restoring humoural balance), by bleeding 'if the age and time permit', and finally by abstention from all food which heated the blood: old hares, salt flesh, red deer, goats, spices, mustard, thick soup, cheese, fish and the like. In addition, the patient should not walk too much, should avoid excessive exertion and emotional upset, and was to eat foods which engendered good blood: mutton, veal, kids, capons; all kinds of fowl (except water fowl); and should drink whey to fortify the liver and the spleen.[112]

Lowe felt it best to avoid surgery at all costs, even when the carcinoma could be excised: 'some counsel to cut it, in such sort, that there remain no root, but my opinion is not to do such things'. He preferred to wet a cloth with the juice of various plants such as morel, plantain, lettuce, sorrel, centaury and shepherd's purse, and lay it on the tumour. He also used 'the urine of a young maid in the same fashion', oil of roses, or burnt lead 'to keep it in one estate, & correct gently the acrimony of the humour'. The desire to be as uninvasive as possible at a time when surgery carried great risks was to his credit, but Lowe's therapy would scarcely have stopped the growth of any tumour, never mind the most virulent. If growth was not checked by application of herbal preparations, he gestured the reader to the cure of cancerous ulcers in Treatise Seven.[113] Here Lowe insisted that their general cure was in purgation and diet, and a potion made of herbi Robeti,

---

was instead dropping the name of 'my good friend Maister James Harvey, chief Surgeon to the Queen's Majestie, who has written diverse learned bookes of Surgery, and in practice has excelled the most part in his time'. Lowe, *A Discourse of the Whole Art of Chirurgerie*, p. 97.

[111] J. J. Keevil, *Medicine and the Navy*, 4 vols (Edinburgh and London, 1958), ii, *1649–1900*, p. 31.

[112] Lowe, *The Whole Course of Chirurgerie*, fourth treatise, chapter 13, 'Of Cancer, Which the Greekes Call *Carsimonia*'.

[113] Ibid., fourth treatise, chapter 13.

scrofuralia, centruodie, Treacle and Mithridate (a complex galenical remedy of animal and herbal components originally intended as a prescription against poisons) 'for they cause venim to come out in the skin'. If the cancerous ulcer was in such a place that it could not be taken away with all its roots, cure was best achieved by incision, cauterisation, or corrosion with arsenic sublimate. In general, only superficial tumours were routinely operated upon in Lowe's time and the rest palliated.[114]

The Fifth Treatise discussed particular tumours, beginning with the head and moving down the body. It is greatly extended in the second edition, with additions on conditions such as alopecia, head lice, cleft lip, sciatica and diseases of the bladder and penis. The only major omission in this treatise is a discussion of lithotomy or cutting for the stone then frequently (if not commonly) performed, often by skilled itinerants.[115] It tends to indicate that Lowe was relatively conservative in his attitude towards surgical intervention. Alopecia – often occurring in the elderly, in cases of rotten fevers, the French pox or leprosy – was seen as a defect of the humour, or vapour fuliginous, of which the hair is made. Where the scalp had been bald for a long time, it was difficult to achieve regrowth, but concoctions suggested for stimulating the follicles were quite as weird and wonderful as modern-day preparations.[116] According to Galen, boar's grease, foxes' dung or vipers (dead presumably) rubbed onto the pate might do the trick; Nostradamus counselled the use of bees dried in an oven until rendered into a powder and mixed with honey or swan's blood, then rubbed into the head; while anointing the bald areas with the milk of a bitch prevented hair from growing back white. Quirkier still was the suggestion of Johanne Lebot who advised distilling the excrements of a red-haired man and anointing the part.[117]

Lowe then dealt with ophthalmy or inflammation of the eyeball, which was to be treated, in the first place, by good diet, emotional equilibrium, and general continence. His recommendations were:

> eat little chiefly at night, abstain from all vaporous things & all evil digestion, fishes, fruits, spices, salt & humid things, abide neither in great darkness, nor too much light, for great light dissipates the spirit ... Endeavour to be laxative, walk not too much, beware of all perturbations of the spirit, from smoke and dust blowing of Alchemy, for both it hurts the eye and consumes the substance,

---

[114] Lowe, *The Whole Course of Chirurgerie*, seventh treatise, chapter 8, 'Of the Ulcers Cancrous and their Curation'.
[115] Hamilton, *The Healers*, p. 60.
[116] Edington, ' "Discourse" of Maister Peter Lowe', pp. 49–50.
[117] Lowe, *A Discourse of the Whole Art of Chirurgerie*, p. 124.

& makes men miserable both in body and goods, hold up your head, and abstain from wine and women and such like.[118]

The bad humour was evacuated and diverted, secondly, by administering pills, clysters, bleeding of the cephalic vein, ventousing the shoulders, frictions on the thighs, legs and extremities, and by opening the veins and arteries of the temples. Thirdly, topical remedies were applied.[119]

In the second edition, the material on the eye is greatly extended. This noblest sense organ, warned Lowe, was 'of so great importance and profit to us, the skilful Surgeon ought to take great care and diligence for the conservation of the same'.[120] There are chapters on its composition,[121] and on 'such common maladies, as befall the whole Eye'.[122] Weak sight was preserved by good diet, through pharmacy and with surgery. To maintain clarity, Lowe recommended that the eyes be bathed in a solution of various herbs including fennel and roses, which should be infused, first, in white wine; second, in the urine of a healthy young boy; third, in a nursing mother's breast milk;[123] and lastly, in finely distilled honey.[124] Surgery on the eye was rather cursorily dismissed.[125]

Since animal components (especially blood) were part of the materia medica of the time, this makes some treatments sound magical in origin: the cure counselled to prevent lashes growing into the eyelid, for example, was to 'anoint the part with the head of a frog', after the hair had been plucked out.[126] Another treatment to the same effect (allegedly 'set down by Alexis of Pimunth, for a rare secret, and learned by him of a noble Lady in Syria') is more reminiscent of the modern-day folk remedy for the prevention of styes:

> take a piece of pure gold that is small and round like a Ring, somewhat crooked at the point, heat it hot in the fire, then rub the inward side of the eyelid gently where the hairs do grow, thereafter anoint the part with oil of Roses, and Violets; if need be, the next day do the like: and if they yet continue do the like, the gold makes that no cicatrice remain.[127]

---

[118] Lowe, *The Whole Course of Chirurgerie*, fifth treatise, chapter 3, 'Of the Tumor in the Eye called *Lippitudo* and *Opthalmia* in Greeke'.
[119] Lowe, *The Whole Course of Chirurgerie*, fifth treatise, chapter 3.
[120] Ibid., p. 140.
[121] Ibid., pp. 140–44.
[122] Ibid., pp. 144–48.
[123] The English surgeon, Joseph Binns, mentioned in note 2 above, still used this as an eye medication in the mid seventeenth century, though he discontinued it in one case because the patient's eye got worse. Beier, 'Seventeenth-Century English Surgery', p. 72.
[124] Lowe, *A Discourse of the Whole Art of Chirurgerie*, pp. 149, 154.
[125] Ibid., p. 155.
[126] Ibid., p. 163.
[127] Ibid., p. 164.

Dilation of the pupils, commonly caused by strokes or falls, was to be treated not only by careful diet, the use of purgatives, bleeding in the arms and corners of the eye and frictions and ventousing of the shoulders with scarification, but also by instilling 'in the eye the blood of a Chicken or Pigeon taken out of the vein under the wing'.[128] Lowe suggested equally quaint cures for nosebleeds. Aside from more conventional remedies like ligaturing of the arms, thighs, testicles and legs, he insisted that 'man's blood dried in powder and put up into the nose is very good'. Similarly, 'The common people do only use for all fluxes of blood at the nose, Hogg's dirt put in cotton or a small linen cloth applied to the nose'; and the smoke of the dung in the nose was equally beneficial.[129]

But possibly the most glaring deficiency of the first edition was the absence of any discussion of cataract at a time when couching was a surgical commonplace. Lowe rectified this in the second edition with two additional chapters, one on cataract, and a second on its cure. It remains an odd omission, though, especially since Lowe had observed the treatment of cataract many times in Paris, and was himself skilled in the operation. He refers to two of his patients in his text:

> The one of a servant of my Lord of Laudum, who had a Cataract five years on both her Eyes, which when I did see, I caused her to stay one year there longer till it became more ripe, then I did couch them both and restored her to her sight. Likewise a Servant of the Lord of Craggie Wallace, who had a Cataract on his Eyes the space of nine months or thereabouts, which was sufficiently ripe, so I did couch it, and restored him in like manner to his sight.[130]

Prior to couching, the patient was prepared by purging and with medicines, syrups, and bleeding if necessary, then he was to rest a day or two, eating only a light diet two days prior to the operation. But it was perfectly permissible (before the use of anaesthesia) for the patient to be fortified before the operation by a slice of toast and a little wine or clean ale. The operation was best performed at eight or nine o'clock in the morning,[131] presumably when the surgeon was at his brightest and the light clear. He was to proceed as follows:

> set him [the patient] on a form or stool in some light place, in such sort that his face be directly before the Surgeon, then one must stand behind him to hold his head fast and stable, and another must hold up his eyelid with his thumb and finger, that the Surgeon may see the better, that being done, the Surgeon

[128] Ibid., p. 165.
[129] Ibid., pp. 181–83.
[130] Ibid., p. 167.
[131] Ibid., pp. 168–69.

shall sit on the same form, or some other seat somewhat higher, directly against the sick, and so near, that the hands of the sick may rest upon the Surgeon's thighs,[132] then the Surgeon shall stab the Needle often in some cloth to make it sound and hot, it must neither be rough nor cold to the membranes of the eye.[133]

The method of rendering the cataract softer to the knife was singular: 'the operator, or some young child who has a sweet breath, shall chew Cinnamon, Ginger, Cloves, or Fennel, and spit it out and breathe three or four times in the eye of the sick, it being open; & that prepares the eye, & makes the Cataract more thin'.[134] Holding the eye firmly, and having 'thrust in the Needle, which must be sharp pointed, a little lower then the middle of the eye, over the membranes named Conjunctiva and Cornea', the surgeon was advised to 'let it go towards the side of the little corner which is near the temples, directly to the foreside of the cataract, not advancing over far'. He continued:

> some do counsel that [the] needle be so conveyed, that it be the thickness of two six pences from the black of the eye, so is it the easier to be couched low, but it is more difficult to pierce: others pass the needle through the middle of the black of the eye, conveying it before the Cataract till it come to the top of it, & taking good heed in picking of it, then by little and little turn the needle gently, until such time as you couch it to the lowest part of the eye, and being there hold it down a pretty space, then retire the Needle the way that it went in.[135]

In this way, the opaque lens was merely pushed into the lower part of the eye with the needle.[136]

Inflammation of the tonsils was treated by an invalid diet, abstention from strong drink, the use of clysters and bleeding in the arm and of the veins under the tongue, the use of ventouses and frictions on the neck, gargarism, refrigeratives (remedies which cooled the body) and dessicatives. Where a tonsil abscessed, the surgeon should open it with a lancet. If they became so enlarged that the patient was in danger of suffocation, then the surgeon was to proceed to tracheotomy, cutting with a bistoury 'between the third and fourth ring', in a manner that 'was practised by Andrew Scot one of the King of Scotland's Surgeons in Paris most cunningly'. But he felt the need to counsel that making the incision was very dangerous, not only because

---

[132] For a contemporary illustration of this stance, see the cover illustration of Conrad et al., *The Western Medical Tradition*, a cataract operation, from G. Bartisch, *Augendienst* (1583).
[133] Lowe, *A Discourse of the Whole Art of Chirurgerie*, pp. 169–70.
[134] Ibid., pp. 170–71.
[135] Ibid. pp. 170–71.
[136] The word couching came from the French *coucher*, to lay down, because the procedure involved pushing the opaque lens down in order to restore vision.

of the flow of blood which usually occurred but because of the nerves in the vicinity.[137]

The surgical repair of a harelip is a fascinating addition to the 1612 edition. It is accompanied by a print of its repair by suture, the source for which was a book by the French surgeon Jacques Guillemeau. Illustrations were added to the second edition of Lowe's text, which appeared under the slightly amended title *A Discourse of the Whole Art of Chyrurgerie*, in an attempt to enlarge and improve upon the first. He poached copies of illustrations from the premier anatomical books of his time: most were taken from Ambroise Paré's *Les oeuvres, avec les figures et portraits, tant de l'anatomie que des instruments de chirurgerie* (1575), and from Jacques Guillemeau's later publication *Oeuvres de chirurgie* (1598). Lowe, who was by no means the first to do this, obviously had a strong desire to spread anatomical and surgical knowledge, though his efforts were limited by the low standard of woodcut in Britain compared with that on the Continent.

For the surgical repair of a cleft lip, the surgeon was advised to follow the standard pre-surgical preparation normally recommended by Lowe – purging the body (if the patient were old enough) and fasting the night before the operation. The recommended surgical procedure was

> with your left hand lift up the one side of the lip: then with a sharp bistoury or lancet curb, cut the outward skin till you come to the middle of the slit, which being done, you shall lift up the other side of the lip and do the like, so both sides shall be altogether like unto a great wound, let them bleed a little to discharge the part and avoid inflammation, which being done, join the sides together so justly as you can, then thrust a needle through both the parts of the lip, taking a reasonable grip, letting there the needle remain: then turn the thread about it after the form you see aged women or Tailors do when they keep the needle on their breast.

Once the wound was sutured, an astringent lotion and compress were applied. It usually knitted in eight to ten days, when the stitches were cut and the needle plucked out.[138]

Also new to the second edition of the *Chirurgerie* were three chapters on teeth. Homing in on a favourite target, Lowe insisted (correctly no doubt) that 'the common Barber Surgeons do commit great error in plucking out of innumerable teeth which might well serve'.[139] Lowe was one of the first people to contradict the then commonly held opinion that worms were the

---

[137] Lowe, *The Whole Course of Chirurgerie*, fifth treatise, chapter 8, 'Of the Tumor of the Amigdalles, called by the Greekes *Parischimia*, and by the Latines *Tonsilla*'.
[138] Ibid., pp. 187–88.
[139] Ibid., p. 189.

3. Portrait of a cleft lip, from P. Lowe, *A Discourse of the Whole Art of Chyrurgerie* (London, 1612).

cause of toothache. He was also trained in the fitting of false teeth.[140] Where enunciation was impeded through teeth loss, the remedy was to

> make artificial teeth of Ivory, Whales' bone or hounds' teeth, which shall be fastened by a wire or thread of gold, passing the wire or thread between the whole tooth on either side next adjacent, then put the artificial tooth in the part, then knit the thread fast through about the ends of the thread, and cut it so near as you can; if any portion rest uncut, pass it between the whole tooth, that the tongue or lips be not hurt by it.[141]

Further treatments for teeth were promised in his forthcoming *Booke of the Infantment*.[142]

Referring to aneurysm, Lowe counselled that it should not simply be slit open because this action usually killed the patient. In his opinion, aneurysms manifesting in the exterior parts of the head, legs and arms might be cured by knitting of the blood vessel, but those in the neck, under the arms and on the flanks where the arteries were very dilated almost always resulted in death within a few days.

> If the tumour be opened the patient dies presently: this happens oftentimes by the unskilfulness of the Barbers and Apothecaries, that meddle therewith, and ruin a number of people through their ignorance, as I have often seen, for such people esteem all tumours, that are soft, to be opened, as common Apostemes.

Indeed, it was hardly a course of action to be recommended.[143] Umbilical hernia, on the other hand, was managed in much the same way as today (though obviously without the benefit of anaesthesia and painkillers). Lowe would reduce it as follows:

> If it [abdominal wall] be rent by great violence or cough, and the intestine comes forth, as happened to an honest Matron in Paris, whom I cured after this manner: first, I enlarged the dilation with a Lancet curb, then reduced the intestine, and used the suture Pellitor,[144] as is set down in the general chapter of wounds, & cured it as other wounds.[145]

---

[140] R. M. Ross, 'Peter Lowe: Founder of the Faculty, Man of Mystery', *Dental Historian*, 28 (1995), p. 6.

[141] Lowe, *A Discourse of the Whole Art of Chirurgerie*, p. 194.

[142] Ibid., p. 198.

[143] Ibid., fifth treatise, chapter 14, 'Of the Tumor Called Auenfrisma'.

[144] This terminology is not used in the general chapter on wounds, but must refer to the retentive suture since Lowe also calls it the 'Suture general called retentive', which, according to him was 'now commonly used in the intestines, stomach, and bladder, and such other membranous parts'. Ibid., pp. 223, 289. It was also referred to as the 'glover's suture', for more on which see below.

[145] Ibid., p. 223.

Scrotal hernia, or the descent of an inguinal hernia into the scrotum, could become immensely huge, and might, if untreated, even descend as far as the knees. Where there was great dilation, and routinely in elderly people, Lowe recommended a truss – 'only the bandage made of cloth with Cotton, Iron, or Steel, as shall be most fitting'. This was merely a palliative treatment, involving the mechanical retention of the hernia by some kind of apparatus in an attempt to keep the hernial sac (a cul-de-sac of the peritoneal cavity) empty. Unfortunately, when the hernia was so huge, it really was not reducible by these means. In late sixteenth-century society, scrotal hernia was more common in 'such people as do ride great horses and are armed ... as I have often seen amongst the French, Allmaine, or Rysters' horsemen: who for the most part have their bandages of Iron, either for one side or for both'.[146] But on castration as a cure for the condition, Lowe was unwavering:

> In this disease there are great abuses committed by a number of unskilful ignorant people, void of all good conscience and fear of God, who for every simple kind of rupture, make incision and cut away the production of the Peritoneum and Testicle: if the rupture be on both sides, they cut off both testicles, which renders a man sterile, and causes the hair of the beard to fall ... Besides that, oftentimes in cutting the sick dies, chiefly when the dilation is great. Sometimes the Intestine sticks to the Peritoneum,[147] which they knit altogether, and cut away the production; after which the sick void [148] their excrements at the mouth, & die most miserably.[149]

Some surgeons, then, simply cut away the protuding lumen, as well as the testicle(s), then closed the abdominal cavity. Significantly, Lowe confessed that he himself had often done this earlier in his career, 'the which I do now repent in committing such a heinous sin'.[150] Indeed, the castration frequently attendant on the surgical treatment of hernia largely explains why the operation remained so long in the hands of itinerant empirics because licensed surgeons were most unwilling to perform it. Always performed as a last resort, the correct surgical procedure was to enlarge the orifice through which the prolapse had occurred, and close the abdominal cavity which is basically the operation as performed today in the absence of additional procedures

---

[146] The hernia truss is illustrated in the second edition of Lowe, *A Discourse of the Whole Art of Chirurgerie*, p. 246. This woodcut is manifestly the mirror-image of a similar figure in Ambroise Paré's *Les oeuvres, avec les figures et portraits, tant de l'anatomie que des instruments de chirurgerie* (1575).
[147] This refers to the retention of the small intestine in the hernial sac by adhesions.
[148] The 'avoydeth' of the original text means void.
[149] Lowe, *A Discourse of the Whole Art of Chirurgerie*, p. 249. This must refer to faecal vomiting following either post-operative peritonitis or a post-operative obstruction.
[150] Ibid., p. 250.

designed to prevent relapse (such as the resewing of muscle bands).[151] Severing of an important blood vessel was all too frequent.[152]

The treatment of wounds is discussed in Treatise Six (including an additional chapter on gunpowder burns in the 1612 edition).[153] Wounds during this period were classified as either simple or composed: the first involved no loss of substance and could be healed by one therapy only; the second did involve loss of bodily matter and required various treatments. However, their treatment was circumscribed: penetrating wounds of the heart, lungs, stomach, bladder and small intestines were generally regarded as fatal.[154] Bleeding was arrested either by healing salves or styptics, or in deep wounds by cautery or suture. Lowe described three kinds of suturing commonly practised in the late sixteenth century: incarnative, retentive and conservative. Incarnative suturing could be executed in a number of different ways. In one method, suturing was begun in the middle of the wound, and an inch observed between stitches. A second method was the 'twisted suture' (as described above for restoring a cleft lip). Thirdly, to repair a large wound, long needles and strong, double thread with a knot in the end were passed into various parts of the wound, leaving an inch between stitches again. Then two small, tubular pieces of wood – 'the greatness of a small goose quill'- were passed beneath the thread on each side of the wound, the lips pressed gently together and the thread knit with double knots, one after the other.[155] 'Quilled' or compound suturing (as it was also known) was used in deep muscular wounds with the aim of preventing the wound from knitting superficially at the surface but remaining open underneath. The first three or four stitches were therefore sunk very deep into the wound and, after the rest of the sutures had been put in, the quill was slipped through the loops of the sutures on one side and then the other in order to keep the centre of the wound and its lips pressed together, and thus prevent undue strain on the threads and painful taughtness.[156]

[151] Sloan, *English Medicine in the Seventeenth Century*, p. 103.

[152] Lawrence, 'Surgery (Traditional)', p. 979.

[153] Here Lowe may possibly have benefited from discussion with his friend, William Clowes, who published a book on the treatment of gunpowder burns and wounds in the year before the first edition of Lowe's *Chirurgerie*. See W. Clowes, *A Profitable and Necessary Book of Observations, for All Those that are Burned with the Flame of Gunpowder, etc. and also for Curing Wounds Made with Musket and Caliver Shot, and Other Weapons of War Commonly Used at this Day Both by Sea and Land* (London, 1596).

[154] Beck, *The Cutting Edge*, p. 12.

[155] Lowe, *The Whole Course of Chirurgerie*, sixth treatise, chapter 1, 'Of the Cause, Signes and Curation of Woundes in Generall'.

[156] Cooper, *Dictionary of Practical Surgery*, p. 591.

4. Portrait of a dry suture, from P. Lowe, *A Discourse of the Whole Art of Chyrurgerie* (London, 1612).

Another method, called dry suturing, was employed where it was important that the scar was as fine as possible – on the face, for example. Since it involved no stitching at all, it has also been referred to as false suture. Little pieces of strong cloth with their edges cut into points like arrow heads were covered with an astringent and glutinous plaster which included whites of eggs, and lain on both sides of the wound, points facing. Once the cloth had dried fast on the skin, 'we put a thread through these points till such time, as we see the lips of the wound to close, and knit the thread with a double knot'.[157] Lowe was not much enamoured of the method of suturing which used 'clasps of Iron sharp pointed, and long, which take the lips of the wound being put together and hold them so'. Much favoured by old practitioners, he considered that this ancient forerunner of stapling encouraged inflammation.[158] It was later replaced by the sticking plaster.[159]

A retentive suture, recommended for superficial wounds, was similar to the uninterrupted stitching employed by glovers. It was made by introducing the needle into one side of the wound from the inside to the outside of the lip, and then into the opposite lip in a similar way, and so on. By the late eighteenth century, it was very unpopular with surgeons because of the number of unnecessary stitches it entailed and for the puckered edge it produced in dragging the edges of the skin together; it was henceforth confined to use in the sewing up of corpses.[160] Lowe had recognised its weakness two centuries earlier: 'for one point slipping, the rest slip also'. Conservative suturing was 'not so hard' or tight; it was used to conserve the separated lips of a wound where there was great loss and laceration of flesh.[161]

In the section on wounds, in particular, Lowe displayed a familiarity with the work of Ambroise Paré. He devoted a chapter to wounds inflicted by a variety of instruments such as bullets and stones, a subject seldom touched by the ancients, except by Celsus 'who tell not with what instrument they were shot'. Introduced in the fifteenth century, gunpowder radically altered the variety and extent of injuries presenting to military surgeons, encouraging

---

[157] This method continued to be used by Scottish surgeons later in the seventeenth century. Alexander Read, surgeon, Fellow of the London College of Physicians, lecturer at the London Barber-Surgeons' Hall, and author of *Surgical Operations* (1672), also used it to stitch small wounds. Sloan, *English Medicine in the Seventeenth Century*, p. 115, quoting R. H. Meade, *An Introduction to the History of General Surgery* (Philadelphia, 1968), p. 17; M. Maltz, *The Evolution of Plastic Surgery* (New York, 1946), p. 192.

[158] Lowe, *The Whole Course of Chirurgerie*, sixth treatise, chapter 1.

[159] Cooper, *Dictionary of Practical Surgery*, p. 592.

[160] Ibid., pp. 590–92.

[161] Lowe, *The Whole Course of Chirurgerie*, sixth treatise, chapter 1.

the development of new techniques.[162] And though sixteenth-century authors had begun to write on the matter, according to Lowe, they were 'of diverse opinions and written in sundry languages', so he decided to give his own view on the basis of protracted experience in the theatre of war. One of the main surgical contentions of the period was that gunpowder produced poisons which were neutralised by cauterising or pouring boiling oil into wounds.[163] The misapprehension probably arose because gunshot wounds were obviously dirty and often foul smelling.[164] Keen to ally himself with Paré who first substituted a cooling salve for this inhumane treatment, Lowe discounted the idea of excessive heat in gunpowder, though not the possibility that it produced poisons:

> some think, that there is venemousness in the powder, and burning in the bullet, which is false ... yet there may be some extraordinary mixture in the powder, which causes venom, for the which we take some other indication, according to the thing.[165]

Trained only by barber-surgeon's apprenticeship, Paré revolutionised military surgery by abandoning the use of boiling oil in the treatment of wounds, substituting a soothing dressing of egg yolk, oil of roses and turpentine. This was the most important advance in wound surgery prior to the development of Lister's antiseptic technique.[166] He also renounced formal Latin to write his books in vernacular French, making the newest anatomical and surgical knowledge accessible to the rank from which he had himself risen, a trend followed by Lowe and his humanist colleagues.[167] Lowe's *Chirurgerie* indicates that he used a similar concoction for gunshot wounds: poultices of bread, milk, yolks of eggs and a little saffron, or mallows, sodden and beaten with wheat flour, oil of roses, hog's grease and saffron. Like Paré, he believed in the use of traditional suppurative applications: his favourite remedy for countering inflammation and correcting putrefaction, was hypericum,[168] turpentine, and yolks of eggs.[169]

---

[162] Lawrence, 'Surgery (Traditional)', p. 970. Hippocrates famous maxim was that 'war is the only real school for the surgeon'.
[163] Wear, 'Medicine in Early Modern Europe', p. 297.
[164] Beier, 'Seventeenth-Century English Surgery', p. 63.
[165] Lowe, *The Whole Course of Chirurgerie*, sixth treatise, chapter 5, 'Of Woundes Done by Gunshot'.
[166] Sloan, *English Medicine in the Seventeenth Century*, p. 106.
[167] C. P. McCord, 'Bloodletting and Bandaging: Barber Surgery as an Occupation', *Archives of Environmental Health*, 20 (1970), pp. 556–57.
[168] Hypericum or Saint John's Wort was a loosening medicine, which relaxed muscles, tendons, ligaments and membranes when distended. Ibid., pp. 210–11.
[169] Lowe, *The Whole Course of Chirurgerie*, sixth treatise, chapter 5.

But it is possibly the trepan – that instrument of torture for relieving pressure in the cranium – which incites most horror in the twentieth-century.[170] The modern form of the instrument with crown-shaped head and central pin, as illustrated in the second edition of Lowe's *Chirurgerie*,[171] was invented by the English surgeon John Woodall.[172] In the view of William Buchan, the eighteenth-century author of *Domestic Medicine* (1769), head wounds were one of the few things which could not be self-treated. The important thing was that it 'touch not the flesh', so the skull had to be uncovered: an incision was made in the scalp, the wound dilated with lint to bare the cranium and the fracture identified by running the surgeon's nail along the bone. This was standard practice. In the sixteenth and seventeenth centuries, trepanning appears to have been performed relatively frequently, especially in cases of depressed fracture where it was necessary to relieve pressure on the brain.[173] Few were skilled in this operation, though Scots who in his opinion were especially adept were the king's surgeons Gilbert Primrose and John Nessmith.[174]

The Seventh Treatise is devoted to ulcers – a far more common complaint among the labouring population of previous centuries. They came in six gloriously named varieties: sanious (with a foul, green-smelling discharge); virulent; filthy; cancerous; putrid or stinking; and corrosive or rotten away; and could be governed by any of the four humours. The prognosis for cancerous ulcers was almost always bad. A fistula was 'a profound ulcer, having the entry, hard, narrow, deep, cavernous, from the which proceeds a matter virulent'. Ulcers that had hard sides were difficult to heal as were ulcers on the extremities. But the real talent lay in diagnosing the type of ulcer and applying the appropriate topical remedies; of course, 'Many ignorant barbers fail herein, thinking one kind of plaster to be good for all sores, in the which they are deceived'. The cure for ulcers was good diet and purging

---

[170] For an illustration, see ibid., p. 318.

[171] See ibid., p. 318, 'A Trepan, with Other Instruments for the Head'. For the subtlety of message relayed to historians by pictorial representations of surgical instruments, see G. Lawrence, 'The Ambiguous Artifact: Surgical Instruments and the Surgical Past', in Lawrence, *Medical Theory, Surgical Practice*, pp. 295–314, particularly, pp. 298–300, 305–7.

[172] Sloan, *English Medicine in the Seventeenth Century*, p. 109. Woodall is most renowned as the author of the standard text for ship surgeons during this period, J. Woodall, *The Surgions Mate* (London, 1617).

[173] Lawrence, 'Surgery (Traditional)', p. 977. It was not until the eighteenth century that a more cautious approach was adopted to trepanation.

[174] Lowe, *The Whole Course of Chirurgerie*, sixth treatise, chapter 10, 'Of Woundes in the Head'.

and bleeding of the offending humour, after which the sore could be healed.[175]

Fistulae near major organs were perilous and mortal; those in the ribs or on the back were difficult to cure; while fistulae in the buttocks were long in healing, though Lowe's good friend, James Henderson, 'a man very expert in the art of Surgery in Scotland' was apparently very skilled in their repair. Anal fistulae had not been successfully treated until the fourteenth century by the French surgeon Henri de Mondeville, who opened the orifice widely and stopped bleeding by pressure (much as today).[176] According to Lowe's recommendation, first the body was purged, and the orifice dilated by tents of gentian and briony,[177] and sponge-prepared cyclamen – normal surgical practice. The next step was to remove the callousness present in a long-standing sore either with caustics and corrosives (like powder of mercury, white soap or arsenic) or by removal with shears or a razor, opening up the fistula. 'If it be profound & such kind of places, where it may be knit, we put a needle with a strong thread through it, and knit the thread every day more and more till it be consumed'. Finally, to appease the inflammation and to cause the scab to fall off, oil of roses was applied with whites of eggs until the fistula voided its discharge.[178]

Fractures, dislocations and embalming are discussed in the Eighth Treatise,[179] which, in the second edition, is preceded by a new chapter on the skeleton. Although almost nothing survives about the teaching of anatomy (or of anatomical demonstrations) by the Faculty in the early seventeenth century, it must be assumed to have been based on Lowe's textbook. The woodcut used by Lowe to illustrate the human skeleton is clearly derived from Ambroise Paré (whose anatomy was Vesalian) but neither the Paré nor the Lowe approach the beauty of the original Vesalian print from the *De humani corporis fabrica* (1543) on which both are based. The first stage in the treatment of a fracture was to extend the limb

> which is done, by laying it on a bench or other place proper, the sick being well situated, there must be two persons to hold the member fractured, the one at

---

[175] Ibid., seventh treatise, chapter 1, 'Of Ulcers in Generall'. Note that the second edition contains no additional chapters.

[176] Sloan, *English Medicine in the Seventeenth Century*, p. 105.

[177] Gentian is a glutinative herb used either to ripen or restore flesh to ulcers, and briony is a suppurating herb, also applied externally to ripen a swelling. Tobyn, *Culpeper's Medicine*, pp. 213–14.

[178] Lowe, *The Whole Course of Chirurgerie*, seventh treatise, chapter 9, 'Of the Ulcer Fistulous'; Beier, 'Seventeenth-Century English Surgery', p. 70.

[179] Note that in the first edition the printer mistakenly duplicated the seventh treatise as a heading twice. The second 'seventh treatise' is therefore, technically the eighth.

the nether part, the other at the upper part, of the which one draws up an other down, to make the extension; if the hands be not sufficient to do this, we take cords or strong cloth and bind fast the member one to the upper part, an other to the nether, which shall be drawn by two men contrary as you have heard.

Treatment was by reduction and fixation. Where great force was necessary, mechanical cases called glossocomions were employed to extend the limbs; the bones were then reduced, and an astringent plaster applied to the fracture prior to bandaging. Splints of card, wood or white iron were used in threes to hold fast a limb. If the fractured limb did not mend in proportion, some surgeons rebroke the limb, but Lowe felt unable to advise rebreaking where the bone had already set, for though guidance had been given on this by both the Persian physician Avicenna and the Italian Guido, 'I counsel no man to try, for better it is to suffer a little deformity of a part, than lose the whole body, to wit, death, which often happens'.[180]

Lowe also describes his method of embalming:

> First we lay the body on a table, and make incision from the clavicles to the os pubis, next lift the sternum as also the musc[l]es of the inferior belly, taking out all which is contained therein as also the brains, having first opened the [brain]pan with a saw, which all shall be presently buried in the earth, saving the heart, which shall be embalmed either with the body, or alone in a box of lead, as the friends shall think good; thereafter, you shall make long deep incisions in the arms, thighs, buttocks and legs and other fleshy parts, chiefly where there are veins and arteries, to the end they may the better void, which being done, you shall diligently wash the three venters, as also the parts incised with strong vinegar, wherein has been sodden Wormwood,[181] Alum [182] and Salt, thereafter with Aquavitae, or fine spirit of Wine do the like, then dry all well with Linen cloths or sponges and fill up the three bellies with powders and a few flocks and sow them up again the incision shall be filled only with the powders and sewed up likewise.

Commonly used preserving powders were roses, camomile, meliot, mint, wormwood, sage, lavendar, rosemary, marjoram, thyme, cypress, gentian, iris of Florane, all dried and beaten to a fine powder, and mixed with nutmegs, cloves, cinnamon, pepper, bengewin, aloes and myrrh. Others (operating on a budget presumably) used only a few of the commonest herbs mixed with a little quicklime, ashes of bean stalks and oak, then rolled the

---

[180] Lowe, *The Whole Course of Chirurgerie*, eighth treatise, chapter 2, 'Of Curation of F[r]actures in Generall'.

[181] Wormwood is a mild astringent, with binding, drying and cleansing properties. Macerated in vinegar, its effect would have been enhanced. Tobyn, *Culpeper's Medicine*, pp. 206, 213, 218.

[182] Alum is also an astringent; burnt alum is mildly caustic. It was also used in styptic powders. Cooper, *Dictionary of Practical Surgery*, pp. 14–15.

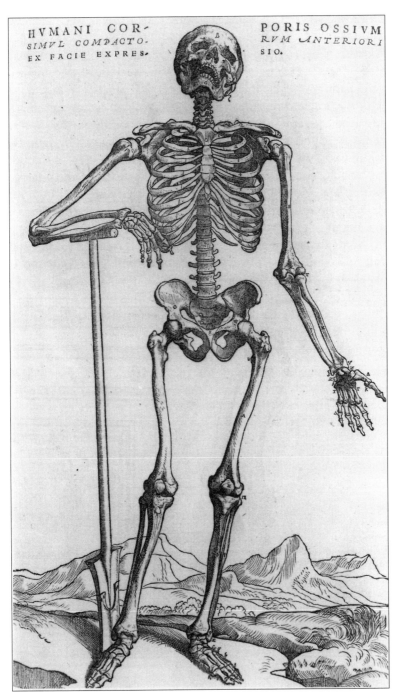

5. The bones of the human body, from Andreas Vesalius, *De humani corporis fabrica* (Basle, 1543).

corpse in a waxed cloth, binding the body with cords before placing it in a lead coffin.[183]

In Treatise Nine,[184] Lowe articulated his views on phlebotomy, as it was known by the Greeks, or venesection, as termed by the Romans – the therapeutic and prophylactic mainstay of surgical practice in his time. Originating in magical and religious ceremonies,[185] blood-letting or 'breathing a vein' involved the removal of about 100 ml. of blood into a dish and sealing the wound by applying pressure.[186] Since illness was caused by an imbalance of the humours, the superfluous humour was removed from the body by drawing blood (frequently). The most common form of surgical intervention, phlebotomy was resorted to in almost every case of disease or injury in this period. There were basically three methods. Blood-letting by direct venesection – the most common method – involved the tying of a ligature round the arm while the patient grasped a thick staff and the vein was incised with a lancet;[187] secondly, there was scarification, sometimes with the use of cupping glasses which, applied hot, drew blood into the vacuum as the vessel cooled; and by the application of leeches to the body, which was more convenient for drawing blood in young children and from places where the lancet or cupping glasses could not easily be used.[188]

In Lowe's opinion, phlebotomy was

> a thing most necessary to be known of all Surgeons not only for the healing of maladies, but also for conservation of the body from sickness, sometime[s] both for healing & preserving, for the which it is most necessary to know the number of the veins, also the true situation, to the end he take not one for an other, because sometime[s] the situation is variable, and in parts not accustomed very uneasy to be opened, we must also know the way to open them and what it is to consider before and after, for the effects, that follow thereupon.

---

[183] Lowe, *The Whole Course of Chirurgerie*, eighth treatise, chapter 4, 'Of the Embalming of Dead Folke'.

[184] This treatise appears greatly extended from seven to twenty-three chapters, but this is mainly because most of the subheadings have been expanded into chapters.

[185] See G. R. Seigworth, 'Bloodletting over the Centuries', *New York State Journal of Medicine*, 80 (1980) pp. 2022–28.

[186] Hamilton, *The Healers*, p. 59.

[187] The staff held by the patient came to symbolise the barber's pole – the red stripe representing the blood that was let and the white the application of bandages. C. P. McCord, 'Bloodletting and Bandaging: Barber Surgery as an Occupation', *Archives of Environmental Health*, 20 (1970), p. 551.

[188] For an account of blood-letting based around the instruments and tools then in use, see A. Davis and T. Appel, *Blood Letting Instruments in the National Museum of History and Technology* (Washington, 1979).

Blood-letting was an all-round cure: not only did it quicken the spirits, but it helped the memory, made the senses more subtle, clarified the voice and sight, was an aid to digestion, purged 'evil blood', and was 'an evacuation natural of the humours of our bodies, by which infinite maladies are cured'. Bleeding was not recommended prophylatically before the age of fourteen or above the age of eighty, though it could be used therapeutically in both those ranges for healing sickness. Lowe's illustration of the veins most commonly opened by surgeons is poached from Guillemeau.[189] Well-nourished, strong bodies were heavily bled; feeble, ill and extenuated bodies, were bled less. Where a person feared fainting, he was advised to 'eat an egg before, or a toast in wine and lie in his bed'.[190]

In order to bleed a patient successfully, Lowe recommended that the surgeon work in a light place – either in natural or candle light; his own sight should be good. His surgical prerequisites were: a ligature an inch broad and sufficiently long to go twice about the arm; a round staff for the patient to hold in his hands; and a little oil to rub on the vein to make it supple and with which to annoint the head of the lancet. The incision could be made either lengthways, breadthways or obliquely but always in the middle of the vein 'not cutting it altogether'! As he put it:

> open the vein softly sliding the point of it in the vein not suddenly, if the vein be not opened at the first time, prick suddenly again a little above or under the first, you may do the same if the hole be not great enough to let out the blood in a reasonable quantity. The vein opened in that manner, the party shall hold the staff in his hand, that he may rest his hand, & the blood come out the better.

Often, in heavy people, streaks of fat protruded at the aperture which were to be replaced. After bleeding, a piece of wet cloth, three or four fingers in breadth was placed on the wound which was bound above and below the elbow. Ventousing was the application of a wide-bellied instrument (or cup) of brass, horn, glass, wood or earth with a short or long neck, to diverse

---

[189] There were forty-one veins for maladies, seventeen in the head, six in the arms (three in each), six in the hands, four in the buttocks (two in each side), and eight in the legs.

[190] Lowe, *The Whole Course of Chirurgerie*, ninth treatise (recte), chapter 1, 'Of Bleeding and Thinges Therein to be Observed'.

[191] Lowe, *The Whole Course of Chirurgerie*, ninth treatise (recte), chapter 1, section 'Things Which are to bee Considered to Bleede Well'.

[192] O. H. and S. D. Wangensteen, *The Rise of Surgery: From Empiric Craft to Scientific Discipline* (Minnesota, 1978), p. 247; R. Porter, *Health for Sale: Quackery in England, 1660–1850* (Manchester and New York, 1989), p. 36; F. N. L. Poynter and W. J. Bishop, *A Seventeenth-Century Doctor and his Patients: John Symcotts, 1592?–1662*, Publications of the Bedfordshire Historical Record Society, 31 (Streatley, 1951), pp. xxx–xxxi; Cooper, *Dictionary of Practical Surgery*, pp. 112–13.

parts of the body to draw some harmful humour from it. It was often applied with scarification, that is, with bleeding (usually by the application of a number of lancet blades fixed in a metal box and worked by springs); otherwise it was called dry ventousing or cupping,[191] in which the glasses were applied to abscesses, ulcers and snake and insect bites with the aim of drawing out morbid matter. They were also used for headache, apoplexy (stroke), stone, pleurisy and pain and inflammation in almost every part of the body,[192] often where counter-irritation was thought necessary.

Leeches, it seems, were fussy about their human hosts: 'The part where we apply them must be clean, otherwise they will not bide'. To encourage them, a surgeon should either prick the part or put a drop of pigeon's or chicken's blood to give them the taste, and where they had to draw a lot of blood 'cut the ends of them with the shears to let the blood run'.[193] They were usually left until fully gorged when they fell off. If they had to come off sooner, this was achieved by sprinkling salt on them. Where they were required to continue sucking and were already fully gorged, their tails were cut off.[194]

The tenth and final treatise explained the different types of remedies commonly used by surgeons to restore a humoral balance, according to their particular quality. Medicines which drew heat were known as attractives; repercussives were cooling; resolutives had a drying effect; while emollients were softening. Supperatives promoted the generation of pus, at a time when 'good pus' was still regarded (erroneously) as essential in the curing of a wound,[195] while mundificatives separated and drew away purulent matter from ulcers and incarnatives dried them.[196] The second edition included an extra chapter on oils, ointments and plasters – Galenic methods of treatment which tended to delay healing.

When Lowe made his will on 13 August 1610, a will 'partly made and given up by his own mouth',[197] the second edition of his *Chirurgerie* had not been

[193] Lowe, *The Whole Course of Chirurgerie*, ninth treatise (recte), chapter 3, 'Of Horsleaches or Bloudsuckers and their Effectes'.

[194] Seigworth, 'Bloodletting', p. 2027.

[195] Encouraging the discharge of this 'laudable pus' – white matter of good consistency which was not foul smelling – was also referred to as bringing the wound to 'digestion'. Beier, 'Seventeenth-Century English Surgery', p. 63.

[196] Lowe, *The Whole Course of Chirurgerie*, tenth treatise (recte), 'Of Antidotaries Chirurgicall which Contayneth Thirtenth Chapters'.

[197] SRO, CC9/7/7, Commissariat of Glasgow, Testaments, Last Will and Testament of Peter Lowe, fol. 24r, 24v. For a written account of this document, see J. Finlayson, 'The Last Will and Testament, with the Inventory of the Estate, of Maister Peter Lowe, Founder of the Faculty of Physicians and Surgeons, Glasgow' (Glasgow, 1898).

published. Curiously, he is mentioned, in the dedication, dated 20 December 1612, as humbly taking his leave at his own house in Glasgow, as if still alive. But Lowe had actually died two days after he made his will.[198] He was buried in the grounds of Glasgow Cathedral where his father-in-law was minister,[199] leaving goods and gear to the estimated value of £436 13s. 4d. Scots.[200] The princely sum of £5486 Scots, owed to him, as a creditor, goes some way to fixing his social sphere of influence within the locality.[201] Local gentry were appointed at tutors to Lowe's son, John, by his first wife,[202] and Glasgow burgesses as tutors to his daughter Christian.[203] His tenement, worth £4000, was left to his son John, as were the proceeds from the sale of 'his whole study with books Instruments & all other things therein', and the best of his moveable goods.[204] His second wife, Helen Weems, was to live in the tenement free of charge, and to have the use of it for four years after his death, so that she could oversee 'the honest Entertainment of his son and keeping of him at the school during the said space'.[205]

Although he had the support of a local physician and apothecary in his battle to erect a regulatory body for medicine in the west of Scotland, it was undoubtedly Peter Lowe who was the main driving force behind the foundation of the Faculty of Physicians and Surgeons of Glasgow. In considering the contribution which, as a Scottish surgeon, Lowe made to the sixteenth-century surgical art, it is important to remember that he was a learned rather than an average surgeon. Though the practice of surgery was taught as a craft skill via the system of apprenticeship, the surgical authors from whom the history of practice can be pieced together were necessarily 'atypical practitioners'. Like Lowe, they were members of an elite group who often

---

[198] SRO, CC9/7/7, fol. 24r.
[199] SRO, CC9/7/7, fos 23r, 24v.
[200] SRO, CC9/7/7, fol. 24r.
[201] For example, he was due £1000 Scots on an unsettled bond by the late Robert, Lord Sempill and his heirs, and Sir Archibald Stewart of Castlemilk, knight junior, Sempill of Mylnebank, Robert Hamilton of Torrence and Humphrey Marshall in Glessine, as his cautioners.
[202] Sir James Sempill of (Weems), James Forret of Borrowfield, James Elphinstone of Woodside and Archibald Muir.
[203] Archibald Muir, the merchant, was clearly a trusted friend and advisor of Lowe's, and was appointed tutor to his daughter Christian, with James Bell and Robert Muir, all Glasgow burgesses.
[204] SRO, CC9/7/7, fos 24v, 25r; Mairi Robinson, *The Concise Scots Dictionary* (Aberdeen, 1991), p. 280.
[205] SRO, CC9/7/7, fos 24v, 25r. John Lowe must therefore have been about ten years of age when his father died.

knew as much about physic as surgery.[206] At a time when – however it is viewed – Glasgow was little more than a small, provincial backwater, Lowe included some of the most eminent metropolitan surgeons among his friends. There is no doubt that he was at the forefront of his art: his book on surgery went to four editions because there was a demand for it.

While it would be churlish to dispute that Peter Lowe was, in the words of the medical historian Douglas Guthrie, 'essentially a kind and humane man',[207] it is as easy to incline to the view of a Mr Edington, writing of Lowe's *Chirurgerie* in the *British Medical Journal* (1922), who said: 'Verily Maister Peter had no small conceit of himself'! But it was, in part, a justifiable conceit: the compilation of a teaching manual emphasises his desire to improve the standard of medical education, which he did by writing in a language which all grades of apprentice could actually understand. His textbook was possibly the most influential book on wound surgery in the early seventeenth century, and opens a window on knowledge and medical practice in Glasgow at the time of the founding of the Faculty.[208] There are signs in the *Chirurgerie* not only of his grounding in medical theory but of his rejection of dangerous procedures (like the repair of hernia) which did not stand the test of empiricism. He created a forum for training and interaction between physicians and surgeons which has survived in a variety of guises until the present day; and finally, if he did not quite succeed in making all practitioners accountable, he significantly transformed the regulatory vacuum which existed in relation to medical and surgical practice in the west of Scotland in the late sixteenth century.

---

[206] Lawrence, 'Surgery (Traditional)', p. 961.
[207] D. Guthrie, 'The Achievement of Peter Lowe, and the Unity of Physician and Surgeon', *Scottish Medical Journal*, 10 (1965), p. 263.
[208] Sloan, *English Medicine in the Seventeenth Century*, p. 107.

3

Surgeons and Barbers

In late sixteenth-century Europe, physicians were accredited and licensed by the universities. Surgeons, on the other hand, were trained and regulated by the trade guilds through a system of apprenticeship in which they were often associated with barbers. This association led to the emergence of an all-round practitioner known as the barber-surgeon who carried out a range of general practice, though the practice of barbery remained a distinct line of work. The work of the average guild surgeon encompassed a range of healing activities: the treatment of wounds and burns, skin rashes and venereal disease, the setting of fractured bones, lancing of infected swellings, application of topical medicines, plasters, poultices and cauteries, as well as more skilled work such as trepanation, the occasional amputation, and assistance at difficult births.[1] The more lowly barber-surgeon functioned almost as 'a sort of body servant',[2] offering a plethora of basic maintenance services from blood-letting, cupping and teeth extraction to the healing of fractures and dislocations and the dressing of ulcers or wounds.[3] But neither surgeon nor barber-surgeon was averse to treating skin conditions and venereal disease, both of which could require internal physicking and were technically, therefore, in the physician's realm.

Throughout Europe since the thirteenth century, barbers and surgeons had associated themselves in guilds. In so organising themselves, surgeons and barbers were simply copying the trend in other trades. In some towns and cities, select companies of surgeons were incorporated but most (finding

---

[1] L. Beier, *Sufferers and Healers* (London and New York, 1987), pp. 10–12; R. Jütte, 'A Seventeenth-Century German Barber-Surgeon and his Patients', *Medical History*, 33 (1989), pp. 184–98; A. L. Wyman, 'The Surgeoness: The Female Practitioner of Surgery, 1400–1800', *Medical History*, 28 (1984), p. 23.

[2] M. E. Fissell, *Patients, Power, and the Poor in Eighteenth-Century Bristol* (Cambridge, 1991), p. 52.

[3] The following characterisation of the barber by the craftsman poet Hans Sachs – a poetic text for a woodcut by the Zurich artist Jost Ammann (here in translation) – is a fairly accurate portrayal: 'I am called everywhere, I can make many healing salves, I can cure new wounds, also fractures and chronic affections, Syphilis, Cataract, Gangrene, pull teeth, shave, wash and cut hair, I also like to bleed'. Quoted in Erwin H. Acherknecht, 'From Barber-Surgeon to Modern Doctor', *Bulletin of the History of Medicine* 58 (1984), pp. 545–46.

survival in isolation difficult) were swallowed up throughout the sixteenth and seventeenth centuries by the more numerous corporations of barbers.[4] In Edinburgh, although joint application had been made by the barbers and surgeons of that city for incorporation, the seal of cause – granted in 1505, and ratified by James IV in the following year – authorised an Incorporation of Surgeons.[5] In London, the Fellowship of Surgeons eventually united with the Barber-Surgeons Company in 1540 (having been closely associated since 1493), as did the surgical community of St-Côme in Paris with the barber-surgeons in 1655. Both occupations gained: incorporation with the surgeons lent prestige to the barbers, while the surgeons acquired all the civic benefits and security accruing to an incorporated guild which thus provided a structure for urban practice.[6] Although barbers were admitted into the Faculty of Physicians and Surgeons of Glasgow from 1602 in a position ancillary, or subordinate, to surgeons, they were accorded greater civic status by the 1656 letter of guildry which established the Incorporation of Surgeons and Barbers. However, occupational distinctions continued to be drawn – especially by the surgeons.

Although physicians, surgeons, apothecaries, barber-surgeons and barbers were united in Glasgow under the charter granted by James VI, barber-surgeons were in the majority in the early years. Peter Lowe was the only surgeon of any distinction, and until the 1620s, Robert Hamilton was the only physician practising in Glasgow (and there were never more than two or three at a time throughout the early seventeenth century). This had certain implications for the organisation of surgeons in Glasgow – a town of only 7000 people in 1600 – because although the town council ratified the Faculty's charter in that year, it became clear that the surgeons needed the support of guild structures in order to reap the full benefits of civic protection. Thus, in 1656, the surgeon, apothecary and barber members of the Faculty acquired a burghal charter or letter of deaconry which gave the new Incorporation of Surgeons and Barbers craft powers within the royalty of the burgh.[7] Such

---

[4] G. Lawrence, 'Surgery (Traditional)', in W. F. Bynum and R. Porter (eds), *Companion Encyclopedia of the History of Medicine*, 2 vols (London and New York, 1993), ii, pp. 968–70; K. Pollak, in collaboration with E. Ashworth Underwood, *The Healers: The Doctor Then and Now* (London and Edinburgh, 1968), p. 96.

[5] H. M. Dingwall, *Physicians, Surgeons and Apothecaries: Medicine in Seventeenth-Century Edinburgh* (East Linton, 1995), pp. 34–35, 53.

[6] J. Dobson and R. Milnes Walker, *Barbers and Barber-Surgeons of London: A History of the Barbers' and Barber-Surgeons' Companies* (Oxford, 1979), pp. 31–35; R. Theodore Beck, *The Cutting Edge: Early History of the Surgeons of London* (London, 1974), pp. 173–86; L. Brockliss and C. Jones, *The Medical World of Early Modern France* (Oxford, 1997), p. 183.

[7] H. Lumsden, 'Bibliography of the Guilds of Glasgow', *Records of the Glasgow Bibliographical Society*, 8 (1930), p. 9.

powers were then essential to the surgeons, although persisting social distinctions made for difficulties in their relationship with the barbers and, as in many other places,[8] it was doomed to end unhappily when the prestige of surgery rose during the eighteenth century.

A Craft Guild

Craft and trade guilds performed a number of important socio-economic functions. They regulated the numbers entering a certain trade, and maintained standards through lengthy apprenticeships, thus ensuring the income of those admitted as freemen. Burghal society distinguished between freemen or burgesses,[9] who enjoyed full burgh privileges and could engage in trade and the manufacture of goods, and the non-free or bondsmen, among whom were classed the apprentices, journeymen, domestics, families and labourers, who did not. Burgesses were subdivided into merchants, who had a monopoly on all kinds of trade, and therefore wielded most power in the town, and craftsmen, who were involved in the manufacturing and retailing and were in the majority. Only freemen, who made up approximately 20 to 25 per cent of the adult population, held the right to undertake commercial activity, to hold property by burgage tenure, and to take part in burgh administration.[10]

The main reasons for constituting a craft guild were to draw up statutes restricting unfreemen from trading; to prevent freemen from performing inadequate work; to regulate the system of apprenticeship; and to establish and preserve a monopoly on the provision of services in that particular field. A guild was formally constituted in Scotland by presenting a petition to the town council,[11] asking it to sanction the statutes of the trade and to authorise their enforcement. Such petitions were granted by the issue of a seal of cause [12] or letter of deaconry – a charter of erection by which an association

---

[8] In Paris, the union of the barbers and surgeons was dissolved in 1743; a new Company of Surgeons was formed in 1745; and in Edinburgh, the barbers separated from the surgeons in 1722 to form the Society of Barbers. Lawrence, 'Surgery (Traditional)', pp. 972–73; Dingwall, *Physicians, Surgeons and Apothecaries*, p. 57.

[9] A freeman or burgess was so called because he enjoyed the freedom of the burgh and of his craft guild.

[10] J. S. McGrath, 'The Medieval and Early Modern Burgh', in T. M. Devine and G. Jackson (eds), *Glasgow*, i, *Beginnings to 1830* (Manchester and New York, 1995), pp. 25–26; J. S. McGrath, 'The Administration of the Burgh of Glasgow, 1574–1586', 2 vols (unpublished Ph.D. thesis, University of Glasgow, 1986), i, p. 56.

[11] The town council comprised the provost, magistrates, dean of guild, deacon convenor or leader of the Trades' House, and merchant and trade councillors.

[12] This was literally a document with the town's seal attached.

was transformed into a corporate body, with the right to elect a deacon and to enact regulations binding on all members.[13] Economic expansion in sixteenth-century Glasgow resulted in nine craft guilds being incorporated; three others were incorporated in the early seventeenth century, including the surgeons and barbers (the latter had not been incorporated beforehand and joined by the surgeons as in many English towns). In joining a craft, members bought into the group protection and benevolent structures of the incorporations. Being more powerful, the merchants had less financial need to be organised (there are few signs of their association in the sixteenth century), but eventually Glasgow was induced to form a merchant guild. The letter of guildry of 1605 created a new stratum of elite craftsmen who vied with the merchants in status. Craftsmen were grouped in the Trades' House, presided over by the deacon convenor, and merchants in the Merchants' House (built in the Bridgegate in 1659), presided over by the dean of guild; both institutions made provision for the poor, widowed and orphaned of their membership. The Dean of Guild Court regulated all aspects of the admission of burgesses and guild brethren and mediated in disputes.[14]

Technically the Faculty was not a craft guild,[15] although significantly, its members referred to their charter, as early as September 1604, as their letter of guild.[16] They also elected a deacon – known as the Visitor because of his powers of visitation – as well as other officers annually like a guild.[17] Similarly, Peter Lowe and Robert Hamilton were both delegates to the deacon convenor's council.[18] According to the constitution of the Trades' House, the surgeons were represented, among a membership of fifty-four, by their deacon (or Visitor) and two assistants, who had rights to vote for the deacon convenor of the house.[19] So, to all intents and purposes, the Faculty operated like a guild from its inception, adopting the organisational structure of the

---

[13] H. Lumsden and P. Henderson, *History of The Hammermen of Glasgow: A Study Typical of Scottish Craft Life and Organisation* (Paisley, 1912), pp. 3–4.

[14] McGrath, 'The Medieval and Early Modern Burgh', pp. 32–33; D. Daiches, *Glasgow* (London, 1977), pp. 14–15. For an account of the operation of the Dean of Guild Court, see A. M. Jackson, *Glasgow Dean of Guild Court: A History* (Glasgow, 1983).

[15] See, for instance, McGrath, 'The Medieval and Early Modern Burgh', p. 26; D. Daiches, *Glasgow* (London, 1977), p. 15.

[16] RCPSG, 1/1/1/1b, p. 19.

[17] Compare the Faculty, for instance, with the College of Surgeons in Edinburgh which was established by a seal of cause of the Edinburgh Town Council in 1505 and subsequently ratified by James IV in 1506. For details, see Dingwall, *Physicians, Surgeons and Apothecaries*, pp. 34–36.

[18] H. Lumsden (ed.), *Records of the Trades House of Glasgow*, i, *1605–1678* (Glasgow, 1910); ii, *1713–1777* (Glasgow, 1934); i, p. 4.

[19] Lumsden and Henderson, *The Hammermen of Glasgow*, p. 167. Other crafts had more elected members.

guild, in regulating apprenticeship and in examining those who wished to enter the incorporation as qualified freemen. Until 1785, when they were given the option of becoming licentiates, all those who entered the Faculty were 'freemen' or full burgess members.[20] Yet, though all were 'free with the vocation', a distinction was made (even before the creation of the Incorporation of Surgeons and Barbers) between those who were permitted to exercise their arts within the burgh of Glasgow and those who were not (burgesses of other towns).[21]

The institution established for their poor by the craftsmen was known as the 'Crafts' Hospital', the 'Trades' Alms House' or Trades' Hospital,[22] in the administration of which the Faculty participated by virtue of its representation in the Trades' House. The Crafts' Hospital contained a hall in which a variety of crafts regularly held their corporate meetings.[23] In 1654 and 1655, Faculty examinations were conducted there, as well as the annual election of its officers,[24] probably forestalling the need for a hall of its own.

The site for the almshouse – in Kirk Street, between the entry of Rottenrow and St Nicholas' Hospital was purchased in 1605.[25] It was agreed that the house should be maintained by contributions of 2d. weekly from every freeman craftsman in the burgh, as well as 13s. 4d. from every member admitted freeman in right of apprenticeship, and half of every fine imposed by incorporations.[26] But this proved too irregular a provision, and from 1609, payments were made by each guild in a block. The surgeons contributed

---

[20] Duncan, *Memorials*, p. 99.

[21] RCPSG, 1/1/1/1b, Transcript Minutes of the Faculty of Physicians and Surgeons of Glasgow, 1599–1688, p. 93, 13 October 1654.

[22] GCA, T-TH 1/60, Foundation Charter of the Trades' Hospital. This was distinct from the Merchant Hospital which predates this and was located in the Bridgegate. W. H. Hill (ed.), *View of the Merchants' House of Glasgow: Containing Historical Notices of its Origin, Constitution and Property, and of the Charitable Foundations which it Administers* (Glasgow, 1866), pp. 15–16.

[23] Lumsden and Henderson, *The Hammermen of Glasgow*, pp. 106, 165; Robert Douie, revised by F. Gibb Dougall, *Chronicles of the Maltmen Craft in Glasgow, 1605–1879* (Glasgow, 1895), pp. 42, 44. An 1804 description of the Trades' Hospital stated that 'In this hospital, which is now in a state of decay, is a hall, where the Incorporations used to convene at their elections and upon other public business, prior to the building of the Trades' Hall in Glassford-street. This room, which is only betwixt twenty and thirty feet in length, contains paintings emblematic of the fourteen professions and six portraits of the most distinguished donors in favour of the charity, besides inscriptions, mentioning many others of its benefactors.' J. Denholm, *The History of the City of Glasgow and Suburbs* (Glasgow, 1804), pp. 177–78.

[24] RCPSG, 1/1/1/1b, pp. 90, 108.

[25] Denholm, *History of the City of Glasgow*, p. 177; *The Merchants House of Glasgow*, p. 16.

[26] Lumsden, *Records of the Trades House of Glasgow*, i, p. 555.

only £3 6s. 8d.,[27] the smallest contribution apart from that of the bonnet-makers,[28] probably because they had few members (eleven, at most, in 1609). The Faculty also paid 5 merks Scots p.a. 'for the use of the poor of the hospital' which in 1656 – the year of its incorporation as a craft guild – was increased to 20 merks. However, the sum was not raised altruistically, but as an inducement for the creation of a third Faculty representative on the deacon convenor's council and more blatantly, 'upon condition ... that the deacons join to intreat the magistrates' concurrence in execution of their laudable acts established in their favour'.[29] Appointed first master of the Crafts' Hospital in July 1605, Peter Lowe took part in negotiations for the purchase of the site. The hospital was finished and ready for inmates by November 1609.[30] Faculty apprentices may have acquired some of their clinical experience when their masters attended the sick poor in the almshouse.

Members of a guild worked together, obeying regulations laid down by their elected deacon and council for monopolistic and protective reasons, and regulating the entry of new recruits.[31] The purpose of a surgical guild was to restrict the practice of surgery to qualified and licensed practitioners, and in so doing to raise standards. In an incorporation governed by collective rules, exercising statutory powers and with entry standards, the opinion of the office bearers inevitably carried greatest weight.[32] As one historian puts it, 'The masters were the incorporation'.[33] Nevertheless, Peter Lowe and Robert Hamilton did their best to establish a democratic tradition at the first meeting of the corporation in the Blackfriar's kirk, on 3 June 1602, when they produced the king's letter of gift before Sir George Elphinstone of Blythswood, Provost, and James Forret, John Anderson, and William Anderson, bailies. Although the powers to regulate medicine in the west of Scotland had been

---

[27] The collector was still paying the same contribution of £3 6s. 8d. in 1655. RCPSG, 1/1/1/1b, p. 114. By 1667, the Faculty's contribution to the Crafts' Hospital had increased to £6 13s. 4d. RCPSG, 1/1/1/1b, pp. 298–99.

[28] Lumsden and Henderson, *The Hammermen of Glasgow*, p. 91; Lumsden, *Records of the Trades House of Glasgow*, i, pp. 24–25. The three main crafts – hammermen, tailors, cordiners – contributed £20 Scots p. a. until 1677, after which payments increased to £40 Scots which continued to be paid until 1729. The maltmen contributed the largest sum, of £30.

[29] RCPSG, 1/1/1/1b, p. 118.

[30] Lumsden, *Records of the Trades House of Glasgow*, i, pp. 3, 7–10, 21. Lowe was discharged, on 1 April 1607, in expenditure of £202 2s. 4d. Scots.

[31] A. M. Smith, *The Nine Trades of Dundee*, Abertay Historical Society, 35 (Dundee, 1995), p. 32.

[32] This follows the discussion in H. J. Cook, *The Decline of the Old Medical Regime in Stuart London* (New York and London, 1986), p. 20.

[33] Smith, *Nine Trades of Dundee*, p. 28.

given to Lowe and Hamilton as individuals, they insisted that elected officials should bear this responsibility.[34] Henceforth, all officers were routinely elected at the end of September. But the self-perpetuating nature of Faculty officialdom, in line with contemporary burgh trends towards oligarchic government, is clear from the frequent repetition of early officers' names.[35]

'Visitor' was a guild term dating from a 1425 statute of James I, which empowered craftsmen to elect a leader called 'a Deakon or Maister man' to ensure that the king's lieges would not be defrauded by incompetent artisans.[36] The Faculty used the terms Visitor and Deacon almost interchangeably – further evidence that it saw itself as a trade guild. The first Visitor of the corporation was Robert Hamilton, who was already in the post by the first public meeting. Elections were closed. The officer was required to swear that he had informed all members of the election, except the 'secluded members' running for office, upon which a new Visitor would be elected by a plurality of votes.[37] The letter of deaconry stressed that the Visitor should be 'one of the most fit, qualified, and worthiest of the said calling, a Surgeon and burgess of the burgh'. It also upheld the Visitor as the final court of appeal in the incorporation, giving him the right to 'judge between master and apprentice, at the bailies' sight, In case any difference of importance arise; and between brother and brother'.[38]

---

[34] RCPSG, 1/1/1/1b, pp. 9–10.

[35] This was inevitable. Besides the founders, there were only a handful of members at the first meeting: Adam Fleming, Robert Allason, William Spang, Thomas Thomson and John Lowe.

[36] James I, c. 39. The power wielded by the various Visitors often led to disturbance in the burghs, particularly between craftsmen and merchants. In Glasgow, as in other burghs in Scotland, the merchants and craft ranks vied for the political control of the town and the management of the magistracy. Often, because they were richer, the merchants had more influence over the election of magistrates than the craftsmen. Therefore, the power of the deacons was checked and revoked in 1426 (James I, c. 86), 1493 (James IV, c. 43), and 1555 (Queen Mary, c. 52), the last stating 'that there be no deacons chosen in times coming within Burgh'. Henceforth, the provost, bailies and council of the burgh were 'to choose the most honest man of craft, of good conscience, one of every Craft to visit their craft that they labour sufficiently, and these persons to be called *Visitors* of their craft, and to be elected and chosen yearly at Michaelmas, and these Visitors chosen, sworn, and admitted to have choosing of officers and other things, as the Deacons voted before'. Although deacons were restored in the following year, by a general charter in favour of all the craftsmen of the burghs and cities within the realm, some guilds – like the Glasgow Incorporation of Maltmen – continued to use the term Visitor. Douie and Gibb Dougall, *Chronicles of the Maltmen Craft*, pp. 15–17.

[37] See for instance, RCPSG, 1/1/1/1b, pp. 179, 258. In 1666, the sum of 6s. Scots was paid 'to the keeper of the door at the election'.

[38] James B. Tennent, *Records of the Incorporation of Barbers, Glasgow, Formerly the Incorporation of Chirurgeons and Barbers* (Glasgow, 1930), pp. 17, 19.

After the physicians returned to the Faculty fold in 1673,[39] to share in the rights of visitation jointly bestowed in 1599 upon surgeons and physicians, a distinction was made between the Surgeon-Visitor and the Physician-Visitor. The word Praeses was adopted for the Physician-Visitor in February 1674.[40] By the early eighteenth century,[41] election of the Praeses was by double ballot in the first round: all the physicians were divided into two lists made by the current office-holder, with the two winners being pitched against each other in the second round.[42]

At the Faculty's second meeting, on 17 June 1602, Peter Lowe, William Spang, Robert Allason and Adam Fleming were elected quartermasters for the following year. Any one quartermaster, convening with the Visitor, was declared a quorum 'for setting down of any good order'.[43] Quartermasters were senior members or masters of the Faculty who were responsible for examining new entrants. It was also their job to prevent people from practising surgery and barbery until found qualified.[44] From 1628, these officials were chosen jointly by the Visitor and the brethren (two by each), and were designated the Visitor's and Craft's quartermasters respectively. The number was increased to six in 1674.[45] Quartermasters were responsible for overseeing the collection of members' quarter accounts – that is, the rates or basic craft taxation collected from each freeman of the guild, out of which were paid all the 'common charges' of the craft, as well as its contribution to the maintenance of the poor in the almshouse. All crafts experienced problems in collecting these payments.[46] Defaulting members (who abound in the minutes of the Faculty) were deprived of their vote in the corporation, and could not hold office. On 8 February 1655, for example, the Faculty ordained 'that the whole quarter accounts be paid upon the day that the visitor is elected', or forfeit their vote in the election.[47] Technically, the payment of quarter accounts was waived only in cases of absence from the country. For instance, in December 1668, the Faculty unanimously freed James Frank,

[39] See above, Chapter 2, 'Surgeons and Physicians in Conflict, 1671–72'.
[40] RCPSG, 1/1/1/1b, pp. 439–40, 477. For a list of Visitors and Praeses (Presidents), see Appendix 4, below, pp. 426–29.
[41] The nature of election procedure in the period between 1688 and 1733 went up in smoke along with the second minute book.
[42] RCPSG, 1/1/1/2, Minutes of the Faculty of Physicians and Surgeons of Glasgow, 1733 to 1757, fol. 7v, 7 October 1734.
[43] RCPSG, 1/1/1/1b, p. 11.
[44] Tennent, *Incorporation of Barbers*, p. 18.
[45] RCPSG, 1/1/1/1b, pp. 40, 484.
[46] See, for instance, Lumsden and Henderson, *The Hammermen of Glasgow*, pp. 86–87, 90.
[47] RCPSG, 1/1/1/1b, p. 99. The Collector's account of 9 November 1658, includes three back payments of the previous four years' quarter accounts. RCPSG, 1/1/1/1b, p. 164.

elder, and Hugh Montgomery (probably one the Montgomerys of the Ards, in County Down) of all their past and future quarter accounts during their residence in Ireland. Non-payment of fines was a big problem and one which the Faculty was sometimes prepared to compromise on, as it did in the same month when James Wilson's bygone quarter accounts were cancelled, providing he paid his freedom fine and quarter accounts in time coming.[48]

But what of lesser officers? The first Visitor, Robert Hamilton, appointed a clerk, a notary, and an officer as administrative officials. Robert Herbertson, notary, was obviously not a practitioner, but it seems likely that George Burnell, the officer, was the 'George Berrell', a burgess's son, who was admitted on 23 May 1605 with licence 'to profess the art of Barbery with simple wounds in the flesh'. Berrell's appointment as officer from 1605 to 1606 tends to support this interpretation.[49] The clerk, as secretary and legal adviser, was usually a qualified notary public (a legal scribe or writer). He would be responsible, as a literate man, for any entries made in the various official books – the statute and minute books. Aside from a basic salary from the incorporation, he also accrued professional fees for registering documents such as indentures and bonds. In 1655 (the year of the earliest extant treasurer's account), the clerk's salary for a half year was £3 6s. 8d. Scots. By 1658 it had risen to £8 Scots; and by 1664 the clerk had a servant or assistant.[50]

The officer was the day-to-day messenger for the guild, delivering notices requiring masters' attendance at general meetings, or, for example, at members' funerals. In effect, he acted as Faculty postman before the advent of affordable postage. In most guilds, the officer received a suit of clothes and a pair of shoes once a year. A junior office, the position was frequently given to the most recently entered member. The officer's salary rose from £1 4s. 0d. Scots in 1655, to £1 10s. 0d. in the following year. The officer's quarterly fee from 21 September 1666 to 4 December 1668 was £3, with an additional £2 8s. 0d. for shoes. The election of John Neill, 'cordoner' or shoemaker, as officer in October 1659, shows an interesting interconnection between guild brethren. By October 1661, the officer's work had reached such dimensions

[48] RCPSG, 1/1/1/1b, pp. 304–5. Hugh Montgomery who graduated MA from Glasgow University in 1649, entered the Faculty in 1664. Duncan, *Memorials*, p. 240. The third Viscount Montgomery, whose heart was partially exposed by a huge ulcer resulting from a fracture of the ribs, had met William Harvey in Edinburgh during the Civil War in 1641. For an account of the meeting see, J. F. McHarg, *In Search of Dr John MakLuire: Pioneer Edinburgh Physician Forgotten for over 300 Years* (Glasgow, 1997), pp. 154–70.
[49] RCPSG, 1/1/1/1b, pp. 10, 19–20.
[50] Douie and Gibb Dougall, *Chronicles of the Maltmen Craft*, p. 54; RCPSG, 1/1/1/1b, pp. 114, 161, 235. The clerk's salary was still £8 Scots in 1667 and 1668. RCPSG, 1/1/1/1b, pp. 297, 299.
[51] RCPSG, 1/1/1/1b, pp. 325–26.

that a 'messenger and servant' was appointed to relay the messages, while the officer dealt with the back payment of quarter accounts and letters of horning. In September 1669, new freeman members who resided in the town were given the choice of serving as officer themselves for a year or paying a fee of £3 Scots.[51] However, as the membership increased, new entrants served only a short period in office before the next entered, so in November 1672, it was ordained that every freemen should pay one rex dollar in officer's fee at their admission. Those admitted since 1665 who had not served as officer were also liable to the collector.[52] For ceremonial occasions, when the Faculty paraded its civic identity in pageantry and ritual, the Faculty employed a drummer. The first – John Jamieson – received 12s. for his services in 1655. By 1674, there were two drummers.[53]

The Faculty did not have a regular meeting-place at its foundation, but met in places like the kirk, the deacon's house or, after 1605, the Crafts' Hospital, the New Kirk Steeple House, the Regality Court and Bailie Hall's shop. It was still sometimes convening in members' houses in 1672,[54] but there is no indication that this was to discuss medical (rather than administrative) matters as the members of the College of Physicians of London had done in the home of Thomas Linacre, a century earlier.[55] It was decided, in July 1673, that meetings should be held every fortnight, seven members and the Visitor making a quorum.[56]

From 1612, the Faculty had a common box with two separate and individual locks in which it kept its most important documents, and the fines from disciplinary cases. The keys were initially kept by two of the quartermasters, as a disincentive to misappropriation.[57] But this responsibility quickly evolved into the office of boxmaster or keeper of the keys, of whom there were two, one responsible for each lock. In most guilds, when a new Visitor was elected, an inventory was made by the boxmasters of the documents and money it contained, often at the Visitor's house. Before the extensive use of banks, the office carried a fair degree of responsibility. The craft's surplus money

---

[52] Douie and Gibb Dougall, *Chronicles of the Maltmen Craft*, pp. 67–69; RCPSG, 1/1/1/1b, pp. 114, 126, 178, 192, 296–97, 325–26, 402–3.

[53] RCPSG, 1/1/1/1b, pp. 114, 485. The drummer was still in receipt of 12s. for his fee in 1667. RCPSG, 1/1/1/1b, p. 297.

[54] RCPSG, 1/1/1/b, pp. 12, 228–29, 252, 390, 657. In February 1672, it met in Archibald Bogle, the Visitor's house.

[55] Cook, *Decline of the Old Medical Regime*, p. 71.

[56] RCPSG, 1/1/1/1b, p. 454.

[57] RCPSG, 1/1/1/1b, p. 25. The box of the Edinburgh College of Surgeons was placed inside a larger chest for additional safety, deposited in the deacon's house. Clarendon Hyde Creswell, *The Royal College of Surgeons of Edinburgh: Historical Notes from 1505 to 1905* (Edinburgh, 1926), p. 16.

was usually invested at interest in bonds which were put in 'the Box' until they were paid or renewed.[58] Many such bonds are recorded in the Faculty minutes in the seventeenth [59] and early eighteenth centuries.[60] In Edinburgh, the surgeons had to forbid their deacon (who acted as treasurer) from lending money to anyone unilaterally without their consent.[61] Similarly, when a series of statutes was set down for the craft in Glasgow, in 1612, it was enacted that 'no Visitor shall depurse any money in any thing concerning the affairs of the craft without consent of two or three of his brethren of craft, otherwise no allowance to be given to him thereof'.[62] But decisions on financial matters were taken only by senior members of the Faculty (and not in the presence of the entire membership) which led to some dissension.

The Faculty's main sources of income were fees from examinations, licensing and entry, a half share in fines for illegal practice, and the quarter accounts. By 1636, there was enough money passing through the box for a treasurer or collector to be appointed. In 1655, £13 4s. 0d. Scots was taken 'from twenty-one persons of quarter accounts being the number of those within the town who are burgesses and freemen as they are in use to pay'; a further £9 10s. 0d. was collected from nine members licensed outside the burgh. This not only gives some idea of the annual income from members' dues, but shows that there were thirty members in 1655, the year before their incorporation with the town. In September 1656, it was decided that the sum paid by each member in quarter accounts was not to exceed 12s. Scots a year, though on collection in November, the quarter accounts of the whole Faculty, 'being nineteen in number',[63] yielded the much higher sum of £18 11s. 8d. Scots.[64] In the 1660s, the quarter accounts rose to 30s. p.a. Members who lived in town could be lax enough in paying their dues but, outside it, members derived less benefit from payment and often let years of debt accrue. In 1661 Robert Rowand, messenger, was given £12 Scots 'to go to the west country to bring in quarter accounts being employed by the Visitor & brethren according to their order'. In the following year, John Weir was paid

---

[58] Douie and Gibb Dougall, *Chronicles of the Maltmen Craft*, p. 53.

[59] See, for example, the first and second collectors' accounts, RCPSG, 1/1/1/1b, pp. 112, 124. Two decisions on long-standing bonds were made in 1667, perhaps indicating that the Faculty was constrained for funds at this point. See RCPSG, 1/1/1/1b, pp. 275–79, 284–85.

[60] See, for example, Lumsden, *Records of the Trades House of Glasgow*, i, p. 27, which shows that the surgeons lent £1000 Scots to the Trades' House in November 1714.

[61] Creswell, *Royal College of Surgeons of Edinburgh*, pp. 16–17.

[62] RCPSG, 1/1/1/1b, p. 27.

[63] This probably refers to the twenty-one members in the previous year less the number of physicians, because the surgeons were, by this time, incorporated with the town.

[64] RCPSG, 1/1/1/1b, pp. 50, 113, 122, 124.

£3 Scots for going to the west, not only to lift bygone quarter accounts but also 'to charge unfreemen'. Some paid up. Dr Wallace in Paisley gave in £9 for six years' back payment of quarter accounts from the time of his booking.[65]

The collector's account for the period between 21 September 1666 and 4 December 1668 was the first in which the income derived from the various Faculty dues and fines was laid out systematically. The total collected in freedom fines (including booking fees) was £217 12s. 8d. Scots. The total in quarter accounts for freemen and others at 12s. p.a. came to £58 10s. 0d. A distinction was made between these and the quarter accounts paid by strangers (who practised outside Glasgow), which amounted to £122 12s. 0d. However, most of this was back payment made by strangers over a period of eight to ten years, showing an improvement in collection of strangers' dues rather than a marked increase in out-of-town membership. At the end of 1668, unpaid quarter accounts amounted to £102 10s. 0d.[66] A deceased member's executors were still liable for non-payment of quarter accounts, though the collectors cannot seriously have held out much hope of their being honoured. From November 1672, a separate list of debtors was kept 'Whereof there is no hope to get any thing recovered of their representers'.[67] There were even problems with non-payment of freedom fines.[68]

Senior officers of a trade guild were obviously accountable to the membership, many of whom demanded collective scrutiny of financial affairs. Being a collector to any of the craft guilds must have been an onerous responsibility, but since the Faculty was a member of the Trades' House, an independent witness from one of the other trades attended the presentation and discharging of the collector's annual account. In November 1656, Robert Carruthers, deacon of the tailors, was present in the Crafts' Hospital to hear Daniel Brown's account. In spite of this, some Faculty members feared that closed accounting encouraged misuse of funds. In February 1661, Mr David Sharp (a pharmacist) declared that he would not keep court with the Visitor and surgeons regarding the auditing of the collector's accounts 'except that the whole brethren were present'. This particular problem arose out of displeasure with its most recent collectors, Robert Harris (1654–55), whom the Faculty paid to have apprehended by a bailie to give an account of his

---

[65] RCPSG, 1/1/1/1b, pp. 202, 204, 208, 218.

[66] RCPSG, 1/1/1/1b, pp. 292, 294–95, 301–3. In the fiscal year 1671–72, strangers' quarter accounts were charged at 16s. and 20s. per annum, but from 1673, country members paid 30s. per annum in quarter accounts. RCPSG, 1/1/1/1b, pp. 411, 462.

[67] RCPSG, 1/1/1/1b, pp. 405–6. Representers were direct heirs of the deceased.

[68] Even the threat of referral to the deacon convenor failed to produce payment within six months from Robert Houston. RCPSG, 1/1/1/1b, pp. 338, 344.

intromissions for 1658 and 1659, and for the poor money before the magistrates; and Daniel Brown (1660–61), who brought the amount in their box 'to [such] a little quantity as changed in its security ... without consent of the calling and put by him to the hazard of the loss of all'.[69]

Apprenticeship and Education

The training of surgeons in the seventeenth century was conducted in Glasgow, as elsewhere, through the system of apprenticeship. A long practical training accompanied by strict examination at various junctures, apprenticeship was, in effect, a course in practical clinical instruction during which the apprentice learned his craft by visiting patients with his master, through observation and by asking questions. In due course, an apprentice would acquire knowledge of materia medica, pharmacy, and the compounding and dispensing of medicines (which, in the seventeenth century, surgeons still administered to their own patients), as well as the ability to perform the smaller surgical operations himself.[70] Usually boarded with their masters, apprentices were excellent labour-saving devices, employed in keeping their masters' shop tidy, running errands, and visiting patients to check on their progress.[71] On the other hand, the master contributed the advantages of his patronage and influence within the community. Guild apprentices came under the protection only of the craft, and not of the seal of cause; their rights of deaconry did not become activated until the end of their apprenticeship. Servants, as a lower status of pre-apprentice assistant (usually appointed for either one or three years at a lower booking fee, and sometimes progressing to apprenticeship), were under the control of the craft only and had no corporate privileges.[72] Maintaining a good relationship with their master was, therefore, of paramount importance.

All European trade guilds laid down ordinances on organisation and management, terms of admission, apprenticeship and training, form of examinations, codes of behaviour, collective responsibility, and competency, in order to protect themselves from unskilled practitioners. Their prime objective was to ensure a steady stream of work for members, which was best achieved by limiting the numbers entering the occupation through insistence on a lengthy training period at no inconsiderable cost. For this reason, most guilds allowed

[69] RCPSG, 1/1/1/1b, pp. 123, 186–87, 205, 238.
[70] RCPSG, 1/5/1, Weir's Faculty Memoranda, fos 20–21; D. de Moulin, *A History of Surgery: With Emphasis on the Netherlands* (Dordrecht, 1988), p. 94.
[71] Fissell, *Patients, Power and the Poor*, pp. 49, 53.
[72] Lumsden and Henderson, *The Hammermen of Glasgow*, p. 9. For the status of a craft servant, see the explanation in Dingwall, *Physicians, Surgeons and Apothecaries*, pp. 67–68.

a master to take on only one apprentice or journeyman at a time; this ensured adequate training and restricted the number of masters. However, this particular standard was not adhered to by the Faculty. The letter of deaconry insisted only 'That no freeman usurp the having of any more apprentices than one ... without express warrant from the visitor and quarter masters', which indicates that permission could be given. Mr Charles Mowat, for instance, booked two apprentice-apothecaries to train with him in April and May 1676.[73] The motivating force behind a guild's statutes and ordinances was self-interest – the promotion and reinforcement of its own special privileges within the town – though the maintenance of standards (especially in the medical field) was of obvious benefit to its inhabitants.[74] Statutes were codified for the smoother running of the association, in June 1602, and entered into the 'Book of the Acts & Statutes of the Said Art of Surgery'.

The 1602 'Act concerning the admission of apprentices & Booking' specified that 'all apprentices to be entered shall remain no shorter space than seven years & the last two thereof for meat and fee'.[75] Most trade apprenticeships of the time were contracted for this duration, with the apprentice earning a small wage, as well as his food, in the last two years served, though he became responsible for providing his own working clothes. The period of apprenticeship was short in comparison with, say, the ten or twelve year stint served by apothecaries in Norwich, but long in comparison with that of continental countries. In the seventeenth-century Netherlands, for example, surgical apprentices trained for differing periods of between three and five years.[76] But the erosion of the more stringent aspects of burgh control and regulation, contingent on the loss of the royal burghs' monopoly on foreign trade in 1672, weakened the burgess-ship system and shortened the duration of apprenticeship. It was no longer so necessary to restrict entry to the craftsman and merchant classes by a lengthy apprenticeship with burdensome fees. Where seven-year apprenticeships had been common in the mid seventeenth century, the duration declined in the latter half of the century. By the 1670s, indentures for five-year apprenticeships with the Faculty were not uncommon,[77] and by the 1730s three-year apprenticeships

---

[73] RCPSG, 1/1/1/1b, pp. 529–30, 533.

[74] De Moulin, *History of Surgery*, p. 67; Tennent, *Incorporation of Barbers*, p. 18; Smith, *Nine Trades of Dundee*, pp. 32–34.

[75] RCPSG, 1/1/1/1b, p. 14.

[76] M. Pelling, 'Occupational Diversity: Barbersurgeons and the Trades of Norwich, 1550–1640', *Bulletin of the History of Medicine*, 56 (1982), pp. 486, 496; De Moulin, *History of Surgery*, p. 113.

[77] See, for example, the booking of William Thomson, who was booked apprentice to Charles Mowat, apothecary, on 9 May 1676. RCPSG, 1/1/1/1b, p. 533.

were as usual as ones served for five years. By this time, the burgess-ship in Glasgow had effectively evolved into a mark of social distinction rather than the necessary qualification it had been in the early seventeenth century to engage in trade and commerce.[78]

Apprentices in Scotland were formally indentured under contract. The master was bound to teach the apprentice his trade, while the apprentice pledged his services, time and paid an apprenticeship fee. But a master's responsibility was not merely educational – an apprentice was fed and clothed at his expense. Indentures specified the length of an apprenticeship and subsequent period as a journeyman. In 1691, for example, John Marshall, surgeon-apothecary in Glasgow bound himself 'not only to teach, learn & instruct the said Andrew Reid in his said art & vocation of surgery & pharmacy', but also to 'do his utmost endeavour to make him a perfect artist therein so far as his capacity can reach during the space foresaid'. His fee was £246 10s. 12d. Scots, payable by Whitsun 1695.[79] An instance of a servant who became an apprentice was William Swan, servant to Andrew Mylne, 'present deacon of the surgeons of Glasgow', who was subsquently booked apprentice to him for seven years in November 1627.[80] And in August 1667, Adam Gray, late servant to the Visitor, was examined and found qualified to practise simple surgery – and to practise as a common barber 'for dressing and polling of heads and beards and In making of periwigs'.[81]

A new apprentice had to be booked into the incorporation. He was taken by his father or guardian to a suitable master where a bargain was made, and a fee paid. The notary acting as clerk to the incorporation then 'booked' the indenture into the books of the craft, in the presence of the deacon and some of the masters. The contract was legalised by the master's payment of a booking fee.[82] Any craft contributions, whether booking fees or entrance fees, were calculated on the basis of civic connection: basically the sons of burgesses or the husbands of their daughters paid at preferential rates. According to the statutes of 1602, a stranger apprentice paid £5 Scots at his entry to the craft and £1 13s. 4d. to the clerk, whereas a burgess's son, paid at the reduced rate of £2 to the box, £1 6s. 8d. to the clerk, and 12s. to the

---

[78] T. Devine, 'The Merchant Class of the Larger Scottish Towns in the Seventeenth and Early Eighteenth Centuries', in G. Gordon and B. Dicks (eds), *Scottish Urban History* (Aberdeen, 1983), pp. 94–95.

[79] SRO, CS232, box 170, Unextracted Processes (Court of Session) 1. Drysdale, Indentures Betuixt John Marshall & Andrew Reid 1691. The last number is in the fold of the manuscript and is unclear.

[80] Lumsden, *Records of the Trades House of Glasgow*, i, p. 124.

[81] RCPSG, 1/1/1/1b, pp. 267–68.

[82] Lumsden and Henderson, *The Hammermen of Glasgow*, pp. 22–23.

officer. The Faculty rate was relatively high, except for barbers who, by an 'Act for admission of Barbers' of 1602, paid only 30s. Scots at their booking, and 12s. to the officer.[83]

The high mortality rate in the seventeenth century led to some casualties of the system. If a master died before the contract was fulfilled, the Incorporation was obliged to find a replacement. In December 1675, Isobel Neilson, widow of David Anderson, shoemaker in Glasgow, petitioned after the death of John Fleming, barber, to whom her son Charles had been recently apprenticed, that the Incorporation had an obligation 'to educate and bring up her said son and teach & instruct him in his art conform to the indenture'. Her son was 'desolate of a master unless his widow Susanna Morrison would take home some journeyman to instruct her said son sufficiently', or he was freed to go to some other master. The Incorporation decided that Susan Morrison should take in a journeyman on the advice of the Visitor and masters who would teach and instruct the apprentice.[84]

Though most apprentices entered the Faculty on the basis of kinship, the evidence shows a reasonable degree of mobility, at a relatively early stage, between Edinburgh and Glasgow. In April 1641, James Young, son of a merchant burgess of Edinburgh, was booked as apprentice to George Michaelson, surgeon burgess of Glasgow for five years.[85] In January 1652, Robert Brown of Leith, was 'booked apprentice with Daniel Brown surgeon to serve him five years as apprentice and two years for meat and fee'.[86] Apprentices also moved after qualification. In June 1658, David Sharp sought admission as an apothecary, presenting an indenture and discharge of apprenticeship from John Foulis, apothecary in Edinburgh,[87] as well as his burgess and guild brother's ticket. Yet, even though qualified, he was for his trial asked to make up a number of remedies, under supervision and to draw blood and apply cauteries 'before two surgeons of the faculty'. Before trial, a candidate was not to practise, under penalty of £10. After trial, his admission was unanimously approved, the compositions 'being very perfectly dispensed' and his having 'made perfect answer'. When James Morton, former servant to Patrick Arthur, Edinburgh apothecary, sought admission in March 1675 'to the

---

[83] RCPSG, 1/1/1/1b, pp. 14–16.
[84] RCPSG, 1/1/1/1b, pp. 518–21.
[85] RCPSG, 1/1/1/1b, p. 56.
[86] Lumsden, *Records of the Trades House of Glasgow*, i, p. 296.
[87] A discharge of apprenticeship from your master was immensely important. An apprentice hammerman, for example, was qualified to enter the craft after seven years of apprenticeship, as long as his master gave him a good discharge. If, on the other hand, he had not made 'thankful service' to his master, he could only enter the craft as a freeman at the stranger's rates. Lumsden and Henderson, *The Hammermen of Glasgow*, pp. 23–24, 26–27.

practise of pharmacy and some parts of Surgery', he produced 'a letter of recommendation in his favour from the faculty at Edinburgh', showing him to have been indentured for four years, and favourably discharged.[88]

In spite of stringent ordinances, the Faculty were prepared to waive fees if it suited them. When Adam Gray, son of a maltman, was booked apprentice with Archibald Bogle, according to an indenture of September 1659, his booking fee was 'forgiven him by the calling as being the son of a member of the faculty'.[89] Similarly, William Bogle, son of a messenger, was booked as apprentice with Archibald Bogle, Visitor, in October 1666, for six years with one for meat and fee, under the following dispensation: 'for his right booking money to the said calling They all In one voice did freely forgive & quit the same to the said William Bogle elder for his undertaking to do them all service he can within burgh in giving of charges of horning or any other service'.[90] Bogle was probably related to the Visitor.

What did a Faculty apprenticeship mean in educational terms? There was little if any provision in the Faculty apprenticeship for formal attendance in classes; the emphasis was always on the individual acquisition of practical skills. Apprentices were required to know a basic body of theory, but no official teaching role was laid down for the Faculty in its charter of 1599. This stated simply that no one was 'to exercise within the said bounds without a testimonial of a famous university where medicine is taught or at least of his majesty's or his spouse's chief physicians'.[91] The onus of teaching therefore fell upon the masters in a surgical apprenticeship, and on the universities in medicine; the Faculty's role was merely regulatory.

Although Peter Lowe demonstrated a basic knowledge of anatomy in his publications, there appears to have been no provision for anatomical demonstrations in Glasgow in the seventeenth century. The founding charters of other surgical companies, including the Edinburgh Surgeons' seal of cause, and the charter of the London barber-surgeons made official provision for

---

[88] RCPSG, 1/1/1/1b, pp. 156–59, 499–500. Note that the Edinburgh association of surgeons was never known as the Faculty, but always as the Incorporation. This was simply the FPSG extending a similar terminology to the Edinburgh Incorporation.

[89] RCPSG, 1/1/1/1b, p. 214. Since his father was a maltman, presumably Gray's stepfather was a Faculty member.

[90] RCPSG, 1/1/1/1b, p. 260. Archibald Bogle entered the Faculty in 1654 and was Visitor on four occasions between 1666 and 1668, in 1669, between 1671 and 1673 and in the year 1674 to 1675. The son of a namesake father who had entered the Faculty in 1649 and served as its collector, the Bogle family was represented in the Faculty through several generations. Duncan states, p. 243, that William was probably the son of Archibald, the Visitor, but the minute book indicates that he was not, though he was probably a relative. Duncan, *Memorials*, pp. 76, 93, 237, 239, 243.

[91] RCPSG, 1/1/1/1b, pp. 2–3.

bodies for dissection.⁹² The Glasgow Faculty had no such right. If any hands-on anatomical teaching did occur in the seventeenth century, no trace of it remains in the official records.

Apart from its box, the Faculty's other common property was manifestly for teaching purposes: it had a human skeleton, a number of books, and 'other Rarities' (probably anatomical specimens) which would have been displayed in a glass-fronted cupboard.⁹³ But, as a corporate body, the Faculty was keen to give its junior members access to the expertise of the seniors. In January 1612, it was ordained that 'the deacon or on[e] of the quartermasters teach upon medicine, surgery, or pharmacy, the nature of herbs, drugs and such like as shall be though[t] expedient by the brethren of the said vocation'.⁹⁴ This is the earliest reference to collective medical teaching in Glasgow. The aim seems to have been to supplement the instruction given to apprentices by their masters with demonstrations and talks given by the Visitor or other senior members of the Faculty, but there is little indication of precisely what this entailed.⁹⁵

Apprentices were examined at three stages during their training – after three, five and seven years' study. After his first examination, an apprentice was obliged to pay £5 Scots for a dinner, 20s. to each of his examiners, and 6s. 8d. to the clerk. In the final examination, he was 'to be examined upon the whole particulars of his art, of the definitions, causes, signs, accidents & cures of all diseases pertaining to his art with the composition of nature & fit medicaments as shall be requisite'.⁹⁶ Precisely what form this examination took in the seventeenth century is unrecorded. However, the Faculty's examination schedule does not appear to have been as comprehensive as that of the Incorporation of Surgeons in Edinburgh, guidelines for which were set out in July 1647. A candidate there had to deliver a general discourse on the entire human anatomy, to inspect anatomical subjects and answer questions on them, and to demonstrate operations on these subjects and answer questions on the procedures.⁹⁷ It does appear that the Faculty's concept of its teaching role was not nearly as well defined as that of the Edinburgh

---

⁹² Dingwall, *Physicians, Surgeons and Apothecaries*, p. 39; S. Young, *Annals of the Barber-Surgeons of London* (London, 1890), p. 589.

⁹³ GCA, T-TH 14 5.2, Minutes of the Incorporation of Barbers, 1707–65, fol. 2r. The books, skeleton and other rarities were allocated to the surgeons in 1707 when the common stock was divided in anticipation of the split between the surgeons and the barbers.

⁹⁴ RCPSG, 1/1/1/1b, pp. 24–25.

⁹⁵ Duncan, *Memorials*, p. 50.

⁹⁶ RCPSG, 1/1/1/1b, pp. 14–15.

⁹⁷ Dingwall, *Physicians, Surgeons and Apothecaries*, pp. 85–86.

surgeons.⁹⁸ Surgical enterprise in Edinburgh was certainly on a larger scale: between 1621 and 1680 the Edinburgh surgeons booked 118 apprentices; in a more or less comparable period, between 1627 and 1684, the FPSG booked but twenty-three, some of them apprentice-apothecaries and apprentice-barbers.⁹⁹

From the tenuous surviving evidence in Glasgow, it seems that the first part of the trial for an apprentice-surgeon was a question-and-answer session on simple anatomy. The second part of the trial – examination of his practical expertise – was conducted during the course of his actual practice. However, the overwhelming majority of candidates in Glasgow presenting for trial in the seventeenth century were examined in pharmacy and surgery, indicating that most wished to practise as general practitioners.¹⁰⁰ Few underwent a trial in surgery alone. George Lockhart, son of a Glasgow apothecary, was examined in August 1668 on several points of pharmacy and found qualified to practise. But for the surgical side of his qualification, he was informed that the next time he had occasion to cure simple wounds, to phlebotomise, to apply cauteries, or to use ventouses and sutures, some of the brethren were to accompany him as observers; if they made a satisfactory report, he would be admitted to practise surgery.¹⁰¹

In November 1674 John Robison, former servant of Mr David Sharp (a senior surgeon-apothecary in the Faculty who acted as an examiner), sought admission to the Faculty as surgeon-apothecary. By way of a test in pharmacy, the candidate had to make up some pills, a syrup, a plaster against rupture and some lozenges. For the surgical component, he was to be examined 'upon that part of the anatomy concerning the head, ears, nose and eyes and the diseases relative thereto *ad usum chirurgicum*'.¹⁰² Robison's examiners reported that they saw him 'weigh the simples and mix the same orderly and after inspection taken by all the members of the faculty present of the same they did allow and approve of each of the said compositions made by him as being well dispensed'. He also gave perfect answers to questions in pharmacy and surgery and was admitted on both counts.¹⁰³ A gap in the

---

⁹⁸ Rosalie M. Stott, 'The Incorporation of Surgeons and Medical Education and Practice in Edinburgh 1696–1755' (unpublished Ph.D. thesis, University of Edinburgh, 1984), p. xxix.

⁹⁹ Dingwall, *Physicians, Surgeons and Apothecaries*, p. 70 (table 3); RCPSG, 1/1/1/1b.

¹⁰⁰ For accounts of general practice in seventeenth-century Edinburgh, see Dingwall, *Physicians, Surgeons and Apothecaries*, pp. 147–84; idem, 'Archives and Sources: "General Practice" in Seventeenth-Century Edinburgh: Evidence from the Burgh Court', *Society for the Social History of Medicine*, 6, pp. 125–42.

¹⁰¹ RCPSG, 1/1/1/1b, pp. 285–86.

¹⁰² This is the first time that the word 'anatomy' is used in the minutes in relation to an examination.

¹⁰³ RCPSG, 1/1/1/1b, pp. 485–86, 488–89.

records makes it difficult to ascertain at what point dissection arrived in the Incorporation's examination but by the time William Stirling was examined in February 1712, he had 'to dissect the eye of an animal' for his trial in surgery. By then, the standard eighteenth-century examination format – a practical essay in pharmacy, 'Extempore questions in Surgery and pharmacy', a dissection, and a surgical discourse on a set subject – was in place.[104]

Ironically, a far clearer picture emerges from the minutes of the pharmaceutical aspects of the seventeenth-century trial. For those seeking licence as apothecaries, formal petitioning served the purpose of laying down in writing the candidate's alleged skills and qualifications. Since a licence was granted on the basis of specific skills and procedures, the petition permitted Faculty examiners to pitch a trial at an appropriate level. In November 1660, Andrew Elphinstone, lawful son of James Elphinstone, former portioner of Woodside, was admitted as a pharmacist having demonstrated his knowledge of an impressive variety of balms and potions, including distilled waters, herbal extracts or decoctions, 'all syrups except purgatives',[105] oil of roses,[106] ointment of roses,[107] oil of St John's wort, ointment of tobacco, and oil of water lilies, pills of washed aloes,[108] stomach pills, not to mention mel

---

[104] GCA, T-TH14 5.2, fol. 13v. The Faculty minutes between 1688 and 1733 were lost in a fire.

[105] For a selection of alterant syrups, see John Quincy, *The Dispensatory of the Royal College of Physicians in London: With Some Notes Relating to the Manner of Composition, and Remarks on the Changes Made in Most of the Official Medicines, from their First Prescribers down to the Present Practice* (London, 1721), pp. 36–54.

[106] Oil of roses or *oleum rosaceum* was made from 4 oz. of exungulated (paring off of the white part of the petal) red roses which were not quite blown, bruised in a marble mortar with a wooden pestle, and 1 lb. of clean olive oil, both exposed to the midday sun in a well-stopped glass vessel for a whole week and shaken every day, then simmered in a heat bath and the oil pressed out. At this point, fresh roses were added and the process repeated. After a third addition, the oil was left to stand for forty days, after which it was ready for use without pressing out the roses. Quincy, *Dispensatory*, p. 159.

[107] *Unguentum rosatum* or ointment of roses consisted of 1 lb. of hog's lard, cleared of membranes and well washed, added to 1 lb. of fresh red roses and left to stand for a week. The mixture was boiled over a gentle fire, and the lard pressed out, and then fresh roses softened and soaked for the same time and boiled and strained as before. Finally, 6 oz. of the juice of red roses, and 2 oz. of oil of sweet almonds were added, and the mixture boiled over a slow fire until all the juice was consumed, then strained to the consistency of an ointment. Ibid., pp. 189–90.

[108] *Pilulae de aloe lota* or pills of washed aloes were made from 1 oz. of aloes, dissolved in juice of roses and then condensed; 3 drams of the Troches of Agaric (a fungus); 2 drams of Mastich (resin of the *Pistacia lentiscus* tree used as a tonic and astringent); and a sufficient quantity of syrup of damask roses to fashion into a mass for pills. Ibid., p. 122; J. Worth Estes, *Dictionary of Protopharmacology: Therapeutic Practices, 1700–1850* (Canton, Massachusetts, 1990), p. 124.

mercurials,[109] sapa,[110] Rob de Berberis,[111] linctuses and electuaries, 'tablets no ways purgative', 'powders ... conform to the prescription of the doctor', various troches or lozenges, and an array of unguents and juleps.[112]

Elphinstone did not stretch himself beyond his abilities: he sought neither to concoct purgatives nor to make up powders without reference to a physician's prescription. For his trial, he was asked to prepare some of the simple mixtures and compounds specified in his petition. In addition, he answered questions posed by Dr Crichton and Mr Archibald Graham on pharmacy, and was admitted to the Faculty 'as a brother to the distilling, dispensing & compounding of the said particulars, only for this time, until such time [as] he attain to more skill'.[113]

Of particular linguistic interest is the subscribed petition of John Ewing in Paisley, who sought admission as an apothecary, asking that the Faculty license him to sell drugs and make up recipes according to a doctor's direction 'which he is to receive from the doctors only in Scots language because he has no other language'. After examination, he was admitted on the condition that he was neither to make recipes nor sell drugs in Glasgow, under pain of £20 Scots.[114]

After passing his master's examination, the newly qualified surgeon-apothecary had to pay £10 Scots for a dinner – clearly more lavish in view of the momentous achievement – in addition to the aforementioned fees.[115] The dinner money was exacted even when the candidate failed; presumably it then became a cause for drowning his sorrows.[116] But admission to the Faculty was granted to no one 'till they produce their Burgess ticket before the deacon and quarter masters according to the act of guildry'.[117] After successful examination, yet another fee had to be paid before the new master was entered in the burgess roll of the town.[118] Only then could the qualified

---

[109] Medicines containing honey for the treatment of syphilitic chancres.

[110] Sapa is the residue remaining after raisins or hard grapes have been evaporated to the consistency of honey. Ibid., p. 172.

[111] Rob de berberis or rob of berberies comprised a pint of clear, strained juice of berberies and ½ lb. of sugar boiled over a slow heat. Quincy, *Dispensatory*, p. 64. The berries of the berberis were used as a tonic and as an antidiarrheal. Estes, *Dictionary of Protopharmacology*, p. 26.

[112] RCPSG, 1/1/1/1b, pp. 181–82.

[113] RCPSG, 1/1/1/1b, pp. 182–83.

[114] RCPSG, 1/1/1/1b, pp. 199–200.

[115] RCPSG, 1/1/1/1b, p. 15.

[116] Duncan, *Memorials*, p. 50. Failure was more likely where a master used his apprentice as a cheap means of undertaking the basic barbering side of his practice rather than teaching him the full range of skills. De Moulin, *History of Surgery*, p. 113.

[117] RCPSG, 1/1/1/1b, p. 13.

[118] Douie and Gibb Dougall, *Chronicles of the Maltmen Craft*, p. 60.

surgeon-apothecary apply for membership of the Faculty. Members were admitted under different categories, on a sliding scale of freedom fines laid down in 1602. A burgess's son paid £40 Scots, with a £10 reduction if he had been apprenticed within the town to a freeman. Sons of freemen of the craft who had been apprenticed to their own fathers or to other freemen in Glasgow could enter for just £20 Scots. A study of the larger Scottish towns has revealed that by the late seventeenth century entry to the burgess-ship by right of paternity was second only to entry by right of marriage to a burgess's daughter or widow as the most common means of registration.[119] At the other end of the scale, a total stranger seeking admission to the Faculty as free man had to pay a phenomenal £66 13s. 4d. for the privilege. Admitted 'as an appendage of surgery',[120] barbers paid £40 Scots (the same as burgess's sons), and 20s. p.a. to the poor.[121] Before the establishment of the Faculty, the barbers had no prior guild status in the town, whereas in England the barbers had the established organisation which the surgeons joined.

Refunds of freedom fines, as of booking fees, were sometimes given for services rendered. When John Hall from Edinburgh was admitted to the Faculty as a barber in September 1671, he paid £40 Scots in freedom fine. But the Faculty, 'having taken to their consideration the respect, Kindness, favour & courtesy the said John Hall had shown to the said calling', unanimously ordained that the collector should refund him £20. He was accorded the special dispensation of paying at the rate of a son of a member. Understandably (given the punitive fine), strangers rarely bothered to become freemen before entering the Incorporation. On 1 January 1669, the Faculty considered the great harm which the poor of their vocation had sustained through strangers coming into town and being admitted as freemen by the Incorporation for little or no fine, although they were neither freemen's sons or sons-in-law, nor had they served apprenticeships in Glasgow. In this way, the gift and patent granted in favour of the calling was being slighted with adverse implications for the poor of the craft. The Faculty enacted that anyone entered as freeman with the craft who lived within the burgh, was to pay 100 merks Scots to the collector for the use of the poor before his admission was booked. The protection from hardship or loss provided by the poor fund was of considerable importance to members. Each stranger

---

[119] Devine, 'The Merchant Class of the Larger Scottish Towns', p. 103.

[120] In the Scottish guild context, this was a specialised craft regarded as a subdivision of a general craft. M. Robinson (ed.), *The Concise Scots Dictionary* (Aberdeen, 1985), pp. 484–85, under 'pendecle'.

[121] RCPSG, 1/1/1/1b, pp. 14–16.

barber was to pay a fine of at least £40 Scots before admission. However, the act was not extended to freemen's children whether they served apprenticeships within the burgh or not.[122]

Much later, in 1816, a committee appointed to codify the Faculty laws observed that before 1783, the apprentices of members of Faculty and sons of members enjoyed an advantage over other apprentices in their admission, and in the term of their apprenticeship.[123]

Internal Guild Discipline

Rules about behaviour were laid down for members' mutual protection. It was stipulated, at the first meeting of the Faculty, that there should be no poaching of patients, and an act ordained that 'none of the brethren take a patient out of another's hand until the time that the said brother be fully satisfied for his pains' or a fine of £40 Scots would be incurred to the box.[124] This was confirmed in the 1656 letter of deaconry, which stated 'That no freeman presume to take another freeman's cure off his hand until he be honestly paid for his bygone pains', unless the latter was unqualified.[125] But it must have been difficult to uphold when patients generally consulted whom they pleased; where the patient had a certain social status; or where a practitioner felt the need to ask a colleague for a second opinion. Indeed, Faculty licensing encouraged consultation with seniors.

A disciplinary case occurred on 22 June 1602, only nineteen days after the first meeting. Thomas Thomson who had sworn that he would discharge his duty along with the rest of the brethren, had been asked to give his assistance in carrying out the Faculty's lawful business, but 'several times has most wrongously contemptuously disobeyed'. The brethren decided that he must forfeit the privileges of membership, but it took time for the wheels of the new association to turn (the first meeting was held three years after the charter was secured), because although John Hall and others were similarly discharged Hall, at least, remained a member. It may simply have been a warning. However, the 'Act for Compearance' (appearance before a court or other authority) which was passed in the same year, ordained that any brethren who failed to obey a summons to a convention would incur a 10s.

---

[122] RCPSG, 1/1/1/1b, pp. 307–10, 365–66.
[123] RCPSG, 1/1/1/5, Minutes of the Faculty of Physicians and Surgeons of Glasgow, 4 May 1807 to 4 December 1820, fol. 170r, 170v.
[124] RCPSG, 1/1/1/1b, pp. 10, 15.
[125] Tennent, *Incorporation of Barbers*, pp. 18–19.

penalty if they did not – in the opinion of the Visitor and masters – have a reasonable excuse.[126]

Various acts relating to the smooth running of the association were formulated on 22 June 1602: one ordained that the names of those who usurped the liberty and privileges of the brethren were to be given to the provost and bailies, in order 'to cause them to convene and to find caution for abstinence conform to his majesty's Commissions and the authority of the town interponed thereto'.[127]

In a small town, reputations could be quickly (and sometimes unjustly) wrecked, and cases of slander were treated as serious 'wrongs'. Those found guilty of damaging and slanderous accusations could be fined, warded or imprisoned or even publicly humiliated in the stocks or by being forced to kneel and beg forgiveness of their victims in court. Where the guilty party was a burgess, he might suffer the severe penalty of having his freedom 'cried down' which could effectively reduce him to poverty. A number of the Faculty's early acts and statutes similarly concerned themselves with slander. One, passed in September 1612, provided for the punishment of any member who blasphemed a brother of the craft publicly or privately or otherwise misused them in word or deed.[128] The 'Act concerning speaking scandalously of the present Visitor', of September 1627, specifically protected the deacon of the incorporation against slander. Any member found guilty after trial would be discharged from the incorporation and fined, as well as having to give satisfaction to the Visitor before being readmitted. In 1653, it was found necessary to reissue a similar bylaw twice in the same month. On 4 April, it was enacted that if any member spoke disdainfully of, or to, the Visitor, he would be relieved (after trial) of his vote with the Faculty, and barred from practice. On the 28th of the same month, it was enacted that if anyone slandered another member, either personally or behind their back, they were to pay 40s. Scots for each offence.[129]

These bylaws were implemented. In September 1667, a complaint was made by Archibald Bogle and William Currie about William Clydesdale, then seeking admission to exercise aspects of surgery and pharmacy. According to them, without any offence given, Clydesdale did 'scold the said visitor by uttering a number of vile expressions as particularly that he was a mere fool

---

[126] RCPSG, 1/1/1/1b, pp. 11–12.
[127] RCPSG, 1/1/1/1b, p. 13.
[128] McGrath, 'Administration of the Burgh of Glasgow', i, pp. 215, 217; RCPSG, 1/1/1/1b, p. 27.
[129] RCPSG, 1/1/1/1b, pp. 36, 71.
[130] RCPSG, 1/1/1/1b, p. 272. Though he had been Faculty Visitor, Clydesdale was always getting himself into trouble. A complaint was later brought against him and his wife for malpractice, for which see Chapter 4, below.

& an ass not worthy to carry office in his place and did call the said William Currie a Warlock & renegade going from door to door'.[130] While confessing the slander against Currie, Clydesdale denied slandering the Visitor. Producing famous witnesses – Messrs Archibald Graham and David Sharp, apothecaries, and Gabriel Cunningham, officer – who proved the article spoken against the Visitor, Clydesdale was fined 20 merks. More punitively, he was barred from being an office-holder or from having a vote within the incorporation until he gave proof of his good behaviour. Two years later he was tried for malpractice. And again, in April 1673, John Fleming was discharged from being a member of the Faculty, and from sitting and voting with other members, because of a past miscarriage which was seen to reflect upon the Visitor.[131]

The Faculty also dealt with sabbath-breaking as part of its disciplinary procedure. In January 1676, the FPSG received information that 'several barbers who are members thereof within this burgh are prophaners of the sabbath by barberising of persons that day'. Finding this a gross and scandalous sin, the Faculty decreed that any member convicted of barberising on a Sunday should be fined £40 Scots on the first two occasions and forfeit his membership. Anyone so foolish as to contravene a third time was denied the opportunity of readmission, being declared 'no member of their said faculty from that time forth as if he had never been admitted and incapable at any time thereafter to be readmitted and his act of admission to be cancelled, severed, & expunged forth of their records as a person unworthy of being incorporated in any society & much less to be a member of their faculty'.[132]

The Split between the Surgeons and the Barbers, 1708–1722

Since 1656, the day-to-day practice of barber-surgeons in Glasgow had been regulated by the Incorporation of Surgeons and Barbers, within which surgeons and barbers coexisted in uneasy coalition. But there was a general divorce in the occupational association between barbers and surgeons in the eighteenth century as the status of surgery improved. In Edinburgh, the barbers split from the surgeons to form a separate Society of Barbers in 1722, and the union of the Paris surgeons of St-Côme and the barber-surgeons in France was finally dissolved in 1743, resulting in separate communities of barbers and surgeons. England followed two years later when the Company of Surgeons split from the barbers. Glasgow was no different, the eventual

---

[131] RCPSG, 1/1/1/1b, pp. 272–73, 450. Details of the 'past miscarriage' are wanting.
[132] RCPSG, 1/1/1/1b, pp. 526–27.

separation between the barbers and surgeons occurring in the same year as in Edinburgh.[133]

The surgeon members of the Glasgow Faculty had effectively operated under two constitutions since 1656 – an unusual situation, the subtleties of which were never fully appreciated. By incorporating the surgeons and barbers as a guild, the town council looked for the same municipal allegiance from them as hammermen,[134] or maltmen, but because of the autonomy granted earlier by its royal charter, the Faculty did not always feel obliged to give it.[135] They particularly came into conflict with the council over their 1679 act restricting the admission of strangers to practise surgery and pharmacy.[136]

It was after the Faculty acquired its first real estate in 1697 that the occupational gulf between the barbers and surgeons in Glasgow really began to widen. The Faculty planned to demolish the property on the west wide of the Tron Kirk, and build anew 'for a public Hall to the faculty for their public meetings, and more particularly for their meetings the first Monday of each month of the year, for communicating to the necessity of the poor gratis, conform to their gift and charter'.[137] The hall was to be shared with the barbers. A petition to the council from the members of the Incorporation, in 1708, confirms the completion of 'their common hall'. The petition requested that the entry beneath the tenement, belonging to the burgh, which was closed at the time be given over to the incorporation for enlarging their laigh [138] houses and adjoining shops.[139]

Barbers had been licensed occasionally by the Faculty since 1602, but they were few in number. However, the 1656 letter of deaconry had afforded them

---

[133] While there are no Faculty minutes for the few decades preceding the split between the surgeons and the barbers, fortunately (as a matter of civic concern) the matter receives an airing both in the burgh minutes and in the early minutes of the Incorporation of Surgeons and Barbers. Note that, contrary to Duncan, *Memorials*, p. 83 n. 1, which discounts the existence of two lots of minutes as 'quite improbable', there are minutes for both the Faculty and the Incorporation of Surgeons and Barbers, though the latter are extant only from the eighteenth century.

[134] This craft incorporation comprised armourers, swordsmiths, blacksmiths, goldsmiths, saddlers, cutlers and shearmakers.

[135] Duncan, *Memorials*, p. 78.

[136] See Chapter 1, 'The FPSG versus the Magistrates of Glasgow'.

[137] Duncan, *Memorials*, p. 81.

[138] A lower building attached to a tenement or main building of several stories.

[139] J. D. Marwick (ed.), *Records of the Burgh of Glasgow, AD 1691–1717*, Scottish Burgh Records Society (Glasgow, 1908), p. 433. Four years later, the town council ordered the entry to be kept open for public entry, this to be one of the terms of the next roup of the Tron. Ibid., p. 480.

civic recognition alongside the surgeons. Never smooth, the professional relationship between the barbers and surgeons began to show signs of severe distress at the outset of the eighteenth century. In 1701, the barbers complained (unsuccessfully) to the Trades' House about the inequities of their position. Two years later, they took their complaint to the town council, which recommended a committee to consider a petition 'by the barbers of this burgh against the surgeons therein'.[140]

In essence, the barbers were complaining about the inequity of their position in the Incorporation vis-à-vis the surgeons. Although they had been accorded full membership of the guild by the 1656 letter of deaconry, at the same time this underwrote the more elevated status of surgeons in ordaining that the deacon or Visitor was always to be a surgeon; barbers were not permitted to hold the office.[141] Furthermore, the surgeons were unwilling to concede any rights to the barbers in the admission of surgeons into the joint incorporation, or in the determination of their freedom fine. They were members of the same incorporation, and they shared the same hall, but the surgeons preferred to regard the two occupational spheres as administratively separate because this reserved them a greater degree of autonomy in operation. In short, the surgeons wished to have the benefit of burgh privileges without conceding any of the independent authority conferred upon them by their original charter.

The nature of the barbers' grievances was discussed by the council in May 1704. The barbers complained that 'of late, the surgeons have committed many unwarrantable encroachments upon the interest of the barbers, contrary to the said letter of deaconry' (as specifically expressed in their petition). They therefore craved the protection of the magistrates and town council, asking particularly for the restoration of their privileges and liberties as laid out in the letter of deaconry, or 'otherwise to fall about some methods for disjoining the barbers from the surgeons'. In November 1703, Mr Peter Patoun, doctor of medicine, Henry Marshall, Visitor, William Thomson, James Weir and David Hall, surgeons, and Thomas M'Allay and Walter Robieson, barbers, appeared before the committee at which the surgeons were asked 'whether they would adhere to or pass from the said letters of deaconry' and to report back.[142] Marshall prevaricated: when the committee reconvened on 2 May 1704, he simply 'declared that he forgot to call the faculty of the surgeons to that effect and craved a new diet for doing thereof'.

---

[140] Duncan, *Memorials*, p. 84; Marwick, *Records of the Burgh of Glasgow, AD 1691–1717*, p. 369.
[141] Office-bearing in the Edinburgh Incorporation of Surgeons was similarly denied to the barbers. Dingwall, *Physicians, Surgeons and Apothecaries*, p. 54.
[142] Marwick, *Records of the Burgh of Glasgow, AD 1691–1717*, pp. 377–78.

Three days later he reported that he had no mandate from the Faculty either to adhere to, or to renounce the letter of deaconry as it then stood, but relayed the surgeons' opinion that 'it was very reasonable and requisite that many things in the letter of deaconry should be amended'. He suggested the appointment of a joint committee of surgeons and barbers to thrash the matter out.

The town council then passed an act declaring that the parties to the letter of deaconry were bound by its rules and conditions, because it had 'proceeded upon the joint application of the surgeons and barbers of this burgh'. Just as important, they also declared that the original gift in favour of the surgeons did not confer any right of deaconry on them or any other civic power. Nor did they consider that the charter afforded the surgeons any loophole. The council therefore ordained 'the said letter of deaconry to be the standard and rule' and disallowed any alterations or encroachments to be made on it. Any violations were to be heard and determined by the magistrates (who were to redress any aggrieved parties and to punish contravenors as they saw fit).[143] But the differences continued.

In September 1706, the town council appointed a committee to listen to the differences between the barbers and surgeons. The committee's findings were ratified by the council a year later, which decreed that the rules in the letter of deaconry 'be inviolably obeyed and observed by them in all time hereafter'.[144] Superseding all former acts of trade on this subject, this act declared, in the first place, that all qualified barbers should have as much right as surgeons to vote for a Visitor (who was, nevertheless, to remain a surgeon). The Visitor was to choose his own quartermasters (three in number), but the trade's quartermasters would be elected by poll from the entire membership. A major concession was the decision that the collector was to be by turns a surgeon and a barber, to be chosen by the whole trade. There was always to be a barber as well as a surgeon in office as boxmaster. The trade was to hold four quarterly meetings per year (as did other trades), of which all members would be advised; they could also be present at the collector's annual accounting (previously attended by the officers only), at the election of any office bearers, at the making of acts of trade, or at any important business concerning the affairs of the trade.[145] However, the act categorically excluded physicians from membership of the Incorporation: 'a

---

[143] Ibid., pp. 378–80.

[144] Ibid., pp. 395–96, 412. This included the provost, bailies, John Anderson of Dowhill, late provost, the dean of guild, deacon convenor, James Sloas, and six others, any seven of them a quorum.

[145] Ibid., pp. 410–11.

Physician praeses is or can be no member of the trade and can neither sit nor vote therein'. Indeed, the Visitor only presided 'conform to the letter of deaconry'. In this way, the council established the pre-eminence of the seal of cause.

It was also stipulated that barbers who were qualified but unentered could have their admission backdated to when they began practising; their apprentices were similarly to be booked from the date of their indentures. This suggests that some barbers had deliberately been refraining from entering the Incorporation until the dispute, and the question of their status, was settled. Although the Incorporation had, at that time, severed its links from the Trades' House, the act stipulated that, should it be reunited, the power to nominate members to the house lay (as it had previously) with the Visitor. In addition, the barbers sought reimbursement of the money expended by them since the beginning of their dispute with the surgeons. It was established that since the surgeons' expenses had been paid out of the public stock of the trade, the barbers were likewise entitled to reimbursement from the public stock, once the expenses had been verified. (They were eventually paid £100 Scots by the collector in December 1707.) [146] The surgeons, however, were protected from interference by the barbers in surgical matters; the barbers were given no interest in the examination of surgeons' qualifications or in any malpractice committed by them; and what privileges they had were not to be prejudicial 'to the liberties and privileges of the surgeons granted to them by their gift from King James the Sixth'.[147]

In spite this, on 1 October 1707 (fast on the tail of the council's conciliatory legislation), the Incorporation of Surgeons and Barbers divided its common stock, two fifths of which was declared to belong to 'the Trade of Surgeons and barbers', and three fifths 'to the Faculty of Surgeons only'. The books, skeleton and other rarities also went to the surgeons. This was done to facilitate the Incorporation's reunion with the Trades' House, David Hall, surgeon, having insisted that 'nothing thereanent might be put to a vote until the Common Stock between the Surgeons and Barbers be first divided'.[148] So, in spite of the act of September 1707 which was so favourable to the barbers, the two occupational groups anticipated imminent separation, anyway.

Indeed, by 1712, the surgeons and barbers were at each others' throats again. In this year, Mr William Stirling applied for admission as a surgeon

---

[146] GCA, T-TH14 5.2, Minutes of the Incorporation of Barbers, 1707–65, fol. 2v.
[147] Marwick, *Records of the Burgh of Glasgow, AD 1691–1717*, pp. 411.
[148] GCA, T-TH14 5.2, fos 1v, 2r. A copy of this agreement also survives in the minutes of the town council and in the minutes of the kirk session.

even though he had not served an apprenticeship.[149] However, under an act of 1679, which stated that 'no person or persons whatsoever shall in any time coming be admitted to practise either of the said arts of surgery and pharmacy within the city of Glasgow but such as either have served their apprenticeship with a free man or member of the faculty for the time for the space of five years', Stirling was not permitted to enter as freeman. Nonetheless, the surgeons were insistent on his admission. In view of his not having served an apprenticeship, Stirling was ordered to pay a punitive freedom fine of 1000 merks (the estimated total of all the dues he would have paid had he served an apprenticeship). The barbers took exception to this, claiming that part of the excess amount should accrue to them as joint partners in the Incorporation. The barbers consequently appealed to the Trades' House for arbitration.[150] On 1 July 1712, the Trades' House found in favour of the barbers, declaring that

> in the case of Mr William Stirling, surgeon, or any others in his case offering to enter with the trade, that the whole joint trade of surgeons and barbers ought to have vote in the admission of any such entrant and in the determination of the quota of the freedom fine of any such entrant, and that the same ought to be applied and joined to their joint stock for maintenance of the poor *in cumulo*.

The surgeons counter-petitioned the town council, on 19 December 1712, 'complaining that the barbers do encroach upon the said surgeons' rights and privileges by pretending right and privilege of a vote in the admission of all entrant surgeons', and in determining the quota of their freedom fine. They asked for an annulment of the decree of the Trades' House and a declaration that only the surgeons had 'the undoubted right to examine, enter and impose freedom fines upon all surgeons'. At this time, the barbers paid £100 into the joint stock at the entry of every stranger barber; the surgeons therefore declared themselves willing to pay just as much in order to obviate the barbers' complaints (the surplus to belong to the surgeons and their separate stock). They also asked that a ceiling of £100 – not to be exceeded without the council's consent – be fixed on the Incorporation's freedom fines.[151]

In October 1713, the Trades' House reconsidered the obscurely worded sentence it had passed in the previous year on William Stirling's petition to be received into the privileges of the Incorporation. Clarifying the position, the deacon convenor of the house, John Graham, declared to the town council that barbers and surgeons together should set entrance fines, which

[149] GCA, T-TH14 5.2, fol. 13v.
[150] RCPSG, 1/1/1/1b, p. 630; Duncan, *Memorials*, p. 87.
[151] Marwick, *Records of the Burgh of Glasgow, AD 1691–1717*, pp. 490–91.

should then be put into their common holding. If either the surgeons or the barbers challenged that ruling they would not be allowed to defray their expenses from the common stock. The town council ratified this judgement in the barbers' favour.[152]

Wrangling erupted anew four years later, and with a ferocity that was to foreshadow the eventual split between the barbers and surgeons. Again in September 1719, the surgeons appealed to the council about a complaint made by the barbers to the Trades' House. Although this petition is not extant,[153] surviving evidence shows it to have been founded upon the Incorporation's Act of 1679 by which membership was confined to sons (including sons-in-law) and apprentices of members. Implying that an entrant became a freeman of his own particular art, the act had specifically mentioned the 'arts of surgery and pharmacy' only. But, following a spate of victories which improved their civic status, the barbers then questioned why (since they were freemen in the same incorporation), their sons and apprentices might not also be eligible to try for admission as surgeons, as long as they were found qualified by the deacons and surgeon-masters for practice.[154] In claiming such a right on the basis of practical, empirical achievement, barbers undoubtedly took their corporate rights a step too far, because this bypassed the need for surgical apprenticeship, but it was a rod that the surgeons had made for their own backs through their insistence in 1712 on the admission of William Stirling although he had not served an apprenticeship. As so often before, the Trades' House decided in favour of the barbers; and, specifically regarding 'a petition from Walter Robertson, barber, his son, that he as a barber's son might be admitted to trial in his skill of surgery and pharmacy, had thereupon taken away all their privileges as surgeons and set barbers wholly on a level with them, and had ordained the deacon of the surgeons to admit him'.[155] As in the case of Henry Marshall in the 1690s, the town was taking upon itself the right to license medical practitioners in Glasgow.[156]

Furious, the surgeons appealed to the town council – which appointed a committee to consider their petition – craving a stay of execution on this

---

[152] Ibid., p. 515. This is also to be found in Lumsden, *Records of the Trades House of Glasgow*, ii, pp. 11–12.

[153] The complaint is not recorded, the records of the house noting simply, on 5 September 1719, that 'The house decerned in the Barbers complaint against the surgeons as is marked on the records'. Lumsden, *Records of the Trades House of Glasgow*, ii, p. 51.

[154] This has been construed from the information in an act determining the difference between the surgeons and barbers, of 7 November 1719. R. Renwick (ed.), *Extracts from the Records of the Burgh of Glasgow, AD 1718–1738* (Glasgow, 1909), p. 72.

[155] Ibid., p. 67.

[156] See Chapter 1, 'The FPSG versus the Magistrates of Glasgow'.

sentence. Having heard evidence from both parties, on 7 November 1719, it upheld the decision of the Trades' House and dismissed the surgeons' appeal on the grounds 'that seeing every surgeon and barber is a freeman of the incorporation and that thereby their sons, sons in law and apprentices have an equal privilege to be admitted members of the incorporation, according to what upon trial they shall be found qualified to practise'.[157]

The town council also gave its opinion that 'most of the differences that have hitherto happened between the said parties is from an undue extension of the rights and privileges conveyed to the surgeons by the gift of King James the Sixth in the year 1599, which both parties endeavour to confound with the letter of deaconry'. But, in truth, the barbers had no claim to the charter, nor did the privileges which it gave to the surgeons and physicians give them any more power as practitioners in the city of Glasgow than it did to those who practised in the neighbouring shires, and certainly none which prejudiced the letter of deaconry. Understandably, the council had merely affirmed the ascendancy of its own gift.

But the town council's decision finally pushed the surgeons' hand: on 19 December 1719, they submitted to the council a document entitled 'Demission and Renunciation of the Letter of Deaconry'. Signed by all the surgeons and apothecaries in Glasgow,[158] the writ unequivocally expressed their intention to renounce and surrender 'all right, privilege and interest which they or their successors had, have or can pretend to, by or from the letter of deaconry granted by the council erecting the surgeons and barbers into one incorporation'. They asked not only that the council record the renunciation in the books of council, but also that it divide the common stock of the Incorporation fairly.[159] In the event, the renunciation was not inscribed into the council's books until 22 September 1722, at which point the council had accepted the irrevocable breakdown in the relationship between the two parties.

The renunciation began by outlining the charter of 1599, and the later incorporation of the surgeons and barbers, which had taken place 'with a view to the interest of both societies'. Despite the 'common benefit and

---

[157] Renwick, *Records of the Burgh of Glasgow, AD 1718–1738*, pp. 68, 72–73.
[158] The document was subscribed by Henry Marshall, William Thomson, Alexander Porterfield, John Boyd, Thomas Hamilton, James Calder, Hugh Sutton, John Melville, Mr William Stirling, Robert Wallace, John Gordon, Robert Hamilton, Thomas Buchanan and Alexander Moffat, surgeons and apothecaries in Glasgow.
[159] Renwick, *Records of the Burgh of Glasgow, AD 1718–1738*, pp. 75–76. Receipt of the renunciation was written into the books on 23 January 1720. It was ordained to lie in the clerk's hands, so that the council could deliberate upon it and give a considered judgment and so that it could be seen by the barbers.

representation of an incorporation in the place', the surgeons insisted that 'their employments were different', and 'had several regulations and acts which [respected] the distinct bodies'. In particular, the fines for surgeons (who conducted most of the Incorporation's business) were heavier, and surgeons' apprentices served only five years, compared with the barbers' seven. But even after their incorporation with the barbers, the surgeons had regulated the admission of servants and new entrants with the surgeons and apothecaries by the act of 1679,[160] being 'very much surprised' when the recent decisions of the Trades' House and the council extended this power to the barbers. Indeed, over the previous forty years, the barbers had frequently been fined for practising surgery and pharmacy 'without any benefit or entering from their being incorporate'.[161] On the other hand, just before the act, in December 1675, the Faculty had been prepared to examine Robert Boyd, a servant trained under the late Dr Thomas Hamilton, physician, who nevertheless craved 'to be admitted to the practise of some points of surgery'. After examination, they had been willing to admit him to open veins, apply cauteries, ventouses, and to cure simple green wounds – all the exclusive province of a surgeon – other than the application of clysters which he was to administer only by a physician's direction.[162] So, at an earlier stage, the surgeons had been prepared to bend this rule for their superiors in the medical hierarchy (though apparently not for the barbers).

Indeed, the records of the early eighteenth-century Incorporation attest to the continuing overlap in the practice of barbers and surgeons until the former were formally separated from the surgeons in 1722. On 18 September 1710, John Pettegrew, barber, had to answer a complaint against him 'for practising Surgery within the City of Glasgow'. He 'judicially confessed his practising frequently within the said City within this year Last', and was fined £9 Scots, to be paid by the following Friday. If he defaulted, he was 'To be secured in prison' until he paid up. On the same day, a similar complaint was made against James Murdoch of Glasgow, who also confessed. He was fined £12 Scots.[163]

The surgeons not only felt that 'we as surgeons and pharmacists have no

---

[160] The date in the minutes of the town council is transcribed as 1677, but the act was passed in 1679. The 1679 act was ratified by the council in 1714.
[161] Renwick, *Records of the Burgh of Glasgow, AD 1718–1738*, pp. 147–48.
[162] RCPSG, 1/1/1/1b, pp. 514–15.
[163] GCA, T-TH14 5.2, fol. 9r. The additional £3 was in satisfaction of a second fine for the unwarrantable practice of surgery, but Murdoch subsequently submitted a petition asking for part of it back. The collector was ordered to return £3 and to keep £9 in satisfaction of both his fines. Pettegrew similarly received back payment of one crown in settlement of his previous fine. GCA, T-TH14 5.2, fol. 10r.

advantage but disadvantage by the letters of deaconry', but also feared that 'the design of the charter from King James is like to be frustrated'. In short, they conceived it impossible to operate under two constitutions: they had been exposed and subjected to all the problems of 'a mixed state, which has been so intimate and perplexing to us and the neighbourhood'. In order to remove their grievances, they chose to renounce their rights to the letter of deaconry, asking that the council accord the Faculty a fair share of the common stock of the Incorporation. The barbers similarly submitted to the council's division.[164] Later, on 27 June 1720, the barbers complained that the renunciation had left them without a representative in the Trades' House. The council therefore gave permission for the deacon convenor (head of the Trades' House) to act as deacon of the Incorporation of Surgeons and Barbers until the problems between the barbers and surgeons were determined.[165]

In 1722, a committee of magistrates eventually divided the common stock of the barbers and surgeons, which had been set down in a contract of 16 September 1708.[166] The Faculty was given three-fifths and the barbers two-fifths. The barbers' share of the corporate stock of the Incorporation was £2116 5s. 10d. Scots. At the same time, the barbers renounced and made over to the Faculty their share in the hall and its adjacent houses. The council was able to purchase the yard which it wanted 'for the accommodation of their new street from the Saltmercat to their other new street from the Trongate to the Bridgate and building of houses', for which it paid the Faculty and Incorporation £216 Scots.[167]

With the division of their common stock, the sixty-six year formal association of the barbers and surgeons in Glasgow came to an end. The union was effectively dissolved by the barbers as they came to appreciate their occupational strength, though it was the surgeons (having more to lose) who eventually took the step of disbanding the joint incorporation. Upon acceptance of the surgeons' renunciation, the barbers were reincorporated under another letter of deaconry on 22 September 1722, and admitted into the Trades' House in the following month.[168] The surgeons adhered to the terms of their original charter.

The Faculty of Physicians and Surgeons of Glasgow acquired the veneer of a craft guild fifty years before the civic incorporation of the surgeons with

---

[164] Renwick, *Records of the Burgh of Glasgow, AD 1718–1738*, pp. 144, 148–49.
[165] Ibid., pp. 86–87. This was renewed again in October 1720, ibid., pp. 100–1.
[166] It followed the guidelines in the agreement of 1 October 1707.
[167] Renwick, *Records of the Burgh of Glasgow, AD 1718–1738*, pp. 144–47.
[168] Ibid., pp. 149–52; Lumsden, *Records of the Trades House of Glasgow*, ii, pp. 76–77.

the barbers. Its charter had been ratified by the town council, its administrative, disciplinary, and regulatory functions mirrored those of the guilds, and it was already represented in the Trades' House. Nevertheless, in 1656, most members of the Faculty (other than physicians) felt the need for a letter of deaconry to formalise its position with the town. As in many other European towns at this time, this resulted in a formal association between the surgeons and barbers in Glasgow.

Group membership conferred privileges and security but also created problems and imposed restraints. Although surgeons, apothecaries and barbers were associated together in one craft guild, they were careful to preserve the distinctiveness of their separate arts; apprentices were therefore booked with particular master-surgeons, master-surgeon-apothecaries and master-barbers. Long-term entry into the Incorporation of Surgeons and Barbers, as into any other craft guild, was achieved by reaching a number of preordained markers; but by-passing the monopolistic guild structures, where an individual had not undergone the standard training, had distinct consequences. Where this was achieved, as in the cases of Robert Boyd and William Stirling, it created precedents for the less skilled members of the guild – namely the barbers – to attempt encroachments on the privileges of the surgeons.

The surgeons (with their more elevated craft status) certainly never fully accepted their union with the barbers and blood-letters, always doing their best to maintain the occupational boundaries between them. These anomalies persisted in the operation of the two groups within the united incorporation: higher freedom and admission fines for surgeons served to emphasise their superior status over barbers, even though by the 1670s the duration of the surgical apprenticeship was commonly only five years, while a barber served a seven-year apprenticeship. The surgeons' ultimate line of defence was always the royal charter of 1599, which they regarded as superior to the civic seal of cause. In the demarcation debate which ensued between the two in the first two decades of the eighteenth century, the barbers were championed by the Trades' House and, ultimately, also by the town council (in defence of its own letter of deaconry). As the prestige of surgery rose, following the general trend throughout Europe, barbery was dissociated from surgery after over 120 years of inferior status in Glasgow – first as a pendicle of surgery in the FPSG and second, after sixty-six years in the slightly more equitable union in the Incorporation of Surgeons and Barbers. Although they united for protective reasons, ultimately the different occupational status of the barbers and surgeons proved too disparate for them to continue together.

4

The Regulation of Practice

In seventeenth-century Scotland, lay interest in medical practice was an accepted feature of daily life.[1] In times of sickness and minor injury the first recourse was to the woman of the house; household remedies were written down alongside culinary recipes – both essential prerequisites of the same domestic economy.[2] Everyone dabbled in healing, and the line between treating the extended family and strangers was hazy. Many practitioners had no formal training in the healing arts; the practice of medicine and surgery was often just a lucrative sideline for men, or a necessary part of the caring role of women. Wise or cunning men and women, astrologers, bonesetters, herbalists, itinerants, tooth-pullers and quacks – working either on a part- or full-time basis – all took their place alongside regular or licensed practitioners.

Healing was indisputably patient- or client-driven at this time rather than 'medicalised'. The client simply paid for the cure of his choice, regardless of whether that healer was officially licensed or not.[3] Personal choice, preference, knowledge, geographic proximity, type of disease, and economic status all played their part in a patient's selection of healer.[4] In actively exercising such choice, patients exerted a direct influence on the social and

---

[1] M. Lochhead, *The Scots Household in the Eighteenth Century* (Edinburgh, 1948), pp. 332–41, 'The Scots Doctor'. For England, see A. Clark, *Working Life of Women in the Seventeenth Century* (London, 1968), pp. 254–58.

[2] See for instance RCPSG, 1/11/33, a late seventeenth- or early eighteenth-century manuscript herbal in the archives of the College.

[3] The seminal work which first outlined the nature of the client and patron relationship is N. D. Jewson, 'Medical Knowledge and the Patronage System in Eighteenth-Century England', *Sociology*, 8 (1974), pp. 369–85. See also, idem, 'The Disappearance of the Sick-Man from Medical Cosmology, 1770–1870', ibid., 10 (1976), pp. 225–44. Much of Roy Porter's corpus has considered the nature of medicine and curing in terms of the marketplace of services. See particularly Roy Porter, *Health for Sale: Quackery in England, 1660–1850* (Manchester, 1989); W. F. Bynum and Roy Porter (eds), *Medical Fringe and Medical Orthodoxy, 1750–1850* (London, 1987).

[4] G. B. Risse, 'Medical Care', in W. F. Bynum and Roy Porter (eds), *Companion Encyclopedia of the History of Medicine*, 2 vols (London, 1993), p. 47.

cognitive structures of medicine.[5] The relationship between client and healer was on a more equal footing than now when practitioners have access to specialised knowledge and terminology not easily shared by laymen.[6] But access to healers was important if only as a palliative because before the twentieth century medicine could do little to alleviate major diseases.

The previous chapters have shown how medical practitioners in Glasgow used the institution of the Faculty of Physicians and Surgeons, and its civic arm, the Incorporation of Surgeons and Barbers, to establish standards in medical practice and to erect barriers against unlicensed medical practitioners. But at a time when medical knowledge and practice was an integral part of learned and popular culture, the Faculty found it extremely difficult to exercise comprehensive control over all healers.[7] However, it did try. While it did not make members of all healers practising medicine within its boundaries, it undoubtedly claimed authority over them all. That it was not uniformly successful shows either that the public had different expectations of the exercise of its authority, that the medical practice of licensed healers was not sufficiently superior to justify its claim to monopoly, or that the powers at its disposal were inadequate.

Licensing and Limitation of Practice

The range in skill and medical training represented among the early Faculty members was extensive. Surgeons, like Peter Lowe, with a university education and substantial surgical experience, were rare.[8] In Paris, where Lowe had trained, surgeons were divided into 'Surgeons of the Long Robe' and 'Surgeons of the Short Robe'. The former – the surgeons of St-Côme among whom Lowe was numbered – tended to perform operations which carried greater risk and required more skill – the excision of tumours, repair of rectal fistulas, plastic surgery on the face and suture of the intestines. The lower surgeons, or 'Surgeons of the Short Robe' – members of the guild of barber-surgeons – undertook general body maintenance in which operative

---

[5] M. E. Fissell, *Patients, Power and the Poor in Eighteenth-Century Bristol* (Cambridge, 1991), p. 51.

[6] A. Wear, 'Medicine in Early Modern Europe, 1500–1700', in L. I. Conrad, M. Neve, V. Nutton, R. Porter and A. Wear, *The Western Medical Tradition, 800 BC to AD 1800* (Cambridge, 1995), p. 238.

[7] R. French and A. Wear, *The Medical Revolution of the Seventeenth Century* (Cambridge, 1989), p. 9.

[8] For an account of Lowe's English contemporaries, the learned humanist surgeons, see V. Nutton, 'Humanist Surgery', in A. Wear, R. K. French, and I. M. Lonie (eds), *The Medical Renaissance in the Sixteenth Century* (Cambridge, 1985), pp. 75–99.

surgery had little place, healing a variety of wounds, for example, as well as operating as specialists in the treatment of hernia repair, cutting for the stone or removal of cataract.[9]

Anyone practising surgery within the Glasgow Faculty's extensive bounds could be summoned and examined. Precise limitations were placed on members' practice, according to the skills they demonstrated competently under examination. According to the charter, successful examinees were to be given 'testimonials according to their art and knowledge, that they shall be found worthy to exercise thereafter'; the procedures which they were authorised to perform were specified in their licences,[10] for instance, phlebotomy or bleeding – a universal medical procedure sanctioned by popular and learned belief, and central to the therapeutics of Galen whose writings formed the basis of the education of learned physicians and surgeons. Possibly the least qualified of any who underwent trial before the Visitor and brethren was one Alexander Lyes who, in September 1628, was found 'only able in practice to vent the blood with a horn'. It was possibly his son, John Lies, weaver, who, in June 1654, was 'licensed and tolerated to draw blood with a horn and such things as pertain thereto alone' – tolerated presumably because he was a weaver.[11] Blood-letting by application of leeches was a separately specified skill which, for example, James Mowat, servant of Dr Rattray, was licensed to exercise in 1656, as well as the application of cauteries and the practice of pharmacy.[12]

Phlebotomy and simple barber-surgery went hand-in-glove. Those who plied the trade of barbering were also admitted into the Faculty's fold, though strictly defined limits were set. The barber was viewed more as 'a purveyor of services, one step removed from a domestic servant' than as part of the medical establishment like the physician and the surgeon.[13] James Braidwood,

---

[9] T. Gelfand, *Professionalizing Modern Medicine: Paris Surgeons and Medical Science and Institutions in the Eighteenth Century* (London, 1980), pp. 21–24; L. Brockliss and C. Jones, *The Medical World of Early Modern France* (Oxford, 1997), pp. 219–25. For a brief general outline of barber-surgeons, see R. E. McGrew, *Encyclopedia of Medical History* (London and Basingstoke, 1985), pp. 30–31. For more extensive discussion, see the work of Margaret Pelling on English barber-surgeons: M. Pelling, 'Appearance and Reality: Barber-Surgeons, the Body and Disease', in A. L. Beier and R. Finlay (eds), *London, 1500–1700: The Making of the Metropolis* (London and New York, 1986), pp. 82–112; idem, 'Occupational Diversity: Barbersurgeons and the Trades of Norwich, 1550–160', *Bulletin of the History of Medicine*, 56 (1982), pp. 484–511.

[10] FPSG, *The Royal Charter and Laws of the Faculty of Physicians and Surgeons of Glasgow* (Glasgow, 1821), pp. 6–7.

[11] RCPSG, 1/1/1/1b, Transcript Minutes of the FPSG, 1599 to 1688, pp. 39, 89.

[12] RCPSG, 1/1/1/1b, p. 130. For more on Rattray, see below.

[13] D. A. Evenden, 'Gender Differences in the Licensing and Practice of Female and Male Surgeons in Early Modern England', *Medical History*, 42 (1998), p. 197.

who entered in March 1636, was limited 'to use only in time comi[n]g barberising, polling, shaving & making of beards without further privilege or points belongi[n]g to their calling'.[14] Robert Muir in the Gorbals and John Liddell were both licensed by the Faculty in February 1657 to exercise 'the art of barberising', while in December 1660, Duncan MacKindo, 'a soldier with the King several times as he alleges', was licensed as a barber until February 1661, by which time he was to have made himself a freeman. The date of the last two entries (which tellingly appear in the Faculty's minutes even after the erection of the Incorporation of Surgeons and Barbers) reveals the unique relationship between the Faculty and the Incorporation. But MacKindo did not bother to become a burgess; he was therefore fined £40 Scots and ordained 'to take in his basins'. The removal of hanging basins – sign of the right to practise – symbolised the withdrawal of privilege. James Wilson was similarly instructed in September 1664 to 'take in and not put out any hanging basins towards the street before his dwelling house or chambers', for non-payment of fines and quarter accounts. As with surgeons, barbers' skills were listed separately on their licences, beard-dressing and periwig-making appearing alongside specified procedures in simple barber-surgery.[15] Supported by a discharge of seven years' apprenticeship and a burgess ticket, William Bogle was admitted freeman barber and periwig-maker in April 1674. But Bogle's additional request to practise such parts of surgery as his master (Archibald Bogle) had been admitted to, could only be granted after trial (a question-and-answer session on simple anatomy, in addition to an examination of his surgical expertise during the course of his day-to-day practice).[16]

George Berrell was the first member on whom a restriction of practice was placed: admitted freeman in May 1605, he was granted liberty 'to profess the art of Barbery with simple wounds in the flesh', under the strict proviso that 'he meddle with [it] no further without special consent to the brethren & deacon for the time'.[17] This licence in barbery and simple surgery (or basic barber-surgery) was the surgical qualification most commonly granted by the Faculty. Precisely what this meant is illustrated in the licensing of Robert Harris, who entered freeman in May 1654. Permitted to practise as a 'simple

[14] RCPSG, 1/1/1/1b, p. 48. His son, James Braidwood, younger, was later admitted in March 1643 under exactly the same terms as his father. Ibid., pp. 59–60.

[15] The barbers bought heads of hair in order to produce the best wigs, but by the early eighteenth century, one general merchant in Glasgow traded in wig remains. Gordon Jackson, 'Glasgow in Transition, c.1660–c.1740', in T. M. Devine and G. Jackson (eds) *Glasgow*, i, *Beginnings to 1830* (Manchester, 1995), p. 78.

[16] RCPSG, 1/1/1/1b, pp. 129, 184–85, 228, 480–81.

[17] RCPSG, 1/1/1/1b, p. 19.

barber surgeon only to meddle with simple wounds', Harris was 'on no terms to meddle with physic, tumours, ulcers,[18] dislocations, fractures nor anything that is composite until he be further qualified'.[19]

The least skilled Faculty members were not only prohibited from undertaking more dangerous procedures, but encouraged to call upon the advice of the better qualified. Examined in October 1628, James Fleming was judged 'qualified to be a barber & to use simple wounds', but was obliged 'to use no further without the advise of the most expert brethren of the calling'. Similarly, the barber-surgeon John Paterson of Paisley in June 1654 was found qualified to cure simple wounds, fractures which did not penetrate the flesh, to phlebotomise and to apply cauteries. However, he was restricted 'in all time coming from using of physic and pharmacy', or administering internal remedies.[20] Fortunately for its members, the FPSG positively promoted the upgrading of qualifications, encouraging them to submit themselves for a further trial when they acquired greater skill. When John Niven in Dumbarton was examined in January 1659, and 'found not qualified as he professed by his paper',[21] in order to encourage him, the Faculty suspended the charge of horning against him, ordaining instead that he 'study further for the space of half a year and he to appear before them for a reexamination between [now] and Lammas next'. Similarly, practitioners who were already Faculty members could enhance their range of practice specified in their licences by submitting for reexamination. After public examination before the Faculty in December 1675 Robert Boyd, servant to the late Dr Hamilton, was licensed to open veins, apply cauteries and ventouses, to cure simple green wounds and to administer clysters on a physician's direction, on the condition that 'when he attains to more knowledge and desires to be further admitted they find it just and equitable that he be reexamined and further admitted according to the knowledge, literature and practise they shall find he shall have then acquired'.[22]

---

[18] Leg ulcers were far more common in earlier centuries, especially among the labouring classes, where they occurred more often in young adults than in the elderly as they do today. I. Loudon, *Medical Care and the General Practitioner, 1750–1850* (Oxford, 1986), p. 76. See also idem, 'Leg Ulcers in the Eighteenth and Early Nineteenth Centuries', *Journal of the Royal College of General Practitioners*, 31 (1981), pp. 263–73; 32 (1982), pp. 301–9.

[19] See Peter Lowe on simple and composed wounds in Chapter 2, 'Operative Surgery'.

[20] RCPSG, 1/1/1/1b, pp. 40, 79, 89, 193.

[21] When a candidate sought admission, he presented a 'paper' or petition setting down the skills he felt himself qualified to practise. In which capacity Niven petitioned for admission is not specified.

[22] RCPSG, 1/1/1/1b, pp. 172, 514–15. Having reentered the Faculty in 1672, Hamilton was Physician-Visitor in the year 1674–75 and died at some time in 1675. Duncan, *Memorials*, p. 242.

To extend the range of practice specified in a licence, the practitioner had first to present a petition to the Faculty. Gilbert Wilson in Strathaven successfully petitioned to upgrade his licence in November 1672, and was found qualified to let blood and apply cauteries 'beyond what he is formerly admitted to'. In addition, he was licensed to administer clysters 'in cases of necessity in absence of physicians'.[23] But qualifying at a higher level led to an increase in the freedom fine paid by a surgeon, though higher fees could probably be commanded to offset it. When, in July 1673, Allan Kirkwood in Darnley was upgraded to encompass the cure of fractures over and above his original licence to cure simple wounds and broken bones, his freedom fine was in consequence increased to £12.[24]

Treatment of fractures and bone-setting was a specialist skill often acquired through informal apprenticeship and, like lithotomy, lent itself to itinerant practice.[25] Treatment of simple breaks and fractures was on a par with the level of expertise necessary to treat simple wounds; practitioners were often licensed to do both. Thomas Harper in Kilmarnock was admitted in May 1668 to practise the cure of simple wounds and phlebotomy at the command of physicians, 'and to cure fractures where there is no loss of substance And to replace the dislocations alone',[26] while on examination in May 1654, John Mathies in Cockhill,[27] was 'found to have little skill of the said arts except only to cure simple wounds and something of broken bones where the flesh is not cut'. Such simple cures were the mainstay of everyday barber-surgery. A fascinating stipulation of Mathies' licence was the permission to undertake cures as long as 'there came not in any complaints against him'. That the Faculty relied not only on its own examination of practitioners but also on patients' testimony illustrates the power wielded by the seventeenth-century patient.

However, such practitioners were commonly licensed for only one or two operations; some surgeons built up special expertise in surgical procedures such as herniotomy and lithotomy, often taking their skills from one community to another. Serious herniae requiring surgical repair were often

[23] RCPSG, 1/1/1/1b, pp. 419–20.

[24] RCPSG, 1/1/1/1b, pp. 110, 455–56.

[25] R. Cooter, 'Bones of Contention?: Orthodox Medicine and the Mystery of the Bone-Setter's Craft', in W. F. Bynum and R. Porter (eds), *Medical Fringe and Medical Orthodoxy, 1750–1850* (London, 1987), p. 160. Although this refers to a slightly later period, it has relevance for this one.

[26] RCPSG, 1/1/1/1b, pp. 83, 282.

[27] RCPSG, 1/1/1/1b, p. 203. The location is uncertain, but is possibly Cockno, a hill in Old Kilpatrick, Dunbartonshire. F. H. Groome, *Ordnance Gazetteer of Scotland: A Survey of Scottish Topography*, 6 vols (London, 1882–1885), ii, p. 274.

6. View of patient's position for lithotomy procedure from John Greenfield, *A Compleat Treatise of the Stone and Gravel* (London, 1710)

treated by itinerant empirics since herniotomy was not commonly performed by the seventeenth-century surgeon (it was far more common to fit the ruptured with a truss). Robert Archibald demonstrated a range of surgical skill, including expertise in herniotomy, being licensed in September 1627 'in particular in the Incision of the stone, Cataract, hernia & all other external acts of surgery'. Cutting for the stone or, in Scotland, for the 'gravel', was an operation performed quite frequently during this period, from which it can only be deduced that the seventeenth-century diet and unpurified water resulted in a high number of bladder stones. A lithotomist usually learned his art by paying a fee to a surgeon already in the business. For this, he would be taught the trade secrets – those individual tricks which a lithotomist kept to himself. Nathan Grey, found qualified 'in particular in cutting of the gravel and stone with the sole cure of the gravel and stone' in May 1654, had clearly served such an apprenticeship, but was restricted 'from curing of humours, fractures, dislocations and all apposite wounds till his further knowledge and after his knowledge to be admitted and approven and allowed to exercise accordingly'.[28]

Like most operations of the time, lithotomy was performed quickly, within four to ten minutes in order to reduce the duration of the acutest pain in the absence of anaesthesia. It was not only essential that the surgeon was strong-stomached, but speedy and dextrous. In the seventeenth century, most lithotomists still used a variation on the operation specified by the ancient Roman writer Aulus Cornelius Celsus. In the Celsian method, the presence of a stone was ascertained by anal palpation. For the operation itself, the patient lay on his back, with his legs drawn up in the jack-knife position, and held down by strong men. The index and middle fingers of the left hand, inserted into the rectum, brought the stone down to the neck of the bladder, while the right hand placed on the lower abdomen was used to bring it down into the grip of the left index finger and prevent it slipping back. At this point, a half-moon incision was made to the left of the anus, its ends pointing towards the hip. A second incision, made at right-angles to the first, allowed for the separation of the soft parts under the skin and the opening of the neck of the bladder. A small stone could then be expelled by the fingers in the rectum, while an instrumental spoon or hook was used to withdraw larger ones. The wound was then allowed to bleed in order to avoid inflammation, and was usually bathed in water tempered with strong vinegar and salt to staunch the bleeding. It was not sutured. Since lithotomy

[28] RCPSG, 1/1/1/1b, pp. 37, 81; De Moulin, *History of Surgery*, p. 235.
[29] De Moulin, *History of Surgery*, pp. 11, 14, 244–45; Sloan, *English Medicine in the Seventeenth Century*, p. 103.

was a risky operation, with a mortality rate as high as 20 per cent according to some estimates, the Faculty would have been keen to regulate it.[29]

Such was the need for the operation that in 1655 Glasgow appointed a municipal stone-cutter, William Souttar, who was initially granted 'liberty to exercise his art of cutting of people of the gravel and preparing of the patients in reference thereto without the city of Glasgow' until such time as he satisfied the town for his burgess-ship. Just over a year later, in August of 1656, Faculty examiners convened for the trial of the Highland lithotomist Iver McNeill, 'who has been in use these ten years or thereby bygone in cutting of the stone'. He was licensed, 'upon sight of several creditable testificates', only for cutting the stone,[30] and became an active member, often appearing on sederunts. Acceptance of testimonials from patients cured, in lieu of examination, is a significant feature of McNeill's licensing.[31] Was there really so great a gulf between the signed assertions of cure waved by quacks and mountebanks, and those on which the Faculty relied?

Continuing in Peter Lowe's footsteps, a few Faculty surgeons like John Panton in Hamilton undertook surgical training in France. Panton appeared before the Faculty in December 1666 charged to refrain from practising physic

---

[30] RCPSG, 1/1/1/1b, pp. 101, 120. The seventeenth century saw the decline of the classically trained Gaelic physicians who had been much less affected by the advance of humanism in the sixteenth century than the Lowlands of Scotland. Gaelic physicians served traditional seven-year apprenticeships under members of hereditary medical familes such as the Beatons and the MacLachlans. The assault on Gaelic society in the late sixteenth and seventeenth centuries resulted in the demise of many of its traditional lines of learning, including the bardic schools and centres of medical education. Therefore, during the seventeenth century, Gaels increasingly considered apprenticeship to apothecaries in Lowland urban settings or a university education if they wished to practise medicine. This theme is expounded in J. Bannerman, *The Beatons: A Medical Kindred in the Classical Gaelic Tradition* (Edinburgh, 1986). Iver McNeill took a Highland apprentice from within his own kin group: on 6 May 1660, letters of indenture were passed between him and Lachlan McNeill in Airdrenish, on behalf of Evir McNeill, younger son to the late Duncan bane McNeill in Drumnamuckloch, for an apprenticeship of seven years. (RCPSG, 1/1/1/1b, pp. 214-15. His booking fee was £2 12s. 0d. Ibid., p. 220.) Indeed, the tradition of appointing Highland lithotomists for the town continued into the early eighteenth century: Duncan Campbell was employed as Glasgow's stonecutter between 1713 and 1714 at a salary of £66 13s. 4d. Scots. J. D. Marwick, *Extracts from the Records of the Burgh of Glasgow, AD 1691–1717*, Scottish Burgh Records Society, 4 (Glasgow, 1908), p. 646.

[31] This differs from practice in Amsterdam, for example, where a lithotomist had to perform the operation in front of three examiners, a doctor of medicine and the equivalent of two bailies, the officials having first satisfied themselves that a stone was present by probing the patient with a metal catheter – the latter to avoid unnecessary intervention. Indeed, by 1728, a fourth examiner had been added in Amsterdam, the professor of anatomy. De Moulin, *History of Surgery*, pp. 244–45.

and surgery within the Faculty's bounds, stating in his petition that he had been practising surgery for some fifteen years since returning from France without becoming a member. He professed 'to cure all sort of wounds apostemations, ulcers, fractures, dislocations, phlebotomies, application of cauteries and [was] in use to give physic to wounded persons when necessity requires' – his own testimony indicating that, like Lowe, he had some experience in military surgery. His claim to expertise was no idle boast: on examination, Panton was not only found qualified but, according to the answers, deemed 'expert' in surgery and duly licensed.[32]

Amputation, however, makes no appearance in the Faculty's licensing restrictions. It seems that the procedure was rarely, if ever, performed by Faculty surgeons at this time, although it had often been executed by Peter Lowe on the battlefield in France and it was certainly practised in Glasgow on a regular basis in the eighteenth century.[33] According to Ghislaine Lawrence, this was not unusual: 'There is ... evidence that only certain licensed surgical practitioners undertook major operations at all, with barber-surgeons unable or unwilling to perform, for example, amputations'.[34] The main problems with the procedure prior to the use of antiseptics were excessive haemorrhaging and the onset of gangrene.

What steps were taken when members failed to observe the restrictions of their licence? The first recorded case occurred in April 1669, when the Faculty met concerning a bill of complaint given in by Margaret Millar, wife of the late John Risk, quarrier in Glasgow, against William Clydesdale, former Visitor (1665–66) and Catherine Muir, his spouse. She claimed that about twenty days previously, John Risk had consulted Clydesdale and his wife about a pain in his chest. They visited him the following day and agreed a fee, half of which was paid in advance, for which Clydesdale promised to cure him of his pain. Clydesdale's wife gave Risk a potion of physic 'in two Cockle shells, a potion of Antimony', instructing their patient to use it 'for purging upward and downward'.[35] The potion was taken on the following day, a Sunday, according to the defenders' direction, but did not begin to work until the Monday at eight o'clock, when 'it wrought the defunct to death'.[36]

---

[32] RCPSG, 1/1/1/1b, pp. 261–62.

[33] MLRBM, 641982, Glasgow Town's Hospital: Minutes of the Director's Meetings, 1732–1764, p. 145.

[34] G. Lawrence, 'Surgery (Traditional)', in W. F. Bynum and R. Porter (eds), *Companion Encyclopedia of the History of Medicine*, 2 vols (London and New York, 1993), ii, p. 977.

[35] This is consistent with the use of antimony in high doses as emetics, and in low doses as cathartics or purges. See J. Worth Estes, *Dictionary of Protopharmacology: Therapeutic Practices, 1700–1850* (Canton, Massachusetts, 1990), pp. 13–14.

[36] RCPSG, 1/1/1/1b, pp. 311–12.

Risk's wife sent for Dr Hamilton,[37] who saw him at 'the height of ... his working', just a short time before he died, his wife helplessly declaring to Hamilton that her husband 'had gotten wrong thereby'. Widowed with two young children to support, she appealed to the Faculty not only to punish and fine Clydesdale and his wife for administering the antimony but to recommend her case to the civil magistrate. The defenders denied the charges. Clydesdale went to the length of stating 'that he never gave physic in his lifetime to the said pursuer's husband except about a quarter of a year since' when he had complained of a pain in his back, for which Clydesdale gave him some oils and pills. Had he actually made up these medicines himself, he would also have been guilty of practising as an apothecary,[38] but supplying was in itself sufficient to incriminate Clydesdale: the administering of pills was contrary to his act of admission. He had no licence to treat medical cases.[39] The Faculty therefore fined him £40 Scots.[40]

The Three Branches

University learning, based on the classical medical texts, was the sole distinguishing feature of physicians who, during the seventeenth century, were nearly always obliged to travel to the Continent for their medical education (other than the few medical students taught in Aberdeen and the medical teaching begun in Edinburgh when Archibald Pitcairne returned from Leiden in 1693).[41] Scots students of physic favoured Paris,[42] Montpellier and Padua earlier in the century, though Leiden,[43] Rheims and

---

[37] This is probably Dr Thomas Hamilton who, at this time, was operating outside the Faculty for which see Chapter 1, 'Surgeons and Physicians in Conflict, 1671–72'.

[38] For an account of the apothecary's art in nearby Edinburgh, see H. M. Dingwall, 'Making up the Medicine: Apothecaries in Sixteenth- and Seventeenth-Century Edinburgh', *Caduceus*, 10 (1994), pp. 121–330.

[39] RCPSG, 1/1/1/1b, pp. 47, 313. Clydesdale had been booked 'barber & a surgeon for curing of simple wounds with out fractures, ruptures, dislocations or like' in March 1636 which he was 'only to profess conform to his knowledge'.

[40] RCPSG, 1/1/1/1b, pp. 314–15.

[41] D. Hamilton, *The Healers: A History of Medicine in Scotland* (Edinburgh, 1987), p. 54.

[42] For an account of medical teaching there during the seventeenth century, see L. W. B. Brockliss, 'Medical Teaching at the University of Paris, 1600–1720', *Annals of Science*, 36 (1978), pp. 21–51.

[43] The theories of the renowned Herman Boerhaave with their emphasis on practical bedside medicine filtered via his Scottish students through to Glasgow as they did to Edinburgh. For Glasgow, see F. A. Macdonald, 'The Infirmary of the Glasgow Town's Hospital, 1733 to 1800: A Case for Voluntarism?', *Bulletin of the History of Medicine*, 73 (1999), p. 91 particularly; for Edinburgh, see G. B. Risse, 'Clinical Instruction in Hospitals: The Boerhaavian Tradition in Leyden, Edinburgh, Vienna and Pavia', *Clio Medica*, 21 (1987–88), pp. 5–9, W. R. O. Goslings,

Angers[44] became more popular from the mid-century. Medical students would commonly begin their education by gaining an MA at a Scottish university, and would proceed to both Leiden and Rheims to take a variety of medical courses and to pass their final examinations.[45] Rheims was favoured for graduation, because medical students did not have to write a dissertation, as in Leiden, but were given a diploma simply upon payment of a fee and an obligation that they would not practise in France.[46] However, acquiring such an education was an expensive enterprise which tended to ensure that students of physic were of a certain social status.

It was the Italian humanist physicians who paved the way for the organisation of physicians into colleges, insisting that only those who studied the ancient texts at first hand could truly diagnose, prognose and cure disease. Physicians began to group together to protect their qualifications, proclaiming that only those who shared their learning and knowledge should treat the sick. Those with an inferior training – surgeons, apothecaries, barbers, and irregulars – should only be allowed to practise under their supervision.[47] Although the evidence marshalled from Glasgow records certainly demonstrates the physicians' inherent sense of superiority, the surgeons also sought to establish themselves as learned practitioners, especially in their moves to distance themselves from the barbers. The incorporation of both practitioners together in one association in Glasgow can only have been assisted by the academic status of Peter Lowe as a surgeon trained at St-Côme.

The Faculty had the right to check that those who sought admission as physicians, or were practising medicine in its bounds, had a medical degree. In August 1657, Messrs James Hamilton, Archibald Graham and John Lowe were ordered, at their first convenience, 'to go to doctor Rattray and crave a sight of his letters of graduation and if he refuse that they may have a sight

'Leiden and Edinburgh: The Seed, the Soil and the Climate', in R. G. W. Anderson and A. D. C. Simpson (eds), *The Early Years of the Edinburgh Medical School* (Edinburgh, 1976), pp. 1–18; and D. Guthrie, 'The Influence of the Leyden School upon Scottish Medicine', *Medical History*, 3 (1959), pp. 108–22.

[44] Fourteen Scottish students studied medicine there in the seventeenth century. E. Pasquier, 'Literary and Historical Notes: Students from the British Isles at the Ancient Faculty of Medicine at Angers', *Notes and Queries*, 164 (1933), pp. 218–20.

[45] H. M. Dingwall, *Physicians, Surgeons and Apothecaries: Medical Practice in Seventeenth-Century Edinburgh* (East Linton, 1995), pp. 101–9. Scottish students at Leiden can be found in R. W. Innes-Smith, *English-Speaking Students of Medicine at the University of Leyden* (Edinburgh, 1932).

[46] H. de Ridder-Symoens, 'Mobility', in idem, *A History of the University in Europe*, ii, *Universities in Early Modern Europe, 1500–1800* (Cambridge, 1996), p. 437.

[47] A. Keller, 'The Physical Nature of Man: Science, Medicine, Mathematics', in J. MacQueen, *Humanism in Renaissance Scotland* (Edinburgh, 1990), pp. 99–100.

thereof to report'.⁴⁸ Sylvester Rattray was a Fifer who entered the Faculty at about this date as a freeman, though his entry is not minuted. No indication as to where he received his medical training survives, though his MD was from St Andrews. He began his practice in Edinburgh, where he took part in an early initiative to found a college of physicians before he moved to Glasgow.⁴⁹ He was responsible for the first medical book written in Glasgow, *Aditus novus ad occultas sympathiae et antipathiae causas inveriendas* (1658),⁵⁰ on the philosophy of sympathetic cures which was later reprinted in Tübingen (1660) and Nuremberg (1662).⁵¹ According to the philosophy of 'sympathetic' treatment, a healing salve or powder was best applied not to the wound but, for example, to the weapon which had caused it. An idea originally derived from the Swiss physician Paracelsus, it received widespread credence in the seventeenth century.⁵² The most renowned treatise on the subject, by Sir Kenelm Digby, who allegedly learned the secret of an effective weapon salve (thought to be copper sulphate) from a Carmelite monk, was published in the same year as Rattray's. And after its foundation in 1660, the Royal Society in London actively supported the search for a single powder to achieve sympathetic cure. Rattray's work dealt with the philosophical principles underlying the sympathetic system, tracing the operation of the occult forces of sympathy and antipathy throughout the natural world. The hidden benefit in this treatment was that after the wound itself had been simply cleaned with water and dressed healing was swifter than after the Galenic application of ointments and powders.⁵³

The Faculty tightened up on the licensing of those practising physic in 1673, the year in which three physicians returned to full membership of the FPSG. Andrew Hamilton in Kilbride was hauled before the Faculty in October

---

⁴⁸ RCPSG, 1/1/1/1b, p. 135.

⁴⁹ Hamilton, *The Healers*, p. 58; Duncan, *Memorials*, p. 237.

⁵⁰ GUL, Special Collections, BG56-l. 11, *Aditus novus ad occultas sympathiae et antipathiae causas inveriendas: per principia philosophiae naturalis, ex fermentorum artificiosa anatomia hausta patefactus* (Glasgow, 1658).

⁵¹ GUL Special Collections, MU53-d. 20, Joh. Andreas Endter (ed.), *Theatrum sympatheticum auctum, exhibens varios authores; de pulvere sympathetico quidem; Digbaeum, Straussium, Papinium, et Mohyum; de unguento vero armario; Goclenium, Robertum . . .; praemittitur his Sylvestri Rattray, aditus ad sympathiam et anti-pathiam* (Nuremberg, 1662). In this edition, Rattray's treatise forms part of a collection of treatises on the subject.

⁵² Aureolus Theophrastus Bombastus von Hohenheim adopted the nickname Paracelsus – 'surpassing Celsus' (the famous Roman medical writer). Wear, 'Medicine in Early Modern Europe', p. 311.

⁵³ Hamilton, *The Healers*, p. 58; A. W. Sloan, *English Medicine in the Seventeenth Century* (Durham, 1996), pp. 115, 142; Duncan, *Memorials*, pp. 197–203. See also J. Finlayson, *Dr Sylvester Rattray: Author of the Treatise on Sympathy and Antipathy* (Glasgow, 1900).

'for practising the art of medicine within their bounds not having letters of graduation or licensed by them'. On confession, he had to give bond promising not to meddle with the practice of medicine until he produced his letters of graduation under pain of £100 Scots. John Soutar in Hamilton was also charged with horning for practising medicine while unlicensed, and 'not being graduate'. He confessed himself in the wrong too, was also fined £100 Scots, and gave bond 'not to practise medicine without the direction of some physician within their bounds'. The fine was commuted to £12 Scots for the cost of taking letters out against him.[54]

Most practitioners who petitioned for admission to the Faculty in the seventeenth century sought licences to practise barber-surgery and pharmacy. In Scotland, the rise of the surgeon-apothecary was not merely an eighteenth-century phenomenon. In England, the Jenkins case had declared in 1602 that 'no surgeon, as a surgeon, might practise physic, no not for any disease, though it were the great pox'.[55] Practising apothecaries were also restricted by the Society of Apothecaries to selling physic prescribed by physicians though they were allowed to dispense medicines listed in the physicians' official *Pharmacopoeia*.[56] In Scotland, though there was a tacit understanding that physicians should prescribe physic to patients – an understanding which the Faculty tried to enforce with its specifications that surgeons be subject to the professional scrutiny of physicians – in reality, the Scottish surgeon's remit was far broader than in England.[57] Surgeon-apothecaries handled most non-critical cases of ill-health, and tended to know a fair amount about medicine. They also suggested therapies which went beyond the popping of pills and medicines and ranged, certainly by the early eighteenth century, into the practice of midwifery. They were concerned more with the practical application of medical knowledge than its theory, and tended to extend beyond their official bounds into the prescribing (rather than simply the dispensing) of medicines.[58] The physicians' concern to retain the administration of medicines firmly within their own

---

[54] RCPSG, 1/1/1/1b, pp. 472, 475–76.

[55] *Annales collegii medicorum*, ii, *1581–1608*, 155b–157a, quoted in Cook, *Decline of the Old Medical Regime*, p. 46.

[56] *A History of the Worshipful Society of Apothecaries of London*, i, *1617–1815* (London, 1963), abstracted and arranged from the manuscript notes of Cecil Ward by H. Charles Cameron, pp. 13–14, quoted in Cook, *Decline of the Old Medical Regime*, p. 47.

[57] For an account of general practice in nearby Edinburgh, see H. M. Dingwall, ' "General Practice" in Seventeenth-Century Edinburgh: Evidence from the Burgh Court', *Society for the Social History of Medicine*, 6 (1993), pp. 125–42.

[58] R. M. Stott, 'The Incorporation of Surgeons and Medical Education and Practice in Edinburgh 1696–1755' (unpublished Ph.D thesis, University of Edinburgh, 1984), pp. xxxiv, xli.

hands was a major area of contention in the seventeenth century, and it was here that surgeon-apothecaries and apothecaries who were prescribing medicines came into conflict with physicians who insisted that they dispense only on a physician's prescription. But, in Scotland, as in England, their fight to maintain a monopoly in the prescription of internal medicines was a lost cause.

With surgeon-apothecaries widely consulted as GPs, physicians were particularly concerned to regulate their practice. Petitioning for entry as freeman in the practice of pharmacy 'and such parts of surgery as they shall find him qualified for', John Whyt in Paisley was licensed in December 1675 'to the making up of receipts according to the physician's order and to sell drugs and compositions and further to open veins by lancet or lock leeches and applying of cauteries in cases of necessity'. He was discharged 'from mixing or making up of compositions without the advise and assistance of a doctor on the place or to meddle beyond this his present admission until he [be] further tried and found qualified by them'.[59] Similarly, Quentin Muir in Paisley was apprehended by letters of caption for 'using and exercising the art of surgery & others within the compass of their gift', and particularly for encroaching upon the rights of physicians in prescribing medicines. Confessing his guilt in September 1655, he declared his resolution to abstain from 'using or prescribing to any patient whatsomever any medicaments' under pain of £40 Scots.[60]

However, in Glasgow, most of the evidence for encroachment upon physicians' rights in prescribing medicines comes from the ranks of the apothecaries. John Miller of Kilmarnock, examined in September 1654 on his skill and knowledge in medical arts, was licensed 'to sell drugs, let b[l]ood, give a clyster, apply a cautery and to compound medicaments, to use pharmacies and give simple potions'. But these were to be given only 'by advice of professors of physic'. In other words, Miller was limited to dispensing (and not prescribing) medicines. This meant that an apothecary could only administer medicines or treat a patient internally under the supervision of a physician. But a few surviving papers relating to the practice of his son, Mathew Miller of Glenlee, apothecary in Kilmarnock, show that he did prescribe and consult as well as dispense medicines. In all probability, so had his father.[61]

[59] RCPSG, 1/1/1/1b, pp. 517–18.
[60] RCPSG, 1/1/1/1b, p. 107.
[61] RCPSG, 1/1/1/1b, p. 90. Mathew Miller entered the Faculty in March 1668, being licensed 'To exercise the art of Pharmacy, the opening of veins by the advice of physicians, the application of cauteries and ventouses, the curing of simple wounds, and the embalming of corpses'. Ibid., p. 280.

In a letter from Irvine, dated 8 October 1682, Mathew Miller was asked to supply medicines, as prescribed by Mr James Irvine, for George Garvan's wife who was 'sadly diseased with stitches all above the short ribs both in sides and back, with a constipation whereby she is swollen ... about the stomach and heart'.[62] Clysters had first been applied according to Dr Johnstoun's prescription, illustrating the contemporary predisposition to consult more than one practitioner until a cure was wrought.[63] It is also an example of the then widespread method of prescribing by letter which obviated the additional expense of paying the practitioner's travel costs.[64] While there is ample evidence for his performing the traditional dispensing role, it is apparent that Miller also consulted. His cousin John Miller, wrote from Skeinstown in May 1668, not only asking the apothecary to come and give his opinion on his two-year-old son, but indulging himself in the tendency, common then, towards lay diagnosis. John Miller intreated his cousin to

> come over to Skeinstown & look upon my son who is in a great sad sickness. It's not an ordinary fever but a universal heat accompanied with much sweating & a deadness of his spirits which is conceived to proceed from his more noble parts withall he is subject to a very great drought wherefore you may bring that cooling syrups both for comforting & strengthening his body upon judge of it.[65]

In the same month, he asked Miller to send 'the physic for purging ... with all several others things that you think needful, with the decoctions what way they should be applied' for his older, nine-year-old son, wishing to know how soon after purging the boy should be bled.[66] The apothecary was not just prescribing but advising on treatment.

If definitive proof were necessary, this comes in a letter of 1 September 1667 from John Campbell in nearby Loudon, who informed Miller that his mother was still in her 'casting', and more or less in the same state as when the apothecary left her. According to Campbell 'they desire her to get half an ounce of cordisidroune[67] and a quarter of an ounce of

---

[62] EUL, Special Collections, La. II. 126, no. 4. Irvine was not a Faculty member.

[63] In all probability, this is John Johnstoun, the father of the son of the same name who was second Professor of Medicine at the University of Glasgow (1714–51). Duncan, *Memorials*, p. 252, states that the first Dr Johnstoun's place of practice has never been ascertained. This letter shows that if Garvan had not travelled to Glasgow to consult the doctor, Johnstoun may have practised in Ayrshire.

[64] For a discussion of consultation and prescription by letter in the Edinburgh context, see Dingwall, *Physicians, Surgeons and Apothecaries*, pp. 164–80.

[65] EUL, Special Collections, La. II. 126, no. 6.

[66] EUL, Special Collections, La. II. 126, no. 12. The letter is dated at Skeinstown, 7 May 1668.

[67] Cordisidroune was lemon peel. While it sounds like a primaeval steroid, 'cordisidroune' was etymologically *écorce de citron*.

cinnamon [68] and the rest of the things you did prescribe your self to her and send them with the bearer and any other thing you think fit for her' – definitive proof, if it were needed, that the apothecary was prescribing.[69] Seeking 'some thing that may be effectual against wind & vapours that proceed from my stomach', John Hamilton later wrote to Miller in August 1682 that 'our minister advises me to get from you the powder of peony, & some water he spoke of which you sent him, to be taken with sugar candy in a spoonful of the said water every morning. So I desyre you may send this or any thing else you Judge fittest for the like case'.[70] Powdered root of peony was taken to alleviate pains in the belly (especially by women as a purge after childbirth).[71] Miller's prescription apparently did the trick. Having finished everything sent him except the powder, John Hamilton perceived himself better from them.[72]

Surgeons and physicians were concerned above all that apothecaries knew their place. Although the 1656 letter of deaconry had incorporated the apothecaries (as well as the surgeons and barbers), the FPSG deemed it necessary by August 1667 for those admitted as apothecaries to consent 'to be exempt from ever bearing place or office within the said faculty as visitor, quartermaster or boxmaster'. They were, however, obliged to give their best counsel and advice to the Visitor on pharmaceutical matters when asked.[73]

John Lennox, merchant in Greenock, appeared before the Faculty in September 1673, seeking admission to the practice of pharmacy. The Faculty undertook a trial of his qualifications and licensed him 'to make and sell common oils', various plasters [74] and ointments,[75] *pulvis sanctus*,[76] and such like compositions conform to the dispensations'. But he was discharged from practising medicine and surgery except under the instruction of a graduate

[68] Cinnamon was often put in water as a drink.
[69] EUL, Special Collections, La. II. 126, no. 14.
[70] EUL, Special Collections, La. II. 126, no. 16.
[71] Nicholas Culpepper, *The British Herbal and Family Physician to Which is Added a Dispensatory for the Use of Private Families* (Halifax, n.d.), p. 255.
[72] EUL, Special Collections, La. II. 126, no. 18.
[73] RCPSG, 1/1/1/1b, pp. 268–69.
[74] Namely, *emplastrum de minio* and *emplastrum de emelelotum simplex*.
[75] Namely, *unguentum de Egiptiacum, unguentum de althew* and *unguentum album*.
[76] It is not clear precisely what *pulvis sanctus* (holy powder) is. It appears in none of the early modern pharmacopoeias in this form. We are grateful to Walter Sneader for suggesting that it may be a rare form of either *pulvis Jesuiticus* (Jesuit's powder) or *pulvis cardinalis* (cardinal's powder), that is powdered cinchona bark. For which see W. Sneader, *Drug Prototypes and their Exploitation* (Chichester, 1996), pp. 108–12. See also his earlier *Drug Discovery: The Evolution of Modern Medicines* (Chichester, 1985).

physician or surgeon under penalty of £40 Scots.[77] Even more stunning, however, was the proviso made at the trial of Hew Hunter on 13 May 1670. After petitioning for admission as freeman, Hunter was found qualified to practise the art of pharmacy under stipulation that 'if it shall be found hereafter that the said Hew shall be able to practise the art of surgery at any time hereafter he is to be admitted thereto providing he have the approbation of a physician of his ability to practise the same'. Such was the physicians' desire to control general medical practice that it extended (at a time when the relations between the surgeon and physician members of the Faculty were more than strained) even to their deciding Hunter's future suitability to practise surgery.[78]

Two years later, in March 1672, James Weir, servitor to Dr John Colquhoun, 'professor of medicine',[79] petitioned for admission on trial to exercise surgery 'having learned the same (as he declared) from his father yet living'. Here a physician was probably employing a junior surgeon to carry out minor aspects of external practice. But there was clearly some doubt as to whether Weir learned the trade from his father; the inference is that he learned at the elbow of his master who, as a physician, was not technically qualified to teach him. Nevertheless, upon examination in points of surgery, Weir was found qualified to practise and subsequently admitted.[80]

The following house account from John Melvill, surgeon in Easter Kilpatrick, Lanark,[81] to the Laird of Barns, hints at the wide-ranging nature of the surgeon-apothecary's practice during his treatment of the laird's son's sickness in the period between 1700 and 1709.

While he designated himself 'surgeon', and undertook embalming (the normal province of the surgeon), most of the account is a list of doses of physic-potions, pills, and powders (the province of the physician), plasters and electuaries (the province of the apothecary).[82] The surgeon was

[77] RCPSG, 1/1/1/1b, pp. 462–63.

[78] RCPSG, 1/1/1/1b, pp. 339–40, 462–63. See Chapter 1, 'Surgeons and Physicians in Conflict 1671–72'.

[79] John Colquhoun was one of three physicians who petitioned for admission to the Faculty in 1671 under the right (as specified in the charter) of being in possession of a university degree. He was appointed Rector's Assessor in the University in that year which, in conjunction with his medical degree, may be the origin of this title. Duncan, *Memorials*, p. 242.

[80] RCPSG, 1/1/1/1b, pp. 392–93. Evenden cites seven cases in seventeenth-century London of candidates applying for a surgical licence, having learned their surgery from physicians. Evenden, 'Gender Differences', p. 204.

[81] Melvill did not feel the need to enter with the Faculty until 1718. He died shortly afterwards. Duncan, *Memorials*, p. 247.

[82] GCA, TD589/20/3, Family correspondence and papers of the Hamiltons of Barns. The account was discharged at Cocknock, 5 October 1709.

		£	s.	d.
1700	To your son a plaster to his knee		14	0
	Two purging potions		8	0
	Two doses of purging pills	1	8	0
5/4/1703	One dose of the same		14	0
	A vomiter		12	0
	The same		12	0
25/3/1705	A pot pectoral electuary	2	0	0
	Two doses former physic	2	8	0
to	Another plaster to his knee	0	14	0
1707	A vomiter	0	12	0
1/5/1708	An electuary pot	2	0	0
1709	Six doses pacific pills	1	9	0
	A vomiter to Mrs Mary	0	12	0
	Materials for three clysters to your son	3	0	0
10/5/1709	A large pectoral lohoch pot [83]	2	0	0
9/6/1709	A mounted Ardgullion	0	12	0
	Two ounces antihectic powders	0	12	0
	One pound and nine ounces pectoral mixture	2	18	0
15/6/1709	Materials for ale	2	0	0
	Box pectoral pills	4	0	0
17/6/1709	A cordial julep	1	4	0
	The same	1	4	0
22/6/1709	The same	1	4	0
	The embalming your son's body to a large Cerecloth	60	0	0
	Oils and powders strong and sweet	12	0	0
		110	16	0

[83] For a pectoral lohoch pot, see Culpepper, *The British Herbal and Family Physician*, p. 292.

manifestly administering internal medicines and not simply applying topical remedies.[84]

Standard entrance for an apothecary in the late seventeenth century was by an essay or trial in compounding medicines. Having petitioned for admission to the practise of pharmacy, to open veins and to apply potential cauteries, Thomas Smith of Glasgow was ordered in April 1673 to make *diacatholicon* or the universal purge, stomach pills, the magisterial stomach plaster, troches of gum arabic, and ointment of marshmallows within the next five weeks. On 4 June, his compositions were deemed 'rightly dispensed', and Smith was found qualified to exercise pharmacy. However, his examiners – David Sharp and Archibald Graham – refused to admit him to exercise phlebotomy or to apply ventouses and potential cauteries 'until such time as it be seen by two or more of the surgeons that he can practise certain parts thereof'.[85] It was surely a serious anomaly of the Faculty's licensing system that lithotomists could be licensed on mere testimonial while licence to phlebotomise – a commonplace of medical and surgical practice – was far more strictly enforced.

At the very bottom of the occupational pyramid of practitioners were members licensed only to sell drugs and make ointments. When Gilbert Neilson in Strathaven applied for membership on a number of counts, he was admitted, in May 1669, to only a few. Licensed to sell drugs in his shop, he was not to compound any without the advice of an apothecary or assistant (though not specifically a physician). And while he could cure simple wounds by the application of green salve,[86] he was forbidden to phlebotomise, apply leeches or the 'use of issues in legs & arms by incision' without advice from those who were expert in these arts. Neither was he to give 'vomiters of graciola' – hedge hyssop, a violent purger of phlegm and bile used in cases of dropsy, gout and sciatica, or externally applied on ulcers – directly or indirectly without direction from a physician.[87] John Weir in Cambusnethan, licensed in January 1659 only 'to make green salve', was

---

[84] In spite of the distinction between physician and surgeon, surgeons were also administering internal remedies. Pharmacy was integrated into the teaching of surgery, and surgeons wrote about internal remedies in surgical textbooks, as Peter Lowe's *The Poore Man's Guide* would have shown, had he lived to have it published for which see above, pp. 44ff.

[85] RCPSG, 1/1/1/1b, pp. 447, 451–52. If found duly qualified, Smith was to be admitted to practise further parts of surgery without any addition on his entry fee.

[86] Green salve or 'green treat' is often mentioned in the records, and was composed of herbs which coloured the salve a green colour.

[87] RCPSG, 1/1/1/1b, pp. 318–19; J. Archer, *Every Man his own Doctor* (2nd edn, London, 1673), p. 75.

declared 'incapable to use any part of surgery or prescribing of physic & specially to meddle with quick silver'.[88]

Powers of Visitation

The fifth clause of the Faculty's charter gave the Visitor powers to inspect all drugs sold in Glasgow.[89] This may have included, as it did in London, the right to inspect prescription receipts so that the Faculty could monitor irregular practitioners.[90] Physicians in London also had the power to examine apprentice-apothecaries who sought admission as freemen to the Society of Apothecaries.[91] Actively exercising its right of inspection, on 6 August 1657 the Glasgow Faculty ordained that a visitation would be held on the 17 August following 'of all the pharmacists' shops of these within this burgh that are freemen and members of the faculty'. Four members were appointed to conduct it – the Visitor (James Thomson), a surgeon, Dr Crichton, physician, Archibald Graham, pharmacist, and Thomas Lockhart, surgeon-apothecary.[92]

In view of the number of irregulars peddling drugs within its bounds, and to protect the livelihood of its apothecary members, the Faculty also tried to maintain a monopoly on drug provision. It seems to date from 1659 when the restriction appeared several times in licences. In January 1659, Thomas Younger, resident in Ardgowan was ordered to desist from the practice of surgery and pharmacy, and from 'prescribing or giving of physic to any patients within any of the shires of the west parts of this nation'. Although he was found qualified to exercise surgery outside Glasgow and its suburbs, significantly, the Faculty 'did order him in the future to buy his drugs from such of the faculty as he should find to have the same as occasion offered'. When John Forster in Auchinclerk was licensed to exercise surgery, in January 1659, he was similarly instructed to 'buy his drugs from freemen of the faculty as he has to do therewith & occasion offers'.[93] And finally, when in September John Gardiner was found to be making and applying plasters without licence (and thus taking trade from apothecaries), he willing consented that if he did so again 'his shop should be closed up and that any

---

[88] RCPSG, 1/1/1/1b, p. 171.
[89] FPSG, *The Royal Charter and Laws*, p. 9. Technically, any which had not been subject to such vetting could be confiscated.
[90] A similar point is made in Cook, *Decline of the Old Medical Regime*, p. 87.
[91] Underwood, *Worshipful Society of Apothecaries*, p. 47.
[92] RCPSG, 1/1/1/1b, p. 134; Duncan, *Memorials*, pp. 235, 237–38.
[93] RCPSG, 1/1/1/1b, pp. 168–70.

drugs therein not found sufficient by such sworn men as they shall appoint should be confiscated'.[94]

In 1661 the Faculty also ordered eleven men not to sell rat poison or arsenic,[95] because they were contravening the sixth clause of the charter which specified that 'none shall sell Rat's poison, as Arsenic or Sublimate, under the pain of one hundred Merks; except only the Apothecaries'.[96]

Licensing Outwith the Burgh of Glasgow

A perusal of the licentiates and of the bonds of desistance from practice signed by irregulars and failed examinees shows that the Faculty was actively examining by the late seventeenth century beyond the boundaries of Glasgow.[97] In February 1655, the year before the erection of the Incorporation of Surgeons and Barbers, the Visitor charged 'all the people that professes physic or surgery within the town of Ayr' with letters of horning.[98] Unlicensed practitioners there petitioned for admission. In the following month, James Tobias, surgeon in Ayr, son of John Tobias, also surgeon there, not only sought admission for himself, but asked more quixotically that 'his said old aged father who was not able to travel might be licensed to exercise his said art of surgery during his lifetime they both being such as are both to be refractory to good order of the said faculty'. This must have been contrary to the accurate conduct of examinations. Nevertheless, the son was examined and licensed to exercise his art within the town of Ayr and other places outside Glasgow and its suburbs, the Faculty also declaring its intention:

> to extend to his said father during his lifetime upon a testificate to be reported to the present Visitor under the present or last magistrates of Ayr & these there of their brethren burgesses of the said burgh who are authorized to exercise the said art their hands of his qualification age & inability to travel & which being done the said John Tobias is to give an act extended to him for the effect aforesaid.[99]

---

[94] RCPSG, 1/1/1/1b, p. 249.

[95] RCPSG, 1/1/1/1b, p. 206. The eleven were, Thomas Lockhart, David Scharp, James Mowat, John Lamb, elder and younger, Andrew Gibson, Hendrie Craig, Adam Ritchie, George Boyl, Alexander Mackinnis and George Lamb.

[96] FPSG, *The Royal Charter and Laws*, p. 9.

[97] See Appendix 3, below, pp. 423–25, for the bonds of desistance (from practice) given to the FPSG between 1657 and 1701.

[98] RCPSG, 1/1/1/1b, p. 99.

[99] RCPSG 1/1/1/1b, pp. 103–04.

This is yet another example of the Faculty licensing on the basis of third-party testimonial. It clearly felt that the expertise of the father was demonstrated well enough in the way he had taught his son.

More significant was the meeting in May 1659 convened to discuss Gilbert Kennedy, surgeon in Maybole, who had petitioned for a licence to practise surgery on the basis of his apprenticeship to David Kennedy, surgeon in Edinburgh. He was duly licensed provided that within twelve months 'he exhibit and produced before them a sight of his Indenture and a discharge thereof otherwise these presents to be Null'. The Faculty's willingness to license merely on the basis of his having served an apprenticeship effectively put him on a par with physicians who simply had to present their degree scroll to secure admission; and that he was allowed a year to produce his documentation shows it reasonably satisfied with his capabilities. But the Faculty's decision on this occasion was contradicted in a similar case only fourteen years later. When, in March 1673, it considered the case of James Robison – a soldier recommended to them for licensing by the incorporation of Surgeons of Edinburgh – it decided, after due deliberation, to reject the Surgeons' recommendation. Consequently, Robison was banned from practising the art within their bounds until his qualifications had been examined.[100] The reasons for this apparently contradictory action are probably two-fold. Kennedy had evidently served a guild apprenticeship in Edinburgh, but Robison may not have done. The Faculty's insistence on examining Robison personally may also have been directly related to the tightening of their regulatory process after the parliamentary ratification of 1672. It may also have indicated a wish not to subordinate themselves to the Edinburgh Surgeons.

Obviously the Faculty's main concern was to ensure that its licensees were suitably qualified to practise. But many healers, particularly outside Glasgow, hoped to continue practising without the financial burden of joining the Faculty. The fine for practising while unlicensed rose in the early 1670s from £40 to £100 for each transgression.[101] Officially extending the terms of its legal gift to include apothecaries, the ratification of 1672 also led to a spate of hornings against those, such as John Tod, apothecary in Kilmarnock, practising pharmacy whilst unentered. After examination he was found sufficiently qualified to practise pharmacy and admitted under the express provision that he should not practise in the town until a burgess and guild brother, under pain of £100.[102]

---

[100] RCPSG, 1/1/1/1b, pp. 176–77, 444–45.
[101] It first appears in 1671. See Robert Naper below.
[102] RCPSG, 1/1/1/1b, p. 397.

However, for many practitioners of the healing arts it was simply of no consequence whether they entered the Faculty or not. Robert Houston was fined £20 in November 1668 'for learning of others to make drugs, syrups & others this half year bygone'. Even healers who had been apprenticed to a Faculty member tellingly failed to become members on qualification. William Sempill of Dalmook, for instance, who had served his apprenticeship with James Frank, then Visitor, at least a decade earlier, was summoned to enter with the Faculty in February 1672. 'Found qualified after examination upon some points of Surgery', he was admitted under the proviso that he should not practise in Glasgow until admitted burgess or guild brother, under penalty of £100 Scots.[103]

The admission from the mid seventeenth century of a small number of physicians and surgeons from the larger towns within the Faculty's bounds, such as Ayr, Kilmarnock and Paisley, enabled it to regulate medicine more effectively throughout the full extent of its jurisdiction. In the year the physicians returned to full membership of the Faculty, powers of inspection were delegated to physicians outside Glasgow. Dr Michael Wallace in Ayr accepted 'the power of visitation for the shire of Ayr including Kyle, Carrick and Cunningham' in March 1673, being held accountable for his actions and for any financial dealings, particularly the money he collected from those licensed. Wallace soon got to work. Charges were sent west to him in April 1673, with instructions approved and subscribed by four representatives of the Faculty. But it was a burdensome task and by July of the same year, the Faculty had appointed him an assessor or helper – Dr Bell in Kilmarnock.[104]

Irregulars and Unqualified Practice

The west of Scotland, like the rest of Europe at this time, had a flourishing irregular medical culture.[105] However, the extent of irregular practice in the west can only be gauged in relation to what was then considered regular practice which, in corporate terms, was always more a question of underwriting privileged status. Regular practitioners claimed to be able to deliver a better service to the patient and so justify their licence.[106] In reality, the

---

[103] RCPSG, 1/1/1/1b, pp. 289, 391–92. Frank was Visitor between 1660 and 1662, and between 1663 and 1664.
[104] RCPSG, 1/1/1/1b, pp. 443–44, 449–50, 453–54, 456. The four Faculty subscribers were the Visitor, Archibald Bogle, Bailie Hall and David Sharp.
[105] H. Marland and M. Pelling (eds), *The Task of Healing: Medicine, Religion and Gender in England and the Netherlands, 1450–1800* (Rotterdam, 1996), p. 21.

therapies plied by unlicensed practitioners were often just as effective as those of the regulars at this time.[107]

Accounts of the prosecution of quacks and unqualified practitioners, which are recorded only from 1655, show that irregular healers came from an immense range of occupations from shoemakers and gardeners to merchants. John Gairner, a Glasgow merchant, summoned in November 1672 for practising surgery and pharmacy without licence, had flagrantly set up shop in close proximity to the surgeons' corporation which can only be an indicator of how little regulatory control the Faculty actually had. But he simply got off with a payment of £6 Scots to the messenger. Some twenty years earlier, however, in June 1654, another merchant, James Scot, had managed to secure a licence 'to cure simple wounds and to make and compose pills according to his skill, practice and experience'.[108]

Although it tried to maintain its monopoly in the licensing of regular practitioners, the Faculty clearly licensed some empirics, the criteria for whose entry into the fold were never precisely delineated. In general, the Faculty merely tried to sideline the whole issue of irregular practitioners. For instance, when John Reid in Beiryairdes was charged in March 1667 with practising surgery and applying treatment to different kinds of wounds, cuts and sores, 'being an old man and the said calling not desirous to trouble him provided he does not meddle with their calling in the future', the Faculty declared itself satisfied with his bond not to practise unless authorised.[109]

However, if, upon trial, candidates failed to qualify, they were fined and expected to sign bonds of desistance – pledges to desist from practising medicine, surgery and pharmacy within the Faculty's bounds. John McClae, schoolmaster, appeared before the Faculty on 25 May 1655, 'declaring being spoken to that he would not meddle hereafter with any point of surgery or prescribe any physic'. If he defaulted, the Faculty expressed its intention to proceed against him with full vigour of the law. James Dougall, gardener, also promised in January 1657 to 'abstain in all time hereafter from using & exercising any part of surgery or prescribing of any medicaments or physic to any person whatsoever either within this city or without whereby he may fall under the censure of the surgeons of the said burgh'. But, the Faculty was not beyond licensing gardeners within strict limitations if they had knowledge of plants and herbs. William Currie in the parish of Douglas was licensed to use herbal remedies in September 1667. He presented a certificate

---

[106] Digby, *Making a Medical Living*, p. 25.
[107] Pelling, 'Appearance and Reality', pp. 250–79.
[108] RCPSG, 1/1/1/1b, pp. 88, 420, 545.
[109] RCPSG, 1/1/1/1b, pp. 263–64.

from his minister 'craving licence to practise the use of herbs for external application only', and was admitted only to apply herbs externally, but not to practise medicine, surgery or pharmacy. Alexander Wilson, gardener in Glasgow, transgressed a bond not to practise surgery in September 1676. Others were tolerated under supervision. In October 1666, John Logan, a Gorbals hammerman, was licensed to make use of green salve and to apply a cautery, 'provided he do it by advice of one approven by the faculty for that end or licensed by them'. He was to pay dues 'so long as they shall tolerate him to do anything relating to their calling'. In February 1659, William Fleming, gardener in Hamilton, was licensed to make green salve and to distill rose water only, while another gardener, Andrew Ralstoun, was found qualified in October 1671 only 'for making green treat of herbs and other ingredients thereof and to cure green wounds where there is no loss of substance and to cure simple fractures and dislocations'.[110] The immense range of practitioner licensed by the Glasgow Faculty which extended from gardeners at the bottom to physicians at the top was perhaps only equalled in renaissance, and early modern, Italy.[111]

In general, the college dealt more (as did most colleges) with illicit practice than malpractice, though a few cases were concerned with gross incompetence. The above-mentioned Alexander Wilson was indicted three years earlier, in April 1673, 'for giving physic & drugs not licensed & whereby doing much harm'. Robert Marshall, maltman, complained that Wilson had administered several things to the late spouse of Robert Sheap 'whereby her death was occasioned'. The patient was pregnant, though we are not told what the gardener thought he was treating her for – a possible sign of socially embarrassing symptoms. Wilson openly confessed that he gave her 'plantain water, roots of comfrey[112] and of red roses with some tormenting roots and also that he gave her a handful of camomile to bathe her feet in and that he desired her to set her feet to the fire and to suffer as much heat as she could but simply denied that ever he knew her with child'. Plantain was commonly resorted to in 'diseases of the ... privities', its juice stopping 'all manner of fluxes, even women's courses, when they flow too abundantly'. It was often used for old, established ulcers that were hard to cure, as well as 'for cankers and sores in the mouth or privy parts of man or woman'. According to Nicholas Culpeper, an unlicensed apothecary and influential author of

[110] RCPSG, 1/1/1/1b, pp. 105–6, 127–28, 172, 260–61, 270, 357–58, 370–71, 545–46.
[111] See K. Park, *Doctors and Medicine in Early Renaissance Florence* (Princeton, 1985), and D. Gentilcore, ' "Charlatans, Mountebanks and Other Similar People": The Regulation and Role of Itinerant Practitioners in Early Modern Italy', *Social History*, 20 (1995), pp. 297–314.
[112] Roots of comphrey were put on wounds and abscesses as an emollient. Estes, *Dictionary of Protopharmacology*, p. 53.

medicinal herbals, both plantain and comfrey stopped rather than provoked menstruation. A decoction of the root of comfrey boiled in water or wine helped 'the fluxes of blood or humours by the belly, women's immoderate courses', while, by encouraging sweating, the juice of the tormentil root, expelled poison and contagious diseases such as the pox or measles; it was deemed 'no less effectual to cure the French pox than Guiacum or China'.[113] Nor was this the only complaint against Wilson. He had also given medicines for the gravel to John Crosbie in Gallowgate which allegedly caused his death too. Wilson was found guilty of illegal practice, fined £40 Scots and ordered to subscribe a bond never to meddle with any part of medicine, surgery or pharmacy in future.[114]

Lairds, or more often their wives, often attended to the ills of the poorer members of their community. In December 1671, the Visitor summoned Mr Robert Naper, brother to the laird of Killmakell, asking whether he would enter the Faculty, but he refused, answering that he did not intend to enter as a tradesman although he had acquired his burgess-ship. The inference that it was beneath his social status to enter as tradesman is quite apparent. He preferred, instead, to give bond to the Visitor not to make up any compositions or physician's receipts, nor to practise any part of surgery or pharmacy under penalty of £100 Scots; he pledged that he had not done either since beginning his practice. When Naper failed to appear at the Faculty's next meeting on Christmas day, the Visitor and Bailie Hall were authorised to apply to the magistrates if he refused to comply.[115]

Although the phenomenon was more common in the following century, the Faculty also took action against army surgeons returning to Glasgow who wished to practise on the strength of their qualification before military examiners. The Faculty ordered a charge of horning in January 1676 against one Hallthorne, 'a soldier in the late disbanded companies', for practising surgery while unlicensed.[116] But the soft touch was employed where possible to gently persuade a person as in October 1665, John Hall and James Frank were appointed merely to speak to John Robison regarding his irregular practice of surgery. He was warned either to desist from so doing or to enter as a member with them.[117]

---

[113] Culpepper, *The British Herbal and Family Physician*, pp. 100, 262, 341.
[114] RCPSG 1/1/1/1b, pp. 448–49.
[115] RCPSG, 1/1/1/1b, pp. 384–88.
[116] RCPSG, 1/1/1/1b, p. 524.
[117] RCPSG, 1/1/1/1b, p. 249.

Prosecution

When those cited before the Faculty refused to comply with its directives, some were prosecuted, a procedure not undertaken lightly. A civil process in Scotland began with a summons executed by the pursuer's messenger on the defender. The case then proceeded with written replies and answers, the court often making intermediate pronouncements called interlocutors prior to the final judgement. Lawyers had a vested interest in being long-winded to push up their fees, so pleadings were often inflated in the final stages by tedious repetition of earlier material. Processes were consequently unwieldy and drawn out, resulting in a final decree which often stretched to a hundred pages or more. All the problems of a system in dire need of reform dogged the Faculty's processes in the courts.

Once a judgement had been obtained, the party in whose favour it had been received could proceed to enforcement of the fine. After messengers had made four separate charges at three days' interval, the transgressor could be 'put to the horn' or outlawed as a rebel for disobedience to the crown. Horning could not proceed on the decree of inferior courts (sheriff court, baron or regality courts or burgh court). If the horning was 'unrelaxed' for the duration of a year and a day, the creditor was entitled to interest on the original sum and to letters of caption (signet letters addressed to messengers requiring them to charge either sheriffs, bailies of baronies and regalities or provosts and bailies of burghs, to apprehend the debtor until he was obedient to the letters of horning). An appeal against letters of horning could be obtained by the defender, either before or after execution, by letters of suspension.[118]

An estimate of the cost of the Faculty's legal actions against irregulars and unlicensed practitioners can be gauged from the first surviving collector's account of 1655 in which John Daniel was paid £10 13s. 4d. Scots 'for charging several persons', and £12 14s. 0d. 'for the registration of the letters of horning denouncing of the persons to the horn, raising of caption any six doubling thereof conform to an account produced'. That it had its job cut out is also apparent from an ordinance of 21 September 1658 to Messrs Archibald Graham and David Sharp 'to speak [to] all those censurable within the town between [now] and the next meeting and to report'. In June 1662, thirty-six people were denounced at the cross of Edinburgh at a cost of £10 16s. – 6s. per individual. The denunciation itself cost a further 12s.[119]

---

[118] Society of Archivists, Scottish Documents (Glasgow, 1980s), unpublished typescript, pp. 83–89.

[119] RCPSG, 1/1/1/1b, pp. 114, 161, 208–9. Where the defender could not be charged in person,

On a very rare occasion, an irregular was imprisoned by the local magistrates: on 29 August 1672, for example, one Gavan in Greenock was imprisoned in the Tolbooth of Glasgow at the Faculty's instance for transgressing its acts and for practising surgery and pharmacy as an unfreeman.[120] But many others had previously so transgressed and had not been incarcerated. Perhaps in making an example of Gavan, the Faculty was demonstrating a tightening up of its licensing policy before the parliamentary ratification of 1672, when the original provisions were extended to apothecaries and barbers. The Faculty agreed to set Gavan at liberty, on condition that he bind himself not to practise surgery or pharmacy until made freeman, under pain of £100 Scots.

On occasion, irregulars were able to keep practising where they would probably have been stopped because they were protected by members of the local elite. For example, John Hall in Paisley, who had set himself up as a licensed surgeon in that town and put out sign, appeared before the Faculty on 11 September 1673 to answer charges of practising all three aspects of the healing arts – medicine, surgery and pharmacy – unlicensed within their bounds. The Faculty asked him for testimonials from the ministers, elders or magistrates of the various places in which he had lived since coming to the country, as well as an indenture and discharge of his surgical apprenticeship.[121] Hall produced passes from various parts of England over the past ten years, but no certificates or testimonials. Considering 'his presumption not only to practise Surgery confessed by him unlicensed but also to set out a sign of his said art at Paisley without their liberty', the Faculty fined him £40 Scots. He was discharged from practising any of the said arts in future until he produced certificates from the respective places of his residence, and a trial was to be taken of his qualification. But an influential patron – Lord Dundonald – seems to have interceded on Hall's behalf and in December 1673, all further execution against him was suspended.[122]

Seeking admission to practise aspects of surgery, Hall eventually produced

---

proclamation was made at the market cross of the main burgh in his county of residence. If he was out of the country, the defender would be charged at the market cross in Edinburgh and then in Leith.

[120] RCPSG, 1/1/1/1b, pp. 394–95.

[121] Similar testimonials from both medical and church personnel played their part in the licensing of practitioners in England. For this see J. R. Guy, 'The Episcopal Licensing of Physicians, Surgeons and Midwives', *Bulletin of the History of Medicine*, 56 (1982), pp. 528–42. However, Evenden contends that 'very little proof of actual skill in terms of successful "cures" was needed, at least for male candidates', but that this was far from the case for women. Evenden, 'Gender Differences', p. 206.

[122] RCPSG, 1/1/1/1b, pp. 456, 460–61, 477.

a testimonial from Dundee and burgess tickets from a number of other burghs in December 1674. Repeating his request two years later, he was on that occasion tried and admitted 'to the cure of simple wounds where there is no loss of substance and where there is neither nerves, arteries or tendons cut and likewise to open veins by a physician's direction and to apply ventouses cauteries & lock leeches'. However, if he attempted a cure beyond his expertise then the Visitor or any surgeon member of the Faculty was permitted 'to take the patient off his hand', because of his ignorance and because he was 'usurping also upon that whereto he is not admitted and has not knowledge thereof'. Hall remained a constant irritant to the Faculty. In November 1676, he was 'warned to the next meeting for keeping of a servant that practises in the art of surgery and barberises'.[123]

Female Practitioners

Female practitioners appear only in prosecutions for unlicensed practice because there is no evidence that women were admitted to apprenticeship with the Faculty.[124] The only women readily tolerated were midwives. As the London College of Physicians had held a meeting in January 1583 specifically to discuss 'the means by which we could prevent itinerant and inexperienced old women from practising medicine', and in 1641 the Edinburgh surgeons drew up a petition to prevent women practising surgery, the FPSG also targeted individuals whom it wished to deter.[125] The first women irregulars make an appearance on 17 March 1656 when the brethren appointed a delegation to speak to the magistrates about the execution of their letters of caption against Catherine Robison and Elspeth Murray.[126]

In a seventeenth-century climate, attempts to exclude women are, to some extent, understandable when the need for a practitioner to have a good all-round education could scarcely be questioned, and most women (other than the rich and leisured) still found it difficult to gain access to such an

---

[123] RCPSG, 1/1/1/1b, pp. 537–38, 547–48.

[124] Though a female surgeon practised in York in the 1570s with the sanction of the company. See M. C. Barnet, 'The Barber-Surgeons of York', *Medical History*, 12 (1968), p. 27.

[125] Pelling, 'Thoroughly Resented? Older Women and the Medical Role in Early Modern London', in L. Hunter and S. Hutton (eds), *Women, Science and Medicine, 1500–1700: Mothers and Sisters of the Royal Society* (Stroud, 1997), p. 67; R. A. Houston, 'Women in the Economy and Society of Scotland, 1500–1800', in idem and I. D. Whyte (eds), *Scottish Society, 1500–1800* (Cambridge, 1989), p. 136.

[126] RCPSG, 1/1/1/1b, p. 117. The delegation comprised the Visitor, Dr Crichton, Archibald Graham, Daniel Brown and Thomas Lockhart.

education. It has also been argued that restricted access to education, and low literacy, have been used 'to explain, justify and perpetuate women's lower status in Scottish society'.[127] It would have been particularly difficult if the Faculty enforced a literacy requirement in Latin, as did the Edinburgh surgeons, because few had access to a classical education.[128] Nonetheless, licensed or unlicensed, women played their part in bone-setting, midwifery, and in surgery and physic, particularly among the poor, serving informal apprenticeships with those prepared to teach them the necessary skills. In addition, women became experienced in nursing through domestic and private practice. Gentlewomen's assistance of the poor is known to have been encouraged, and there was a market in literature to help them.[129] But an understanding of the roles played by women is complicated by the fact that distinctions between the different kinds of healers per se were often not strictly delineated in this period.

The year 1656 saw a crackdown and an increase in prosecution of women by the Faculty in the year before the surgeons were granted their letter of deaconry. In March 1656, the collector 'acknowledged his receipt of £20 from the lady [blank] who was incarcerated by letters of caption for her bygone transgression'.[130] This entry is interesting for a number of reasons. It shows (as mentioned above) that if transgressions were severe enough, the Faculty could appeal to the magistrates for the imprisonment of an irregular, though it must have been unusual for a lady to be imprisoned. Secondly, the use of the term 'lady' shows that she was a woman of status in the community, like the wives of the aristocracy, of wealthy burghers and sometimes of clergymen who were expected to help their sick social dependents. But women were seldom defended in their exercise of skills which they had often acquired through many years of practice.

Elite female philanthropists treating their own households and the poor of their neighbourhoods were joined by a group of women from the lower classes – the so-called 'wise women' – who used what empirical knowledge they had acquired to assist the sick. Midway were the wives and daughters of surgeons and apothecaries, who acquired some knowledge of surgery and

[127] Houston, 'Women in the Economy and Society of Scotland, 1500–1800', p. 136.
[128] Dingwall, *Physicians, Surgeons and Apothecaries*, pp. 61–62. No literacy requirement is mentioned in the FPSG records, though literacy would have been presumed upon indenture.
[129] See, for instance, *An Alphabetical Book of Physicall Secrets, for all those diseases that are most predominant and dangerous (curable by art) in the body of man. Collected for the benefit, most especially of householders in the country . . . as likewise for the help of such ladies . . . who of charity labour to doe good. Whereunto is annexed a small treatise of the judgement of urines* (London, 1639), for Walter Edmonds.
[130] RCPSG, 1/1/1/1b, p. 117.

medicine by working in the family shop; they could possibly make the greatest claim to the term 'professional healer', and widows sometimes carried on their husbands' businesses, as they did in England.[131] But, as Doreen Evenden has demonstrated for London, they were very few (and most practised barbery rather than surgery).[132] In May 1657, Elspeth Murray, widow (mentioned above), appeared before the Visitor and masters of the craft having been incarcerated by letters of caption 'for usurping & taking upon her to prescribe and give physic & use the art of surgery'. Liberated on favour, she had agreed – under caution – not to practise medicine or surgery upon 'any person whatsomever except to those of her own family', under penalty of £40 Scots. There are also cases of mothers practising irregularly with their sons: in January 1676, for example, the Faculty ordained that Mr James Snodgrass in Paisley and his mother be charged with horning for their practice of medicine.[133]

Remarried by August 1657 to John Wallace, wright, Elspeth Murray was soon in trouble again. She was asked whether or not she had, since the previous May, given physic to any one, particularly to one James Watson, and

> openly confessed the same in so far as he having had as she called it the gentlemanly sickness She gave him first Julep & then ointment of olibey,[134] galtsam & quicksilver all mixed together, a quarter of ounce of every one of them'.

Her use of quicksilver shows that she clearly appreciated the use of metallic remedies against venereal disease. Though she was fined £40, the Faculty nevertheless had some sympathy for Elspeth Murray (who had already been imprisoned once): in view of the fine and her confession, it entreated the magistrates to intervene in preventing execution of the decree of 19 May. Finally, on 16 October, the Faculty accepted £10 in satisfaction of the entire fine. The next day, Murray appeared to free John Rollo, merchant, who had stood caution for her. But further transgressions were to cost her the full £40.[135]

---

[131] A. L. Wyman, 'The Surgeoness: The Female Practitioner of Surgery, 1400–1800', *Medical History*, 28 (1984), pp. 22–23. In Bristol, women were still being permitted to carry on their deceased husbands' practices in the early eighteenth century. Between 1700 and 1750, a phenomenal 50 per cent of surgeons and apothecaries who died in the course of their career were succeeded by their wives but this high succession rate was increasingly curtailed by the mid century. Fissell, *Patients, Power and the Poor*, p. 64.
[132] Evenden, 'Gender Differences', p. 196.
[133] RCPSG, 1/1/1/1b, pp. 131 (quotations), 524.
[134] This is probably a corruption of *olibanum*, the Latin for frankincense. For which, see Christophe Wirtzung, *Praxis Medicinae Universalis: or A Generall Practise of Physicke . . . in the Germane Tongue, and Now Translated into English . . . by Jacob Mosan* (London, 1598), the third or Latin index containing all the Latin and Greek names of simples (no pagination).
[135] RCPSG 1/1/1/1b, pp. 135–37, 152–54.

Although women had been admitted as freemen into provincial barber-surgeons' guilds in fifteenth-century England, by the sixteenth and seventeenth centuries women who practised physic and surgery were tolerated less openly. In Scotland, the attitude was generally hostile. The main reason was economic: beyond the monopolistic control of the guilds, it was cheaper to go to female practitioners for a cure because the most lucrative areas of medical practice were dominated by men. As A. L. Wyman put it: 'The bad reputation attached to the women stems largely from the strong hostility of the men trying to maintain their own monopoly and status in their closed corporations'.[136]

The case of an itinerant woman practitioner – Margaret Granfield, wife of Englishman David Torrell – illustrates the differing degrees in the Faculty's tolerance of irregulars. A complaint was brought against Granfield by Janet Anderson, widow of the late James Rodger, merchant of Glasgow. The aggrieved often initiated proceedings against the incompetent in the hope of recruiting the body corporate as an ally when they were getting nowhere themselves with a complaint.[137] Anderson stated that on 6 August 1657 her husband had been heavily and mortally diseased, and having heard 'that the said Margaret had given out herself as a most expert physician, he caused her go for her who came to visit him'. After examining the terminally ill patient, and in the graphic original language having 'groped his pulses', she not only told him that his disease was curable but promised to cure him in fifteen days. Credulous, Rodger agreed to pay Granfield £3 10s. 0d. sterling for which sum 'she promised to cure him perfectly'.[138]

To accomplish this, Granfield sent the patient some julep in a can with one or two other things most of which, still unused, was exhibited before the Faculty. But in spite of three or four visits by Granfield, during the course of which 'she told that there was no death working with him', Rodger died five days later. His wife subsequently petitioned that Granfield 'be declared incapable of the forsaid science and art of medicine and surgery and punished for the forsaid wrong'. Insisting that Granfield take back her medicines (which she had thus far refused), she demanded reimbursement of the fee paid, appealing for justice in front of the Visitor, members of the Faculty and a bailie (the presence of a town magistrate showing that the proceedings were legitimised as a court.)[139]

Having been apprehended by officers, Margaret Granfield was present with

[136] Wyman, 'The Surgeoness', pp. 26–28, 30.
[137] Pelling, 'Older Women and the Medical Role in Early Modern London', p. 68.
[138] RCPSG, 1/1/1/1b, pp. 138–39.
[139] RCPSG, 1/1/1/1b, pp. 139–40.

her husband. After hearing the complaint, 'She denied the whole expressions contained in the narrative', though she did acknowledge receiving the money and tendering medicines to the deceased as contained in a recipe given in by her. The patient had been plied with two pints of seck, two pints of claret wine, two loaves of sugar, 3 oz. of syrup of hour head or horn head, 3 oz. of maidenhair with some syrup of gilliflouris. A decoction of the fernlike herb maidenhair was generally used to counter jaundice or in cases where there was a stoppage of urine to dissolve a kidney stone. Syrup of clove gilliflowers – a commonplace in every apothecary's shop – was taken to expel poison and to ameliorate 'hot pestilent fevers'.[140] Rodger 'had a hard cough, convulsive gasping, [was] swollen in his belly', and eventually died of a hydropsy in which his legs had swelled downwards. Being questioned as to who was present at the striking of the original bargain, Granfield indicated Mrs Anna Crawford, daughter of the deceased laird of Carssmuir, as sole witness, had been there. If she swore on oath that she had spoken the words libelled, then not only was Granfield content to undergo the censure of the Faculty but would also refund the fee. But asked whether she had either a licence to exercise medicine or a testimonial from any authority to travel through the country by which her honesty and good carriage might be known, she conceded that she had not.[141]

Taking all this into consideration, the Faculty fined Margaret Granfield £40 Scots for her transgression, and 'for her usurping to exercise the said calling within their bounds without any testimonial'. In addition, it also ordained that 'she shall not prescribe or give any physic to any patient whatsomever within this city or the bounds committed to the faculty's trust' or practise the arts of surgery or pharmacy under penalty of £40 Scots. Granfield and her husband, David Torrell, agreed not to meddle in physic or surgery within the Faculty's bounds. But Granfield was shrewd, pleading that 'the truth is she was altogether ignorant of her being censurable'. As a stranger, she humbly craved that the Faculty would refund her fine since she was 'most willing to submit herself to them and all judicatories of the land', and it took a sympathetic stand, instructing the collector to surrender the fine to Dr Crichton and Mr Archibald Graham to dispose of as they saw fit. But Thomas Lockhart, one of the quartermasters, contested the decision, asking that it be settled to the benefit of the Faculty.[142] Certainly, the occupation seems to have encouraged English empirics to Scotland. The collector

---

[140] Culpepper, *The British Herbal and Family Physician*, pp. 154, 204.
[141] RCPSG, 1/1/1/1b, pp. 140–42, 151.
[142] RCPSG, 1/1/1/1b, pp. 142–46, 151, 16 October 1657.

noted 6s. in 1661 'for charging of James Fairie, smith in England, who has given bond to abstain from our calling'.[143]

Another female practitioner, Janet Hood, stood indicted in October 1671 of giving physic to Barbara Hart, servant to James Muir, merchant, as a result of which she died. Hood confessed that the night before she died she had given Hart Venice treacle or *theriaca Andromachi* (a theriac-like preparation of over fifty ingredients which contained 1.3 per cent opium or up to 15 mg. morphine),[144] as well as some julep on the following day. To clarify the matter further, the Faculty sent for Muir who declared that after Hood gave his servant physic 'she from then forth vomited until she expired'.[145] The deceased had obviously been given an exceedingly potent emetic. Hood was not only practising internal medicine irregularly, but may possibly have administered an abortifacient. At best, this incident feeds the popular stereotype during this period of women as poisoners.[146] After due consideration, the Faculty commended Hood to the magistrates 'for her condign punishment and for procuring her never to give or prescribe any physic to any persons whatsoever at any time hereafter she being neither burgess or guild brother or having right thereto by her husband nor having entries nor privilege within the burgh of Glasgow'.[147] This statement confirms that women could have certain rights of practice if their husband had a shop but, as in Edinburgh, women were not admitted to the Incorporation.[148]

The Faculty minutes between 1688 and 1733 perished in a fire, but legal records make it clear that the wrangling between the Faculty and individuals who infringed its monopoly on licensing continued. Some of these individuals were women. Inscribed in the general register of deeds on 5 November 1697 is a bond at the instance of the Surgeons of Glasgow against Isabell Crawford, wife of the late William Young, dyer in the Gorbals. She was prohibited, at the instance of the 'Faculty of Surgeons', from practising surgery and pharmacy unless she gave proof of her qualifications and knowledge in the said art, and consequently undertook to grant bond 'for preventing of any further trouble that I may incur through my suspect

---

[143] RCPSG, 1/1/1/1b, p. 206. On 30 May 1653, the Faculty voted Muir fifty merks. For an account of the Cromwellian regime in Scotland, see F. D. Dow, *Cromwellian Scotland, 1651–1600* (Edinburgh, 1979).

[144] Estes, *Dictionary of Protopharmacology*, pp. 201–2; Culpeper, *Complete Herbal*, p. 478.

[145] RCPSG, 1/1/1/1b, pp. 371–72.

[146] See M. Pelling, 'Compromised by Gender: The Role of the Male Medical Practitioner in Early Modern England', in H. Marland and M. Pelling (eds), *The Task of Healing: Medicine, Religion and Gender in England and the Netherlands, 1450–1800* (Rotterdam, 1996), p. 105.

[147] RCPSG, 1/1/1/1b, p. 372.

[148] Dingwall, *Physicians, Surgeons and Apothecaries*, p. 68.

practice'. Crawford bound herself to desist from the practice of surgery and pharmacy under penalty of £40 Scots.[149]

Nevertheless, female irregulars continued to practise medicine in Glasgow and beyond into the early eighteenth century. Elizabeth Dunlop, widow of the late George Hutcheson, merchant, 'being formerly prohibited from the practise and Exercise of Medicine Surgery and pharmacy within the City of Glasgow or precincts of their Respective shires of their freedom', appeared in front of the Incorporation of Surgeons and Barbers in March 1710 to be prohibited once again.[150] Moves towards the wider professionalisation of medicine, especially the advent of the man-midwife in the eighteenth century, only exacerbated the marginalisation of the female practitioner.[151]

In most European countries by the early seventeenth century, physicians, surgeons and apothecaries were organised in their own groups which, to a certain extent, delineated the nature of their practice. In Glasgow, where all practitioners were organised in one body, the boundaries of practice may have been somewhat more fluid, though physicians tried to maintain an influence on the practice of surgeons and apothecaries. Nonetheless, it is clear that the Faculty's main activity was the licensing of general practice and that its day-to-day licensing and admission policy was dominated by the surgeons.[152]

While the relatively humble origins of medicine form much of its fascination, the earlier period has been portrayed as 'a series of battlegrounds, largely lacking in standards or centralised control, and yet at the same time over-rigidly stratified into the three parts of practice represented by apothecaries, surgeons and physicians'.[153] The day-to-day reality of practice however was far less rigidly defined.[154] Regular practioners worked alongside irregulars and many traditional healers practised without the legitimisation of a licence. Indeed, the entire debate about regular practitioners (university or apprenticeship-trained) versus irregulars or quacks (with no formal training) arose out of the desire of nineteenth-century medical reformers to discredit the standards of previous centuries.[155] It says far more about the stance of

---

[149] SRO, RD3/93, Register of Deeds, p. 153.
[150] GCA, T-TH14 5.2, Minutes of the Incorporation of Barbers, 1707–1765, fol. 8r.
[151] Digby, *Making a Medical Living*, pp. 20, 259–78.
[152] L. McCray Beier, 'Seventeenth-Century English Surgery: The Casebook of Joseph Binns', in C. Lawrence (ed.), *Medical Theory, Surgical Practice: Studies in the History of Surgery* (London, 1992), pp. 53–54.
[153] M. Pelling, 'Medical Practice in Early Modern England: Trade or Profession', in W. Prest (ed.), *The Professions in Early Modern England* (London, 1987), pp. 90, 94.
[154] Fissell, *Patients, Power and the Poor*, p. 62.
[155] Porter, *Health for Sale*, pp. 222f.

the nineteenth-century commentator than of the seventeenth-century practitioner.[156] In the seventeenth century, regular and irregular healers practised within the Faculty's bounds, offering a range of services to a variety of clients.

[156] See, for instance, R. Porter, 'The Language of Quackery in England, 1660–1800', in P. Burke and R. Porter (eds), *The Social History of Language* (Cambridge, 1987), pp. 73–103; Bynum and Porter, *Medical Fringe and Medical Othodoxy*, introductory discussion.

5

Enlightenment

Glasgow's strength lay in its sea trade. Before and after the Union of 1707 trade routes extended to all the important ports in northern Europe and the New World. Histories of Glasgow never fail to remark on its access to the sea, describing the goods, the tonnage carried and the types of ships making their way to Greenock or Port Glasgow. Images of tall, square-rigged, three-masted ships, with wooden sterns high over the waves and flags flapping briskly in the breeze symbolise the city's early dependence on import and export, trade and manufacture.[1] None of the recent histories has, however, done justice to the ideas, knowledge, sciences, politics and religious influences that benefited from quick transport by sea.

As the seventeenth century faded into a new one, being a port town on the west coast of Scotland didn't only bring the tobacco trade,[2] or linen and sugar production.[3] Nobody measured or taxed the invisible goods that appeared with the ships. Intellectual expansion, combined with the tough staying power of Scottish commerce, made possible the golden era of the Scottish Enlightenment, rooted not so much in the Union with England as in a deep interest in science and philosophy. Moreover, contrary to some recent historical opinion, the fertile soil on which this genius flourished was not the cleansed one of *literati* moderation and civic humanism. That was only an elegant transmutation of more deep-seated beliefs. The raw body of Scotland's intellectual industry was tempered by Calvinism, forged in the truly gruesome trials endured by Covenanters and Presbyterians. Presbyterianism was remarkable in Europe, along with other radical offshoots of the Reformation, for preserving democratic principles. Toughness, industry, persistence, intelligence and frugality gave the Scottish people the edge, derived, as some foreign visitors observed, from the Calvinist ethos of austerity and

---

[1] This picture graces the cover of the most recent history of Glasgow. See T. Devine and G. Jackson, *Glasgow*, i, *Beginnings to 1830* (Manchester, 1995).

[2] For the importance of the tobacco trade, see T. Devine, *The Tobacco Lords: A Study of the Tobacco Merchants of Glasgow and their Trading Activities, c. 1740–90* (Edinburgh, 1975).

[3] G. Jackson, 'Glasgow in Transition, *c.* 1660 to *c.* 1740', pp. 81–85, and R. H. Campbell, 'The Making of the Industrial City', pp. 189–97, in T. Devine and G. Jackson (eds), *Glasgow: Beginnings to 1830*.

discipline. In this the Scots were like the Swiss, as even Sir Walter Scott remarked.[4]

With episcopal power demolished in 1690, and the crown pacified, regeneration focused on civic renewal. In this context both local developments and international trends have to be considered to understand the place of the FPSG. Its role was strong in that it controlled the only legitimate port of entry to medical practice outside the MD. It licensed surgeons, while not itself engaged in teaching. By 1700 it recognised it could harness the resources of the university, at a time when Glasgow and Edinburgh were both the poor relations of the richer, more progressive European centres of learning and their burgeoning civic, commercial and scientific enterprises. This is why the story of medical education and the shaping of the profession cannot be told without considering the interplay of civic institutions or the impact of European scientific networks on Scotland.

The Presbyterian elite in Scotland was in a prime position to engage, commercially and scientifically, with the educational advances that were a legacy of the Reformation, revived in Scotland under the pressures of maintaining religious freedom and despite political upheaval. Education had long been a Protestant priority and the 'new learning', which challenged scholastic traditions of knowledge, grew strong as the Scottish intelligensia engaged with the sciences to which medicine undeniably belonged. Thus medical training struck a note of modernisation, encouraging renewal at home by importing the sciences accessible abroad. Without international contacts the FPSG investment in a new style of medical education would have been both parochial and deficient.

Scottish regeneration after 1690 derived mainly from belief in educational betterment and a hard-headed pushing of town renewal. The most striking characteristic of the success Scotland made of its medical training was its melding of universities, medical incorporations and teaching hospitals, a prime achievement when measured against developments in England. This distinguished the model of the Scottish 'medical school' from education in London, or, for that matter, Vienna or Berlin, with Paris being the closest prototype. London medical training was carried out in hospitals,[5] not by the availability, as in both Glasgow and Edinburgh, of university teaching, apprenticeship and, by the end of the century, hospital clinical instruction.

---

[4] H. Utz, *Schotten and Schweizer: Brother Mountaineers. Europa entdeckt die beiden Völker im 18. Jahrhundert* (Frankfurt, 1995), pp. 77, 129.

[5] R. Porter, 'Medical Lecturing in Georgian London', *British Journal for the History of Science*, 28 (1995), pp. 91–99.

The ability to choose combinations best suited to the – often itinerant – medical student or apprentice enabled the educational system to operate on a free trade basis. This was undeniably advantageous to the FPSG. In a very economical if parasitic venture, the medical incorporation utilised the expansion of university teaching to change the context of its medical education. It did so from the beginning of the eighteenth century through university educational diversification and then in the nineteenth through the extramural colleges, such as Anderson's Institution, which by the 1820s was able to challenge the universities. During the lifetime of its affiliation with teaching institutions the FPSG very cannily coopted lectures and facilities from different institutions while it held sway over surgeons' licences. Both teaching and the standards to which it was held were dependent on the tacit approval of the Faculty to a surprising degree. This dominant position lasted until mid century in regard to Glasgow University, when teachers like William Cullen and the Hamilton dynasty tipped the balance in favour of the MD degree. As medical and scientific education at university became good enough to attract surgeons *and* physicians it came to dominate the market; but this was an ongoing struggle only significantly resolved by the 1858 Medical Act.

In the first part of the eighteenth century teaching agendas were not determined by universities but relied heavily on surgeons drawn to the FPSG qualification. The University and Faculty were jointly interested in importing new ideas and techniques, above all in accrediting the teaching of skills in new fields such as botany, chemistry, anatomy and midwifery. In a commercial city few would resist a 'free trade' system of maximising opportunity, especially if it culminated in the prestige – and additional income – attached to university teaching. This made possible, and indeed necessary, the rapid development of scientific education.

Surgeons began to utilise university instruction along with physicians in Glasgow in 1714, before the town council in Edinburgh inaugurated the university medical faculty in 1726. In both places the universities and the incorporations pursued a modern agenda in teaching despite occasional disagreement. Contrary to European practice, the Scottish universities welcomed surgeons as lecturers as well as students although, as in Europe, degrees were not granted, as the accreditation for practice for surgeons in Britain remained the licensing examination, tied to the separate institution of the medical incorporation. Thus institutions with diverse functions, the universities, the royal colleges (including the FPSG), and the hospitals, were fused together in the Scottish model – much earlier than elsewhere – to create a 'free trade' commercial pattern of medical training. As residential requirements did not apply until the end of the century, clients were free to come and go, paying teachers by ticket and being accredited by certificate, examination or testimonial.

A monopoly was maintained within this very flexible system only through local jurisdiction. The local 'trials for qualification' influenced teaching and this gave the medical incorporations their power over other institutions. Those outside this system were, however, not hindered from study – not even empirics, although these could meet their nemesis if fined for breaching licensing practice.[6]

Within the profession medical competence was primarily a matter of peer review of those 'on trial for their qualifications'. The 'trial' was the port of entry to legal medical practice in each jurisdiction. Thus a flexible commercial system was adopted to control qualifications and competence at home with a remarkable openness to itinerant medical students and apprentices. The success of the lecture system in Scotland ensured its prominence in the medical marketplace, latterly attracting students from all over Europe. As a result of such versatility, one could select lectures or instruction from university to private lecture to extra-mural school in Glasgow or Edinburgh, the sequence being left to the 'buyer'. No set curriculum was required. In London, surgeons', physicians' and especially apothecaries' guilds banked on exclusion (an enduring thorn in the side of the Scottish practitioners going south). The gentlemen's club of privilege was ruled by the few in London while elsewhere the modern practical work of training depended on versatility.

One could argue that William Hunter's success in 'free enterprise' lecturing in London in the 1740s broke open the English system by making a success of the Scottish model. He sold modern anatomy and midwifery teaching to anyone who paid, circumventing the self-selection of the 'closed-shop' of hospital ward 'clerks'. Hunter was trained in the west of Scotland, under John Gordon and William Cullen. He had experienced the rising success of 'free trade' teaching and the active search for modern training on the continental model. Adam Smith described this system well when he championed competition among private and university lecturers as the best way of securing a high quality of teaching.[7] By that time, 1774, William Cullen was seeking a regulated curriculum, Smith holding out against him on free market principles.

Medical life within the fold was equivalent to FPSG medical management. The local professional body cast medicine in its own image. It cleverly used

---

[6] See for example, the case of A. Turnbull, Glasgow City Archives A 2: 25, decreet absolvitor in part and for expenses, the Magistrates and Town Council of Glasgow against the Faculty of Physicians and Surgeons of Glasgow, 1794.

[7] A. Skinner, 'Adam Smith and the Role of the State: Education as a Public Service', in S. Copley and K. Sutherland (eds), *Adam Smith, Wealth of Nations: New Interdisciplinary Essays* (Manchester, 1995), pp. 70–96.

its civic position to align itself with social needs and thus helped shape developments that sustained its professional role. Piety and industry, commerce and philanthropy combined to create the main institutions that were to advance the medical profession.

Presbyterian Influence

Glasgow was known as 'the most evangelical city in Scotland', its piety suffusing all social classes, especially among merchants and craftsmen. Local patriotism was infused with a strong sense of religious justification and devout introspection. George Drummond, Lord Provost of Edinburgh from 1725, wrote in his dairy:

> Begun the day with God, my employment last night, and this morning, was to examine the growth of grace in my soul ... but I cannot see much of it, I dare not say that sin is getting ground, yet, I as little see, grace growing, – prayer, in the morning, run in the strain of this tryal but I had not the Lord, I was not breathed on ...[8]

In this vein he urged himself onward in his great municipal works not least that of building the Edinburgh Royal Infirmary. He wrote about the Infirmary in 1738: '. . . I of late began to be afraid, vanity and not regard to God is become the spring of my activity'.[9]

Such a disciplined and passionate Calvinist conscience had driven forward civic and educational good works as Presbyterians rebuilt the civic fabric more than a decade before the Union. Most visible was university regeneration, Presbyterians taking firm control with William Dunlop (Principal of Glasgow University in 1690) and his brother-in-law William Carstares (Principal in Edinburgh University in 1703).[10] Both were vital to university regeneration. John Stirling, Dunlop's successor as Principal of Glasgow, was yet another pious Presbyterian bent on improvement. Gershom Carmichael, the Regent who introduced the secularising texts on natural law to Glasgow, again exemplified the revolutionary strength of spiritual introspection and its quests for truth. As Robert Wodrow recorded, 'in his advanced years he was singularly religious, and I know he was under great depths of soul-exercise'. Wodrow continued, 'he was of very great reputation, and was exceedingly valued, both at home and abroad, where he had considerable

---

[8] A. Chitnis, 'Provost Drummond and the Origins of Edinburgh Medicine', in R. H. Campbell and A. Skinner, *The Origin and Nature of the Scottish Enlightenment* (Edinburgh, 1972), see p. 92 for this quote.

[9] Ibid., p. 96.

[10] A. Chitnis, *The Scottish Enlightenment: A Social History* (London, 1976), p. 131.

correspondence with learned men'.[11] Wodrow knew the type, belonging to it himself. These 'dark men', rejectors of Descartes, despising Catholic metaphysics, were modernisers, shrewd about stepping carefully into the future. As university librarian Wodrow added significantly to university holdings, discounting only the *Ethics* of Aristotle and 'Cartesius [and] his gang'. Seen as 'mean' before Wodrow's time, in 1692, the library was 'overflowing with books' by 1704.[12]

The FPSG had passed a rule of strict Sabbath observance in 1675,[13] while medical doctors had no compunction about associating closely with 'ousted' ministers, an instance of which concerns Robert Wodrow. Wodrow was one of Scotland's earliest historians, recording the sufferings of the church in the 'Killing Times' before 1690. He was half-brother to John Wodrow, Praeses of the FPSG. Robert Wodrow died in 1734, his church history vilified by Episcopalians and Cameronians alike, this very fact pointing to his tenacious fairness. He was a true patriot in writing about the Union of 1707: 'How we are to get rid of it I do not see'.[14] He felt the patronage of the landed to be servile and his history vindicated the right of the congregation to elect their own, giving the world 'some view of what this Church underwent for religion, reformation, rights and the cause of liberty'.[15]

Robert Wodrow's writings described a time, as his biographer put it, 'when persecution was at its worst, and when his father's house and kin were immersed in it'.[16] His story is important because it shows empirical science gaining ground in Glasgow at the beginning of the century and because he, his father and his brother were deeply interested in medicine and botany. John financed a botanical garden for the use of candidates taking the FPSG examination so important to the usual licensing as 'surgeon-apothecary'. There are few records on early eighteenth-century science in Glasgow, but the story of the Wodrow family helps explode the notion that Enlightenment was the creation of moderates and *literati*. Interest in modern science was prevalent from the beginning of the century and not a product of secularisation.

The close relationship of medicine and Presbyterianism under siege is evident in the dramatic circumstances of Robert Wodrow's birth, which 'gives some view of the violence of the times'.[17] His father, James Wodrow,

---

[11] R. Wodrow, *Analecta or Materials for a History of Remarkable Providences Mostly Related to Scotch Ministers and Christians* (Edinburgh, 1842–43), iv, p. 96.

[12] Ibid., p. 126.

[13] RCPSG, 1/12/2/2, Dr William Weir's Notes on Faculty Minutes, 1599–1859.

[14] W. J. Couper, *Robert Wodrow* (n.p., 1828), p. 5.

[15] Ibid., p. 12.

[16] Ibid., p. 2.

[17] Ibid., p. 16, and the following quotes from pp. 61–64.

had a price on his head as a Covenanting minister when, in 1679, his mother 'fell in travail of me'. She was fifty-one and in peril, when 'an intimate of my father's', the 'worthy and excellent' Dr Thomas Davidson, was called for. All agreed she was 'under prospect of death' and James Wodrow was called to her 'about dusk', but was recognised, and a 'party of the guard came to the house to seize him'. He was about to be found, but Elizabeth went into labour just then and the soldiers retreated. 'Dr Davidson came in at this juncture to visit my mother, and had a man-servant carrying a lanthorn before him, it being now night.' The doctor acted quickly and clothed Wodrow in the servant's clothes, enabling him to escape. When the soldiers returned, searched, and even stabbed the bed with their bayonets, 'he was out of their hands and out of the house'.

Dr Davidson was probably a member of the medical incorporation.[18] The other 'intimate' of Wodrow, Dr Thomas Kennedy, who attended at James Wodrow's death in 1707, was certainly a member. Kennedy graduated MD from Leiden University in 1682 with a dissertation on the nutrition of the foetus, a subject under much scientific scrutiny at the time, and was an examiner for one of the first medical doctorates at Glasgow University in 1703.[19] He donated books to the FPSG library.[20] James Wodrow had himself studied medicine diligently for several years, 'particularly botany and anatomy'.[21] He was the first professor of divinity, appointed in February 1692 in the Presbyterian aftermath of the Revolution,[22] and so was hardly an outcast in the town – nor, as is so evident, was science scorned in these circles. Wodrow was installed expressly at the wish of 'my father's old and intimate friend' William Dunlop,[23] the Principal, and was in an influential position to promote change. In 1704 the physic garden was started under advice from the FPSG, a story to be told fully in a later chapter. John Wodrow, Robert's half-brother and James's third son, studied medicine and became an influential member of the Faculty. He showed the family's love of medicine and botany by planting a physic garden.

---

[18] The minute books of 1688 to 1733 were lost in a house fire.
[19] Duncan, *Memorials*, p. 248, and GUABRC, 26631, *Faculty Meeting Minutes, 1702–1720*, p. 11 (29 September 1703).
[20] Duncan, *Memorials*, p. 216.
[21] Ibid., p. 67.
[22] R. Wodrow, *The Life of James Wodrow, Professor of Divinity in the University of Glasgow, from MDCXCII to MDCCVII*, edited principally by John Champell (London, 1828). This book was published by the minister Dr John Campbell of Edinburgh who found it in MS amongst R. Wodrow's papers. The original bears the date 5 February 1724. The following material is taken from this book.
[23] Ibid., p. 116.

John Wodrow was the son of James and his second wife, Janet Luke. John Luke, Janet's father, was 'of the oldest standing in trade and business of many in Glasgow, and by blood and affinity related to most people of any fashion and continuance in that place: the Andersons, Grahames, Gibsons, Campbells, Crosses etc'.[24] He was 'amongst the first' to bring sugar manufacture to the city, and, by the account of Robert, 'he favoured, supported, and stuck by Presbyterians in their sufferings and suffered by impositions and finings very much'.[25] Luke had eighteen children, all of whom were pious and successful, and related in property holdings and religious beliefs to the Wodrow clan. Robert was librarian to the university together with his father's successor, John Simson, who was under fire for his liberal theological views. Robert's son, James, was librarian from 1750 to 1755. The 'minister of Eastwood' wanted 'to make religion and usefulness to mankind [a] constant study', the wish he recorded for his brother and his children.[26] John Stirling, the Principal, too, was close to the Wodrows, being the godparent to three of Robert Wodrow's children. Stirling secured the funds for the regius chair of medicine, supported the physic garden and appointed a lecturer in anatomy.

Robert Wodrow had a collection of 500 or 600 specimens 'of one thing and another relative to natural history'.[27] 'From the first he seems to have taken to natural history' and corresponded with the Keeper of the Ashmolean Museum in 1709.[28] He took a scholarly interest in the Balfour and Sibbald collections and was devoted 'to every subject connected with science or general literature'.[29] He owned a copy of the *Auctarium Musaei Balfouriani, e Musaeo Sibbaldiano; sive enumeratio et descriptio rerum rariorum*... (Edinburgh, 1697), which was inscribed in his own handwritings as '*ex dono authoris amici plurim colendi*'.[30] He was part of a small circle of eminent people interested in collaborating on manuscripts and exchanging specimens. Other members included Robert Sibbald, James Paterson, Eduard Lhwyd (the above-mentioned Keeper of the Ashmolean) and Sir Hans Sloane.[31]

---

[24] Wodrow, *The Life of James Wodrow*, p. 120.
[25] Ibid., p. 121.
[26] Ibid., p. 196.
[27] Couper, *Robert Wodrow*, p. 5.
[28] Ibid., p. 4.
[29] 'Prefatory Notice from the Diary', by M. L., in R. Wodrow, *Analecta*, iv, p. xiii.
[30] GUL, Special Collections, Mu 7- i. 41: *Auctarium Musaei Balfouriani, e Musaeo Sibbaldiano; sive enumeratio et descripto rerum rariorum ... quas Robertus Sibbaldus ... Academiae Edinburgenae donavit ...* (Edinburgh, 1697).
[31] A. D. C. Simpson, 'Sir Robert Sibbald: The Founder of the College', *Proceedings of the RCPE Tercenary Congress 1981* (Edinburgh, 1982), p. 79.

When only nineteen he recorded in his journal 'chemical experiments, observations with the microscope, cures for some common diseases, memorable interferences of Divine Providence, meteorological observations and interesting anecdotes, etc'.[32] In 1698 use of the microscope was a sign of serious scientific study. Chemical experiments, too, were indicative of modern scientific training and not yet available at university. The identification of plants and plant substances was just becoming an educational issue.

Wodrow admired John Ray, the botanist whose *The Wisdom of God Manifested in the Works of Creation*, was published in 1691, then reprinted in 1692, 1701, and 1704, which was perceived as Ray's 'most popular and influential achievement'.[33]

> Dec. 31, 1700. This afternoon **** told me that Mr Ray was once a fellow in on [one] of the University, and apeared much for the Covenant: and because he would not renounce it, he was afterwards put out of the University, and reduced to some straits; on which a gentleman bestowed £50 per annum, and a country house, upon him, in which he lived hitherto privately; and on this stock has done all these noble things the learned world is soe much endebted to him for. Compare this with the Preface to his Wisdom of God, in the Works of Creation and Providence.[34]

Understanding nature scientifically was equivalent to appreciating God's providence for Creation. These ideas found fertile soil in Glasgow. Men like Wodrow and the commercial leadership of the city were receptive to this religious interpretation binding together improvement and science. Works like Ray's inspired a close adherence to scientific methodology and ultimately spawned improvement in agricultural and chemical manufacture. Analytical empiricism undermined traditional ways of transferring skills, putting emphasis on standardised knowledge (for example the classification of plants) and conceptual understanding (learning anatomy as opposed to learning a specific hands-on skill). Science impacted on craft skills as well as commerce, on surgeons, linen manufacturers, agriculturists, horticulturists, chemists and merchants alike.

Civic renewal in Glasgow emerged from local Presbyterian self-assertion and contained a definitive and positive attitude to empirical science. Such strong evidence for the mutual support of learning in the town at the beginning of the eighteenth century severely challenges the opinion that the Scottish Enlightenment only emerged around 1750 and was a product of secularisation. The history of the Faculty strongly suggests science was

---

[32] Wodrow, *Analecta*, quoted in the 'Prefatory Notice from the Diary', p. xxxiii.
[33] C. Raven, *John Ray, Naturalist: His Life and Works* (Cambridge, 1950), p. 452.
[34] Wodrow, *Analecta*, quoted in the 'Prefatory Notice from the Diary', p. xxxiii.

incorporated very early in the Glasgow context. This strong tendency to further empirical science propagated so assiduously by Glasgow Presbyterians distinguished a very early turn to educational renewal and shaped the future of the medical profession.

Men like Gershom Carmichael, Robert Wodrow's teacher and in charge of students for thirty years, were engaged in a 'cheerful commerce' that united Protestant political thinking with an affirmative view of nature. The pessimism inherent in Calvinist views on man's 'fallen nature' was giving way. Natural law theory which Carmichael championed saw nature as benevolent and acknowledged 'natural' reason as good. Thus the works of Samual Puffendorf and Hugo Grotius impacted on Scottish thinking and Scottish students.[35] Later secularising tendencies, legal and cultural, served to curtail the voice of religion as arbiter of moral and civic conduct, but it had unlatched the door already to scientific method and common sense.

The universities and civic reorganisation thus surged to life as men exiled for political and religious dissent built a new educational and scientific framework after 1690. Peace meant being able to scupper overlords, which had meant, in political terms, strife. The leaders of university reform were Presbyterian and if they were pro-science and pro-commerce so was the temper of Glasgow. Their experience abroad taught them where to look for knowledge to import and how to revitalise medical training. In Scotland before 1700 there had been nothing much in the way of science. It might have remained a closed and secret world of the 'adept', akin to the 'virtuosi' of Oxford or Cambridge,[36] had not Protestant dissent welcomed the 'new learning' and Scotland's commercial outlook made a practical thing of it.

Cullen and the European Context

Robert Boyle is often named as a virtuoso of chemistry, employing the 'scythe of scepticism' without reference to the anti-authoritarian thrust of Protestant science. William Cullen's achievements in chemistry, too, have been taken out of context,[37] creating the impression that Cullen, like Boyle, managed to pull discoveries like rabbits out of a hat. But both men were credited with advancing scientific progress because they adopted an observational methodology that was critical of received knowledge. Like Protestant dissent, this

---

[35] J. Moore and M. Silverthorne, 'Gershom Carmichael and the Natural Jurisprudence Tradition in Eighteenth-Century Scotland', in I. Hont and M. Ignatieff, *Wealth and Virtue: The Shaping of Political Economy in the Scottish Enlightenment* (Cambridge, 1983), pp. 76–83.

[36] R. G. Frank, Jr, 'Science, Medicine and the Universities of Early Modern England: Background and Sources', *History of Science*, 11 (1973), pp. 194–216.

[37] Chitnis, *The Scottish Enlightenment*, pp. 164–67.

was a political statement. Presbyterians and Puritans and other 'separatist' despoilers of the union between Anglican episcopacy and the crown, were not to be enticed back to signing 'articles' of any kind. By Cullen's time these religious roots had been secularised, but the tap root had given the growth.

Cullen began his medical lectures in Glasgow by explaining why he was not following Hermann Boerhaave's *Institutes and Aphorisms*, the recently installed epitomes of medical knowledge, because he had developed a more valid scientific agenda of his own, thus aligning himself with the spirit of scepticism and dissent.[38]

Boyle's work was part of the complex 'new learning' networks of Protestant reform. Science was not 'English',[39] nor was it 'Dutch', really, despite the oft-cited 'Dutch connection', so important to Scotland.[40] These places were significant only because Protestant dissent had engendered there a freeing of minds unhampered by orthodox restraints. Holland was popular because it was a place of freedom, a place where radicals and exiles gathered, tolerating scepticism as well as religious diversity. The 1660s brought with them the exploration of the body beneath the skin, the microscope aiding a talented group in Amsterdam and Leiden including Jan Swammerdam, Frederik Ruysch and Regnier de Graaf, who inaugurated modern anatomical dissection.[41] Looking at the natural world, such as in Swammerdam's work on insects, was charged with the optimism of understanding and appreciating the function and beauty of nature as scrutinised by the eye of man. Swammerdam in particular waxed lyrical in describing the intricate beauty and artistry of nature, binding this, his pleasure in science, to God's many good works and his great art.[42] The quest to learn from the Book of Nature was writ large, and was common to many of the outstanding naturalists of the time. John Ray's much read book, *The Wisdom of God Manifested in the Works of Creation*, exemplified this school of devout science whose leaders were the major figures shaping discovery, not only in anatomy or botany

---

[38] J. Thomson, *An Account of the Life, Lectures and Writings of William Cullen MD* (Edinburgh and London, 1854), i, p. 26.

[39] Paul Wood makes the point of how restricting 'national' views of scientific achievement are, especially in the so-called 'scientific revolution', see P. Wood, 'Scientific Revolution in Scotland', in R. Porter and M. Teich (eds), *The Scientific Revolution in National Context* (Cambridge, 1992), pp. 263–64.

[40] F. J. Cole, 'The History of Anatomical Injections', in C. Singer (ed.), *Studies in the History and Method of Science* (Oxford, 1921), ii, pp. 297ff.

[41] E. G. Ruestow, *The Microscope in the Dutch Republic* (Cambridge, 1996), pp. 46ff.

[42] Ibid., pp. 134–44.

but also in chemistry and medicine.⁴³ As John Ray wrote: 'No knowledge can be more pleasant to the soul than Natural History: none so satisfying, or that doth so feed the mind'.⁴⁴

Thus Protestant science pioneered changes across several disciplines. Robert Boyle's circle included men like Thomas Sydenham and John Locke, one a medical man, the other a philosopher. Sydenham, to whom Locke was apprenticed in medicine,⁴⁵ was a Puritan (as was Boyle). Locke, of course, became very influential in Scotland for his empirical philosophy. Empiricism was a common epistemological tool, used by Sydenham, Boyle and Locke, in chemistry, medicine, and philosophy became revolutionary. However, observational science was soon engaged in its own book-burning, adamant about replacing tradition of authorised teaching. Sydenham's attitude was typical and gave no ground. When asked to recommend medical textbooks he replied: 'Read *Don Quixote*, it is a very good book, I read it still'.⁴⁶

The mission to reform learning needed the natural sciences. In this Sydenham was a key player. His influence is often forgotten, but the clinical teaching so proudly ascribed to the initiative of John Rutherford in the Edinburgh Royal Infirmary, and the clinical teaching in Glasgow, which was just as effective, both owe a great debt to English Puritan radicalism. Of course, this too was mediated over time, but it anchors the notion of Scotland's medical Enlightenment deriving from the 'new learning'.

Sydenham shared Robert Boyle's deep religious sense and devotion to Baconian methods.⁴⁷ He worked in London on the nature of disease, revolutionising learning by rejecting textbook authority, making nature the author of his observations. The epistemological point he was making concerned a 'modern' problem, not a classical one: what disease teaches us as we see it, not as others have recorded it; from nature, not from tradition. The radicalism of empirical investigation provided the rationale for clinical training, the 1676 *Medical Observations* becoming its paradigm. In some ways a more difficult enterprise than securing anatomical knowledge, and done without hospital practice, Sydenham's clinical methodology nonetheless took root.

---

⁴³ On this question see R. Hooykaas, *Religion and the Rise of Modern Science* (Edinburgh and London, 1972), pp. 144–49; see also C. Webster, *The Great Instauration* (London, 1975) pp. 1–2.

⁴⁴ H. Phillips, *History of Cultivated Vegetables: Comprising their Botanical, Medicinal, Edible and Chemical Qualities, Natural History, and Relation to Art, Science and Commerce* (2nd edn, London, 1822), i, who quotes John Ray, p. 2.

⁴⁵ K. Dewhurst, 'Thomas Sydenham (1624–89): Reformer of Clinical Medicine', *Medical History*, 6 (1962), p. 108.

⁴⁶ Ibid., p. 111.

⁴⁷ Ibid., p. 108.

Sydenham's clinical method directly influenced Jena and Halle, where Friedrich Hoffmann (1660–1742) and Georg Ernst Stahl (1659–1734) took up observational science. Their teaching and the publication of their many *Observations* in medicine, and chemistry predated Herman Boerhaave's work.[48] Their works were influential in clinical teaching, in the form of observational case histories, taking their cue from Sydenham. They reproduced case histories from practice, abandoning citation of cases from medical textbooks. Neither Hoffmann nor Stahl practised in hospitals, their clinical knowledge deriving from extensive private practice. It is to their cases and theories that Cullen and Joseph Black made reference so often. Stahl in particular questioned both contemporary and received scientific theory, emphasising scepticism and medical reform. He challenged the recent theories on the origin of fever, contradicting the idea that it arose from the friction of motion (in the blood). He also questioned material explanations in biology, thus questioning the main principle of mechanist medicine. Cullen's later interests in nerve theory go back to his readings of Stahl and Hoffman and their explanations of holistic function.

Stahl was similarly critical of contemporary chemistry. Not surprisingly, he turned to experimental chemistry, claiming to learn from nature. Stahl's observations in both disciplines, like Sydenham's, became exemplary publications reaching many editions. Thus the influence of the 'Dutch connection' on Scotland must be revised in favour of a Protestant network drawing on many centres of learning.

Cullen's ideas were largely worked out before he taught medicine in Glasgow in November 1746. Cullen began his lectures by arguing against using textbooks such as the *Institutes and Aphorisms* of Herman Boerhaave.[49] In his chemistry lectures, begun even earlier, in 1744, he wrote:

> It may be alleged I have done wrong in setting the chemical part of philosophy in opposition to the mechanical ... The elements of bodies are vastly too small for us to discern their form and size, we have nothing yet with respect to these but conjectures, & the application of such conjectures & those of the mechanical

---

[48] See, for example, G. E. Stahl, *Observationum chemico-physico-medicarum curiosarum-menses Julius-December* (Frankfurt, 1697); G. E. Stahl, *Experimenta, observationes, animadversiones, CCC numero, chymicaet physicae, qualium alibi vel rara ... commentatio aut explicatio invenitur ...* (Berlin, 1731); F. Hoffmann, *Friderici Hoffmanni observationum physico-chymicarum selectiorum libri 3, in quibus multa curiosa experimenta et lectissimae virtutis medicamenta exhibentur ...* (Halle, 1722); F. Hoffmann, *Observationum physico-medicarum sylloge comprehendens quam plurima medicinam tam theoreticam quam practicam elinicam ac forensem nec minus philosophiam naturalem utramque insigniter illustrantia et confirmantia medici veterani phoenomena experimenta coll. ...* (Frankfurt, 1736).

[49] Thomson, *An account of the Life, Lectures and Writings of William Cullen MD*, i, p. 26.

philosophy has hitherto been very unsuccessful in explaining the properties of bodies which chemistry considers.[50]

These comments acknowledge the recent controversies raging over the work of Stahl, Boerhaave and Hoffmann, and can be understood only in that context. Friedrich Hoffmann had visited Robert Boyle in Oxford in 1684 after travelling through Holland.[51] In his *Medicina rationalis* (Halle, 1718–40), a book much read in Scotland, Hoffmann cites Sydenham.

The six beds of the university St Caecilia Hospital in Leiden, the mythical cradle of bedside teaching, were to have been the model for Archibald Pitcairne, Alexander Monro and others in founding modern medicine in Scotland. The 'sofa' of medical chairs held down by Hermann Boerhaave (institutes of medicine, botany and chemistry) is said to have been seminal to Scottish medical teaching methods. But Boehaave's consolidation of his 'sofa' was late, between 1709 and 1724, many scientific achievements in chemistry, botany and clinical observation were already under way or now in dispute. His was an achievement of eclecticism, unproductive for research, but suited to being a teaching tool.[52]

For Scottish medicine to emerge and compete in the European forum it needed the international networks built to bolster the new observational sciences. This 'new learning' provided the material which, by 1690, Scotland wanted for its own.

Botanical Networks

Increased trade with overseas swamped the tidy knowledge of home remedies. Plants and plant substances were entering Europe with each vessel. For instance in 1698 James Cunninghame, a surgeon with the East India Company, brought back natural history specimens and paintings of 800

---

[50] As quoted from A. Donovan, 'William Cullen and the Research Tradition of Eighteenth-Century Scottish Chemistry', in R. Campbell and A. Skinner, *The Origin and Nature of the Scottish Enlightenment* (Edinburgh, 1972), p. 103. Donovan sees this as Cullen using the 'scythe of scepticism' without tracing Cullen's remarks in the context of the mechanism-organism debate as described above.

[51] M. Neuburger, 'Some Relations between British and German Medicine in the First Half of the Eighteenth Century', *Bulletin of the History of Medicine*, 17 (1945), pp. 217–18.

[52] Everyone could agree on Boerhaave's *Institutes*. They are not pioneering, but consolidating achievements. See R. James, *The Modern Practice of Physic; as improv'd by the celebrated professors, H. Boerhaave and F. Hoffman . . . being a translation of the Aphorisms of the former, with the commentaries of Dr Van Swieten, so far as was necessary to explain the doctrine laid down, and of such parts of Dr Hoffman's works, as supply the deficiencies of Boerhaave. And render the whole practice of physic compleat: wherein, the various diseases to which the human body is subject, are distinctly consider'd, whence the diagnostics and prognostics together with the*

plants in their natural colours.⁵³ Cunninghame's collection became part of Sir Hans Sloane's vast documentation of natural science. Sloane's connections with Scotland are not well researched, but he had local knowledge and close ties to the west of Scotland, his family being of Scottish descent. Both private and public gardens were scrambling to come to terms with the growing habits, conditions of light and heat, the possible extracts in drugs and narcotica, or even the edibility, of botanical imports. The impact of this trade looms large, from the tulip craze in Holland to Scottish imports of Dutch onions, to books written by sea captains on how best to preserve plants on long sea voyages (pack them, bare rooted, in wooden boxes, in wet moss and keep them wet).⁵⁴ Exotica were available, marketable and saleable but, as yet, changelings. Medicinal uses of indigenous plants were being replaced by the demand of the public for panaceas like tea and coffee, or, for example, 'Peruvian bark'. Tracts were numerous on their efficacy or danger. John Coakley Lettsom, a Quaker and a student of William Cullen's, wrote his dissertation in Leiden in 1769 which was published in English as *The Natural History of the Tea Tree: With Observations on the Medical Qualities of Tea* in 1772.⁵⁵ The private gardens of lowland Scotland were home to the wondrous potato, not yet a field crop, but dear and extravagant, gracing the tables of the aristocracy as a novel delicacy.⁵⁶

Nomenclature and classification were an urgent scientific problem throughout the eighteenth century, acute in the beginning as well as the end of the century, despite Linnaeus. The problem and its solution in botany had significant repercussions on medical science, as learned physicians and surgeons almost invariably shared in the scientific expansion of botany and materia medica. A book of 1804 still lists the other forms of plant classification used together with the Linnaean method, those of Robert Morrison (*c.* 1680), John Ray (two methods, of 1682 and 1703), Tourneforte (1694), Paul Hermann (*c.* 1695), Hermann Boerhaave (1710), Christian Knaut of Halle (1716) and Christian Gottlieb Ludwig of Leipzig (1757).⁵⁷ Except for Ray and Hermann, these men were all physicians. Carl Linneus' system of nomenclature was

method of cure are regularly deduc'd, and the prescriptions adapted thereto from Boerhaave's Materia Medica, are added to every aphorism (London, 1746), in two vols. The FPSG bought books by James for the Library in the 1750s.

⁵³ G. Saunders, *Picturing Plants: An Analytical History of Botanical Illustration* (London, 1995), p. 80.

⁵⁴ T. Davies, *Some Instructions for Collecting* (n.p., 1790), p. 4.

⁵⁵ W. T. Stearn, *The Influence of Leiden on Botany in the Seventeenth and Eighteenth Centuries* (Leiden, 1961), pp. 36–37.

⁵⁶ R. Salaman, *The History and Social Influence of the Potato* (Cambridge, 1985), p. 389.

⁵⁷ S. Barton, *Elements of Botany: or Outlines of the Natural History of Vegetables* (London, 1804), pp. 4–9. Appendix on methods of classification.

set out in his *Systema naturae* (1735) and *Species plantarum* (1753).[58] When James Sutherland compiled the *Hortus Medicus Edinburgensis: or A Catalogue of the Plants in the Physic Garden at Edinburgh* of 1683, he used the system 'of our most learned and incomparable countryman Doctor Morrison'.[59] Morrison's system was one of the most difficult.

Public gardens were as essential as private collections, fulfilling a didactic and scientific purpose. They were systematic, accessible to the public and provided instruction by way of specimens. Private collectors had the means to augment the teaching aims of the public gardens and often had well-placed contacts in major overseas trading companies. This interaction was crucial. Sir Hans Sloane and his support for the Chelsea Physic Garden is but one instance. Sir Joseph Banks's contribution of 500 seeds to the same garden from his voyages is another.[60] Knowledge of indigenous plants flowed into teaching; and this was expertly supplied by men like John Ray or Sir Robert Sibbald. The Edinburgh Botanical Garden originated as a systematic repository of 'simples' of local origin, the term designating plants with medicinal value.[61] In the 1670 physic garden 'simples' were to be grown and made available to the medical profession. Only in 1675 when Sibbald and Andrew Balfour were appointed 'visitors' did they ask 'to embellish the garden and import plants from all places'.[62] When the catalogue was written in 1683, it was with the exchange of plants in mind, and by then Sutherland could describe in detail how they had been derived from the Levant, Italy, Spain, France, Holland, England and the East and West Indies.[63] The prime purpose of the collecting activities of the botanical garden in Amsterdam was also medicinal, its catalogue appearing from 1697–1701 under a title reflecting the new proximity between 'simples' and the exotic: *Horti medici Amstelodamensis rariorum plantarum descriptio et icones*.[64] Other aspects of horticulture had as great an impact in these seminal years, as F. A. Roach explains in *Cultivated Fruits of Britain*:

> The late seventeenth century marked the beginning of a period of renewed interest in gardens, not only those laid out for the royal palaces and many large private

[58] Stearn, *The Influence of the Leiden on Botany*, pp. 31–35.
[59] M. V. Mathew, 'James Sutherland (1638(?)–1719): Botanist, Numismatist and Bibliophile', in H. M. Wright (ed.), *The Bibliotheck* (Glasgow, 1987), p. 6. These are the words of Sutherland himself from his introduction.
[60] Phillips, *History of Cultivated Vegetables*, p. 12.
[61] Stearn explains the term in his *The Influence of the Leiden Botanical Garden*, p. 21: 'simple' (Latin, *simplum*), meaning a 'medicinal herb'.
[62] Mathew, *James Sutherland*, p. 6.
[63] Ibid., p. 5.
[64] Saunders, *Picturing Plants*, p. 72.

houses built at this time but also the gardens attached to the more humble dwellings of country people ... The walls around the gardens of the houses of the rich were clothed with trained trees of apples, pears, plums, cherries, peaches and other tree fruits, while trees trained as espaliers or in bush form, the apples often being on dwarfing rootstocks, were planted alongside the paths and walks. Soft fruit bushes, especially red and white currents and gooseberries, were planted out in beds ...[65]

Looking at pictures of both the physic garden and the college garden at Glasgow University one is struck by just this combination of fruit trees, shrubs and 'simples', these last in a special garden. Whatever the extent or success of an individual garden, planting it heralded a modern programme of medical teaching, particularly if medical interests were associated.

These medical interests went beyond the botanical. The Oxford Physic Garden, founded 1621, did not originate with the medical faculty and was 'not established with primarily medical aims in mind'.[66] Its influence as a scientific centre, with extensive catalogues and trading of plants, began in the period after the appointment of the Scotsman Robert Morison (1669) as Professor of Botany, and was continued by the Bobart family, originally from Brunswick, who advanced the cultivation of exotica and the cause of classification.[67] Johann Jacob Dillenius, a German renowned for his work on cryptogams, extended its scientific reputation.[68] His position was secured by the eminent botanist William Sherard, whose generous bequest saved Dillenius from university parsimony. The Chelsea Physic Garden was the place where the medicinal knowledge of plants merged with the new botanical interests, first pursued by Hans Sloane. This was the agenda followed by the happy – and early – combination of medicinal interests with the study and planting of exotica in Edinburgh. James Sutherland's 1695 appointment as professor of botany joined the plantsman to the teacher of materia medica.

The 1704 foundation of the physic garden in Glasgow also fitted these objectives. Glasgow's trade routes to the Americas were expanding, and, as W. T. Stearn has pointed out, there were three important waves of botanical imports coming to Europe just then. Canadian and Virginian herbaceous plants had reached Europe between 1620 and 1686; South African plants entered from 1687 to 1772, as did North American plants from the eastern

---

[65] F. A. Roach, *The Cultivated Fruits of Great Britain: Their Origin and History* (Oxford, 1985), pp. 54–56.
[66] C. Webster, 'The Medical Faculty and the Physic Garden', in *The History of the University of Oxford* (Oxford, 1986), v, p. 712.
[67] Ibid., pp. 713–15.
[68] Ibid., p. 719.
[69] W. T. Stearn and W. Blunt, *The Art of Botanical Illustration* (London, 1994), p. 330.

areas of settlement.[69] Glasgow assumed its place in the new scientific and educational enterprises connected with physic gardens, which became indistinguishable from botanical gardens. This melding of interests formed the backbone of the botanical sciences, tied to the universities and to the medical incorporations by teaching. In Glasgow – and elsewhere – botany and chemistry were taught by university medical graduates – excepting only the outstanding William Hooker – until the nineteenth century.

This marriage of botany to medicine in the very practical arena of university posts and support of physic gardens had an important impact on the development of the medical sciences. Classification and the means of identifying diseases was hugely influenced by botany. Importing new plants and understanding where they were to fit into the descriptive methods of identifying them to others (colleagues as well as pupils) strained inadequate and local nomenclature.

The 'catalogues' of indigenous and 'exotic' plants produced by men like John Ray or James Sutherland, and highly valued in Glasgow,[70] were more than botanical in scope. Their pivotal value for medicine lay in the problem they were tackling. No longer were such scientific activities trying to describe a 'simple' so that it could be found and used, as in the standard herbals. They were trying to differentiate 'specific entities', a broad scientific enterprise that Sydenham had tried to tackle in his *Observations* of 1676. Like botanists, he was trying to read the 'signs' which would identify discreet types. Secondly, and plants are prototypical for this, seasonal growth emphasised morphological change: a problem familiar to medical men as they observed the *process* of how illness developed.[71] Thus 'classification' was by no means only a botanical pursuit, it was also a common one concerned with morphology, specificity and classification. Botany was linked closely with 'physic' by more than the use of plant remedies. The new style 'physic garden' was no longer the herbal garden of the monks. Instead, it fulfilled a multiplicity of purposes: domesticating imports for pharmacological use; providing a 'laboratory' to investigate and identify plants; and setting new research agendas.

The catalogues published as part of the work of the new style physic gardens and the experts needed to fill the posts of keepers enhanced the great push toward systematising the sciences. The habit of recruiting talent internationally was very noticeable, again undermining narrow nationalist

[70] Principal Stirling of Glasgow sought to recruit Sutherland. See A. D. Boney, *The Lost Gardens of Glasgow University* (London, 1985), p. 27.

[71] An analysis of this epistemological change is contained in W. Lepenies, *Das Ende der Naturgeschichte: Wandel kultureller Selbstverständlichkeiten in den Wissenschaften des 18. und 19. Jahrhunderts* (Munich and Vienna, 1976), p. 74, but our interpretation differs from his.

interpretations. One example was G. D. Ehret, an outstanding botanical artist of the time. Born in Heidelberg, he was the protégé of the medical doctor Johann Jacob Trew in Nuremberg. Moving to Paris, England, Holland and then back to England where he stayed after 1736, Ehret produced for Linnaeus, in Leiden, the definitive drawings introducing the binomial sexual classification of plants, which were then pirated throughout Europe, before he settled to teaching and working in London.[72]

There are many other examples of the growing demand for international scientific exchange. Scotland was not slow in using its trade routes to build modern scientific outlets for education and commerce. Establishing botanical gardens, in the brief period between 1695 and 1704,[73] that linked with the definitive intellectual developments was a sign of its alertness. Botany was a crucial first step, showing Scotland's active participation in scientific networking. In 1709 Boerhaave attracted many students to Leiden in botany chemistry and medicine. In the new University of Göttingen, founded in 1734, Albrecht von Haller achieved an international reputation for, among other things, his stewardship of the botanical gardens from 1736–53. His great work in systematising botany, surgery, anatomy and medicine survives in his unsurpassed bibliographies.

In Vienna, the medical reformer Gerhard van Swieten established a separate physic garden in 1754, annexing for medicine what was a Habsburg-financed botanical extravaganza, the gardens and sumptuous publications cared for by Nicolas von Jacquin. Paris was no different. Like Vienna it was the centre of a powerful and moneyed court. Its many gardens satisfied the rich and benefited middle-class academicians and medical men. Paris and Vienna attracted trained minds to their horticultural and botanical bounties. The interest in cultivating and managing plants was a modern medical pursuit, turning to scientific purpose the personal patronage of the rich and their need for ostentatious display. The sciences slowly constructed a new man-made nature, composed of classification systems and theories. Glasgow and Edinburgh were deeply involved, joining the culture of science to educational strategies. Their physic gardens and their schools of medicine had as their first clients surgeons and physicians – and a growing number of other men and women, who wanted to benefit from the new remedies and paid for tickets to attend courses and lectures.[74]

[72] Stearn and Blunt, *The Art of Botanical Illustration*, pp. 159–62.
[73] The date 1695 rather than 1670 is taken for Edinburgh because our emphasis is on the combination of teaching use and botany; the dates given imply the involvement by surgeons.
[74] R. Stott, 'The Battle for Students: Medical Teaching in Edinburgh in the First Half of the Eighteenth Century', in *Edinburgh's Infirmary: A Symposium Arranged under the Auspices of the Scottish Society of the History of Medicine* (Edinburgh, 1979), pp. 4–5.

The International Agenda in Anatomy

The same spirit of exhilarating scientific exploration characterised anatomy. From 1700 to 1800 a leap was made from the infrequent, ritualised and moralising dissections common since anatomical theatres were first founded to scientific, and morally neutral (for doctors) anatomy lessons. Anatomical teaching was systematised, much as botany was, and became a subject taught at university.

It was of critical importance that surgical teaching in Glasgow and Edinburgh liberated itself from the constraints of the local craft tradition, with which it was already uneasy. Although the incorporations remained professionally and territorially jealous of their hegemony, what they were aspiring to establish in teaching terms, after 1690, bore little resemblance to homegrown practices. Dissection was rapidly becoming, after 1700, part of practical training, but was not practised in Scotland. In anatomical teaching in Edinburgh after 1704, for instance, the Master Surgeons of the Incorporation demonstrated on various parts of the body in succession over nine days.[75] Students were not given the opportunity to practise dissection, or to learn how to make preparations. Instruction was infrequent, with the dissection being demonstrated only once. Matters did not improve by the granting of dissection rights on the unclaimed bodies of infants and children from the foundling hospital.[76] To learn systematic anatomy and to operate on the cadaver (an innovation used in Paris) or to learn midwifery, it was necessary to go abroad.

Change came about in Scotland between 1714 and 1720 when John Gordon taught anatomy at the university in Glasgow and Alexander Monro held his first lectures in Edinburgh. Lectures implied a systematisation of knowledge not used in apprenticeship learning, or in the occasional dissection of amputated limbs.

In Leiden and Amsterdam anatomical dissection was entering a new era as chemical expertise expanded. The ability to inject preparations or to preserve them in spirits was the product of Dutch science. For the system of modern anatomical teaching to emerge, this ability to 'freeze' visual instruction needed to be joined to systematised learning. Scientific progress was impacting on education in much the same way as in botany, the difference being an already nascent culture of private lectures in hospitals. In London William Cheselden was amongst the first to inaugurate private

---

[75] 'Teaching of Anatomy at Edinburgh', *Annals of Medical History*, 8 (1926), pp. 325–26.
[76] Ibid., p. 326.

surgical lectures (1711), moving them to St Thomas' Hospital in 1718 and delivering four courses per year.[77] This 'private' teaching developed by surgeons attached to the major hospitals became greatly influential under various teachers at St Bartholemew's and Guy's Hospital. John Hunter later taught about a thousand students at St George's Hospital, an institution favoured by Scots.[78] The importance of these lectures was recognised abroad. Lorenz Heister, the great textbook author and surgeon, went to London in 1710 (after medical studies in Leiden and Amsterdam) 'collecting everything new in the several branches of Physic'.[79]

From about 1690 to 1730 anatomical and surgical teaching in Paris underwent a startling revision. Labelled the 'Paris manner of dissection' this involved teaching anatomy and surgery with manual exercises on dead bodies. The supply was not abundant, but it was sufficient for many students to do preparations, try out operations and sometimes dissect. With this approach established, the other strength of Paris teaching was the wealth of lectures and training in the different areas of surgery and midwifery available there. The surgeons and physicians of Paris were leading in specialist areas (lithotomy for example) and opened their houses commercially to the influx of foreign students. The high concentration of patients in city hospitals, and the method of working directly on cadavers, attracted the diligent and ambitious among medical men. These came from all over Europe. Peter the Great of Russia and Frederick Wilhelm I of Prussia paid for their military surgeons' education there.[80] Many relied on private means, such as the Swiss Albrecht von Haller and Johannes Gesner, and many Germans, English and Scots, converged on Paris as well, either as 'pensionaires' of the renowned teachers, or as students permitted to attend the hospital or private teaching.

A medical 'grand tour' evolved, shuttling enterprising medical men willing to invest in 'gathering in the new', as Heister put it, between London, Holland and Paris. Its main axis was Leiden and Paris, its subsidiary stops London and, latterly, Edinburgh. Anatomy and surgical operations were at the top of this agenda and Glasgow tapped into its potential throughout the eighteenth century. Both Scottish cities established their own medical teaching empires because they transferred knowledge and skills learned by travelling

---

[77] R. Porter, 'Medical Lecturing in Georgian London', *British Journal for the History of Science*, 28 (1995), p. 98.

[78] Ibid.

[79] Neuburger, *Some Relations between British and German Medicine*, p. 225.

[80] H.-U. Lammel, 'Zur Stellung der Pensionärchirurgen an der Berliner Charité', pp. 59–68; M. B. Mirskij, 'Medico-chirurgische Ausbildung in Rußland', pp. 159–79, both in G. Harig (ed.), *Abhandlungen zur Geschichte der Medizin und der Naturwissenschaften*, 57 (Husum, 1990).

this circuit to institutions established at home. These exchanges of knowledge formed the backbone of the new system favoured in Scotland. Leiden supplied a model for systematic teaching, while in Paris and London ample opportunity was available to practise surgical techniques.

These itineraries served other purposes as well. Books unavailable or costly in Britain were bought abroad. Walter Sinclair, an Edinburgh medical student who went to Leiden in 1736 and then to Paris in 1737, devoted 10 per cent of his expenditure to books, and other Scots students did likewise.[81] Exposure to an international culture and its ways of doing things broadened horizons: 'there is a concourse of students from all parts, even from Muscovy'. Medical teaching was intensive with up to four lectures a day plus (for 'private colleges') the time needed to write out dicates'.[82] One student, Alexander Coventry, who went from Glasgow to Edinburgh vividly described his attendance at medical lectures in 1785:

> My time was fully occupied. I took notes from all the lectures, generally the leading topics, which I filled up at my lodgings, which kept me from bed till two in the morning. Then Webster at seven, Cullen at nine, Infirmary at twelve, Munro [Alexander Monro, secundus] at one, and twice a week Clynical Lectures at three, besides our notes and visits at the Clynical Ward.[83]

Scottish students who were planning to go to Paris after Leiden secured private language teaching in French while in the Dutch city. They worshipped in the French church, 'combining devotion with practising their French'. Laguage proficiency was at a premium, as the French surgeons lectured in the vernacular. The German-speaking Swiss Johannes Gesner heard the lectures in Paris in French, but kept his diary of what he learned in Latin. Gesner had sufficient private means to seek the best medical education, and he chose to study in Zurich, Basle and Leiden, but could not dissect until he went to Paris in 1727.[84]

Anatomy had a broader remit than dissection, including learning how to make preparations. Surgical studies, too, while including dissection, combining systemised learning with practical operations, first on corpses and then, if the student was judged to have had sufficient experience, on the teacher's own patients. Preparations could be made or imported and animals were used to demonstrate basic medical knowledge. The first preparations

---

[81] K. van Strien, 'A Medical Student at Leiden and Paris: William Sinclair, 1736–38', *Proceedings of the Royal College of Physicians of Edinburgh*, 25 (1995), i, p. 297.
[82] Ibid.
[83] Van Strien, *A Medical Student at Leiden*, i, p. 297; p. 300, for 'practising French'.
[84] U. Boschung, *Johannes Gesners Pariser Tagebuch* (Bern, 1985), p. 111.

1 The City of Glasgow, *c.* 1820, by J. Clark

2 Broomielaw, Shipping, by J. Fleming

3 View of the Hunterian Museum, by J. Fleming

4 The Old Town's Hospital, c. 1849, by T. Fairbairn, to the far right

5 The University of Glasgow, c. 1693, by John Slezer

6 The Second Faculty Hall (1791–1860), photographer unknown

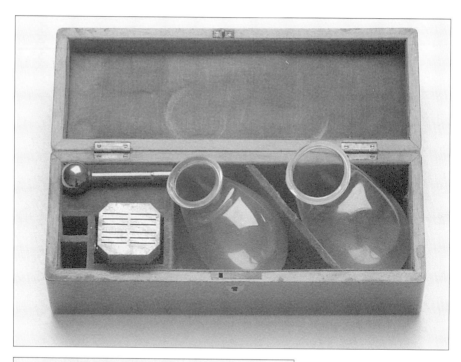

7 Cupping set, *c.* 1790 presented to the Royal College of Physicians and Surgeons of Glasgow by Dr J. Menzies Campbell DDS

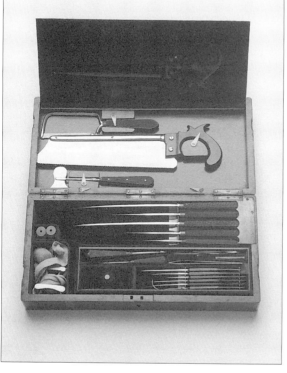

8 Amputation set, *c.* 1840–1850 made by S. Maw and Sons of Aldergate Street in London

9 William Cullen (1710–1790), oil on canvas, protrait by the Scottish School, undated.

10 William Hunter (1710–1790), oil on canvas, portrait by Allan Ramsay, c. 1785

11 Robert Watt (1774–1819), artist unknown

12 John Moore (1729–1799), oil painting, copy by J. Barr of the original by Sir T. Lawrence

13 Joseph Black (1728–1799), oil on canvas, by Sir Henry Raeburn, undated

14 John Burns (1774–1850), oil on canvas, by John Graham-Gilbert, 1832

15 Glasgow Royal Infirmary by David Allen

16 Blythswood Place, St Vincent Street, by J. Fleming

documented as having in been made in Scotland were skeletons.[85] In 1815, more than a century after their introduction as a technical innovation, the surgeon John Cross wrote of the value of preparations as teaching aids when comparing Paris with London:

> In regard to skeletons, dried diseased bones, course blood-vessel preparations, dissected or corroded, nothing is wanting ..., but, with the exception of foetuses and monstrosities, this museum does not contain thirty preparations preserved in spirit, whatever of morbid or natural parts. There is not one moist preparation of the minute structure of the organs of hearing, the eye, the nose, the viscera, etc.; nor a mercurial injection of the absorbents or excretory duct of the testicle. To conclude, however, that the anatomical lectures at Paris, are what those delivered in London would be without preparations and dissections preserved in spirit, would be erroneous, because, for the former, much more trouble is taken to make recent dissections for the lecture, and there are more able assistants and better dissectors ... But it is impossible wholly to supply the place of minute injections and preparations. How are we to shew the progressive steps in the formation of bone, without their assistance; the commencing vascularity of soft cartilage, the shooting of considerable vessels into it, when bony matter is beginning to be deposited? The minute structure of the eye may be described, but can never be demonstrated without injected preparations.[86]

Anatomical instruction was in its infancy at the beginning of the eighteenth century. At its best it was systematic, as graphic as possible, detailed and three-dimensional. Thanks to new teaching aids spatial relations and manual precision were taught. Although valued for their verisimilitude, expensive folios could not supplement work on the corpse or in making preparations to teach surgical cuts. Textbooks like Vesalius's *De humani corporis fabrica* failed in comparison to visual and manual 'hands on' instruction. Gesner declined to buy the new edition of 1725.[87]

Scots went on the same educational itinerary as other medical men. Albrecht von Haller recorded his work in Paris, Leiden and other cities in 1727–28.[88] So did Adam Murray from Edinburgh, who wrote to his elder brother William in 1724 that two years study was essential: 'As to my coming

---

[85] V. Tansey and D. E. C. Mekie, 'A Collection of Curiosities (1699–1763)', in *Catalogue of the Museum of the RCSEd.* (Edinburgh, 1982), p. 257.

[86] J. Cross, *Sketches of the Medical School of Paris: Including Remarks on the Hospital Practice, Lectures, Anatomical Schools and Museums; and Exhibiting the Actual State of Medical Instruction in the French Metropolis* (London, 1815), p. 31.

[87] Boschung, *Johannes Gesners Pariser Tagebuch*, p. 112.

[88] E. Hintsche (ed.), *Albrecht Hallers Tagebuch seiner Studienreise nach London, Paris, Strassburg und Basel, 1727–28* (Bern, 1968); and *Albrecht Hallers Tagebücher seiner Reise nach Deutschland, Holland und England, 1723–27* (Bern, 1971).

home, you may be sure I'll make all possible dispatch, but before Whitsunday 1726 I don't see how I can finish what I must necessarily ply here [Leiden] and in France'.[89] Murray took the same educational route as Haller, from Leiden to Paris. The attraction in Paris was dissection and Murray described succinctly what he needed to do, writing that he 'shall be close employed about dissections (sic) in the Hotel Dieu Operations, etc., till some time in April, minding all the while nothing but nasty butchery work'.[90]

The scion of a well-known Swiss family, Johannes Gesner, who became Zurich's most famous naturalist, kept a diary of his stay in Paris, on the 'nasty butchery work' which he and his brother, and his countryman, Albrecht Haller, engaged on in the years 1726–27. Paris was shameless in dispensing with cultural inhibitions regarding dissection and provided the most extensive practical surgical expertise in Europe.

In 1686, *l'année de la fistule*, Louis XIV was cured of an anal ulcer by his first court surgeon Charles François Felix, who was rewarded handsomely and subsequently ennobled in 1690. This gave tremendous prestige to surgeons, who afterwards no longer needed to incorporate solely as a craft guild, being given the status of a free trade (1699). They separated formally from the barbers in 1743.[91] In 1725 regular lecture courses began at the college of St-Côme with the appointment of five demonstrators to teach specialist courses. The third and fourth demonstrators were responsible, respectively, for a course of anatomy using a corpse and a course of surgery in which operations were undertaken on the cadaver. This teaching brought them into conflict with the Paris Medical Faculty, including the famous anatomy teacher Jacques-Bénigne Winslow, but in general boosted the already dominant place of Paris in providing for dissection and surgical operations. This significant expansion placed the surgical and anatomical teaching of Paris ahead of Leiden. For a time Leiden continued to be influential through its anatomy teaching and mounting of preparations. The coronary and vascular systems and other soft tissue parts could only be made visible through the injection methods developed by the Dutchmen Johannes Rau, Frederick Ruysch and Bernhard Albinus. But even Bernhard Siegfried Albinus went to Paris in 1718 to expand his anatomical knowledge, studying with Jacques-Bénigne Winslow.[92]

---

[89] T. C. Smout, 'A Scottish Medical Student at Leiden and Paris, 1724–26', *Proceedings of the Royal College of Physicians of Edinburgh*, 24 (1994), i, p. 102.
[90] Ibid., p. 103.
[91] Boschung, *Johannes Gesners Pariser Tagebuch*, p. 30.
[92] Ibid., p. 68.

Gesner enrolled in the courses of Henri-François Le Dran (1685–1770), a leading surgical teacher who lectured on surgery, but, above all, 'while appointed to the Charité', demonstrated surgical operations on the dead body.[93] He also offered his students, who lodged with him, training in how to make their own preparations. This, too, would have meant being supplied with a corpse. Indeed this was the appeal and the danger of private teaching in Paris. Paying for both the teaching and a corpse was expensive. The journals of Gesner reveal an intense dedication, however, to all aspects of anatomising, showing it as a group activity where everyone was making the most of the police turning a blind eye. The courses were only one aspect of what could be learned; as operations were shown by Le Dran on the corpse, its head, legs and arms were cut off and given to students to practise dissection and preparation techniques. Heads were removed by Gesner and other medical friends; the Gesner brothers cooked their head to identify and study the bones, in student digs on Christmas Eve, hoping to defy detection.[94] Albrecht Haller lists two heads in 'two round boxes' among items sent home from Paris.[95]

The greatest difficulty seems to have been learning the techniques of making preparations. A crucial thing to learn, because teaching was now coming to depend on more than skeletal preparations, was injection technique. These had been pioneered in Holland under Ruysch, but the raw material was apparently easier to acquire in Paris, and crucial to acquiring expertise. Both Gesner and Haller failed in their attempts to inject inner organs with solutions made of wax, turpentine and oil.[96] Gesner recorded that his injection slipped and the coronary vessels became distorted. His description details every minute action, as did Haller's notations, because they intended to make use of them later, since both wanted to teach.[97] Together with the obvious care taken to learn making preparations, the journals show an assiduous search for ways of illustrating the internal structure of the body. The anatomical museum of Duvernay and the wax preparations in the collection of the surgeon Desnoues were studied. The wax preparations were so lifelike that they made a strong impression, the art of making them was still a secret and their use was to be potent in the future. The Hapsburg emperor and enlightened reformer, Joseph II, paid for the collection used in the military surgeons' school in Vienna (the

[93] Ibid., p. 94.
[94] Ibid., p. 105.
[95] Ibid.
[96] Ibid., p. 108.
[97] Ibid.

Josephinum) towards the end of the century. The intricacies of the nervous and vascular systems were easily visible in wax, which was particularly appreciated after the miseries encountered by students trying to do preparations themselves.

These anatomy courses set the pattern for teaching in Scotland. Osteology, myology, angiology and neurology were followed by more specialised subjects.[98] The circulatory system and the lymphatic system were scrutinised, followed by lectures on sensory perception and the inner organs. Sections of the brain were examined, as this was a speciality of Winslow.[99] The surgical courses were just as intensive, consisting of sixteen lectures and twenty-two demonstrations. The demonstrations were of surgical operations performed on cadavers. Students could also watch operations in the hospitals. Autopsies were the norm at the Charité where it was found that in four out of five cases the operation could not have helped the patient to recover.[100]

The need to acquire skills in visualising and presenting the complex inner structure of the body originated in Leiden, Paris and (to some extent) London. Scotsmen were assiduous in utilising learning proffered in these centres, the palette of skills acquired being then transferred to Glasgow and Edinburgh. Early examinations required the dissection of animal parts, but the new specialised techniques of making preparations culminated in producing a spatial map of the body. In the end this was the decisive challenge. The trading affinities of Glasgow and Edinburgh with France, the Netherlands, Russia and its union with England set a fast pace for amalgamating the new. The advances just outlined in such detail give a measure of how quickly technologies and medical science moved. The context was international and the medical arena that of the enterprising bourgeoisie. The sources for the rise of Scotland's superior medical education, acknowledged by so many in the nineteenth century, were these.

The European parentage of Scottish intellectual and professional proficiency is typified in William Cullen. He was first a product of the intellectual culture of the west of Scotland and shows, more than anyone, its burgeoning influence in the larger arena of the pan-European development of medical education.

---

[98] Ibid., pp. 112–15.

[99] Ibid., pp. 117–24.

[100] Ibid., p. 147. For the foregoing information, see pp. 127–29. An exact plan of teaching worked out for Le Dran shows his operations and their teaching purpose in minute detail.

Cullen and European Medicine

'I was in Paris when the news of your establishment in the University of Glasgow reached me – my joy was too great to be silent ... Yet not on your account alone did I rejoice – ever affectionate to Alma Mater I have been long ambitious of her making that figure in the Medical that she has already done in almost every other science.'[101] These sentiments are those of an Irish medical student, Alexander Haliday, in 1751, when William Cullen was appointed Professor of Medicine in Glasgow, after having taught there for seven years. In 1756 Cullen went to Edinburgh as Professor of Chemistry, transferring to the medical chair in 1766.

Cullen's career in Scotland best exemplifies its mid century genius in linking practical skills and theoretical science. He best represents the coming of age of Scottish medical education. In his years in Glasgow he adapted *virtuosi* scholarly pursuits to a teachable curriculum. He helped transform the skills-based learning of apprenticeship tradition to an instrument capable of appreciating medical research. This was crucial because it changed the emphasis of medical education from competence in skills to cutting-edge involvement with educational change. The pursuits of the few and the learned were joined to accommodate those who in increasing numbers attended university lectures. In Scotland this meant something different to the exclusive academic practices of, for example, Oxford and Cambridge. Tickets for lectures were increasingly available to those interested. Matriculation was not a barrier, the 'non-toga' students being a recognised group.

Right from the beginning of his teaching career Cullen advocated empirical observation, but only if it could be used to sustain a viable theoretical superstructure. This marked Cullen out as an advocate of the European systematisers of medical knowledge. A student recorded the following directives given by Cullen for his course on diseases (clinical pathology) which were typical:

> The Characters delivered in the Nosology are from matters of fact, and independent of all theory; and I am now to unite 'em in a theoretical way, which is the best proof of a natural Class of diseases, usefully constituted for practice.[102]

Cullen and those he inspired by his university teaching were creating a new educational agenda. They reworked modern knowledge to suit the student. This was not learning by rote, but a testing of currently available

---

[101] GUL, Special Collections, MSS Cullen 1–249, no. 92, Letter Alexander Henry Haliday to William Cullen, Belfast, 2 May 1751.
[102] GUL, Special Collections, William Cullen, Lecture Notes, MS Gen. 685, ii, p. 216.

knowledge in the crucible of theory 'usefully constituted for practice'. Teaching had a new focus: that of bringing 'unproven' observations to the attention of the profession. As John Crellin has remarked: 'A sympathetic awareness of students' needs characterises all Cullen's lectures, yet in assessments of his influence this has received little attention, despite his fame as a system builder'.[103]

Cullen's sympathy for teaching was different from that of the many other good teachers Scotland produced at this time. He went beyond an ordinary teacher's coverage of the subject and teased out its uncertainties and puzzles. He went even further: criticising contemporary medical teachers, such as Boerhaave, Stahl, Hoffmann, to teach his students that even the most modern science must be sceptically assessed and if possible superseded. On countless occasions his students were instructed why it was best to reject the past:

> Formerly Chemists thought gold a very necessary Medicine, as having a particular property of strengthening the vital powers of our system, but [this] is entirely hypothetical and we are now perfectly assured it was either Superstitious Ignorance or direct fraud, that kept it in use.[104]

Cullen also staked out new ground when reviewing the important issue of the (physical or immaterial) links between sensory knowledge and physiological process:

> There must be a power giving a new force besides the intellectual Operations, and this we call the Action of the Brain ... There are Cases of Impressions exciting motions without the Intervention of Sensation either reflex or simple and without your Intervention too of Will. Sometimes indeed Consciousness is destroyed by habit ... But we may admit actions it seems without the Intervention either of Sensation or Will. We shall not enter on the Disputes of the Stahlians here.[105]

This is a good example of how Cullen used his teaching to explore the frontiers of both philosophy and medicine, paying a good teacher's tribute to a student's inquisitive mind. He was determined that teaching include research questions.

While Cullen is indeed not known for any specific 'discovery', he completely transformed how medicine was taught. Learning was no longer tied to minimum local licensing requirements. He strengthened the prestige of university lectures by offering students a unique opportunity to learn the

---

[103] J. K. Crellin, 'William Cullen: His Calibre as a Teacher, and an Unpublished Introduction to Treatise of the Materia Medica, London, 1773', *Medical History*, 15 (1971), p. 80.

[104] GUL, Special Collections, MS Ferguson 51, The Medical Properties of Chemical Bodies, As Taken from Dr Cullen's Lectures, 1765, pp. 65–66.

[105] GUL, Special Collections, MS Gen. 685, p. 514.

theoretical implications of practical skills. This speeded up the demise of the apprenticeship system and showed that university courses were capable of teaching practical medicine. After Cullen the significance of the *virtuoso* scientist began to pale, his place being taken by the professional teaching his subject.

Cullen's practice near Glasgow in the 1730s had prepared the ground for his teaching of medicine, chemistry and botany. His work was never that of a lone pioneer. Instead he developed strong links in both the FPSG and the university, gathering around him in the process the founders of two medical dynasties associated with the west of Scotland. If the most famous is that of the Hunter brothers, the three medical professors named Hamilton were equally influential although their impact was local. These relationships, despite separation, remained extremely close, disseminating influential teaching beyond Glasgow.

William Cullen's career touched on the many medical men who used education as a key to success. In a later chapter we look at John Gordon, Cullen's friend within the FPSG, and his lifelong devotion to better skills. Cullen too was part of this trend. The surgeon Robert Wallace, writing to Cullen's biographer on Glasgow medical teaching before Cullen, was right in his perception that the 'Institutes' (physiology) were not being taught. There was then little need, as most clients of medical lectures were seeking surgical and pharmaceutical skills compatible with FPSG examination criteria.

Indeed Cullen was intending to be a surgeon-apothecary. He was indentured as an apprentice to John Paisley,[106] and applied to be examined by the FPSG on 1 March 1737.[107] He did not, in the end, become a surgeon, as there is no record of his completed examination (nor is there one, incidentally, of his licensing by the Barber Surgeon's Company before he became surgeon's mate on a ship), but he did enter the Faculty in 1744 as a physician after receiving the MD degree from Glasgow in 1740.[108]

Cullen's departure from a conventional career as a surgeon may have been connected with his voyage to Jamaica. Sir Hans Sloane had undertaken an earlier voyage in 1687 with the avowed purpose to advance natural knowledge and its usefulness to medicine, 'the best physicians having travelled to the places whence their drugs were brought, to inform themselves concerning them'.[109] *The Natural History of Jamaica* came out in two volumes, the first

---

[106] Thomson, *An Account of the Life, Lectures and Writings of William Cullen, MD*, p. 3.
[107] RCPSG, 1/1/1/2, Minutes of the FPSG, 1733 to 1757, fol. 21r, 21v. There are no minutes for 1 March 1737 (only for 7 February and March), and no examination of Cullen is recorded.
[108] RCPSG, 1/1/1/2, fol. 86r.
[109] Thomson, *An Account of the Life, Lectures, and Writings of William Cullen, MD*, p. 45.

published in 1707 and the second in 1725. Cullen journeyed to Jamaica in 1729 and stayed for two years. He must surely have used Sloane's books, which were among the best guides to contemporary medical investigations into plants and drug use. Joseph Pitton de Tournefort's great *Voyage into the Levant* had been published in Amsterdam (in French) and in London (in English) in 1718 and was a notable contribution to the collection and investigation of unfamiliar plants and drugs derived from them. Sloane, Tournefort, the Prussian physician Andreas von Gundelsheimer, Stahl's predecessor as first physician to the Prussian king, John Ray, and the Keeper of the Ashmolean Museum, the Welshman Edward Lhuyd, were all great collectors and engaged in the enterprise of scientific classification. Tournefort's and Ray's classification systems for plants were among the most widely used of their time. Of Tournefort's collections it was written: 'His Cabinet... was a second Ark, to which the Creatures, animate and inanimate, were come to own [to acknowledge] themselves as it were the Tributaries [those who pay tribute] of him who had brought them together; for each Piece, according to M. Tournefort, had its Quota of Proofs to pay in'.[110] John Ray, in reviewing Sloane's catalogue of Jamaican plants in the *Philosophical Transactions* (1696), called attention to Sloane's prowess in reading, considering and comparing what had been written and relating it to current knowledge.[111] His Jamaican volumes were more than botanical studies: they were intensely observational – 'everything interests him, and, as he describes it, we share his interest'.[112] This unfettered interest and equally valuable desire to explain and make useful – combining of the empirical and the analytical – must have influenced the young Cullen deeply. There was also a strong Celtic connection. John Ray, Edward Lhuyd, and Sloane all had avid interests in the language, the literature and flora and fauna of Scotland and Ireland. Lhuyd wrote his *Archeologica Britannica* (1702–7) 'applying [the] empirical analytic criteria of [the] science of that day to language, and ... recognising the Celtic family of languages'.[113] Lhuyd is seen as having 'awakened from the grave the hardy language that was extinguished'.

Cullen's Jamaican experience and the reading he must have done for it may well explain why he turned to integrative, comparative and systematic methods in medicine. Physiology and chemistry, the subjects to which he turned after returning to Scotland, show the strong influence of natural history. Physiology, plant science and chemistry were integrated in his

[110] 'The Life of M. Tournefort', preface to *A Voyage into the Levant* (London, 1718) i, p. xvi.
[111] E. St J. Brooks, *Sir Hans Sloane: The Great Collector and his Circle* (London, 1954), p. 67.
[112] Ibid., p. 75.
[113] E. P. Hamp, 'Background and European Setting', in C. Ó Dochartaigh (ed.), *Survey of the Gaelic Dialects of Scotland* (Dublin, 1977), i, p. 17.

teaching. Like Tournefort, Sloane and Gundelsheimer, Cullen espoused systematics as indispensable in science. The Quaker physician John Lettsom, who was part of the group around the innovator Linnaeus in Leiden, wrote to Cullen in 1798 when his *Materia Medica* finally appeared: 'I have read with singular pleasure the immense and luminous work of the *Materia Medica*, which will produce a new era in this department, where the truly useful will be discriminated, and the voluminous Inertia discarded'. Lettsom then added, surely grasping the formative influence of Cullen's voyages, that he must not forget 'to mention a Gentleman, who was once thy patient, from one of the Granadilloes of the West Indies, his name Maze ... whilst I write this, I doubt whether he will ever arrive alive in Scotland, but should he enjoy that happiness, he will solicit thy aid'.[114]

Scientific improvement rests on men who have more confidence in their own abilities than in tradition. The culture of medical improvement was not far removed from the experience of other self-made men. The famous advocate and founder of the *Edinburgh Review*, Francis Jeffrey, was of 'undistinguished birth' and set the Tory Sir Walter Scott's teeth on edge with his liberal views, even though Scott, significantly, acknowledged his expertise.[115] He was 'the son of a clerk-depute', a middle-ranking civil servant. Cullen's father was factor to the Duke of Hamilton; William and John Hunter were the sons of a grain merchant who later purchased a farm, Long Calderwood; the Hamiltons, Robert and Thomas, were descended from the Rev. William Hamilton, minister of Bothwell Kirk, all untitled property owners married to educated daughters of the prospering middle classes. The sons could not expect an inheritance, though they were educated and urbane. They were the heirs to frugality and industriousness. As the Reverend William Hamilton wrote to his son, Thomas, in a poignant letter, for he was dying and would not 'ever see [the] face [of his son] in this world':

> do not neglect your duty. Fourthly, Beware of taking up with Bad Company, so particularly such as are given to Cursing, Swearing, Drinking or Whoring for they will bring down the wrath and curse of God upon you so it utterly ruin your reputation, as to your business no one will employ or trust themselves in your hands ...

> Fifthly, See that you diligently attend your business for it is by that you must make your fortune in the world and if you neglect your Business, ye have nothing to expect in this place wherefore be as saving of any little thing ye have as possible.

---

[114] GUL, Special Collections, MS Cullen 225, Dr Lettsom to Dr Cullen, London, 3 August 1789.
[115] 'Francis Jeffrey', in T. Thomson (ed.), *A Biographical Dictionary of Eminent Scotsmen* (London, 1875), half-vol. iv, p. 392.

> Thomas I must now bid farewell perhaps never to meet again. I have given you your education as liberally as was in my power. All that I require of you is that you will diligently pursue and observe the directions I have given you ... Then you will conscientiously fear God and keep His Commandments.[116]

This letter sent Thomas Hamilton on his way to London with John Hunter in the autumn of 1748,[117] both keen to improve their knowledge. His brother William had established a Scottish bridgehead in London with the help of the family of Dr James Douglas at Red Lion Square where William had entered the world of high-risk experimental medical science. He had trained with Cullen and had brought with him the teaching interests that Cullen had fostered. Hunter's sense of purpose was gleaned from Cullen's high ideals, as Hunter made clear in 1745 when he wrote to Cullen about both of them preparing medical lectures, his in London and Cullen's in Glasgow:

> Well, how does the animal economy appear to you, now that you have examined it, as one may say, with precision? I have good reason to put the question to you, because, in my little attempts that way, since I begin to think for myself, Nature, where I am best disposed to mark her, beams so strong upon me, that I am lost in wonder, and count it sacrilege to measure her meanest feature by my largest conception. Ay, ay the time will come when our philosophers will blush to find that they have talked with as little real knowledge, and as peremptorily of the animal powers, as the country miller who balances the powers of Europe!

These men who broke the mould of teaching in Glasgow, Edinburgh and London, who competed there with the best and most ambitious medical men, were treated in London and even in Edinburgh as upstarts. Cullen and the Hunters were lampooned for a supposed lack of Latin,[118] much as the nouveau riche were deplored for a lack of social graces. In the broadside *Dr Puff* Cullen was taken to task for lecturing in English in Glasgow, the implication being that he would lose Edinburgh's higher class of student with such common parlance. They complained, too, that he did not have the approval of the Edinburgh medical clique.[119] John Hunter, equally, was under fire, when already established as a leader in investigative medical science, for the sin of an 'ungentlemanly' lack of Latin. He was robust in his defence, as he was when, as a medical student, they wanted him to acquire

---

[116] GUL, Special Collections, MS Gen. 1356/2, Letter of William Hamilton of Bothwell to Thomas Hamilton, Bothwell, 18 October 1748.

[117] J. Hunter: Bicentenary Celebration: The Hunters and the Hamiltons: Some Unpublished Letters, *Lancet* (1928), p. 354.

[118] *A Letter from a Citizen of Edinburgh to Dr Puff* (Edinburgh, 1764), p. 11.

[119] Photocopy of an article on Cullen with no bibliographical reference extant, RCPSG, 1/3/8/6, p. 28.

the polish of Oxford: 'They wanted to make an old woman of me or that I should stuff Latin and Greek at the University; but these schemes I cracked like so many vermin as they came before me'.[120]

The University of Glasgow educated Francis Jeffrey, who was elegant in Latin and Greek,[121] but also caught in the conflict, an archetypal one, between the industrious middle class student and those who modelled themselves on the idle, their criteria being not useful knowledge but the ennui of good taste. Jeffrey wrote to his sister Mary about his studies in Oxford:

> Is there anything, do you think, Cara, so melancholy as a company of young men without any feeling, vivacity, or passion? We must not expect, here, that warmth and tenderness of soul which is to delight and engage us; but let us at least have some life, some laughter, some impertinence, wit, politeness, pedantry, prejudices – something to supply the place of interest and sensation. But these blank parties! oh! the quintessence of insipidity. The conversation dying from lip to lip – every countenance lengthening and obscuring in the shade of mutual lassitude – the stifled yawn contending with the affected smile upon every cheek.[122]

The stifled yawn is a far cry from the zest of cracking 'so much vermin'. Scottish culture was Calvinist and pulled no punches. When it took on the foibles of society it pounced on the greedy, or the fashionable and the extravagantly profligate, and held these up, much in the tradition of the mirrors of virtue, showing them up as bizarre idle aberrations. Tobias Smollett, who trained with John Gordon in Glasgow, was skilled in lampooning both cultures. John Moore, his fellow apprentice with the firm of Gordon and Stirling, also took to the pen in his London years. The adroitness of the Scottish portrayal of men and manners may hark back to a self-consciousness induced by being in 'foreign parts', among the English. But even such a respectable figure as William Hamilton, son of the professor, was vilified on the streets for being Scottish.

An anecdote from his letters gives a glimpse of these sensitivities. The reference to 'Sawney' is to the legend of the cannibal of the Galloway coast,[123] a man cast out because he was thought to eat human flesh, as did Sweeney Todd, the butchering barber. William recounts the following incident in a letter to his mother from London of 1 January 1778.

> I am now at no loss in the streets of [London] tho at first I certainly looked strange, for a woman one day spying me look[ed] me in the face and said Oh Sawney is this you [?] I was so angry at the speech that I should have struck her

---

[120] G. C. Peachey, 'The Homes of the Hunters', *Lancet* (1928), pp. 362–63.
[121] Thomson, *Biographical Dictionary of Eminent Scotsmen*, p. 390.
[122] Lord Cockburn, *Life of Lord Jeffrey* (Edinburgh, 1852), i, p. 39.
[123] R. Holmes, *The Legend of Sawney Bean* (London, 1975), pp. 66–75.

had she been a man and I had an excellent stick in my hand. What irritated me most was her finding me out to be Scotch by my face when I thought I looked very like an Englishmen. This is the only affront I have met with since I have come here. Indeed the Scotch at present are rather respected than otherwise.[124]

The Medical Literati

The end of the eighteenth century saw the triumph of medicine as a cultural phenomenon. The great collections of William Hunter and Sir Hans Sloane were housed in 'domes of Enlightenment', inviting the public to engage with science and revere education. William Hunter's collections of anatomy specimens, coins and artefacts (and even the skeleton of an elephant) could be seen under the elegant cupola of the Adam brothers' revival of Greek classicism. Sir Hans Sloane's bequest to the nation was marvelled at in Montague House, then moved behind the majestic Doric columns of the British Museum. Even Anderson's College in Glasgow displayed its anatomy and natural history collections under a dome.[125] Collecting the natural world specimen by specimen to be scrutinised in museum exhibits made universally available the painstaking scholarly efforts of men like Hunter and Sloane. These 'domes of Enlightenment' were architectural paeans to the need to classify and understand the natural order.

As secularisation edged religious belief from centre stage, 'man' was scrutinised as part of this 'natural order'. Tobias Smollett, friend to William Smellie and John Moore, explored the 'science of man' in his satirical novels,[126] often portraying behaviour as the product of ill-health. The novel *Humphrey Clinker* has been described as 'Dr Smollett's travelling clinic'.

In some of his other novels Smollett had his characters entertain the reader with the philosophical topics of the day. One of the great debates of the time was over materialism. De la Mettrie's *L'homme machine* had appeared in the 1750s, with *avante-garde* philosophers, among them Voltaire, disputing the existence of any invisible agency. These criticisms were meant to discredit religious belief both Catholic and Protestant. Philosophers were keen to applaud rationalism and discount both orthodox Christianity and its evangelical offshoots. Neo-Platonism fed the rivers of all who objected to material reductionism. *Peregrine Pickle*, published in 1750, has been described as 'a

---

[124] GUL, Special Collections, MS Gen. 1356/36, Letter of William Hamilton to his Mother dated London, 1 January 1778.

[125] T. A. Markus, 'Domes of Enlightenment: Two Scottish University Museums', *Art History*, 8 (1985), p. 174.

[126] D. Bruce, *Radical Doctor Smollett* (London, 1964), chapter 4, 'The Mechanism and Necessity of Our Natures', passim.

large and bulging carpet-bag of a novel' containing a 'great deal of satire on the Neo-Platonists and the theory of innate ideas'. Smollett fell in line with one of the most cherished ideas of the time: that the soul has no knowledge except for that which it acquires through the senses and in this showed his materialist leanings.

But Smollett was versatile and didn't bow, either, to the philosopher's desire to be consistent. He wrote for a living, and was more likely to finish off a theme with 'a quick astringent certainty'.[127] Writing was a commercial venture, producing a commodity. Smollett sank his life and wit into publications. Henry Grahame writes of him:

> His pen was busy, leaving little time to see his friends, for the income from the Jamaican estate had dwindled. He had to labour at a huge compilation of *Universal History*, projected by a band of booksellers, for which his hacks supplied raw material; to make a *Compendium of Voyages* in many volumes, with the assistance of his myrmidons; to translate *Gil Blas*; and to struggle, with the aid of a Spanish dictionary, a grammer, and Jervas's English version, at a translation of *Don Quixote* – besides preparing for the press Smellie's work on midwifery and completing a *History of the German Empire*.[128]

No wonder when Smollett was on a surprise visit to Scotland, his own mother failed to recognise him. But 'relaxing into a smile', in an instant the old lady knew who he was. She is said to have cried out, 'If you had continued to glower, you might have imposed upon me for a while longer, but your roguish smile betrayed you at once'.[129]

John Moore, Smollett's fellow apprentice under John Gordon, tapped the same commercial vein with his travel books and novels, but was less given to morbid insight or dalliance with 'a panorama of the victims and profiteers of the flesh'.[130] John Moore was cut from solid oak compared with the aspen-leaved sensitivity of Smollett. He was less inclined to lampoon, having 'the same solid basic notion of behaviour' that critics attributed to Jane Austen. He did not equivocate; he did not veil in sentimentality, but, inviting empathy from his readers, asked them to feel as their own the misfortunes of others and enjoy happiness where it was deserved.[131]

Moore was the son of a Presbyterian minister in Stirling, and the family,

---

[127] Ibid., p. 45.
[128] H. G. Graham, *Scottish Men of Letters in the Eighteenth Century* (London, 1901), p. 307, where Graham quotes from letters of Smollett.
[129] Ibid., p. 308.
[130] The apt phrasing comes from Bruce, *Radical Doctor*, p. 47.
[131] J. Moore, *Mordaunt: Sketches of Life, Characters and Manners in Various Countries* (London, 1965), ed. with an introduction by W. L. Renwick, p. xv.

against the fond hopes of John's grandmother, soon acknowledged that medicine would need to substitute for a profession he would not follow, the ministry. His mother was Marion Anderson, the daughter of a man of considerable property, who had inherited an estate called Dovehill near Glasgow. In 1737 Moore's father died and the seven children were brought back to Glasgow by their mother. Moore was apprenticed to John Gordon, 'by the advice of his relatives and his own predilection for the medical profession'.[132]

Moore was educated at the Grammar School and Glasgow University where he studied arts. Even though his widowed mother possessed an independent inheritance, her children had to make their own way in the world. John Moore was booked as apprentice to the firm of Gordon and William Stirling on 3 December 1744 for five years.[133] Being apprenticed to John Gordon ensured a good start, as he was described as 'a surgeon of extensive practice, and a citizen of great and well merited popularity'. In later years Gordon asked him to join his practice as a partner, where he dealt with all aspects of medicine, 'a custom very common in Scotland, where the extensiveness of the practical embraces the several branches of physic, surgery, pharmacy and midwifery'.[134]

He was admitted to the FPSG on 3 September 1750 when he was examined by William Stirling, John Gordon, Andrew Morris and Andrew Craig.[135] On 7 February 1751 his 'trials' were completed and he was admitted a freeman member of Faculty to practise surgery and pharmacy.[136] He continued to be active in the incorporation, being elected Visitor on 3 October 1757 and collector in 1763.[137] Moore not only trained as an apprentice but also attended lectures in the university.[138] There he heard the lectures of William Cullen, Robert Hamilton and George Montgomerie.[139] Moore joined William Cullen in 1769 as consultant at the bedside of the young Duke of Hamilton on the recommendation of this most eminent of Scottish professors. His father-in-law was the Glasgow Professor of Divinity, John Simson, brother of the liberal mathematician Robert, who fell foul of orthodox church opinion and

---

[132] R. Anderson, *The Life of John Moore, MD* (Edinburgh, 1820), p. iv.
[133] RCPSG, 1/1/1/2, fol. 86v.
[134] Both quotes Anderson, *Life of John Moore*, introduction, pp. iv, vii.
[135] RCPSG, 1/1/1/2, fol. 127r.
[136] RCPSG, 1/1/1/2, fol. 131r.
[137] RCPSG, 1/1/1/3, fol. 29.
[138] This is glossed over by H. L. Fulton, 'John Moore, the Medical Profession and the Glasgow Enlightenment', in A. Hook and R. Sher (eds), *The Glasgow Enlightenment* (East Lothian, 1997), pp. 178–79.
[139] Ibid., p. 177.

was brought before its disciplinary councils.[140] John Moore married Simson's daughter Jean in 1751.[141] Later in life he wrote to her about the strict observance of the Sabbath. 'He could never think of Glasgow, when absent, without calling up a mental picture of Sunday morning in that city, and all the good families proceeding to kirk looking as if they were going to the gallows and knew they richly deserved their fate'.[142] But this rejection of Calvinist sobriety did not mean a rejection of Glasgow. Moore had engraved on his tombstone 'born in Glasgow in Lanark in Scotland' and his funeral followed the best of Scottish traditions, although he died in his house near the Thames in Richmond.[143]

In 1772, when Moore embarked on his travels on the Continent as medical adviser and moral watchdog for young Douglas, the eighth Duke of Hamilton, he envisioned a literary future, planning to use his observations for publication. Significantly they were conceived as descriptions of the society and of the manners of the countries he visited: France, Switzerland, Germany and Italy.[144] Soon he risked a different medium, taking his theme of morals and manners into the realm of fiction. These novels, *Zeluco: Various Views of Human Nature, Taken from Life and Manners, Foreign and Domestic* (1786), *Edward: Various Views of Human Nature Taken from Life and Manners, Chiefly in England* (1796) and *Mordaunt: Sketches of Life, Character and Manners in Various Countries, Including the Memoirs of a Lady of Quality* (1800), became his most famous works. As their subtitles reveal, Moore was one of the main contributors to, and creators of, the genre of moralistic character novels, best-sellers in a market keen on edification and improvement.

Moore's works of fact and fiction were known for their levity and appraisal of the 'lives and manners' of society at home and abroad.[145] In Smollett's case the theme of 'lives and manners' took on a satirical bent, using the grotesque as a foil for the normal, but he is recognised, equally, as a great realist.[146] While Smollett is perhaps more fun to read, his comic portraits providing an entertainment moralists rarely risk, his were typically improving

---

[140] J. K. Cameron, 'Theological Controversy: A Factor in the Origins of the Scottish Enlightenment', in R. H. Campbell and A. S. Skinner (eds), *The Origins and Nature of the Scottish Enlightenment* (Edinburgh, 1972), pp. 118ff.

[141] C. Oman, *Sir John Moore*, p. 3; the marriage took place in 1753.

[142] Ibid., p. 9.

[143] Ibid., pp. 298–302.

[144] *A View of Society and Manners in France, Switzerland and Germany* (1779) and *A View of Society and Manners in Italy* (1781).

[145] Oman, *Sir John Moore*, pp. 55, 59.

[146] G. Rousseau, 'From Swift to Smollett: The Satirical Tradition in Prose Narrative', in J. Richette, J. Bender, D. David, M. Seidel, *The Columbia History of the British Novel* (New York, 1994), p. 127.

novels, representing 'a complex system of morality and didacticism' that makes him one of the most profound moral commentators of his age.[147] It was this undercurrent of serious purpose that held Moore to him because, in a labour of love, Moore edited Smollett's collected works.[148]

William Hunter once told his nephew and heir Matthew Baillie, who he trained in his vivid techniques of lecturing, that: 'If you know the subject really well you will know it whether the preparation be present or absent'.[149] Smollett, Moore and Smellie used this Scottish vividness to literary effect, incorporating portraits of patients, doctors, and the dramas of illness, death and advice, either taken or not, into fiction. In their works Smollett and Moore pursued the empirical mode, and used it to make their moral points about folly, infamy and human foible. Reality, viewed through the lens of satire or 'candid integrity', the point of comparison between Jane Austen and Moore,[150] was the Glaswegian's answer to the Gothic imbroglio of senseless love-affairs often maligned as 'the bad taste of lady novelists and womanish readers'.[151]

The rubbishing of 'womanish readers' came from another descendant of Glasgow's medical elite, the grandson of William Cullen's practice partner Thomas Hamilton. His son William fathered two men of letters, the philosopher and educationalist Sir William and his brother Captain Thomas Hamilton, the author of *Cyril Thornton* and *Men and Manners in America*. Reality, once more, was the strong suit of this Scottish writer. 'Its power lies in its reality', wrote one reviewer,[152] while another wrote: 'The style is flowing and nervous, and tolerably correct; the descriptions are graphic, sometimes powerful; the characters for the most part ably drawn'.[153] The book remained in print until 1880. America seemed to shock Hamilton. He wrote to William Blackwood, the publisher: 'I was always, as you may remember, a bit of a Whig; I shall come home a Tory. God defend England from any experimental imitations of American democracy'. In depicting the Glasgow of his youth, however, he earned his literary laurels: 'while ... the manoeuvrings of manners perhaps seem somewhat crude and clumsy compared to the handling of similar situations by Jane Austen ... Hamilton recreates ... more

---

[147] Ibid.
[148] Anderson, *The Life of John Moore*, p. xxxiii.
[149] W. MacMichael, *The Gold-Headed Cane* (Springfield, 1953), p. 166.
[150] Moore, *Mordaunt*, introduction, p. xvii.
[151] T. Hamilton, *The Youth and Manhood of Cyril Thornton* (Aberdeen, 1990), ed. by M. Lindsay, where he cites a letter of Thomas Hamilton's to his publisher Blackwood, p. x. Lindsay gives as a source NLS, MS 4017/127.
[152] Ibid., p. xi.
[153] Ibid., p. xiii.

vividly than any social historian the long-vanished days of the tobacco lords whose blunt-spoken energies laid the foundations of Glasgow's future industrial success'.[154]

Even when skimming the surface of these mirrors of manners and mores, packaged in fiction, satire, or foreign travel, the habits of a good education and training in medical observation shine through and delight. The medical Enlightenment in Glasgow had come full circle, from depicting God's wisdom in Nature to incorporating scientific research in university teaching to medical men engaging the great philosophical and moral questions of the time.

These themes gave them a place in the Enlightenment pantheon of high culture and good investments. Their legacy is perhaps best expressed by John Moore's son, Sir John Moore of Corunna, when he was ten years old, writing to his mother at home: 'Compliments to my Brothers, and tell them to apply well, for without talents a man is despised when he comes abroad. I have seen some English that were ridiculed by every body, though they were rich, because they knew hardly anything but nonsense'.[155]

---

[154] Ibid., p. xvii.
[155] J. C. Moore, *The Life of Lieutenant General Sir John Moore, KB*, 2 vols (London, 1834), i, p. 251.

7. View of the Cathedral and Infirmary from Five Mills, drawn and engraved by Joseph Swan.

6

Botany, Anatomy and Chemistry

Historians who have credited the universities with the leading role in educational change have missed the independent power of closely-knit medical elites pursuing professional aims across institutional boundaries. Local city politics played a fundamental role in determining which educational structures evolved successfully. Until university medical education came to dominate in the nineteenth century, medical schools were only as good as the multiple institutions and the diverse specialist courses, from private to extra-mural, that had been encouraged to proliferate within city bounds. Refurbishing university medical chairs was deftly turned to the advantage of surgeons needing applied skills and practical learning, steered by the corporate aims of the Faculty of Physicians and Surgeons of Glasgow.

This chapter tells a very broad story, from the many gardens planted in Glasgow to grave-robbing and anatomical museums, all instrumental in training medical men. Chemistry too belongs to this kaleidoscope of resources financed by public and private initiative. Because practising physicians and surgeons needed expertise in materia medica (the art of preparing efficacious drugs) they were equipped to tackle applied research in chemistry. William Cullen and Joseph Black, for example, used chemical laboratories financed by the university to advance experiments on latent heat and soil composition which were well outside their teaching briefs. They formed the nucleus of a community of skilled chemists derived from the ranks of medical doctors and FPSG members, including William Irvine and John Robison. These men could trade on their scientific expertise to the extent that they were being head-hunted for projects as far afield as Spain and Russia.[1] Their medical colleagues, men like the surgeon-apothecary Ninian Hill, who operated his own laboratory, also extended the reach of experimental chemistry. His friendship with Black helped launch major discoveries. Because licentiates of the FPSG were for the most part accredited as 'surgeon-pharmacists' their knowledge and facilities thinned

---

[1] For William Irvine see A. Kent, 'William Irvine, MD', in idem, *An Eighteenth-Century Lectureship in Chemistry* (Glasgow, 1950), p. 147; for John Robison, see E. Whittaker, 'John Robison, MD', in Kent, *An Eighteenth-Century Lectureship*, p. 127.

the demarcations between university chemistry, mixing and supplying drugs, improving the use of chemicals for the bleaching fields or as agricultural fertilisers to vanishing point. The emphasis on applied skills so strongly advocated by the FPSG was pronounced in the disciplines it influenced. In 1847 Sir William Hooker, recently made director of Kew Gardens, built a widely praised 'economic museum' of plants illustrating their potential commercial applications.[2] He had spent twenty years teaching in Glasgow, where the emphasis on the industrial applications of science surely influenced his belief in the importance of turning knowledge to commercial account.

The Glasgow medical school emerged from the many coalitions of public institutions and private capital investment available in an industrialising city. An early example of this was the way in which medical chairs and lectureships were provisioned. The university recruited its medical lecturers from FPSG members while candidates for the FPSG licence made up a large portion of its paying 'students'. Botany and anatomy were the more viable part of university medical teaching, despite the founding of chairs in traditional subjects, mainly because the 'new sciences' were useful to practising surgeons. The medical chair and the composite 'botany and anatomy' chair, endowed in 1714 and 1720 respectively, were undersubscribed in comparison with the anatomy and botany lectureships with which they were associated. Botany had been taught since the surgeon John Marshall's instalment as Keeper of the Physic Garden in 1704, well before the chair was inaugurated. John Gordon, a prime mover in the inner circles of the FPSG, taught anatomy from 1714, as an adjunct to the medical chair. The ability of chairholders to engage other lecturers to fulfil needs central to medical practice highlights the composite nature of teaching. This became something of a tradition in Glasgow. Midwifery, for example, was taught from 1757 to 1815 without a separate chair, as were surgery, materia medica and chemistry.

The university inaugurated chairs of a traditional nature when John Johnstoun became Professor of Medicine in 1714 and Thomas Brisbane, another physician, was appointed to the chair of 'botany and anatomy' in 1720. Both came from city practices. Of Johnstoun we know little, but Brisbane came from an established Glasgow medical dynasty. He was the son of Matthew Brisbane, the first physician to be appointed Rector of the University, and the first physician in a family of ministers. Matthew Brisbane was admitted to the FPSG in 1671, having received his doctorate at Utrecht.[3]

---

[2] M. Allan, *The Hookers of Kew, 1785–1911* (London, 1967), p. 156.
[3] Duncan, *Memorials*, pp. 62, 112.

He is mentioned in Glasgow city accounts as town's physician.[4] Thomas's son John was the author of an illustrated book on anatomy,[5] and was influential enough to be among the founding members of the Literary Society of Glasgow, together with William Cullen and Robert Hamilton, although, unlike the others, he was not a professor at the University.[6] John Brisbane was admitted to the FPSG as an honorary member on 4 December 1769, having presented his book to the Faculty.[7]

Both Thomas Brisbane and Johnstoun have unjustly been labelled as 'inert' teachers.[8] Both were in physician's practices of considerable prominence, their livelihood dependent on their consultation practices, not on university salaries. They might have taught students not by lecturing but through the continental practice of taking students on case rounds in private practice and discussing medical theory 'at table' in the lecturer's home.[9] Joseph Black described this practice when he was teaching in Glasgow: 'The Professors get £10 a quarter for Diet and Lodging and in other houses they take seven or eight [students]'.[10]

Johnstoun and Brisbane were both Presidents of the FPSG, Brisbane from 1734 to 1737 and Johnstoun from 1737 to 1739. They would not have been opposed to appointments in anatomy that supported the aims of their professional incorporation. Johnstoun and Brisbane were both *ex officio* members of the FPSG examination boards. Both professors are on record as diligent in examining the few candidates for university MDs in their time. Brisbane was elected Dean of the University Faculty in 1734.[11] Both he and Johnstoun presided over the MD examinations of the following men: James

---

[4] Ibid., p. 245. See also R. P. Ritchie, *The Early Days of the Royal College of Physicians, Edinburgh* (Edinburgh, 1899), pp. 147–48.

[5] J. Brisbane, *The Anatomy of Painting: or A Short and Easy Introduction to Anatomy* (London, 1769).

[6] *Notices and Documents Illustrative of the Literary History of Glasgow during the Greater Part of Last Century* (Glasgow, 1831), p. 123.

[7] RCPSG, 1/1/1/3, Minutes of the FPSG, 1757–1785, 1 May 1769.

[8] The accusations originate in documentation for Thomson's *Life of William Cullen, MD* (on p. 24), published in the nineteenth century, and, by Thomson's own admission, after those who were living witnesses of the period had died. The evidence for 'inadequate' teaching surfaces in a letter by 'Dr Wallace' saying that 'Dr Brisbane never gave lectures' and that 'Dr Johnstone [sic] did not give lectures', see GUL, MS Gen. 1508, letter to William Thomson, 25 September 1844.

[9] J. Geyer-Kordesch, 'German Medical Education in the Eighteenth Century: The Prussian Context and its Influence', in W. F. Bynum and R. Porter, *William Hunter and the Eighteenth-Century Medical World* (Cambridge, 1985), pp. 177–204.

[10] Edinburgh University Special Collections, GEN874/v/13–14, Joseph Black writing to his father, 4 October 1763.

[11] GUABRC, 26645, Dean of Faculty's Meeting Minutes, 1732–68, 26 June 1734.

Arbuckle (19 June 1724), George Montgomery (19 October 1732), who was later instrumental in initiating clinical teaching at the Town's Hospital, Samuel McCormick (11 March 1736), John Shiells (4 August 1737) and Thomas and Joseph Johnstoune (16 May 1739). Other MD awards were also made, on the basis of testimonials, overseen by the medical professors, a practice strongly discouraged by their successors in the 1760s. Joseph Black wrote:

> We have had some very warm debates of late upon some instances of this kind. I believe that if a young man who came now in the same way obtained his degree at all, it would cost some of us more trouble than we would wish to undertake, and I can assure you that however well qualified he would run a great risk of being disappointed. It is insisted upon that a lad who comes with that view should both bring testimonials of his standing and good character and stay with us the whole of session; the reasons given are that the manner of teaching here, particularly the different Sciences, is in the way of a digested and systematised set of Lectures, with examination upon them which last the whole Session, viz from the middle or end of October to the middle or end of May.
>
> Every single Lecture therefore is an important part of the whole; and those who take only a part of the course, especially if they come towards the end, when the Plan, the Principles, and Introductory Parts have been already delivered, can reap no advantage.[12]

The supposed lack of teaching by Brisbane and Johnstoun came from an accusation made in a nineteenth-century biography of William Cullen, written to accentuate the latter's importance. The lack of lectures on internal medicine, to be distinguished from the teaching needs of surgeons, was probably due to the university's meagre granting of MD degrees in the first decades of the century, the first awarded in 1703 and 1711.[13]

In 1720 anatomy and botany, both of which had been taught before at university level, were elevated to professorial teaching status but, in the event, fell victim to the accepted hierarchies between physicians and surgeons. Thomas Brisbane, the incumbent, was a physician and would have rejected dissection, which he abhorred, as incompatible with his status. His anatomical teaching was carried out by John Paisley, a prominent city surgeon. Brisbane personally undertook his botanical obligations, actively overseeing the physic garden and gaining from the sixteen years that botany had been supervised by the surgeon, John Marshall.

---

[12] EUL, Special Collections, GEN 874/v/13–14, Letter of Joseph Black to his father of 4 October 1763, Glasgow.

[13] Coutts, *A History of the University of Glasgow*, pp. 481ff.

Gardens and Plants

In 1704 the physic garden was laid out. In Edinburgh James Sutherland was at work creating the teaching gardens there, and, indeed, his transfer to Glasgow was sought.[14] Glasgow's commercial and intellectual elite was keen to keep pace with their trade partners in cultivating foreign plants and evolving a systematic method of identifying local flora. Such an investment would not only serve teaching but underpin advances in treatment, in effect giving Glasgow a stake in developments underway in Edinburgh and abroad. Sir Robert Sibbald and Sir Andrew Balfour, under the auspices of the newly founded College of Physicians of Edinburgh (1681), were issuing a pharmacopoeia, an official compilation of approved medical substances. The enterprise was a difficult one as it sought to distinguish useless from efficacious remedies, despite vested interests or popular lore. It first appeared in 1685. In later years a Lord Advocate, denying Glasgow's medical incorporation the title 'Royal College', believed the work of a College of Physicians to be its supervision of an official pharmacopoeia, and this, of course, was overseen in Edinburgh not Glasgow. The listing of acceptable drugs and how to make them was an obvious advance in regulating medical care. Prussia had published its official pharmacopoeia in 1689 and its Collegium Medicum remained responsible for it. But official sanction of how drugs were to be made required the education of medical practitioners.

Establishing a physic garden was crucial to these developments. Andrew Balfour had planted his own garden with exotica. Many of his plants and those of Patrick Murray, his student, were the first of their kind in Scotland. Murray's botanic garden contained 1000 species of plants, which were transferred to Edinburgh on his death, since he had bequeathed them to Balfour.[15] Balfour's knowledge was incorporated directly into the Edinburgh pharmacopoeia, as he was responsible for the arrangement of its section on materia medica.[16]

'Druggists' gardens cultivated the traditional medicinal herbs, roots, trees and shrubs, while botanical or 'physic' gardens were devoted to the scientific study of plants, both native and exotic.[17] The new physic garden gave Glasgow

---

[14] A. D. Boney, *The Lost Gardens of Glasgow* (London, 1988), p. 29.

[15] T. Thomson, *A Biographical Dictionary of Eminent Scotsmen* (London, 1875), half-vol. i, p. 64.

[16] Ibid.

[17] K. Maegdefrau, *Geschichte der Botanik* (2nd edn, Stuttgart, 1992); *Botanical Gardens*, Ciba Symposium, 11, no. 1 (1949); C. E. Raven, *John Ray, Naturalist: His Life and Work* (Cambridge, 1950); *Hortus Eystettensis: zur Geschichte eines Gartens und eines Buches* (München, 1989); J. R. Green, *A History of Botany in the United Kingdom from the Earliest Times to the End of*

medical men the opportunity to study three innovations: drugs won from cultivating non-native plants, the new systemisation of plants and their nomenclature (as its layout of beds indicates) and the effects of nutritional and culinary regimens.[18] Nutritional studies had recently become the subject of medical research. Friedrich Hoffmann, the Halle University physician, was pioneering the chemical study of food substances.[19] Glasgow's investment in a physic garden signalled its participation in this agenda.

The scientific methods of semiotic differentiation and species systemisation were now being made accessible in herbaria and compendia. James Sutherland produced his *Hortus Medicus Edinburgensi: or A Catalogue of the Plants in the Physic Garden at Edinburgh* in 1683. His catalogue and that of Jan Commelin, the *Horti medici Amstelodamensis rariorum plantarum descriptio et icones* (1697–1701) were surely used in Glasgow. These gardens dispersed plants, seeds and botanical skills, including the new agricultural plants, maize, the potato and the tomato.

> Plants and gardens became status symbols, and exotic ornamental plants themselves became part of a much wider trade in luxurious commodities. The cargoes of the ships of the British East India Company and the Dutch equivalent returning from China, India and the Cape included plants as well as silks and calicoes, lacquer-work and porcelain.[20]

The FPSG was actively consulted in the founding of the physic garden. The Principal, John Stirling, asked the 'physicians and surgeons of the city' to choose a site to the east of Blackfriar's church.[21] The man chosen to be the 'overseer' or curator of the physic garden, John Marshall, was a city surgeon member of the FPSG.[22] His brother Henry was an innovator involved in a test case over widening FPSG membership and was Visitor at a crucial juncture.[23] Through Henry's marriage, John was related to the Earl of Wigton.[24] The

---

the *Nineteenth Century* (London, 1914); *Hortus Botanicus: The Botanic Garden and the Book*, exhibition catalogue compiled by I. MacPhail (Morton Arboretum, 1972), introductory essay by J. Ewan, pp. 5–36.

[18] Boney, *The Lost Gardens of Glasgow*, p. 31.

[19] C. Teucke, 'Nahrungsmittel in der Frühen Neuzeit an der Schnittstelle zwischen Alltagswissen und Nahrungsforschung', *Braunschweiger Veröffentlichungen zur Geschichte der Pharmazie und der Naturwissenschaften*, 37 (Braunschweig, 1996), pp. 49–53.

[20] G. Saunders, *Picturing Plants: An Analytical History of Botanical Illustration* (Berkeley, California, 1987), p. 71.

[21] Boney, *The Lost Gardens of Glasgow*, p. 31; GUABRC, 26631, Faculty Meeting Minutes, 1702–20, 7 September 1704.

[22] Boney, *The Lost Gardens of Glasgow*, pp. 48–50; Duncan, *Memorials*, pp. 246 and 248; Coutts, *History of Glasgow University*, pp. 185–86.

[23] Duncan, *Memorials*, p. 114.

[24] Ibid., p. 246.

Marshall family subsequently intermarried with the Horsburgh and Cowan families. Alexander Horsburgh was active as a town surgeon, holding office in the FPSG and sitting on its examinations board.[25] Robert Cowan, another descendant, was the first holder of the chair of medical jurisprudence in Glasgow from 1839 to 1841.[26]

The FPSG allocated £10 per annum, recorded in 1760, to another teaching garden. This garden belonged to Dr John Wodrow. It was situated near what is now St Andrew's Square.[27] The physic garden and Wodrow's garden were just along the street from each other. Dr Wodrow was the son of the Professor of Divinity, James Wodrow, who had studied medicine 'very diligently for several years, particularly botany and anatomy, and was a pretty exact botanist'.[28] His other son Robert further recorded:

> I know about 1686 he had thoughts of going over to the universities abroad and taking his degrees in medicine. I remember when very young, about the years 1685 and 1686, he would take me with him to the fields, when walking, and cause me to pull different herbs in the fields and from the corns, and tell me their names and uses. And when old, he would still know from his physician Dr Kennedy what the manner of his prescriptions for him or his children was, and sometimes would reason with him, and propose alterations in what he ordered for us.[29]

John, born in 1695 by James's second wife, Janet Luke, was president of the FPSG three times (in 1739–41, 1749–51 and 1753–55). His garden was not, however, maintained after his death in 1769. In 1779 it seems to have been in disrepair, a victim of the city's rapid need for new housing.[30]

On the maps of the city the walled enclosure of the teaching gardens at the university and, down the High Street and Saltmarket, nearer the Clyde, that of Dr Wodrow, are not hard to locate.[31] The keys to their gates were

[25] RCPSG, 1/1/1/2, Minutes of the FPSG, 1733 to 1757, fol. 13r.

[26] Duncan, *Memorials*, p. 156; Boney, *The Lost Gardens of Glasgow*, p. 49; A. Crowther and B. White, *On Soul and Conscience: The Medical Expert and Crime* (Aberdeen, 1988).

[27] Boney mentions Wodrow's garden, but did not write about it as another place to educate students and apprentices and, indeed, others interested in materia medica and plants. (On Wodrow's garden, see p. 283.) Boney repeats the view that the university and the FPSG were rivals, an interpretation we would like to revise.

[28] R. Wodrow, *Life of James Wodrow, AM, Professor of Divinity in the University of Glasgow from MDCXCII to MDCCVII* (Edinburgh, 1828), p. 67.

[29] Ibid.

[30] Senex (Robert Reid), *Glasgow, Past and Present: Illustrated in Dean of Guild Court Reports and in the Reminiscences and Communications of Senex, Aliquis, J.B*, 3 vols (Glasgow, 1884), ii, p. 251–52.

[31] J. N. Moore, *The Maps of Glasgow: A History and Cartobibliography to 1865* (Glasgow, 1996), p. 40 (map of 1778); Senex, *Old Glasgow and its Environs* (Glasgow, 1864), opposite p. 1 (map of 1760); Boney gives locations and relocations of the physic garden.

available to students for the purposes of study. From these teaching Edens it was more difficult to steal liquorice roots, or 'liquory sticks', as did 'Senex' (Robert Reid), the chronicler of Glasgow's past, when he was a boy.[32]

More ordinary gardens filled Glasgow as well; these were so much part of the city that in 1736 it was written about in idyllic terms as 'surrounded with cornfields, kitchen and flower gardens and beautiful orchards, abounding with fruits of all sorts, which, by reason of the open and large streets, send forth a pleasant and odorous smell'.[33] A glance at town maps of this period shows the wynds and 'closes' (small back lanes) giving way to the numerous white patches marked 'garden grounds'.[34] The 'liquory sticks' mentioned above were stolen from these more common gardens, in this instance one rented by a druggist on the Trongate, kept 'partly as a vegetable garden, and partly for raising medicinal plants and roots suitable for a druggist's shop'.[35] Next to the druggist, an orchard was located, also used for raising vegetables, and this was not far from the large garden kept by Hutcheson's Hospital until it was sold in 1788.[36] The new Royal Infirmary of 1794 also maintained a garden. The managers discussed its expense in 1808, dismissing the present gardner, and adopting new plants for maintaining it and 'keeping the walks in order'.[37] Dr John Moore's garden and the gardens of Mrs Fleming and Mrs Balmanno, too, are to be found on city maps, and these were not the gardens of scientific pursuit but working gardens, used to effect when apprentices sought the tutelage of those versed in the healing properties of plants.

When looking at the diversity of gardens, distinctions begin to blur, as knowledge about plants shades from the culinary to the medicinal. John Reid, in the *Scots Gardener* of 1683, distinctly encouraged a knowledge of herbs amongst the nobility and gentry.[38] In the gardens of everybody, as popular songs name them, were the familiar stalwarts, the 'sweet herbs' used for cooking or for physic: pennyroyal, clary, rosemary, sweet-basil, fennel, besides the usual sage, mint, and wild marjoram.[39] The druggists, apothecaries and surgeons of the city either had their own gardens or access to gardens such

[32] Senex, *Glasgow, Past and Present*, ii, pp. 100–1.
[33] H. G. Graham, *The Social Life of Scotland in the Eighteenth Century* (London, 1950), p. 131, who is quoting 'Glasgow's first historian', M. Ure, *History of Glasgow*.
[34] J. N. Moore, *The Maps of Glasgow*, passim.; both Graham, *The Social Life of Scotland*, and Senex, *Glasgow, Past and Present*, provide evidence for the many gardens in Glasgow.
[35] Senex, *Glasgow Past and Present*, ii, p. 100.
[36] Ibid., p. 101.
[37] HB, 14/1/2, GRI Manager's Minutes, ii, p. 174.
[38] Graham, *The Social Life of Scotland*, p. 16.
[39] Ibid., p. 6.

as those planted for hospitals,[40] and in these grew the 'simples', the non-exotic medicinal components of the local pharmacopoeia. 'There were found the hyssop, camomile, and hore-hound, cat-mint, elacampine, "blessed thistle", "stinking arag", rue and celadine, which were in constant request in time of sickness.'[41] Remedies were ubiquitous and, if a cue is taken from William Cullen's consultation letters written to patients, bridged the many crafts of preparation from herbal drugs to cooking nutritious soups.[42]

Until about the middle of the century gardens in Scotland lacked turnips, onions and potatoes, and other vegetables which were later to become so common. These 'exotica' travelled from the gardens of the rich and titled to the common kitchen garden, merging the properties of healing and nutrition. One physician who encouraged this trend was John Hope, King's Botanist in Edinburgh from 1761, and later Regius Professor of Botany and Medicine, who had learned his botany as a medical student at Glasgow University. He published papers on the practical cultivation of rhubarb, once an expensive medicine, and made its planting plentiful in gardens. Considerable energies were also spent by Hope in working out a systematic order for vegetables, once more attesting to the importance, in medicine, of regimen, and its dependence on the knowledge made available to clients about new plants and their uses. This pursuit of science was invariably cultivated by alliances between 'amateurs' among the nobility, such as the botanically versed Earl of Bute, an acknowledged lover of vegetables and trees, who financed gardens (those of Hope in Edinburgh, for one) and books, such as the engravings of the 'vegetable system' done at great expense for him by John Hill. *The Vegetable System* was 'of great importance because it [gave] for the first time in the vernacular a comprehensive treatment of the plant kingdom, on a lavish scale and with coloured illustrations, adopting the Linnaean generic names and introducing binary nomenclature'.[43] Importing trees was another favourite activity of botanically interested members of the aristocracy, such as the second Duke of Atholl, who was the first man to plant the larch on a wide scale for forestry. The fourth Duke 'was so smitten with the larch that he planted 17,000,000 of them'.[44] The rich traded on the professional

---

[40] We can safely assume the Town's Hospital and the GRI bought from local suppliers or, for a time, had their own 'kitchen gardens', as this was common practice.

[41] Quoted in Graham, *The Social Life*, p. 6, who quotes from Moncrief of Tippermalloch's, *Poor Man's Physician* (Edinburgh, 1712).

[42] J. Geyer-Kordesch, 'New Beginnings: The Importance of Medical Case Histories in the Scottish Enlightenment', unpublished talk given at the RCPSG in May 1992.

[43] 'A Magnificent Collection of Botanical Books', in *Sotheby's Florilegium Catalogue* (London, 1987).

[44] H. Johnson, *The International Book of Trees* (London, 1973), pp. 92–93.

interest of physicians and surgeons, and supported men from artisan backgrounds switching to botanical pursuits. This predilection for working with plants led one authority to write that 'botanist' and 'Scotsman' were almost interchangeable.[45] The surgeon Archibald Menzies discovered and described the new trees from the Pacific north west while he sailed with Captain Vancouver on his voyage of 1792.[46] On the other hand, David Douglas, to whose labour in gathering seeds by the chestful is owed much of the magnificent treelife of Britain, was a stonemason's son whose botanical talent was spotted by the celebrated William Jackson Hooker.[47] Thus the enterprises of botanical teaching, the compilation of reference works, the planting and care of gardens, and the medicinal use of plants as drugs or nourishment spanned the whole of Scottish society.

One garden in the centre of the city illustrated the mercantile rather than the teaching use of plants. Like many of its kind it points to the difference between learning about plants and their medicinal applications and marketing them to mediciners and the public. This garden, a largish plot of land glorious with 'medicinal plants and roots',[48] belonged to Mrs Balmanno, an enterprising woman who owned her own shop and married the employee who gave her his name. This was the kind of shop to which everyone came for needs from beauty aids to sweets, to ink;[49] and, on the medical side, for consultations and drugs.[50] Mrs Balmanno's trade was carried out under the sign of the Golden Galen's Head, an apt appellation for what was an old-style apothecary's shop with over-the-counter trade. The shop sold drugs, and offered advice on their application, but was not a surgeon-apothecary's practice. Its part in the town's economy was through its status as a supplier. Mrs Balmanno's son John did not enter the trade but studied medicine instead.

John Balmanno was one of Glasgow's most prominent physicians, and twice President of the FPSG (1802–4 and 1812–14). He was a physician to the Glasgow Royal Infirmary and the successor to Robert Cleghorn as physician superintendent at the Glasgow Asylum for the Insane. His doctorate was awarded from Edinburgh in 1798. As a physician, John Balmanno was not dependent on the garden or shop for his livelihood. He sat reading medical journals in his mother's old chair overlooking the Laigh-Kirk Close and the

---

[45] Ibid., p. 46, and on the employment found by Scots as gardeners, Graham, *The Social Life of Scotland*, p. 514.

[46] Johnson, *International Book of Trees*, pp. 46–47.

[47] Ibid.

[48] Senex, *Glasgow, Past and Present*, ii, p. 115.

[49] J. D. Hadcraft, 'John Balmano's Receipt Book', *M & B Pharmaceutical Bulletin*, copy in the University of Strathclyde Archives, S 178.

[50] Senex, *Glasgow, Past and Present*, ii, p. 115.

interior of the shop, and providing medical advice gratis to the poor. 'Indeed, the lower orders of our population at that time were infinitely obliged to Dr Balmanno for his kindness and liberality on all occasions where his services were required by the needy.'[51]

Keeping a shop as a physician was against the rules of the FPSG, which would have required Balmano to be classed as a surgeon. Thus the surgeon William Kennedy took on the druggist's trade of Mrs Balmano and kept the Golden Galen's Head in business in the nineteenth century.[52] John Balmanno's gentlemanly status, however, resulted in boosting the business of his colleagues. His fellow members of the FPSG and surgeons to the Glasgow Royal Infirmary, James Monteath and William Couper, opened a rival drugs business at the north-east corner of Stockwell Street and realised the trade potential of a new age by being both a wholesale and retail concern.[53] Monteath was a leading city surgeon who gave lectures in Glasgow on midwifery, and a power in the councils of the FPSG. William Couper was also a city surgeon and involved in the expansion of the chemical industry, in partnership with Charles Tennant when he set up his bleaching business at St Rollox.[54] The Monteath and Couper retailing company was latterly in competition with the Glasgow Apothecaries' Hall in Virginia Street. This company sent travellers through Scotland 'supplying the country practitioners with the best of medicines, to the greatest advantage of all ranks in our native land'.[55]

The licences granted by the FPSG were primarily those termed 'surgeon-pharmacist', a standard part of the examination requiring the mixing of drugs. The resources of gardens were part of medical commerce, each type feeding both teaching and supply needs. Materia medica and botany could not have been taught without the specimen use of plants. Gardeners were often on the lists of those prosecuted by the Faculty for illegal practice. For instance, on 5 December 1743 a number of gardeners were summoned for practising medicine, surgery (this was usually a case of bloodletting) and pharmacy. Donald McLauchlan, gardener to the Laird of Drumakill, Alexander Sinclair, gardener to John Buchanan in the parish of Kilmarnock, John Thoburn, Tenant in Faslane, John Ramsay, gardener to the Duke of Argyll, William Todd, gardener to the Laird of Macfarlane, Walter Cameron,

[51] Ibid.
[52] Ibid.
[53] Ibid., p. 116.
[54] W. Alexander, 'Charles Tennant', in *Enterprise: An Account of the Activities and Aims of the Tennant Group of Companies First Established in 1797* (London, 1945), pp. 117–34.
[55] Senex, *Glasgow, Past and Present*, ii, p. 116.

innkeeper at Aldacyly, and Lawrence Pate, gardener to the Laird of Bonhill, were all named as in violation.[56]

Like the diversity of plantings, teaching was an enterprise of high aims and mixed success. When John Marshall died in 1719 the 'new' subjects, botany and anatomy, were given the status of a professional chair. A.D. Boney has made clear in his study of the physic garden at the University that, although the garden had a mixed history of care, teaching was consistent. This is particularly evident from the payment made to one of the gardeners to go into the countryside to gather teaching material for students.

The Professor of Botany and Anatomy, James Jeffrey, detached the teaching of botany from his own schedules. He wrote to the university, 'I beg leave to refer you to Dr Brown'.[57] Dr Thomas Brown entered the FPSG in 1799 and was surgeon to the Glasgow Royal Infirmary between 1804 and 1810 and later its physician from 1824 to 1828.[58] He belonged to an established Glasgow family, married Marion, the sister of Francis Jeffrey of the *Edinburgh Review*, and inherited the estates of Waterhough and Langside. His father, also Thomas Brown, a surgeon in London, built Langside House, near the supposed site of the battle, which his son sold in 1852.[59] His large collection of minerals, fossils, and antiquities was given to the universities of Glasgow and Edinburgh in equal shares.[60] His uncle John entered a partnership with Robert Carrick of the Ship Bank.

A compilation of students attending Brown's lectures from 1799 to 1810 shows him beginning with fifteen and ending with fifty-five.[61] Brown's lectures 'required large numbers of flowering plants, both for demonstration and for examination (each student being issued with a specimen of each plant under discussion)'.[62] Because the physic garden was then insufficient in providing for the rising student numbers, the gardener William Lang was sent into the fields to gather specimens:

> It is required of me to collect elsewhere whatever plants may be necessary for carrying forward the lectures. For which purpose I have to traverse the country round in search of plants: and that, Gentlemen, not on a particular occasion but almost every day during the course ... Because a certain number of different plants, all in flower, must be had for each lecture, And oftentimes after, I have travelled to a wood or waterside two or three miles from Town ... And,

---

[56] RCPSG, 1/1/1/2, Minutes of the FPSG, 1733–57, fol. 77r.
[57] Boney, *The Lost Gardens of Glasgow*, p. 246.
[58] Duncan, *Memorials*, p. 267.
[59] Ibid.
[60] Coutts, *A History of the University of Glasgow*, p. 503.
[61] Boney, *The Lost Gardens of Glasgow*, p. 247.
[62] Ibid.

Gentlemen, as the number of students last season was upwards of thirty, it became necessary for me to provide upwards of thirty specimens of each individual plant demonstrated. And as several hundred Genera and Species were examined last Season, the Botany Garden not furnishing near one hundred in perfect conditions. A great proportion of my time must be occupied in this manner.[63]

The quality and thoroughness of Glasgow teaching is once more vouched for, a point often neglected because institutional histories do not take the complex nature of teaching in this period into account. Earlier botanical instruction was also systematically available, only the loss of records making it appear patchy.

Thomas Brisbane, Thomas Hamilton and William Cullen all expressed concern over the condition of the physic or 'Botany Garden' showing them taking responsibility for teaching. They probably did not rely entirely on its specimen collection because of all the other gardens in Glasgow. Especially the 'private' gardens should not be underestimated as resources for teaching. Robert Cleghorn, the physician and FPSG activist, started a Botanic Garden for the use of students at Langside Cottage,[64] where the widowed mother of Thomas Brown spent her summers.[65] There was also a market dealing in herbs in Glasgow in 1760.[66] The private garden of Thomas Hopkirk of Dalbeth contained a great number of plants, 3000 of which formed the basis of the Sandyford Garden, the successor to the physic garden on the College grounds.[67] The herbarium of William Hamilton, the professor of botany and anatomy, provided study of a different kind. He apparently invested considerable private money in the physic garden. He wished to bequeath £200 to form a new Botanical Garden, along with other property, its worth assessed by William Couper, one of the city's pre eminent surgeons. In the end this property was sold on the open market, giving a good indication of what he had invested in botanical teaching:

> According to a contemporary advertisement, the sale of the greenhouse and effects was to take place in the Physic Garden, the sale to include the Hot-House, Hot-bed, Frame etc erected by Mr Hamilton, also a collection of curious Exotic Plants, among which is a Banana or Bread Tree.[68]

[63] Ibid., p. 246, as quoted there.
[64] G. Thomson, 'Robert Cleghorn MD', in A. Kent, *An Eighteenth-Century Lectureship in Chemistry* (Glasgow, 1950), p. 169.
[65] T. Annan, *The Old Country Houses of the Old Glasgow Gentry* (2nd edn, Glasgow, 1878), p. 159. See also M. Greene, *From Langside to Battlefield* (Glasgow, n.d.). The geriatric unit of the Victoria Infirmary now occupies the site of Langside Cottage.
[66] G. Eyre-Todd, *History of Glasgow: From the Revolution to the Passing of the Reform Acts, 1832–33* (Glasgow, 1934), iii, p. 296.
[67] Boney, *The Lost Gardens of Glasgow*, p. 256.
[68] Ibid., p. 207.

William Hamilton is one of Glasgow's unsung medical heroes. His obvious command of the main teaching subjects in medicine, botany (materia medica), anatomy, midwifery and surgery, would have rivalled Cullen's, but he died young. His notes for his botanical lectures survive.[69] They indicate his appreciation of the Linnaean system, but show him adopting a teaching strategy in line with his audience of apprentices and surgeons as well as students. He explained plants and their identification through the more obvious characteristica of their parts. His lecture headings show he taught the history of botany, speaking on Ray and the Europeans who advanced botanical learning.[70]

The Faculty's role in the development of herbal gardens for medical purposes is clearly illustrated by William Jackson Hooker's remarkable tenure as Regius Professor of Botany at Glasgow University. Hooker, who is now remembered mainly for his hugely successful custodianship of Kew Gardens, first made his reputation in Glasgow – and with it the reputation of the recently established Sandyford Botanic Gardens, the precursor of the present gardens at Kirklee.[71] Yet the university was only one of the controlling influences on Sandyford. Although it was a much needed replacement for the university's venerable physic garden, which had been badly affected by encroaching industry, the initiative for the establishment of a new garden in the city's western suburbs had been taken by Hopkirk with the help of a group of prominent citizens who included businessmen as well as academics.[72] Their aim was to provide amenity for the new suburbs, whose gardens were meagre compared with those that had graced the older part of town in the seventeenth and eighteenth centuries and to promote botanical study and research.[73]

The Faculty's interest in these developments was natural, in view of the financial contributions that it had evidently made to the upkeep of private herb gardens such as John Wodrow's, not to mention the active involvement of some of its members in the cultivation of these gardens – not least the surgeons who kept 'druggists' shops.[74] Surprisingly, the Faculty minutes contain no mention of Sandyford at the time of its opening near the west

---

[69] GUL, Special Collections, MS Hamilton 112, book 2 (1782).

[70] Ibid.

[71] J. Davies, *Douglas of the Forests: The North American Journals of David Douglas* (Edinburgh, 1980), p. 12.

[72] E. Curtis, 'An Historical Introduction' to a supplement on the natural history of Glasgow Botanic Gardens, *Glasgow Naturalist*, 23, part 3 (1998), p. 43.

[73] J. Hutchison, *The Associations between the Faculty of the Physicians and Surgeons of Glasgow and the Herbal and the Botanic Gardens*, RCPSG, 1/13/7/9, p. 168.

[74] Ibid., p. 166.

end of Sauchiehall Street in 1819, but in 1821 a committee of the Faculty met the directors of the gardens and undertook to acquire twelve shares at a cost of 120 guineas – and instead of paying the capital sum agreed to make an annual contribution of thirty guineas, a commitment that was honoured until 1887.[75] In return, Faculty members were to enjoy the privileges of subscribers and the Faculty was to nominate annually three of its members as joint directors, each with one vote.[76] This allowed the FPSG to share in the responsibility for the training of doctors and the practice of medicine, in which, of course, a knowledge of herbal remedies played an important part.[77] To that end the garden contained a lecture room for the use of the Professor of Botany, who was obliged to deliver at least one course of lectures each season.[78]

The interests of botanical teaching for medical students could not have been better served than by Hooker, whose tenure as Professor of Botany almost exactly coincided with the two decades of the Sandyford gardens' existence. Although the gardens had been established under his predecessor, Robert Graham, it was under Hooker's 'skilful guidance' that they attained international stature.[79] Thousands of new plants, including many new species, were acquired on his initiative – he prevailed, for example, on the director of Kew Gardens to ship surplus plants north by the Leith smack. Visiting botanists declared that the little eight-acre garden would not suffer by comparison with other European gardens.[80]

Hooker's success as a lecturer was equally remarkable, particularly since he had no experience of teaching before coming to Glasgow and indeed had spent the preceding period as a brewer in Suffolk. Although he was an established botanist, having devoted himself to natural history ever since discovering a rare moss hear his boyhood home in Norwich, his accomplishments had been in the field of botanical exploration rather than in education (his expeditions were adventurous – he was rescued from a burning ship on the voyage back from Iceland and arrested as a spy in the Highlands).[81] Nevertheless, he became one of the university's most accomplished lecturers. His popularity grew, like the garden, from modest beginnings and for a time

[75] Ibid.
[76] *Companion to the Glasgow Botanic Garden or Popular Notices of Some of the More Remarkable Plants Contained in it* (Glasgow, 1818), p. 19.
[77] Hutchison, *The Associations*, RCPSG, 1/13/7/9, p. 165.
[78] *Companion to the Glasgow Botanic Garden*, p. 20.
[79] D. McLellan, *Glasgow Public Parks* (Glasgow, 1894), p. 102; M. Hadfield, *A History of British Gardening* (3rd edn, London, 1979), p. 296; Davies, *Douglas of the Forests*, p. 13.
[80] Allan, *The Hookers of Kew*, p. 79.
[81] M. Hadfield, *A History of British Gardening* (London, 1979), p. 316.

he had a smaller following than any other medical professor. But before long an increase in his fees and a threefold increase in the endowment of the chair reflected a marked rise in numbers. While this may have been partly due to changes in curriculum regulations,[82] there can be no doubt that Hooker had become a star attraction. Many outsiders attended his lectures, including officers from the barracks three miles away, who were drawn by his striking presence, graceful style and innovative visual aids such as huge coloured drawings of plants.[83] Hooker's sons William and Joseph also attended the lectures as young boys and were apparently inspired with an interest in natural history.[84] William, who had a special interest in ornithology, later graduated MD at Glasgow University and became a member of the FPSG, while Joseph eventually succeeded his father as director of Kew Gardens.

Hooker also took his students botanising in the countryside around Glasgow and, once a year, in the Highlands and Islands. Wearing top hat, high collar, fobs and seals and watch chains, he covered huge tracts of land every day, reducing some of his students to utter exhaustion.[85] Students, who were sometimes invited to breakfast after the lecture, were also given the run of Hooker's private herbarium in his home in Woodside Crescent, which also attracted botanists from all over Europe.[86] For the benefit of his students, too, Hooker wrote textbooks, beginning with *Flora Scotica* in 1821, which was a revelation to botanists. His later *British Flora*, which reached a wider audience, reflected the new interest in the distributional problems of native plants.[87]

Hooker's career in Glasgow may in some respects have been more pleasing to the Faculty than to the University, with which he was frequently at odds. The Senate, it has been suggested, frowned on his work at Sandyford because it was not under the direct control of the University.[88] Moreover, in 1827 Hooker baulked at the proposal to devote the fees for the C.M. degree to the defence of the graduates in surgery against the Faculty's legal action to

---

[82] Coutts, *A History of the University of Glasgow*, p. 532.

[83] Allan, *The Hookers of Kew*, p. 79.

[84] D. Murray, *Glasgow and Helensburgh: As Recalled by Sir J. D. Hooker* (Helensburgh, 1918), p. 7.

[85] T. Whittle, *The Plant Hunters* (London, 1970), p. 118.

[86] D. Patton, 'The British Herbarium of the Botanical Department of Glasgow University', *Journal of the Glasgow and Andersonian Natural History and Microscopial Society*, 17 (1954), p. 10.

[87] H. R. Fletcher and W. H. Brown, *The Royal Botanic Gardens Edinburgh* (Edinburgh, 1970), p. 111.

[88] Allan, *The Hookers of Kew*, p. 86.

stop them from practising in Glasgow without taking their own license. He strenuously objected to the siphoning off of £30 from his salary since botany, as he pointed out, was not even in the surgical curriculum.[89] It was, however, in the medical curriculum and that of the FPSG.

Hooker was faced with a typical problem of the early nineteenth century, separate disciplines detaching from an umbrella subject. What made botany viable on its own was not what made it interesting to training for the medical profession. Hooker was astute enough to realise how tight was the hold of the curriculum on teaching positions. Both his sons studied medicine in Glasgow. William, his namesake, taught materia medica at Anderson's College from 1838 to 1839. He could have travelled the world, but was genuinely attached to Glasgow, about which he wrote in 1837: 'much as I like rambling, home is home after all, and though I might like *travelling* abroad, I daresay that I would like settling abroad still less than settling at home'.[90] But this was not to be, as William became seriously ill. Sir William turned to his friend William Connell, whose firm had commercial and shipping ties to Jamaica, to send William there to recoup his health. True to his Glasgow training, William Hooker was involved in studying infectious disease when he died.[91]

Jackson Dalton Hooker, too, finished his medical studies. He followed his father to become Director of Kew Gardens. The man with whom Hooker disputed the curtailing of his salary over dinner, John Burns, he recommended for membership of the Royal Society in London. In 1829 William Jackson Hooker, his long-time mentor Dawson Turner, and Dawson's son-in-law, Francis Palgrave (of *Treasury of Poetry* fame), along with the Glaswegians Charles Macintosh, inventor of the famous raincoat, and Thomas Thomson, professor of chemistry at Glasgow University, pledged 'from personal knowledge' that Burns was 'likely to become an active and useful member of the Royal Society'.[92] Burns was elected on 1 April 1830. In 1841 when Hooker was leaving Glasgow for Kew, he shared an upset with Burns because both of them were Episcopalians.[93] They stood accused of not complying with the ancient rule to sign the Confession of Faith.

Burns' election to the Royal Society underlined the unity in the disciplines that shaped modern science. William Hooker, the great botanist, Thomas Thomson, the renowned chemist, and Glasgow's most famous – and first – professor of surgery, John Burns, were acknowledged for their contributions to their fields. But above all they and their students saw them as good teachers.

[89] Coutts, *A History of the University of Glasgow*, p. 551.
[90] Allan, *The Hookers of Kew*, p. 97.
[91] Ibid., p. 104.
[92] J. Burns, Certificate of Election, vii, 295, the Royal Society (London).
[93] Allan, *The Hookers of Kew*, p. 111.

Anatomy Teaching

'Learning that the ship had struck the rocks, and was fast sinking [the governess], awakened by the shock, rushed to the berth of her little charge, whom she snatched from her slumbers, and hastily proceeded to dress. The child, though surrounded by half frantic passengers hurrying upon deck, knelt down, and with clasped hands, frequently repeated her morning prayers. In the same attitude of devotional fervour was Dr Burns last seen.'[94]

The John Burns who so calmly went to his death in the wreck of the *Orion* off Portpatrick in 18 June 1850 was 'one of the most respected of our fellow citizens ... few men in Glasgow were better known or more universally admired'.[95] A high Tory and a convert from the Presbyterian Church of Scotland to Episcopalianism, he became the first Regius Professor of Surgery in Glasgow University in 1815. His career was based on an early dedication to anatomical research and he championed high standards in surgery and midwifery throughout his life. Yet Burns's early anatomical work almost cost him his career when he walked the legal tightrope of body 'snatching' before the reforms of the 1832 Anatomy Act. His expertise fuelled his unremitting zeal to secure a rigorous medical curriculum and the advancement of the profession. In this he was like another great, and just as historically forsaken, Glasgow teacher and surgeon, John Gordon. The lives of these two men spanned the 150 years when improvement and reform were nailed to the mast of the medical profession, mainly by surgeons.

John Burns and Gordon, to both of whom we return throughout the next chapters, owed the eminence of their professional careers to the activities of the FPSG which they did so much to shape, although the course of time Burns contributed to antagonism between the FPSG and the University over who could license surgeons when he initiated the first degree ever awarded in surgery by a British university. He himself took the CM in 1817.[96] His fellow surgeon and the first Regius Professor of Midwifery, James Towers, did the same. Characteristically, both of them were active in the FPSG as well as mainstays of university teaching and hospital medical care. Burns's career represents a culmination of the surgeons' quest for an upgrading of surgical skills that made anatomy teaching a priority in the eighteenth century.

The *Orion* floundered 'not fifty yards from the entrance' to Portpatrick,

---

[94] Report of the *Glasgow Herald*, 28 June 1850.

[95] *Glasgow Herald*, 24 June 1850.

[96] W. I. Addison, *Roll of Graduates of the University of Glasgow, 1727–1897* (Glasgow, 1898), p. 80.

with a 'grating sound, like as if a powerful saw was rapidly eating through the bottom'.[97] The many who died were not pulled under by heavy seas: 'the sea was as smooth as glass, it was perfect calm, there was not the least fog'.[98] Burns was returning to Glasgow with his niece from medical lobbying activities in London by way of Liverpool. His brothers, George and James, were the owners of the steamboat company of Burns and Laird,[99] which became the Cunard Line in the twentieth century. By marriage Burns was related to the west-coast shipping enterprise of David MacBrayne. Burns seamlessly integrated with the commercial elite of Glasgow. His own business interests extended from his early anatomy school to shares held in water and transport recorded in his estate.[100]

Burns's death in his seventy-sixth year brought to a close his considerable political activity in safeguarding the interests of Glasgow University, the value of surgical degrees and their recognition nationally. His social and commercial ties provided openings for his success in the lobbying arena and profound professional and religious convictions gave him the heart to engage the fight. In April 1830 he is mentioned as writing, together with James Jeffray, the professor of anatomy in Glasgow University, a memorial criticising Warburton's Anatomy Bill.[101] In 1826 he was called to testify on medical education before a select committee of the House of Commons.[102] In 1842 he fought against the transfer of degree-granting powers from the universities to the colleges of either Physicians or Surgeons;[103] and again, in 1845, Burns was requested by the University Senate to watch carefully yet another Bill before Parliament on regulating the medical profession. He was instrumental in its downfall, from Glasgow University's point of view a success, for he 'was heartily congratulated on the ability which he had shown in opposition to the bill'.[104] His political lobbying was the outcome of a lifetime spent in teaching and research, and as a member of the FPSG and the university. His university associates valued his contribution to their interests sufficiently to commission an oil portrait, which now hangs in that 'Valhalla' of the University of Glasgow, the Bute Hall.[105]

[97] *Glasgow Herald*, 24 June 1850.
[98] Ibid.
[99] P. Mathias and A. W. H. Pearsall (eds), *Shipping: A Survey of Historical Records* (Newton Abbot, 1971), no. 35.
[100] NAS, SC 36/48/37, Register of Inventory, 7 June 1850 to 19 May 1851, Glasgow Sheriff Court, pp. 496–97.
[101] Coutts, *A History of the University of Glasgow*, p. 518.
[102] Ibid., p. 563.
[103] Ibid., p. 564.
[104] Ibid., p. 530.
[105] Glasgow University, Hunterian Portrait, GLA, HA 44171.

'The late Dr Burns', said the *Glasgow Herald*, 'gave himself up with great zeal to the study of anatomy, especially to that department which is styled relative or surgical anatomy. He afterwards began to give instruction to others, and was the first private teacher of anatomy in Glasgow. His lecture room was originally at the head of Virginia Street, at the north west corner behind the present Union Bank.' The *Herald* continued: 'In these days ... bodies for dissection ... were all obtained by the students robbing the churchyards. Mr Burns being detected in some scrape of this sort, the magistrates agreed to squash proceedings against him on condition that he gave up lecturing in anatomy. This he agreed to do, but his younger brother, Allan, took up the lectures on anatomy, while John began to lecture on midwifery. The lecture room of the two brothers was now removed to a brick flat, built in the remains of the Old Bridewell, on the North Side of College Street where it still remains'.[106]

Allan Burns, John's brother, has been called the 'the first imaginative experimentalist to deal with the coronary circulation'.[107] The 1964 reprint of his *Observations on Some of the Most Frequent and Important Diseases of the Heart* names him as 'of major importance in anatomy, pathology, surgery, and medicine; and this in the face of obstacles and frustrations that would have turned back any but the most dauntless'.[108]

The teaching done in the College Street rooms was private, at least as to income and inclination, but was soon recognised for accreditation through the FPSG examining boards. In 1800 attaining a recognised medical qualification depended on course certificates and these in turn had to be acceptable to the examiners for a university MD or to the FPSG for a licence. Where certificates came from or how they were evaluated was only then on the verge of regulation. Which courses to take and their number was purely the choice of the client, with no requirement to study in one place. By 1802 the university was beginning to set minimum requirements, including a year's minimum attendance in Glasgow.[109] This changed the balance of free market education. John Burns's career stood at the cusp of critical changes in standardising medical qualifications and in the typically Scottish conflation

---

[106] *Glasgow Herald*, 24 June 1850.
[107] A. Fishman and D. W. Richards (eds), *Circulation of the Blood: Men and Ideas* (Oxford, 1964), p. 222.
[108] A. Burns, *Some Observations of the Most Frequent and Important Diseases of the Heart* (reprint, London, 1964), Introduction, p. v.
[109] GUABRC, Senate Minutes, 1/1/2, 30 March 1802.

of the MD and the surgical licence. He was Regius Professor of Surgery when the Senate of Glasgow University granted him an MD in 1828. His early qualifications were those of a surgeon, being awarded a licence of the FPSG in 1796 at the age of twenty-two.[110] His first steps to a successful career were in line with FPSG control of hospital medical positions. After completing his studies, in 1792, he became the Glasgow Royal Infirmary's first surgeon clerk.[111] He then moved in rapid succession to other stepping-stone posts, first as apothecary's and surgeon's clerk in July 1795 (to May 1796).[112] John Burns was subsequently elected (by signed lists), on 26 January 1797, one of four surgeons for the year, to continue in office until the first Monday of February 1798. He was also elected surgeon in 1808.[113] Burns was appointed in June 1800 to lecture on surgery and midwifery at Anderson's College, a post he did not relinquish until he resigned in 1817. Granville Sharp Pattison, the friend of Allan Burns, and owner of the latter's admired specimen collection, succeeded him at the extra-mural school.[114] Anatomy was one of the golden threads in Burns' rise to fame and his educational reforms.

In 1814 there were in Glasgow some 'eight hundred students of anatomy'.[115] These high numbers stretched across university, private and extra-mural courses. They were also prone to fluctuate and show that medical education, as a business venture, could carry risks for the independent teacher.[116] But anatomy was no longer an optional part of any branch of medical education. Anatomy was basic: 'For anatomy clears up doubts, purges the mind from all visionary or fanciful prejudices, penetrates and discovers the fallacy of various hypothesis and injurious systems'.[117] What is more, by the end of the eighteenth century it bound together the separate classes of practitioner:

> Surgery, that inestimable art, on the skilful administration of which, the lives of all ranks ... frequently depend, would be totally incapable of performing its salutary functions without the safe guide of anatomical science. From the common operation of bleeding, to the higher and more difficult operations of surgery, the knowledge of anatomy guides the hand of the artist, to avoid fatal errors, and

[110] Duncan, *Memorials*, p. 178.
[111] Thomson, *A Biographical Dictionary of Eminent Scotsmen*, p. 252.
[112] HB, 14/1/1, GRI Managers' Minutes, i, pp. 160–61.
[113] HB, 14/1/2, GRI Managers' Minutes, ii, p. 166.
[114] J. Butt, *John Anderson's Legacy: The University of Strathclyde and its Antecedents, 1796–1996* (East Linton, 1996), pp. 36–37.
[115] J. Herrick, 'Allan Burns, 1781–1813: Anatomist, Surgeon and Cardiologist', *Bulletin of the Society of Medical History of Chicago*, 4 (1928–35), p. 463.
[116] J. Christie, *The Medical Institutions of Glasgow* (Glasgow, 1888), pp. 14–15.
[117] W. Rowley, *On the Absolute Necessity of Encouraging, Instead of Preventing or Embarrassing the Study of Anatomy* (London, 1795), p. 4.

ensure probable success ... An ignorance in anatomy, therefore, must produce ignorance in the practice of surgery, midwifery and physic.[118]

Anatomy had not been so well placed at the beginning of the century, its later success owed in large part to the professional aims of FPSG surgeons.

Early Teaching in Anatomy

Anatomy was crucial to the 'trials' by which surgeons qualified for the FPSG licence. In 1714 the FPSG took the important step of encouraging attendance at university-taught courses held by one of their surgeons as a preliminary to their examinations. Course teaching took skills training away from the master-pupil contract and established systematised, publicly accessible lectures. It undermined protective or 'secret' trade practices.[119] The 'closed-shop' of the craft visibly declined as private and university teaching attracted an increasingly itinerant profession from 1700, a trend underway in London and different contexts of university cities, such as Leiden and Paris.[120] Leading surgeons welcomed the opening of the profession, the Glaswegian surgeon John Gordon,[121] being quick to attach the needs of the FPSG for bettering surgical training to the refurbishment of 'internal' medicine within the university. When his FPSG colleague, John Johnstoun, a physician, was appointed to the chair of medicine in Glasgow in 1714, Gordon took up lecturing anatomy in the university. Physicians, according to incorporation rules, were banned from manual 'handiwork' including dissections. In university teaching physicians explained anatomy but the manual work was done by the 'demonstrator'. In a quick move to upgrade education, Glasgow and Edinburgh surgeons instigated anatomy lectures taught and demonstrated by themselves. Surgeons in Scotland stepped into lectureships, eventually having demonstrators of their own. John Innes, a Highlander, was dissector for Alexander Monro for twenty years, for example, and wrote

---

[118] Ibid., p. 5.
[119] Chris Lawrence remarks on Alexander Monro primus breaking with 'secret' formulas by making known his techniques for injecting preparations. C. Lawrence, 'Alexander Monro Primus and the Edinburgh Manner of Anatomy', *Bulletin for the History of Medicine*, 62 (1988), pp. 193–214. The same mission to unmask the 'arcane' or 'occult' is visible in the writings of H. Boerhaave, F. Hoffmann and G. Stahl.
[120] For international comparisons and their impact, see the previous chapter.
[121] On J. Gordon see Duncan, *Memorials*, p. 251; Coutts, *A History of the University of Glasgow*, pp. 484–85; Boney, *The Lost Gardens of Glasgow University*, pp. 65–66. RCPSG references to activities of Gordon, RCPSG, 1/1/1/2, fol. 20r (regarding the 1736 amendments to the Faculty laws, regulations and acts and successively after that in examination and governing matters).

his own teaching text, *A Description of the Human Muscles*, which was then reedited by Robert Hunter of Glasgow in 1822. Hunter was the extra-mural Lecturer in Anatomy and Surgery at Anderson's College and a FPSG member.[122] This instance of the demonstrator publishing an anatomy textbook shows how common the interest in the subject had become.

Anatomy teaching on a regular basis started in Glasgow in 1714 when John Gordon taught apprentices for FPSG qualifying examinations on the university's premises. This accommodated all concerned: the university as the appropriate venue for lectureships; the FPSG as the major regulatory agent whose candidates for licenses were thus cared for; and medical students seeking instruction. Lecture courses were systematic and universal in a way the craft indenturing of apprentices was not. To have a surgeon who was a principal in a leading Glasgow firm actively support a public form of teaching outwith craft regulations heralded the Faculty's firm will to modernise.

The surgeons in Glasgow shifted to university lecturing earlier than in Edinburgh, although no one denied the importance of apprenticeships or the examinations held by the surgical incorporations. At Edinburgh Alexander Monro taught anatomy as a member of the Town's College in 1726, twelve years later than John Gordon, although some earlier lecturing was undertaken under the auspices of the Royal College of Surgeons. John Monro and his son were leading figures in the Edinburgh College of Surgeons. While they too indentured apprentices, they were obviously also determined to defy the restrictive craft practices of the guild traditions of skills training.

To learn as an indentured apprentice meant a direct, personal tie to a master's practice. Craft skills were specific, taught so that each special skill could be perfected, for example the excision of tumours, applying ligatures, cutting of the stone or amputations. Before the eighteenth century, the FPSG licensed practitioners in a specific skill of this kind. With anatomical lectureships a more unified homogeneous standard could be introduced. This was crucial for moving the skills demanded by the surgeon's licence away from piecemeal local instruction to a common level of skills. Systematic learning, the use of new textbooks and a unifying nomenclature were now to change the face of the profession. Uniformity in teaching made possible a common understanding of the internal structure of the body. As this knowledge grew it imposed conformity on anatomical atlases. The exploration of the body through engravings, dissections and preparations secured anatomy its formative role.

[122] J. Innes, *A Description of the Human Muscles with their Several Uses, and the Synonyma of the Best Authors: A New Edition with Notes, Practical and Explanatory by Robert Hunter* (Glasgow, 1822).

The rare public dissections used in teaching anatomy before systematic teaching took hold did not hone the manual skills essential to good surgery. Mainly public spectacles, they were a relic of earlier religious customs attached to anatomical theatres, staged to remind onlookers of the brevity of life and man's unavoidable encounter with death. In stark contrast to lectures – taking these to include teaching in universities or hospitals or as private courses – public dissections did not teach skills. Nothing was preserved from the corpses used in public dissections, few anatomical or pathological collections were enhanced through them, nor were they used for preparations, making anatomical plates or for learned papers.[123] Useful anatomical teaching in this period was usually only accessible through the hospitals of Paris or London.

As anatomy lectureships took hold, this new repertoire of medical knowledge shaped the agenda of examining boards and, indeed, what practitioners were expected to know. It thoroughly undermined the authority of the craft guild. To make modern anatomy teaching useful for surgeons it had to be visually mimetic, applicable in surgery, standardised in nomenclature and resourced through reference works and textbooks. Systematic lectures connected these requirements together like the spokes of a wheel. The mission to modernise moved forward on the bandwagon of this complex undertaking, not on what single institutions alone had to offer. This was one major reason why travel and trade routes remained crucial: they assembled the scientific jigsaw of an ever expanding teaching enterprise. Locating new textbooks and learning to do preparations was part of this puzzle of piecing together the bigger picture of anatomy.

Expenditure on teaching aids was a crucial input in this process of unifying surgical education via anatomy courses. The FPSG fully underwrote these changes to surgical education; indeed, it provided active support in buying specimen collections and supplying textbooks through its library. The didactics of the classroom and the textbook became an essential prerequisite in training the mental imaging and the manual dexterity needed by the surgeon. As Susan Lawrence remarked, when a student came to dissect he could then recall the anatomical knowledge deeply impressed on his mind by a series of class sessions where the senses (particularly visual) had been rigorously disciplined.[124]

Dissection (of animal parts) and examinations in anatomy became the

[123] R. Kilpatrick, *Nature's Schools: The Hunterian Revolution in London Hospital Medicine, 1780–1825* (unpublished Ph.D. thesis, Cambridge, 1988), p. 96.
[124] S. Lawrence, 'Educating the Senses: Students, Teachers and Medical Rhetoric in Eighteenth Century London', in W. F. Bynum and R. Porter (eds), *Medicine and the Five Senses* (Cambridge, 1993), pp. 154–78.

major 'trial for qualification' for surgeons accredited by the FPSG. They were introduced by the FPSG sometime between 1688 and 1733 (the Minute Books for that period were destroyed by fire), John Gordon being one of the examiners.[125] When Nathan Wilson, surgeon in Greenock, sought admittance to the Faculty on 4 August 1735, his 'trial' or examination was based on the ability to prepare medicines (an ointment, pills and a plaster), and on his skill in dissecting a heart while explaining the circulation of the blood. Other examinations followed this pattern. Thomas Simson had to dissect an eye and explain how vision functioned; John Love, a Greenock surgeon and later an anatomical lecturer himself, dissected the lungs and gave a discourse on respiration. John McFarland dissected the liver and John Muir the brain.[126] All these candidates were examined in the 1730s showing that anatomy had become central to gaining accreditation by the FPSG.

Both John Gordon and John Paisley, his successor, taught in classrooms at the university. Paisley taught in the Old Humanities Classroom, which was expressly designated for use as an anatomy room.[127] While no lecture notes survive to reveal what Gordon or Paisley taught, their lectures would have been similar to those in Edinburgh. Both were apprenticeship masters and active in the incorporation, like John and Alexander Monro. Chris Lawrence described the situation well when talking of Alexander Monro's efforts in Edinburgh: 'Monro had to build a course in a city lacking both a strong anatomical tradition and, until 1726, a medical faculty. The one factor in his favour was that he had reasonably large potential audiences in the local surgical apprentices'.[128] At Glasgow Gordon was already teaching local surgical apprentices and was as aware as Monro of the need to pay attention to changing patterns in medical education and the sciences that fed it.[129]

Gordon's and Paisley's methods for teaching anatomy were probably similar to Monro's and they would have found the course set up by Alexander Monro in Edinburgh much to their liking. Monro, while not an innovator himself, polished and refined the methods he had acquired by watching others. From 1712–13 Alexander Monro served an apprenticeship in his father's surgical practice, undertook training in London, with eight dissections recorded in his commonplace book from 1717–18. Moving on to Paris and Leiden the next year, he again did dissections, learning how to

[125] Gordon's activities in the FPSG predate the extant minute books, he being, for example, the apprenticeship master of William Smellie and Tobias Smollett.
[126] RCPSG, 1/1/1/2, fos 13r-17v.
[127] Coutts, *A History of the University of Glasgow*, p. 488.
[128] C. Lawrence, *Alexander Monro Primus*, p. 197.
[129] J. Butterton, 'The Education, Naval Service and Early Career of William Smellie', *Bulletin of the History of Medicine*, 60 (1986), p. 17.

do preparations in Holland. He studied Frederick Ruysch's methods of injecting specimens (known then as the 'Ruyschian art') and heard the lectures of Boerhaave and Bernard Albinus, the latter, perhaps, Europe's leading anatomist, whose anatomical atlas and specimen collection were renowned. In 1719 Alexander Monro returned to Edinburgh and began teaching a systematised series of lectures which were the basis of his later course at the university (after 1726).[130] It is not known where these lectures were held but presumably they were private lectures approved by the incorporations of surgeons.[131] In 1747 Alexander Monro wrote a treatise for his son Donald. Its purpose was to explain, in a step by step approach, how he had built up his lectures,[132] and specifically to show how to combine verbal explanations with visual aids. All the lectures on bones, tissues, organs, nerves and the physiological functions of the body were carefully presented in sequence and made visually mnemonic by the use of preparations.

Reflecting difficulties of cost and acquisition, only two cadavers were used in the whole course.[133] This however seems unimportant in an overall teaching strategy built on using multiple approaches to instil knowledge about the inner spatial structure of the body. Private teaching, on the increase in London since 1703,[134] and available in Leiden and Paris, meant individual surgeons had an eye on learning new teaching techniques. Teaching by multiple methods was more useful and economical. The diversity of representational techniques included, for instance, first making watercolour sketches, done quickly before putrefaction set in. These were essential as reference points for the slower process of producing etchings. The laborious process of making preparations was useless without honed dissecting skills. These examples demonstrate the many specific skills involved in creating an anatomical teaching framework. Textbooks were a spin-off of these endeavours, a repository of the knowledge acquired by different hands and different processes. The expensive and difficult collaboration between artists and research anatomists explains the rampant pirating of good anatomical

---

[130] The following interpretation follows Lawrence, *Alexander Monro Primus*, pp. 193–214.
[131] For the Edinburgh Incorporation of Surgeons, see H. Dingwall, 'Original Edinburgh Fellowship Examination', *Journal of Royal College of Surgeons of Edinburgh*, 36 (1991), pp. 357–61.
[132] Lawrence, *Alexander Monro Primus*, p. 197.
[133] Ibid.
[134] Kilpatrick, *Nature's Schools*, p. 97.
[135] M. Kornell, *The Ingenious Machine of Nature: Four Centuries of Art and Anatomy*, exhibition catalogue by M. Cazort, M. Kornell and K. B. Toberts, Ottowa, National Gallery of Canada (1996), nos 72–74.4, pp. 186–90; idem: 'The Study of the Human Machine: Books of Anatomy for Artists', ibid., pp. 43–70; for Cowper, pp. 59–60, 64, 66; idem, 'Cowper, William (1666/7–1710)', entry for the New Dictionary of National Biography, forthcoming.

plates.¹³⁵ The history of anatomy teaching has not been served well by dividing it into parts, art history or printing techniques or charting the rise of 'hands-on' dissection. All of this is important but has to be seen holistically, as teaching methods depended on a considerable network of newly acquired expertise. The later richness and heavy investment in anatomical and medical 'museums' has its roots in the teaching enterprise initiated with the beginning of the eighteenth century. What began as science quickly fed into teaching.

In London the Scotsman James Douglas employed a plethora of artists and engravers.¹³⁶ His work was fundamental to William and John Hunter's interests in anatomy, including comparative anatomy. Early in the century Douglas was working on his osteology and giving private lectures on anatomy,¹³⁷ while William Cowper was perfecting the description of the muscles which appeared in book form in 1694. His *Anatomy of Human Bodies* became a standard work, the FPSG buying its reissue by C. B. Albinus when it came out.¹³⁸ Anatomy teaching underpinned many other surgical and medical uses, for example diagnostics and pathology.¹³⁹ In clinical teaching in Glasgow in the 1780s autopsies were frequent and used to audit diagnostic skills.¹⁴⁰ Matthew Baillie, part of the Scottish 'mafia' in London, famously published the first comprehensive atlas of pathology. Baillie's access to the collections of his uncles William and John Hunter:

> furnished the necessary material not only for his beautiful copper engravings, but also for *The Morbid Anatomy of Some of the Most Important Parts of the Human Body* (first edition 1794), to which the atlas formed a supplement. This book is the first text of pathology devoted to that science exclusively by systematic arrangement and design.¹⁴¹

The skills involved in making preparations were considerable and implied a good understanding of chemistry. The acrimonious dispute between Alexander Monro secundus and William Hunter, touching the work of important pupils like William Hewson, over the lymphatic system in the 1750s had, at its core, experiments with 'vessels filled with quicksilver by extravasation'

---

[136] H. C. Brock, *Dr James Douglas's Papers and Drawings in the Hunterian Collection, Glasgow University Library: A Hand List* (Glasgow, 1994), pp. 12–15.

[137] Ibid., pp. 17–19.

[138] On W. Cowper see M. Kornell's entry in the forthcoming *New Dictionary of National Biography*, a draft of which was kindly lent us.

[139] The most famous example is M. Baillie, *The Morbid Anatomy of Some of the Most Important Parts of the Human Body* (London, 1793).

[140] For example see notes on autopsies in GUL, Special Collections, MS Hamilton 83, Common Place Book, vol. 2.

[141] E. R. Long, *A History of Pathology* (Baltimore and London, 1965), p. 93.

done in connection with his lecture courses, as Monro was at pains to explain.[142]

The first Alexander Monro's advocacy of scientific anatomy was applauded in Glasgow, with city surgeons contributing early to his new scientific journal, the *Medical Essays and Observations*, founded in 1733. The respected FPSG surgeons Peter Patoun and John Calder published their findings there. John Paisley contributed no less than seven articles on anatomical, pharmacological and obstetrical subjects. Paisley was John Gordon's successor as anatomical lecturer. John Love, a candidate for Paisley's anatomical lectureship when the latter died in 1740,[143] published on a drug he thought beneficial for curing cancer.[144] This journal was the first periodical to devote itself solely to medical scientific observations, paralleling the *Philosophical Transactions* in intent, but focusing the discipline of study more narrowly.[145] When Alexander Monro made public his improvements on injecting techniques in the *Medical Essays and Observations* of 1732–37,[146] the Glasgow surgeons were no doubt already familiar with them.

Anatomy Teaching after 1740

By 1740 the FPSG possessed a collection of anatomical preparations in their Hall. John Love was able to borrow them:

> upon an Inventory thereof to be made up and delivered to him and his signing and delivering to the Collector an obligement therto to make the same forthcoming

---

[142] M. H. Kaufman and J. J. Best, 'Monro Secundus and Eighteenth-Century Lymphangiography', *Proceedings of the Royal College of Physicians of Edinburgh*, 26 (1996), p. 78.

[143] *Scots Magazine*, 2 (May 1740), p. 238.

[144] A number of publications of Paisley, Love, and Gordon can be found in GUL, Special Collections, Bh10-k. 9–14, vol. i–vi. These include various contributions by Paisley, 'The Dura Mater Ossified, and other Morbid Appearances', in ibid., *Medical Essays and Observations Published by a Society in Edinburgh*, 6 vols (Edinburgh, 1752), ii, pp. 267–71; 'An Account of an Extraordinary Worm', in ibid., pp. 284–89; 'A Mortification Cured by the Peruvian Bark', in ibid., iii, pp. 39–42; 'A Hydrocephalum, with Remarkable Symptoms', in ibid., pp. 305–11; 'Coagulated Blood Extravasated upon the Uterus and the Thickness of the Womb in a Laborious Birth', ibid., iv, pp. 355–61; 'The Dissection of a Calculous Person', in ibid., v, pp. 283–87; 'A Dropsy and Large Vesicae in the Ovarium' by Mr Paisley, in ibid., pp. 298–98. Also, 'Observations of the Effects of Lignum Guiacum in Cancers' by Mr J. Love, in ibid., v, pp. 82–84.

[145] The description of the journal's aims taken from M. DeLacy, 'Influenza Research and the Medical Profession', *Albion*, 25 (1993), p. 45. Others date the beginning of the journal to 1732, *Medical Essays and Observations Published by a Society in Edinburgh*, 3 vols (Edinburgh, 1732–37).

[146] A. Monro, 'An Essay on the Art of Injecting the Vessels of Animals', *Medical Essays and Observations* (Edinburgh, 1732–37) i, pp. 94–111.

to the Faculty when they please in the same Case and Condition they are now in.[147]

Love's use of preparations documents the continuity of lectures in the old humanities classroom refitted for anatomy teaching when Paisley began his lectures in 1730.[148] Robert Hamilton and John Crawford also petitioned to use the old classroom for teaching and dissecting.[149] In 1741 and 1742 Love and Crawford once again inventoried the anatomical collection of the FPSG. Their petition contains a reference to dissection, significant because it shows that cadavers were being used for teaching students and apprentices, probably also for preparations, as these would benefit the teaching collection. The University Senate in 1745 resolved 'the encouragement of Anatomy and preventing Mobbs etc'.[150] Mobs or riots seemed to have swept through the university in 1744, 1745, 1747 and 1748, and are mentioned specifically in connection with Robert Hamilton's appointment. Glasgow was subject to the social turmoil caused by public disquiet over body acquisition as were other places, but this very protest shines a light on the modernising activities of the town's medical men. It is well to note that dissection was being practised in Glasgow in 1744, two years before Hunter began his anatomical lectures in London, to the extent that it caused public disorder. Despite the FPSG condemning grave-robbing publicly, its members were practising dissection and doing autopsies on patients. On 5 June 1744 the Faculty passed an 'Act against Violating the Graves of the Dead'. It stated:

> In abhorrence and detestation of the crime of violating the Graves of the Dead Do hereby Revive & Confirm their former acts against it particularly an act Declaring all members of faculty guilty thereof to be incapable of being any longer Members of the said faculty and to forfeit all privileges they or theirs might Claim by their having been members thereof. And that all apprentices so Guilty shall be forever excluded from being members of said faculty tho otherways Intitled by their services.[151]

A year later, on 1 July 1745, the Faculty appointed the Praeses, the Visitor, Dr Robert Hamilton and Hector McLean as a committee to wait on the magistrates of Glasgow 'and with them Concert on proper measures to be taken for suppressing the raising of dead corpses' and to report their resolution to the Faculty.[152]

[147] RCSPG, 1/1/1/2, fol. 55v.
[148] Coutts, *A History of the University of Glasgow*, p. 486.
[149] Ibid.
[150] GUABRC, 26639, UMM (Senate), 1730–49, pp. 191–92.
[151] RCPSG, 1/1/1/2, fol. 83v.
[152] RCPSG, 1/1/1/2, fol. 90v.

In 1749 the windows of the university buildings on the High Street were smashed with many people injured in a disturbance rooted in suspicions that students were raising corpses.[153] The contraband in exhumed bodies seems to have become part of commerce. Illegality was no hindrance in a city where His Majesty's excise officers were often the foe. Bodies shipped from Ireland were not a rarity, as some broadsheets reported for 8 December 1826. In boxes marked as stationery, but leaking sawdust, six corpses were found; while four more were discovered five days later. The commentary on this macabre trade was laconic:

> It is to be regretted that importations of this kind are not more carefully managed. Since dead bodies are necessary for dissection, it is much better that they be brought from Ireland, than our own church-yards should be robbed. Such packages might easily be wrapped up in such a manner as to avoid suspicion, and prevent those popular effervescences that are sure to follow detection.[154]

Legally only those condemned of heinous crimes punishable by death were to be dissected. James Glen, senior, was convicted of 'the cruel murder of his male child ... barbarously throwing it into the Forth and Clyde Canal, near Port Dundas'. He was twenty-two years old and 'after hanging the usual time, his body was cut down and conveyed to the Dissecting Room'.[155] There are other sad and raucous stories about the 'resurrectionist times'. It appears:

> trap-guns were set to scare the violators of the so-called last resting places of the dead; but in spite of all the dangers the outrages were numerous. Once instance is recorded of a student in Glasgow being killed by stumbling over one of these guns ... When he dropped dead, his fellow students were horrified, but the fear of discovery forced them to adopt an extra-ordinary method of taking away the body ... they passed slowly along the streets towards their lodgings, shouting and singing as if they were three roysterers returning from a carouse.[156]

Robert Hamilton and his brother Thomas, in succession professors of botany and anatomy from 1742 to 1781, all in all nearly forty years, coordinated much of the anatomical and botanical teaching for both the FPSG and the university. They supervised, updated and catalogued library and specimen collections. Robert Hamilton improved the university's collection by travelling to London in 1746 to procure 'Anatomical Preparations and other things very necessary for his teaching Anatomy successfully'. The Senate approved this desire to buy the best and appointed John Carrick to teach the anatomy

---

[153] G. MacGregor, *The History of Burke and Hare and of the Resurrectionist Times: A Fragment from the Criminal Annals of Scotland* (Glasgow, 1884), p. 30.
[154] GUL, Special Collections, Bh 14-X5, N. 46, 'Horrible Seizure of Dead Bodies'.
[155] GUL, Special Collections, Eph. G/70, 'Trial and Execution of James Glen'.
[156] R. Alison, *The Anecdotage of Glasgow* (London, 1892), p. 263.

class in his absence.[157] John Carrick entered the FPSG in 1746 as a surgeon and was the brother of one of the partners of the famous Glasgow Ship Bank.[158] The Carricks were related to John Moore, a friend of Cullen's, who was a partner in John Gordon's practice.

John Carrick was called upon by the Senate in 1747 to give advice, together with William Cullen, on buying apparatus for teaching chemistry. Their recommendations were approved and the money 'was to be laid out by the direction and at the sight of these two Gentlemen'.[159] The same year, 1747, saw a committee appointed 'to make the present Anatomy Class fit for teaching Chemie in and to consider what place is proper for the Anatomical Class'.[160] Carrick's budding scientific career was tragically cut short by an early death in 1750. In 1754 a committee was convened by the FPSG 'to Consider the new method proposed of Examining and Entering freemen surgeons'.[161] John Moore, Thomas Hamilton and John Crawford, all deeply committed to medical improvement, sat on the committee, but its proposals were neither adopted nor recorded.

In 1746 William Hunter began his private lecture courses on anatomy in London which are so often seen as introducing new teaching methods. These were almost certainly influenced by Glasgow anatomical teaching in the early 1740s, especially given his and William Cullen's close relationship with the Hamilton family, and Hunter's working partnership with Thomas Hamilton. Thus Hunter's often quoted remark that neither Paris, where he 'learned only with his ears but not his eyes',[162] nor that Alexander Monro's classes in Edinburgh [163] had added much to his knowledge, reflect well on Glasgow: he may have seen the equivalent or better amongst his colleagues in his native city.

Certainly the close interchanges between Hunter's London teaching and that at Glasgow University for candidates taking the FPSG licence as well as the MD deserves recognition. Indeed Glasgow teaching was as receptive as a sponge is to water for useful tips from the education of its medical scions in centres of learning. The letters of Thomas and William Hamilton, father and son, epitomise capitalisation in these skills. Anatomical, surgical and midwifery teaching at Glasgow University benefited from 'the Scottish

[157] Ibid., pp. 216–17.
[158] C. Munn, *The Scottish Provincial Banking Companies, 1747–1864* (Edinburgh, 1987), p. 42.
[159] Ibid., p. 229.
[160] Ibid., p. 230.
[161] RCPSG, 1/1/1/2, fos 158–59.
[162] F. Beekman, 'William Hunter's Early Medical Education', *Journal of the History of Medicine*, 5 (1950), p. 83.
[163] Ibid.

connection'. In the two decades in which William Hunter's school of anatomy became famous, in the 1770s and 1780s, its key figures had Scottish roots or went back to Glasgow. Indeed the letters, mainly of William to his father, but also other letters of various correspondents, show a lively trade in observations medical and surgical.[164] Of Thomas Hamilton it is known that he had an extensive practice 'connected with many of the most respectable families in Glasgow and its neighbourhood'.[165] He was sought after as a skilful surgeon, notwithstanding a reputation as 'a man of great hilarity and genuine humour, his company courted by all who relished wit and good fellowship'.[166]

William Hamilton, his son, was by all accounts an extraordinary scion of this medical family. Thomas Hamilton was connected with William and John Hunter 'by early friendship and a constant intercourse of good offices'.[167] William Hunter was apparently happy with William's professional ambitions, and 'particularly pleased with them in the son of his old friend, that, after the first season, he invited Mr Hamilton to live in his house, and committed the dissecting room to his care'.[168]

This acknowledgement of William Hamilton's qualities was due as much to Hunter's medical principles as to family ties. William Hunter writes to Thomas in 1780:

> Your son has been doing everything you could wish, and from his own behaviour, has profited more for the time than any young man I ever knew. From being a favourite with every body, he has commanded every opportunity for improvement that this great town afforded during his stay here; for every body has been eager to oblige and encourage him. I can depend so much on him, in every way, that if any opportunity should offer for serving him, whatever may be in my power I shall consider as doing a real pleasure to myself.[169]

William Hunter was instrumental in helping William Hamilton's appointment to the chair occupied until his death in 1781 by his father. When consulted about the appointment William Hunter wrote to the Duke of Montrose: 'That from an intimate knowledge of Mr Hamilton, as a man, and as an anatomist, he thought him every thing that could be wished for

---

[164] GUL, Special Collections, MS Gen. 1356/1–78.
[165] R. Cleghorn, 'A Biographical Account of Mr William Hamilton, Late Professor of Anatomy and Botany in the University of Glasgow, Read 6th November 1792', *Transactions of the Royal Society of Edinburgh*, 4 (1798), p. 38.
[166] Ibid.
[167] Ibid., p. 36.
[168] Ibid., pp. 36–37.
[169] Ibid., p. 37 (quoted from letters that Cleghorn had to hand before 1792).

in a successor to his father, and that it was in the interest of Glasgow *to give him* [emphasis in original], rather than his to solicit the appointment'.[170]

Hunter's approval of William Hamilton named the qualities the new breed of top medical teachers needed to have. The ability to teach was foremost amongst them as Hunter notes in a letter of December 1778 to Thomas Hamilton:

> He [William Hamilton] is now on the direct road for acquiring knowledge, as director in the dissecting room. It obliges him to apply, because he is to answer any question, and solve any difficulty that may occur, and, which is best of all, he is to demonstrate all parts of the body again and again to students. This is the most instructive province, and a fine introduction to giving lectures, as it gives facility in public speaking, and a habit of demonstrating distinctly and clearly, both of which are easily acquired when we are young; and yet, for want of that very opportunity, are possessed by few.[171]

Clarity in explaining and the anatomist's specific knowledge of the body characterised a good teacher.

William Hamilton himself recorded what he learned from William Hunter and his brother John in a series of highly descriptive letters to his father. On December 19, 1777, William wrote to his father that he had not 'waited on Dr Hunter yet, nor Mr Hunter'. He continued:

> I find it is not consumary to dissect without attending the class and that I may take out a ticket now which will serve me till the same lecture in the next course, as the Dr has a month of this course to go through. Mr Hunter has only one set of lectures in the winter. He only reads twice a week so that if I enter now I will only lose four and twenty lectures and he does not finish till April, this loss Mr Young has offered to make up from his notes. But of these things I can write you more fully to-morrow after seeing the Hunters whom I delayed calling on today ... I find the rates are three guineas for one course, and five for two courses, bodies are at present difficult to get. There are no whole ones in the dissecting room at present, I went there along with G. Reid. He introduced me to Cruickshanks [sic]. I have seen several Edinburgh acquaintances here and some Glasgow ones.[172]

On 22 December 1777, William wrote:

> I would have written to you on Saturday night but went to Dr Hunter's and after the class I delivered my letter, the Dr asked me and G. Reid who was along with me to come and eat an oyster with him and we sat until eleven. The Dr's class

---

[170] Ibid., p. 38.
[171] Ibid., p. 37.
[172] GUL, Special Collections, MS Gen 1356/33, copy of a letter from William Hamilton to his father, 19 December 1777. (Original of this and following letters in RCSE, CR HUN: J49. c. 170).

meets from half after one until half after three and on Saturday he has two meetings a night, one from six till eight at which I was. This week we meet for Tuesday and Saturday twice a day, he cannot meet at night as Mr J. Hunter lectures in his theatre then. I went to-day to see the Dr; we gave the money to Mr Cruikshanks, but he said he would speak to me about that to-morrow when he gave me my ticket. I offered him the dissections fee but he put off the acceptance of it in the same way. The Dr is particularly hurried this week as he is afraid his body won't keep, he is on surgery just now, he explained lithotomy and passing [the] catheter today. Bodies are vastly scarce at present, some of the men have been taken up and tried, but I hope this will soon be over. I waited on Mr J. Hunter today, he said he was vastly happy to see me and enquired very particularly about your health and both he and the Dr begged their compliments to you. He would take no fee from me and said he was happy to have it in his power to be of use to your son, he asked me to dine with him tomorrow and I go to his class tonight at seven. The Dr is an exceedingly good lecturer and vastly plain, he tells me a story with good humour, one or two of which he introduced into his lecturers on Saturday. He was vastly chatty with us at supper, G. Reid and he seem to be very intimate, he has breakfasted once or twice and dined with him.[173]

The themes established in these first letters were carried through in the rest of the correspondence. The intimacy with the Hunters, together with named friends from Glasgow, such as G. Reid and J. Young, William's constant companions, continued. The medical aspects of the correspondence centre on William Hunter's lectures and on what he was learning: dissection and the preparation of specimens. This would, of course, lay the foundation of his future teaching methods in Glasgow. William wrote:

am close employed in the dissecting room. I have got a leg and thigh from the body the Dr showed the operations upon and I had it injected, I should have done it myself but Mr Home said, as it was rather putrid, it was not proper for a first attempt as I should be apt to burst the vessels, and he did it and very well injected it is. I have been employed upon it since Sunday and I have not got to the knee yet. Bodies are vastly scarce, two resurrection men are taken up and all the burying ground is watched so that I am afraid we shall have very little dissecting for some time. There is nothing but an arm and my leg in the dissecting room at present.[174]

These themes continue in the letter of 3 January 1778, where William again wrote about his gaining knowledge of hands-on dissection and making

[173] GUL, Special Collections, MS Gen 1356/34, copy of a letter from William Hamilton to his father, RCSEng., Hamilton Letters, William Hamilton and Thomas Hamilton, 22 December 1777.
[174] GUL, Special Collections, MS Gen 1356/37, copy of a letter from William Hamilton to his father, 1 January 1778.

preparations from the material: 'I have got all the thigh and leg dissected. I keep on all the muscles and dry them with the vessels. I paid half a guinea for it and the injecting'; and then William commented: 'Mr Cruikshanks has been showing the nerves. He does it on just such a preparation as yours and which appears to have been kept for a considerable time'.

William Hamilton was learning dissecting skills and how to do preparations. These are the first of many that he used for his Glasgow teaching. Two of the letters (25 December 1777 and 1 January 1778) contain descriptions of the healing of fractured bones and the lectures on abscesses and hydrodropsy. He also noted the later lectures on the nerves and midwifery. The detailed notes Hamilton took on his work with the Hunters and under Thomas Denham and William Osborn for midwifery (to which we return below) are among his papers in Glasgow.[175] William recorded the dissections done at Hunter's anatomy school in the winter of 1779–80.[176] In all there were sixty-seven. William lists them as twenty-two males, nineteen females, eleven male children and fifteen female children. Adults were dissected to learn the muscles (twenty-three), for operations (eight), to understand the circulatory system (two), and for making preparations (three). Seven bodies were used for the lecture course that winter.

William Hamilton's teaching in Glasgow mixed detailed notation of cases with teaching and consultation with colleagues. Hamilton's teaching record in surgery was practical and explicit. It covered 'wounds, tumours, ulcers, fractures and dislocations',[177] amongst other subjects, indicating he taught the major requirements of an FPSG examination. He also taught clinical cases, as his notes show, which included discussions of the operation and treatment of particular patients, such as Marion Davidson (on tumour) or David Christie 'This case appears to be catarrh'.[178] His 'Heads of Lectures on Anatomy, Physiology and Medicine' contain the outline of his course, including 'the General Account of the Body' to 'Physiology and Pathology of the Brain and Nerves' where he covered respiration and digestion'.[179] He also lectured on venereal diseases, the growth and nourishment of the body, and then again on surgery.[180] In his surgical lectures he cautioned that the practitioner must be sure of the disease before the operation, question

---

[175] GUL, Special Collections, MS Hamilton 120/1 and 120/2: 'Dr Osborn's Lectures on Midwifery', London 1780; MS Hamilton 120/2 and 120/1: 'Dr Denman's Lectures on Midwifery', London 1780.
[176] GUL, Special Collections, MS Hamilton 83, p. 117.
[177] GUL, Special Collections, MS Hamilton 87.
[178] Ibid.
[179] GUL, Special Collections, MS Hamilton 90 (dated 27 November 1780).
[180] GUL, Special Collections, MS Hamilton 91.

whether it be removable and whether, if cured, a worse one would ensue. When the arguments for or against came out equal, he counselled to let it be.[181] The notebooks for the anatomical course are as comprehensive as those on surgery. All the major areas are listed in succession.[182] The quality of Hamilton's teaching can be assessed in the following advice to his students:

> You must see that the knowledge of the actions of the body in a state of health can be the only key to the actions in a state of Disease. Unless we know the natural [state] we can never know the deviation from it. Our anatomical knowledge besides assisting us in determining the seat of the disease is likewise of use in explaining the cure. In almost all diseases the powers of the body attempt to restore themselves to a state of health, in many the cure is entirely effected by them. These efforts must be understood by the Physicians, that they may aid them when too weak and that he may be on his guard not to counteract them when properly exerted.[183]

As Cleghorn's obituary of Hamilton on this evidence justifiably explained:

> While he practised extensively as a surgeon, his skill in anatomy made him be consulted by many surgeons, older than himself, before they performed operations; and, in a few years, those who had been his pupils, practicing in distant parts of the country, consulted him on similar occasions. Besides anatomy, he taught botany and midwifery; which last he practiced with such success, that he was called to almost every difficult case in Glasgow.[184]

In regard to William Hamilton's clinical knowledge, Cleghorn comments:

> He kept a regular account of all uncommon cases, accompanying the conclusion of each with remarks suggested at the moment, forming, at the end of each year, a general table of the diseases which had prevailed during the difficult seasons ...
>
> This plan facilitated his practice, and was highly gratifying to his patients, by convincing them, that their former complaints were distinctly remembered. But he had a higher object in view than the assisting of his own memory, or the gratifying of particular patients. His object was to have published a System of Surgery, illustrated with cases, of which several are fully and accurately drawn up.[185]

The *System of Surgery* would certainly have made William Hamilton's name, it being, as Cleghorn indicates, a textbook representing the best of the new art. It would obviously have combined the practical art of anatomy, so well learned with the Hunters, with its clinical alter ego, diagnosis and

[181] Ibid.
[182] Unfortunately Hamilton records only headings of what he taught. He did not write out all his lecture material.
[183] GUL, Special Collections, MS Hamilton 90, 'Notebooks for the Anatomical Course'.
[184] Cleghorn, *Mr William Hamilton*, p. 39.
[185] Ibid.

treatment of disease. This indeed is what the Hunters had envisioned, to enlarge the art of medicine by relating the science of the body to the evidence of how a disease develops. The *System of Surgery*, although it was not published, was at least part of William Hamilton's teaching. It served his students, who, as Cleghorn makes clear, sought his practical guidance and secured excellence for the medical school.

Library Resources

John Paisley the anatomy lecturer was also the FPSG's librarian.[186] He is said to have opened his extensive collection to students, in particular those of his friend Cullen, and to members of the Incorporation and their apprentices. Paisley died in 1740 and in September 1741 the Faculty asked Dr George Montgomery 'to Visit the Library of books belonging to John Paisley a former Collector and to Inspect if the faculty books missing and mentioned in sederunt first January last be amongst them'.[187] The FPSG had increased its holdings in books on botany and anatomy in 1736.[188] The main drive to stock the library came in 1745, however, when the Faculty asked Robert Hamilton, its President in 1745–47 and the Professor of Botany and Anatomy to substantially upgrade its holdings. The University Senate had also called on Robert Hamilton soon after his appointment (in 1742) to catalogue the books in the university library, which he undertook with Mr George Rosse in 1744. Present holdings were to be listed, including all treatises bound in volumes, and suggestions were to be made for new acquisitions.[189] In 1748 the task was recorded as completed.[190] When the Faculty asked Hamilton to act as their specialist agent, he was informed:

> that the Memoirs of the Academy of Sciences in sixty volumes octavo doun to the year 1739 were soon to be auction'd at Edinburgh and considering it would be an advantage to their Library that this full copy or Edition was purchased and the faculty's present incomplete set of twenty volumes quarto sold off. The faculty now that Doctor Robert Hamilton is at Edinburgh Impower him to sell their present copy and buy the other with the price of the quarty Edition and to apply what more is got for the quarto edition than given for the octavo for such other

---

[186] Duncan, *Memorials*, p. 251.
[187] RCPSG, 1/1/1/2, fol. 59v.
[188] E. Blackwell, *A Curious Herbal Containing Five Hundred Cuts of the Most Useful Plants Which are Now Used in the Practice of Physick*, 2 vols (London, 1737 and 1739); W. Cowper, *The Anatomy of Humane Figures Drawn after Life*, revised and edited by C. B. Albinus (2nd edn, Leiden, 1737); RCPSG, 1/1/1/2, fos 18v, 19v.
[189] GUABRC, 26639 UMM (Senate), 1730–49, p. 181.
[190] Ibid., p. 259.

books as he shall judge proper for the faculty and the faculty appoint the praeses [president] to intimate this present [situation] to Doctor Hamilton.[191]

Robert Hamilton was obviously being entrusted with both the medical libraries in Glasgow. He oversaw a period of extensive refurbishing. Robert's brother and successor at the university, Thomas, also became a valuable adviser to both libraries. He was commissioned by the FPSG in 1753 to purchase in Edinburgh 'the Dutch Edition of the French Memoirs'.[192] The FPSG took care to purchase the books produced by its famous native sons, subscribing to 'Hunter's Anatomy of the pregnant uterus presently advertised' in August 1752, some twenty years before publication, and buying 'Smellie's Cutts' on 2 February 1756 for 46 Shillings.[193] The FPSG acquired new 'standard works' in medicine as they appeared, so Le Clerc's *History of Physick* and Albrecht von Haller's edition of Boerhaave's medical writings.[194] In 1767 they purchased Friedrich Hoffmann's complete works, which may have been the Geneva edition.[195] The 'Mr Hamilton' mentioned as overseeing the transaction with Mr Daniel Baxter was still Thomas Hamilton, since 1757 the Professor of Botany and Anatomy.[196]

In 1762 the FPSG once more engaged Thomas Hamilton, wishing to allot part of their funds 'to purchase a small Library for the use of their apprentices'. He was appointed library keeper to buy books to the amount of £10. The list of books was to be drawn up by a committee including 'Dr Gordon, Dr Montgomerie, and Dr Black' (viz. John Gordon, George Montgomery, Joseph Black) along with the original committee of John Gibson, the Visitor, Thomas Hamilton and Ninian Hill.[197] All this suggests library holdings being coordinated between the FPSG and the university. Thomas Hamilton's stewardship of the apprentices' library in particular highlights this close association, as his students in anatomy and botany were in all probability mainly candidates for the FPSG licence.

This concern to underpin teaching with reference works and text books continued. It was taken to new heights when Robert Watt compiled his own teaching library to complement his clinical teaching course.[198] Out of this

[191] RCPSG, 1/1/1/2, fol. 87v.
[192] RCPSG, 1/1/1/2, fol. 148v.
[193] RCPSG, 1/1/1/2, fol. 168v.
[194] RCPSG, 1/1/1/2, fos 169r, 169v; D. Le Clerc, *Histoire de la médecine* (The Hague, 1729); A. von Haller, *Methodus studii medici*, 2 vols (Amsterdam, 1751).
[195] F. Hoffmann, *Opera omnia physico-medica*, 6 vols, in 4 (Geneva, 1740).
[196] RCPSG, 1/1/1/3, p. 138.
[197] RCPSG, 1/1/1/3, pp. 93–95.
[198] F. Cordasco, *A Bibliography of Robert Watt, MD: Author of the Bibliotheca Britannica* (New York, 1980), p. 9.

grew the first great compilation of books in print in Great Britain and Ireland. 'The most monumental bibliographical catalogue ever singly planned and achieved' was the formidable *Bibliotheca Britannica* with its 50,000 entries.[199] The *Bibliotheca* grew from Watt's original intention to provide his students with access to the original writings of medical contemporaries and past masters. They were 'to draw [their] own conclusions'. His *Catalogue of Medical Books* (Glasgow, 1812) supplemented his lectures on 'The Principles and Practise of Medicine'.[200] As he explains in his preface:

> To obtain correct views in medicine, it is necessary to have recourse to original authors, to such as write from actual observation, who have seen and treated the diseases they describe.
>
> Many students, however, are neither possessed of such works, nor have access to them. To remedy this defect, the present plan of establishing a Library is undertaken, and it is hoped that it will meet the approbation of those for whose benefit it is intended.
>
> In my Lectures on the Practice of Medicine, after considering the history and treatment of each disease, I give a list of the best authors who have written on that subject, and I now put it in your power to peruse these authors, to examine their facts and opinions, and to draw your own conclusions.[201]

Students of Watt's lectures could borrow books from his library in Queen Street, two at a time to be returned within two weeks. They forfeited all privileges if they wrote in them or left them at a shop in a public place and they had to pay five shillings 'to assist in making additions to the collection, and in keeping the Books in repair'.[202] Watt's collection contained some 1066 books, but there is no record of his gifting them to the Faculty's library.[203]

Robert Watt held private lectures, being neither a teacher at the university nor at an extra-mural college. His was a case of direct accreditation through the FPSG, as its members could hold courses acceptable to the FPSG examining boards. Watt's teaching and scientific work unites the diverse legacies of medical education coming together around 1800. He was particularly strong on the close affinity between anatomy and 'internal' medicine. Watt also pioneered the use of statistics to audit the results of medical treatments. Watt's *Inquiry into the Relative Mortality of the Principal Diseases of Children*

---

[199] Ibid.
[200] Available today in facsimile.
[201] R. Watt, 'An Address to Medical Students', in *A Catalogue of Medical Books* (Glasgow, 1817), pp. 5–6.
[202] Watt, 'Regulations', ibid., p. 16.
[203] A. Goodall and T. Gibson, 'Robert Watt: Physician and Bibliography', *Journal of the History of Medicine*, 18 (1963), p. 42.

was published in 1813, the year after his *Catalogue of Medical Books*. He studied the effect of smallpox vaccination, compiling mortality statistics, a method not far removed from his bibliographical interests.

He examined fifteen manuscript volumes of registers of burials in Glasgow,[204] finding that, although smallpox vaccination reduced child mortality, overall death rates remained consistent, as high as one-half of all children dying before the age of ten. As the great statistician Walter Farr acknowledged, Watt proved that smallpox was the greatest cause of child death in Glasgow and that these deaths were reduced to a fifth by vaccination, but 'the children died in nearly the same numbers as before, but of other forms of disease'.[205]

Watt was admitted to the FPSG in April in 1799 as a licentiate. Medicine was not his original field of study, coming from a farming family and having worked as a ploughman and cabinetmaker. He entered the Greek and Latin classes at Glasgow University in 1793, transferring to Edinburgh to study moral and natural philosophy, but also took a class in anatomy in 1796.[206] Rather than becoming a schoolteacher or a minister, Watt finished his medical studies and began practise in Paisley, his partner being the surgeon James Muir. He was active in medical politics, helping to found the Paisley Medical Society in 1806. On 5 January 1807 Watt was admitted as a full member of the FPSG practising in Glasgow. His first book on diabetes and diseases in general was published the following year. The university lecturers Robert Cleghorn, who taught chemistry, and Thomas Brown, who was teaching botany, wrote him testimonials. He used these for a quick, but mostly frowned upon, method for gaining an MD, in this case from King's College, Aberdeen. Watt bought a large house in Queen Street, testimony to his success, and was 'an assiduous attender at meetings of the Faculty of Physicians and Surgeons and appeared often in its minutes'.[207] He was elected Boxmaster,[208] and joined the Library Committee. Between 1814 and 1816 he became physician to the Royal Infirmary, a founder member of the Glasgow Medical Society and president of the FPSG. Watt's understanding of the need for up to date libraries and the dissemination of medical learning in journals also extended to his advocating the use of visual aids. For him these were complimentary learning tools, physical specimens in 'museums' increasing the practical knowledge of anatomy and physiology.

'It has been too clearly proved, that every attempt to explain disease,

[204] F. Cordasco, *A Bibliography of Robert Watt, MD*, p. 11.
[205] Ibid., p. 13.
[206] A. Goodall and T. Gibson, *Robert Watt: Physician and Bibliography*, pp. 36–50.
[207] Ibid., p. 41.
[208] Effectively keeper of the FPSG document chest.

without the aid of Anatomy and Physiology, has been vain and fruitless.'[209] Watt's advice to his students in the *Catalogue* clearly delineates how medicine now defined its foundation subjects. Medicine was described by Watt 'as an extensive circle'.[210] 'I have it in my power, through the kindness of my friend Mr Allan Burns, to show you, from his Museum, specimens of many of the most remarkable organic affections'.[211] Watt was signalling what his contemporaries were avid to explore, the impact of anatomy on clinical medicine. Specimens and preparations, in the hands of men like Allan Burns, whose talents in dissection were of almost fabled repute in Glasgow, were honoured as refined tools, advancing diagnostic precision. Post-mortems, too, were now becoming routine. Watt quotes from William Hunter's *Introductory Lectures* to make this point, validating the impact of research anatomy on the training of doctors:

> Were I to guess, says Dr Hunter, at the most probable future improvements in physic, I would say they would advise from a more general knowledge, and a more accurate examination of diseases after death. And were I to place a man of proper talents in the most direct road for becoming truly great in his profession, I would choose a good practical Anatomist, and put him into a large hospital to attend the sick and dissect the dead.[212]

We will return to the subject of clinical teaching in a subsequent chapter on the Glasgow Royal Infirmary, an ideal locus for examining diseases and honing medical skills. Here it is sufficient to note that the teaching attached to hospitals was yet another component in a system of vital parts welding together the Glasgow medical school. As the reference libraries expanded, so did the use of collections. Watt advertising the availability of Allan Burn's specimens for use in his teaching points to their significance. The 'museums' were a vital part of hands-on visualisation, and Glasgow was rich in this material. But as specimens derived from anatomical research became a part of teaching collections, their value changed, becoming part of the popular currency of teaching.

Attendance at anatomy lectures became part of the 'irregular' practitioner's education, spotlighting its enormous popular success. Irregular healers were the less affluent. Yet as standards in medical education rose – and indeed were pushed upwards by the ambitious elite within the FPSG – the question of too narrow a point of entry arose. In 1754 a change in entry standards was voted down by the majority of FPSG members, probably with an eye to

[209] Cordasco, *A Bibliography of Robert Watt, MD*, p. 9.
[210] Ibid., p. 10.
[211] Ibid., p. 11.
[212] Ibid.

keeping the surgeon-apothecary eligible. The details of a 1786 law case brought against Andrew Turnbull for unlawful practice, long after he had failed the FPSG examination,[213] however, shows where the incorporation drew the line. Interestingly enough, Turnbull had attended the anatomy lectures of Thomas Hamilton at the university for two or three years, while openly admitting he had not been even an apprentice. On being asked about his study of medical textbooks, Turnbull rounded neatly on his FPSG 'colleagues', answering 'that he met with so many cures of diseases that he had no occasion to read any'.[214]

The evidence used against Turnbull to disqualify him conveys, for this period, the high standard of FPSG examinations. The questions asked were extensive and detailed on anatomy, required pharmaceutical expertise, surgical knowledge, a grasp of disease nomenclature, and knowledge of basic medical treatment such as blood-letting and bone-setting. When 'irregulars' like Turnbull were made to justify their qualifications, the FPSG enforced its regulatory role. The poor often turned to these 'cheaper' practitioners. At the opposite end of this medical continuum were the well-educated surgeons like John Moore who typified the quality of medical men preferred by the FPSG. These doctors were the product of the upgrading of the many resources needed for the study of medicine.

John Moore, gifted with a ready wit and 'courted by a numerous and respectable circle of acquaintance' was a successful general practitioner,[215] the counterfoil to Turnbull, but a man the poor could not normally afford. His educational and career choices reflect how the medical elite rode the cresting wave of higher standards in medical education. Moore was apprenticed to Gordon on 3 December 1744 for five years, gaining further experience as a surgeon's assistant in army service. Returning home to practice in the 'business' of his mentor, John Gordon,[216] he sought admission to the FPSG on 3 September 1750. Successful in his 'trials' in 1751, he was licensed to practise surgery and pharmacy. By 1757 his skill and repute saw him elected Visitor of the Faculty. In October 1763 he was elected collector.[217] Moore accepted apprentices, one of whom was Robert Montgomery in 1767.[218] While studying, Moore attended the lectures of William Cullen, Robert Hamilton

---

[213] GCA, A2:25, 'Decreet absolvitor in part and for expenses, the Magistrates and Town Council of Glasgow against the Faculty of Physicians and Surgeons of Glasgow 1794', pp. 17–21.
[214] GCA, A2:25, p. 19.
[215] J. Strang, *Glasgow and its Clubs* (London, 1856), p. 45.
[216] RCPSG, 1/1/1/2, fol. 86v (for indenture).
[217] RCPSG, 1/1/1/3, fos 29, 99.
[218] RCPSG, 1/1/1/3, fol. 148.

and George Montgomery,[219] learning the practice and theory of medicine from Cullen, anatomy from Robert Hamilton and clinical practice from George Mongomery.[220]

Educational Networks

Moore also went to London and Paris. He was fortunate to find, in Paris, a mentor he had previously encountered on the battlefields, the Earl of Albermarle; who was now British Ambassador to the Court of France. He invited Moore to become physician to his household, a position Moore took, but he used much of his time for medical studies. His biographer writes that Moore availed himself

> of the opportunities of improvement to be found in attending the hospitals and lectures of the French metropolis, which at that period, had deservedly the reputation of being the best school of medicine and surgery in Europe.[221]

Moore, writing to William Cullen in 1749, stated that he made the two branches of medicine that Cullen recommended 'my principal branch of study since my arrival at Paris, as the most likely to succeed in Glasgow, viz., performing surgical operations and midwifery'.[222] Moore attended at a Paris hospital, which is not named, and studied in two courses under unnamed but 'celebrated' surgeons:

> where under their inspection I have performed every operation several times over upon dead subjects, particularly the stones and some others which I have never performed on a living person, I would fain have had my Coup D'esee [the French is *coup d'assais* meaning 'first attempt'] of this operation over here but find it to my great sorrow Impracticable, I mean upon a living person.
>
> As to midwifery I have attended one course, seen a good many births and performed some myself, have also read upon this subject Mauriceau and La Motte with tolerable diligence. And shall give the finishing stroke under Smellie whom I design to attend in London upon my return. My time at present is occupied in Dissecting and attending the Lectures of the famous Astruck [Jean Astruc] upon the Diseases of Women and Children, a branch of my business which I freely own I have great need to study.[223]

---

[219] H. L. Fulton, 'John Moore, the Medical Profession and the Glasgow Enlightenment', in A. Hook and R. B. Sher (eds), *The Glasgow Enlightenment* (East Linton, 1995), pp. 176–89.
[220] Ibid., p. 179.
[221] R. Anderson, *The Life of John Moore, MD* (Edinburgh, 1820), p. vi.
[222] Ibid.
[223] GUL, Special Collections, MSS 2255/91, William Cullen Papers, Letter of John Moore to William Cullen (Paris, 1749).

Moore proves the value of a 'free market' education, utilising where necessary Glasgow teaching and its networks, but also seeking specialist training elsewhere. As a surgeon he is not the product of a university medical education, achieving a solid reputation primarily through his contacts within the FPSG. Moore's choices were not so different from a typical MD candidate, however, whose educational profile, under the guidance of Cullen, included anatomy and pharmacy, skills the FPSG nurtured for their licence, as well as the requisite 'Institutes', that is, the study of physiology.

Like William Hamilton's journey to London, good examples of the exchanges between learning abroad (including Edinburgh) and teaching at home abound. Cullen advised his student from Belfast, Alexander Henry Haliday, how to optimise his medical studies. Writing to Cullen on 2 May 1751, Haliday mentions that he 'passed four winters happily and I hope not idly in Glasgow', the last two of these in Cullen's classes.[224] He spent his summer 'in the shop of our principal apothecary' in Belfast, in the same way Cullen had in London and, indeed, as required by the FPSG for its surgeon-apothecaries. He continued:

> the next winter was spent in Edinburgh, where I attended with some care Mr Monroe, Dr Rutherford (both his college and his clinic lectures) and, which was of more advantage (in consequence of your advice), Dr Young and the Infirmary, – from the Doctor I heard much and in the other saw somewhat of the genuine appearance of diseases and effects of applications – the summer was employed in essays in practice and in the reading of practical authors, particularly Hoffman [Friedrich Hoffmann] whose writings though diffuse and ill ordered, I have found much more instructive than the clear well digested axioms of Boerhaave – in the beginning of Winter I went to London, where I attended with a pleasure and improvement that well rewarded my care the lectures of Mr Hunter – and for eight months St George's Hospital – from this I passed to Holland in which place and in Paris half a year was employed as usefully as my situations would allow – on my return to London I renewed my attendance at the Hospital, inspected the files of an Apothecary and dissected a little; after some months I had proposed to revisit Glasgow, to conclude my studies where they at first commenced and assumed a character I began to think myself not altogether unqualified for – but the bad state of a tender mother's Health and an only sister's concurred with some other circumstances to hurry me home [to Ireland].[225]

Ten years later William Stark, Thomas Hamilton's student, thanked him for recommending attendance at William Hunter's lectures and practice in 1765. William Stark was a surgeon of promise who died young, at twenty-nine

---

[224] GUL, Special Collections, MSS Cullen 1–249, no. 92, Letter of Alexander Henry Haliday to William Cullen, Belfast, 2 May 1751.
[225] Ibid.

(born in 1742). His work on tuberculosis, the study of cysts, the cause of hemoptysis, and the sections on phthisis and scrofula, which he did under William Hunter's supervision, were incorporated in Matthew Baillie's book, *The Morbid Anatomy of Some of the Most Important Parts of the Human Body* (1794), a landmark in pathology and extraordinarily popular.[226]

William Stark wrote to Professor Thomas Hamilton, on 8 October 1765, about London giving him opportunities to improve himself 'considerably in that study wherein you first initiated me, the Anatomy of the sound and of the morbid parts'.[227] Stark thereafter entered:

> St George's Hospital where I have reaped much benefit. I have seen a considerable variety of diseases, I have learnt to interrogate patients with more propriety, and to remember and class their complaints much better than I could do before.[228]

Besides his enthusiasm for the benefits of clinical instruction, Stark was able to examine bodies and dissect them, working from these observations to hone diagnostic skills. He was present as William Hunter performed 'the chief part of a dissection of the brain', and had the opportunity to see several operations at close hand. But he commented that 'the practice of the physicians seems to be rather timid so that I have not had many opportunities of seeing powerful medicines fairly tried'.[229]

The remaining part of the letter discussed in some detail the 'fine preparations' made by William Hunter that could be seen at his house. Stark wrote of injections that make these preparations especially 'elegant and distinct'. Above all Stark praised the 'fine casts of the Gravid Uterus':

> The plates from these Casts will all be finished before Spring and are soon afterwards to be published. They are in number about thirty and contain a gradation of the different states of the Gravis Uterus from two to three months after impregnation to near the full time. The Doctor's Lectures upon this subject and upon the parts of generation in males and females were truly admirable.[230]

The educational attributes delighting Stark were not dissection alone, nor the provision of a human body for every pupil, but the ability to combine the demonstration on the body with a lecture by an experienced and knowledgeable teacher. Dissection was only one aspect of the new quest to create educational tools, regardless of whether the object was the body as a

---

[226] E. R. Long, *A History of Pathology* (London and New York, 1965), p. 36.
[227] GUL, Special Collections, MS Gen. 1356/72. Copy of letter from William Stark to Thomas Hamilton, 8 October 1765. This year is the correct date.
[228] Ibid.
[229] Ibid.
[230] Ibid.

corpse, or a preparation, or indeed wax sections or a wax model. The sense of control for both the surgeon and the physician lay in the ability to visualise spacial relationships and to be in complete command of medical knowledge.

Collections

The greatest specimen collection in Glasgow, William Hunter's 'Museum', which he had left to the Principal and Faculty of the College of Glasgow in 1781 'for the improvement of students and the use of the public', was extolled by his contemporaries because it gave their lectures visual substance.[231] Wealth, not gained in medicine alone but 'by dealings in government lotteries and by buying and selling of stocks and shares',[232] had given Hunter the means to furnish 'a public school of anatomy' to house his anatomical preparations and library. In 1806 'the first of the British Greek Temples to Arts and Science' was built using the design of the Glaswegian architect William Stark.[233] Glasgow 'desperately wanted the anatomical preparations',[234] although students had only restricted access (perhaps this meant, in reality, supervised or regulated access, as in the British Library today).[235] The Hall of Anatomy contained presses (cupboards) with preparations preserved in spirits. On the table were injected preparations 'elegant' and with the 'colours very fine'.[236] The exhibit of the 'corrosion of the lungs' provides an example, 'the artery red, the veins yellow, trachea green', and 'an elegant and minute corrosion of the liver; it rests on its upper or convex side ... the artery is white, the vena porta green, and the gall ducts yellow, with the cava red'. The number of the preparations seems truly splendid, for example 120 alone of the kidney 'illustrative of it's structure and diseases', and 500 preparations of the gravid uterus.

Stark's building rose in two floors, a shallow dome in a high drum sitting over an octagonal first floor picture gallery with two wings.[237] Thomas Markus has shown its close affinities with Hunter's Great Windmill Street anatomical school.[238] The building in London was Hunter's brick and mortar

---

[231] H. C. Brock, 'Dr William Hunter's Museum, Glasgow University', *Journal of the Society for the Bibliography of National History*, 9 (1980), pp. 403–12.
[232] Ibid., p. 404.
[233] Ibid., p. 406.
[234] Ibid.
[235] Ibid., p. 407.
[236] J. Laskey, *A General Account of the Hunterian Museum, Glasgow* (Glasgow, 1813), p. 47. This was the first printed description of the museum.
[237] T. A. Markus, 'Domes of Enlightenment: Two Scottish University Museums', *Art History*, 8 (1985), p. 170. This seminal article illustrates the importance of Glasgow's investment in popular education and explains its architectural iconography.

vision of how observation, teaching, reading and the sensory immediacy of the arts came to operate together. The museum in Glasgow with its dome and Greek portico conveyed the educational interplay. It invited both the student, the scholar and the public to appreciate the proximity of science and art. Walking from room to room linked natural history and comparative zoology to the minute disclosure of the human form. Very clever appreciation was evoked of the breadth of what research had assembled. From William Hamilton's letters, already cited, we know how laboriously won were these fixed effigies, the wax and mercury preparations of the human body. Their very grotesqueness was soothed and given a classical air under the dome and symmetrical spaces of an architectural citation of Grecian culture, admired by every schoolboy who took Greek and Latin as part of the obligatory arts curriculum. This 'Museum' was the ultimate teaching tool in the practice of medicine.

When the partnership between university medical education and the FPSG began to shift, favouring John Anderson's extra-mural college, the Faculty won additional leverage. As we have seen from the teaching of William Hamilton, his university course covered surgery, anatomy and midwifery. From the Turnbull prosecution it is evident Thomas Hamilton taught the surgeons as well. The whole strategy of courses combining botany, anatomy and surgery points to the instruction of candidates for the FPSG licence. With Anderson's College beginning to operate in 1799/1800, in particular with the appointment of John Burns to teach anatomy, surgery and midwifery,[239] the first serious challenge to university hegemony in teaching requirements for the *surgical* licence was mounted. The challenge was maintained, as the Burns brothers' friend, Granville Sharp Pattison, continued the lectureship at Anderson's in 1818 (Burns resigned in 1817). He was brought down through a scandal involving his colleague Dr Andrew Ure and his wife, but went on to a successful teaching career in the United States.[240] William Mackenzie was his successor, well-known as the founder of the Glasgow Eye Infirmary. When Mackenzie moved to the university in 1828 in ophthalmology, Robert Hunter took his place, transferring from the Portland Street School of Medicine. This institution was founded before 1827 (probably 1826)

---

[238] Ibid., p. 164
[239] Various sources point to his teaching this triumvirate of traditional Glasgow subjects thought crucial to general practice. Both J. D. Comrie, *The History of Scottish Medicine* (London, 1932), pp. 538–40, and Duncan, *Memorials*, p. 185, list him under anatomy and surgery at Anderson's from 1799 to 1815. John Butt, *John Anderson's Legacy*, notes him as teaching anatomy, surgery and midwifery from 1800, pp. 36–37. Butt confirms his resignation in 1817.
[240] F. L. M. Pattison, *Granville Sharp Pattison: Anatomist and Antagonist, 1791–1851* (Edinburgh, 1987).

and closed in 1844.²⁴¹ It also taught the basic medical subjects, 'internal' medicine, anatomy, surgery, midwifery, chemistry, botany, materia medica, physiology and medical jurisprudence. Anderson's taught the full medical curriculum by 1818, as did the Portland Street School. Anatomy, in both schools, was well up in students taking the courses. The certificates given at these schools were accepted by the FPSG, the other surgical colleges, the Apothecaries' Society and the Army and Navy Boards. Figures for attendance given by Alexander Duncan for the Portland Street School and Anderson's College correspond to the teaching at each institution, respectively, by Moses Buchanan. He taught from 1836 to 1840 at Portland Street with student numbers fluctuating between fifty-five to seventy-two. At Anderson's the numbers for anatomy were well over one hundred in each year from 1841–60.²⁴² Anderson's was seen as a 'popular institution' which possessed 'an excellent library and museum, and its students have free admission to the Botanic Gardens'.²⁴³

This colonising of medical education outside the University by the FPSG and the extra-mural schools found expression in Glasgow's second 'dome of enlightenment'. There the medical lecturers of Anderson's College taught anatomy, midwifery, chemistry and the other subjects needed to pass FPSG examinations. The lecture room at Anderson's College was, as Markus remarks, 'forty-five feet in diameter (almost the same as Hunter's "ideal" museum) and its circular space accommodated 500 seats. The new Anderson's College building included conversion for a library, committee rooms and apparatus apartments'.²⁴⁴ The dome housed collections and gave ample space for demonstrations, and for the increasing numbers that came to hear the medical and science classes of Anderson's. The new site on George Street mirrored Anderson's aspirations to become a full-fledged medical school:

> The conversions were carried out by Scott and Watt, who also prepared a scheme for a museum and library in a specially purchased piece of semi-circular ground. Its shape was dictated by the Trustee's desire to remove the dome of their existing building and re-erect it ... Thus for a total expenditure of £9000 they had a splendid new building which not only included lecture facilities, laboratories and demonstration workshops, but also a school of medicine with its own anatomy theatre and anatomical museum.²⁴⁵

The original John Street front 'had Ionic pilasters and classical ornaments;

---

²⁴¹ Duncan, *Memorials*, p. 183.
²⁴² See Duncan, *Memorials*, pp. 183–84.
²⁴³ Medical Times and Gazette, 'Medical Education in Glasgow', 20 November 1862, p. 578.
²⁴⁴ Markus, 'Domes of Enlightenment', p. 174.
²⁴⁵ Ibid., p. 175.

the central space was circular and glass-domed, with a gallery supported by [iron?] columns'.[246] It had been built to cope with Glasgow's upsurge in selling butchered meat and live poultry, macabre when related to human dissection, but evocative too of the popular market, the need to exchange and buy in a busy city. The Greek revivalist mantle clothed modernity, Anderson's medical school and science teaching having strong links with the education of working men. The Mechanics' Institutes were its offspring, including Birkbeck College in London.[247]

James Jeffrey, the Professor of Anatomy, had induced the University to provide him with a state-of-the-art dissecting room. Such a room, not surprisingly, was absent from the plans for the Hunterian, the one point at which it differed significantly from Hunter's Great Windmill Street anatomy school. The school in Glasgow was not concentrated in one place but was spread throughout the city, and the FPSG held many of these strings in its hand because it controlled surgical examinations. Moses Buchanan used wax models, diagrams, plates, and wet and dry preparations in his teaching in the 1840s.[248] James Paterson modelled specimens for teaching at Anderson's,[249] and an applicant for the anatomy lectureship who taught previously in College Street wrote: 'my anatomical museum is now pretty extensive, besides which Dr Hunter's [Museum], the equal of which for illustrating a popular course, I reckon is not in Scotland, wants this entirely at my command, should I receive the appointment'.[250] This Dr Hunter was Robert, Professor of Anatomy at Anderson's College from 1828 to 1841, not William. In 1839 a rather jaundiced John Wood who was teaching popular courses of anatomy notified the secretary of the Mechanics' Institute, J. MacDougall, that he was returning the paintings he had used for his anatomy cum physiology lectures:

> I send back your paintings, which I certainly would not have done had they been of any material use to me – I have had specimens, drawings and apparatus from the most eminent men in Glasgow, but as to your committee I am not sorry to return all I ever had from them.[251]

By 1840 its popularity made anatomy part of the education of working men. The 'extensive circle' that Watt, the farmer's son from Ayrshire, saw

---

[246] Ibid., p. 174.
[247] Butt, *John Anderson's Legacy*, pp. 39–40.
[248] M. Buchanan, *Lecture Introductory to a Course of Anatomy* (Glasgow, 1842), p. 23.
[249] Ibid.
[250] SUA, C. 8.1.25, David Wark to John Leadbetter, President of the Mechanics' Institution, seeking the anatomy lectureship, 23 June 1836.
[251] SUA, C. 8.1.105, John R. Wood to J. MacDougall, 27 February 1839.

medicine to be was linking progressive reform with medical education. This signposted Glasgow's special role in making medical education accessible and vindicated what the 'new sciences' of anatomy, botany and chemistry had promised at the start of the century: that medicine needed to be backed by extensive specialist training.

Chemistry

Ninian Hill was a Glasgow trained surgeon-apothecary. His career demonstrates the success of the integrated educational programme available in Glasgow after the middle of the eighteenth century. The networking of medical education provided the new-skills training that benefited an auxiliary science like chemistry, then in the hands of medicine. Scientific precision in mixing drugs was transferred to the new commercial interest in chemistry. Under William Cullen and the group around Joseph Black chemistry shifted to encompass uses not medical in application. The many industrial follow-ups that emerged from the experiments financed through medical teaching then benefited the creation of a separate chair. The first incumbent was Thomas Thomson. Thomson was born in Crieff in 1773 and acquired a thorough classical education, the background to his numerous improvements of chemical nomenclature. His brother James was one of the editors of the *Encyclopaedia Britannica*. He studied with Black in Edinburgh and came to Glasgow as lecturer in 1817. He was a prolific writer on chemistry and involved with John Dalton's work on atomic theory.[252] Chemistry was singularly successful as a research and teaching subject in Glasgow but came in many guises. At Anderson's College it was popular from the start in 1799, but taught under natural philosophy. The expertise in chemistry in Glasgow stretched across several boundaries, academic, extra-mural, professional (the surgeon-apothecaries) and commercial. This was its particular strength, rooted, however, in structural terms, in the 'hotbed' of competitive medical schools.

Chemistry remained wedded to medicine for some time. Within the expanding sector of extra-mural education chemistry achieved a new following. It soon became a subject attracting students who would never take medical degrees. Chemical skills and knowledge were brokered through several agencies, with the FPSG in the front ranks. It set the standards of skills for its surgeon-apothecaries. Hill's working life as a surgeon-apothecary was largely independent of the university educational system, but his contacts, not least

---

[252] On Thomas Thomson, see J. R. Partington, 'Thomas Thomson, MD', A. Kent, *An Eighteenth-Century Lectureship in Chemistry* (Glasgow, 1950), pp. 176–90.

through the FPSG, connected him to the top experimental scientists in chemistry in the 1750s.

In 1749 Ninian Hill applied to the FPSG for entry as a country licentiate. He was duly examined under the existing rules of the Faculty, writing an essay and submitting it to the 'three or at least two essay masters' appointed. He dissected a heart and explained the circulation of the blood before the examining board. Hill passed his 'trials', paid his entry fee and was licensed 'to practise surgery and pharmacy' except in the city of Glasgow and its suburbs. In 1754 Hill moved from Paisley to Glasgow and was soon brought to book for infringement of the FPSG regulations. He subsequently enrolled as a city surgeon, paid his 'freedom fine' and was once more enabled to 'practise surgery and pharmacy'.[253]

Hill's was a typical pattern of entry into the medical profession. He was the offspring of an established middle-class family, his father being the Reverend Laurence Hill. As Ninian never married, James, his brother, became head of the family and consolidated its fortunes. Ninian Hill was elected 'box-master' of the FPSG in 1755. His brother James became clerk to the Faculty two years later. James studied law and was thus also professionally qualified. In 1758 he became clerk to Hutcheson's Hospital and was made factor to the University in 1759.[254] Medicine and law were fitting professions for the sons of the clergy and the tight kinship system prevalent in Glasgow saw these posts passing on within the family. In the next generation James's son, another Laurence, trained in law, and he too became factor to the University Senate and clerk to the FPSG. Qualifications not nepotism, however, were the essential ingredient of the Glasgow entrepreneurial culture of the middle classes, placing a premium on talent, education and skills. Without them the patronage possible through family connections would soon have been wrecked.

Ninian Hill's surgeon-apothecary's training enabled him to keep pace with the scientific skills and curiosity of men like James Watt and Joseph Black. These men's correspondence in 1768 and 1769 documents their using instruments from Hill's laboratory. At the time both men were also deeply concerned over Hill's health. In 1769 Black wrote to Watt that he was 'grieved to the heart for my friend Ninian Hill'.[255] Although the grief is unspecified, subsequent letters cite an illness, presumably grave, from which Hill

---

[253] RCPSG, 1/1/1/2, fos 118r, 124r.
[254] Coutts, *A History of the University of Glasgow*, p. 268.
[255] E. Robinson and D. McKie (eds), *Partners in Science: James Watt and Joseph Black* (London, 1970), p. 15.

recovered at the end of January.²⁵⁶ Hill can thus be counted among Glasgow's leading medico-chemical circles.

Hill owned one of the few private chemical laboratories in Glasgow, a boon in the years 1757 to 1758 which were crucial for the 'take-off period of the Industrial Revolution'.²⁵⁷ The catalysts were Joseph Black, James Watt, William Irvine and John Robison, all active in the city in these years. Black, Robison and Irvine held chemical lectureships in Glasgow University having trained as medical doctors. Their most famous discovery is of latent heat but it was the tip of an iceberg. Collaboration and the way their paths crossed culminated in discoveries on specific heat, the steam engine, prepared the ground for the theory of combustion, the early use of mineral acids, plant physiology and so much more.

Joseph Black was President of the FPSG from 1759 to 1761 and 1765 to 1766 while William Irvine was President in 1775 to 1777 and again in 1783–85. Irvine was 'greatly captivated with Chemical Science. He engaged with great pleasure and zeal in all examinations which seemed to interest the Professor [Joseph Black]', as his friend Robison said of him in his edition of Black's *Lectures on the Elements of Chemistry* (1803).²⁵⁸ Like Black and Hill, Irvine had access to one of the chemical laboratories in Glasgow doing experimental science. The university laboratory had been fitted out for William Cullen in 1747 when he became Lecturer in Chemistry and Black made sure that University funding kept it in good working order.²⁵⁹ Irvine probably also maintained a laboratory, as his son mentions spending most of his youth with his father there.²⁶⁰ Hill's laboratory, too, must have been used for their common exploration of scientific and industrial uses of chemistry. Hill was in partnership with William Couper, another eminent city surgeon, whose chemical and financial interests were entwined with the marketing of drugs and industrial chemistry. He was an early partner of Charles Tennant and Co., whose establishment became the biggest chemical works in Europe (at the St Rollox Chemical Works). A further early venture was his original partnership of the Glasgow Apothecary's Company,²⁶¹ in which surgeons, such as John Burns held shares.²⁶² Couper was President of the FPSG in 1822 to

[256] Ibid., p. 16.
[257] Ibid., p. 3.
[258] A. Kent, 'William Irvine, MD', *An Eighteenth-Century Lectureship in Chemistry* (Glasgow, 1950), p. 140.
[259] J. Geyer-Kordesch, 'Die medizinische Aufklärung in Schottland: nationale und internationale Aspekte', in H. Holzey, D. Brühlmeier, V. Murdoch (eds), *Die Schottische Aufklärung: 'A Hotbed of Genius'* (Frankfurt, 1996), pp. 91–106.
[260] W. Irvine, Jr, *Letters from Sicily* (London, 1813), p. xl.
[261] Duncan, *Memorials*, p. 263.
[262] SRO, SC 36/48/37, pp. 496–97.

1824. He was also designated Professor of Chemistry in the testament of John Anderson. This attests Couper's thorough knowledge of chemistry since Anderson only designated for appointment those who were in his eyes keen on practically-orientated scientific education.

The generation that followed Hill increased rather than deflected the gains made by chemistry under its medical umbrella. Robert Cleghorn, whose medical interest lay in clinical observation, spent his professional career lecturing to good effect on practical chemistry.[263] The Irish Presbyterian Church encouraged its theological students to attend medical lectures, so Josias Christopher Gamble heard Cleghorn in the 1790s. Georg Thomson writes:

> The lecturer and the subject so fascinated Gamble that he spent his long vacations experimenting in chemistry, particularly in trying to prepare chlorine solutions for bleaching the handwoven linen of his native Enniskillen. After a few years in the ministry of the Irish Presbyterian Church he resigned his charge to become a chemical manufacturer first in Ireland and later in England where, in company with James Muspratt, he became one of the founders of the alkali industry in St Helens, Lancashire.[264]

Cleghorn's fellow chemistry lecturer was Thomas Charles Hope, described as doing 'brilliant research' in Glasgow,[265] later transferring to Edinburgh as Black's successor. Thomas Charles Hope was the nephew of Alexander Stevenson, Professor of Medicine and active in promoting the Royal Infirmary. He held the chemistry lectureship from 1787 to 1791 when he succeeded his uncle as medical professor. Hope's main scientific accomplishment was in chemistry. He discovered strontium hydroxide and why ice floats (because water reaches its maximum density at 4° Centigrade). This is a temperature above freezing and explains why water congeals and sinks, letting warmer and lighter water rise up from below.[266] But Hope chose to invest the prime of his life in teaching chemistry rather than research. After his death the following note was discovered among his private papers:

> Those who devote themselves to the science of chemistry may be divided into two classes: First, Those whose labours are employed in original researches, to extend our knowledge of facts and principles. Secondly, Those whose business it

---

[263] One instance of his teaching success concerns an Irish student.
[264] George Thomson, 'Robert Cleghorn, MD', in Kent, *An Eighteenth-Century Lectureship in Chemistry* (Glasgow, 1950), p. 167.
[265] James Kendall, 'Thomas Charles Hope, MD', in Kent, *An Eighteenth-Century Lectureship*, p. 159.
[266] Coutts, *A History of the University of Glasgow*, p. 498; on Hope see J. Kendall, 'Thomas Charles Hope, MD', in Kent, *An Eighteenth-Century Lectureship*, pp. 157–163.

is to collect the knowledge of all that has been discovered or is going forward, to digest and arrange that knowledge into lectures, to contrive appropriate and illustrative experiments, and devise suitable apparatus for the purpose of communicating a knowledge of chemistry to the rising generation, or others who may desire to obtain it. From my professional situation I consider myself, as Dr Black had done before me, as belonging to the second class of chemists. I consider my vocation to be the teaching of science.[267]

The teaching of science was writ large from the inception of Anderson's College, the great rival to Glasgow University in the practical sciences. Its first teacher of note in chemistry was Dr Thomas Garnett, the pupil apprentice of the surgeon-apothecary John Dawson of Sedburgh in England. He offered 'popular' chemistry illustrated by experiments in the Trades' Hall at 8 p.m. and an additional course of practical chemistry two nights a week.[268] When he left for London in 1799 Dr George Birkbeck (after whom the college in London is named) took up his post in 'natural philosophy', also holding chemistry classes. In 1804 when Birkbeck resigned, James Corkindale was a candidate for his post.[269] 'Corky', as he was known among friends, was one of the most powerful members of the FPSG, with his finger in every pie. He was closely involved in the management of the Royal Infirmary and the founding of the Glasgow Lying-In Hospital. He was President of the FPSG in 1834–36 and was both a trustee and manager at the Andersonian. Corkindale had studied chemistry in Edinburgh and London for five years. He was good in mathematics and natural philosophy, having studied under John Anderson.

Dr Andrew Ure was elected to the post of Professor of Natural Philosophy. He had an excellent reputation as a chemist and held classes three days a week at 8 p.m.[270] Ure promoted close ties between industry and science. He was the author of several standard publications.[271]

The new chair in scientific chemistry, founded in 1830, brought another very successful man into teaching, Dr Thomas Graham, the son of a Glasgow textile manufacturer.[272] Like his teacher, Thomas Thomson of Glasgow University, Graham trained his students in laboratory practical classes.[273] Practical chemistry was on offer, with 'a series of the most important Experiments and Processes of Scientific Chemistry, with Exercises in Analysis and

---

[267] Ibid., p. 160.
[268] J. Butt, *John Anderson's Legacy*, p. 29.
[269] Ibid., p. 35.
[270] Ibid.
[271] Ibid., p. 40.
[272] Ibid., p. 57.
[273] Ibid.

in Pharmacy' to anyone who wished to buy a ticket.²⁷⁴ This course 'in the Laboratory 11, Portland Street' commenced in May, and was held five times a week for three months. The laboratory was opened in the evening for another class 'suitable for Gentlemen in Business'.²⁷⁵ Thomas Graham also taught 'chemistry and mechanics' in the evenings twice a week for six months. These lectures gave 'a Popular View of the Principles of the Chemical and Mechanical Sciences in their most recent and improved state'.²⁷⁶ Dr Andrew Ure, the Andersonian Lecturer in Chemistry, competed in this arena by advertising his 'Mechanics' Class' with 'Important Additions, newly made to the Apparatus and Museum', and could 'give unprecedented variety and splendour of illustration to his Lectures'.²⁷⁷ Those attending the 'mechanics' class' had access to Anderson's 'extensive Library, containing upwards of two thousand volumes, in every department of Science and Literature' and was to be available for the use of students until the commencement of the next year's session (a full year).²⁷⁸ The use of the library for a full year was also part of the benefits of enrolling in the course on natural philosophy and chemistry run by the Glasgow Mechanics' Institution in 1830–31.²⁷⁹

Thomas Graham published twenty-nine research papers while teaching at Anderson's College and did his work on gaseous diffusion and the nature of phosphates there. He moved to University College London and later became Master of the Mint.²⁸⁰ His successors were equally capable men. The attraction of chemistry was reflected in the large attendance figures.²⁸¹

The demand for instruction in the chemical sciences was obviously strong and woven into the fabric of scientific instruction for popular consumption first created for licensing surgeons and now the general practitioner. The requirements for graduation in all of the medical branches and for both the MD and the FPSG licence included chemistry. An essay on Charles Tennant's Glasgow chemical works which quotes a 'classical review of the years 1760 to 1832' describes 'the changes that transformed travel, transport, commerce, manufacture, banking, farming and all the various arts and means of social life' as 'a chapter from the Arabian Nights':

> The blind Metcalf had introduced the art of making roads; the illiterate Brindley the art of building aqueducts; Telford, a shepherd's son had thrown a bridge

---

[274] GUL, Special Collections, Bh12-y. 25, No. 90.
[275] Ibid.
[276] GUL, Special Collections, Bh12-y. 25, no. 63.
[277] GUL, Special Collections, Bh12-y, 25, no. 10.
[278] Ibid.
[279] GUL, Special Collections, Bh12-y, 26, No. 59.
[280] Butt, *John Anderson's Legacy*, p. 57.
[281] Ibid., pp. 43, 52, 53.

across the Menai Straits; Bell, a millwright's apprentice, had launched the first steamer on the Clyde; Stephenson, the son of a fireman, had designed his first railway engine; while a long line of inventors had by their patience and courage and imagination made Britain the workshop of the world.[282]

Charles Tennant was the son of an Ayrshire farmer, one of sixteen children, born in 1768. One of his brothers, Captain David Tennant, was a privateer and fought the French at sea. He refused a knighthood from George IV. When asked why, he said 'I just considered it a little better than a nickname'.[283] Charles Tennant exploited the discovery of chlorine, made in 1774 by the Swedish apothecary Scheele, turning it to manufacturing use by applying a milder solution and creating bleaching powder. His mastery of these techniques came by way of information obtained by James Watt in Paris.[284] When he established the St Rollox works in 1797 his partners included William Couper, the surgeon, who belonged to the inner circles of the FPSG and was involved with creating Anderson's College. Couper's three daughters married the three sons of Charles Tennant.[285] By 1835 the St Rollox chemical works were the most important in the world. 'They covered over one hundred acres and there were upwards of a hundred furnaces and retorts in use. The consumption in coal was 600 tons daily.'[286]

The symbiotic relationship established between medicine and chemistry, from both ends of the medical spectrum, from the surgeon-apothecary to the physician, proved to be a seminal one. Hill's two laboratories, in King Street and at No. 54 Trongate, were held in partnership not only with William Couper but with the surgeon James Monteath. James Monteath brought together medical politics, educational enterprise and the sciences. He was the first President of Anderson's College in 1801 and a founder of the Glasgow Apothecaries' Hall. This company 'was formed in 1805 by a number of medical men entering into partnership for the purpose of getting their prescriptions properly dispensed'.[287] The Apothecaries' Hall was opened in Wilson's Court, 29 Argyll Street, but as it soon evolved into a retailing business and moved in 1810 to one of the old mansions in Virginia Street.[288] James Monteath was President of the FPSG in 1820–22, only to be succeeded in the post by William Couper (1822–24).

---

[282] W. Alexander, 'Charles Tennant', in *Enterprise: An Account of the Activities and Aims of the Tennant Group of Companies First Established in 1797* (London, 1945), pp. 122–23.
[283] Ibid., p. 120.
[284] Ibid., p. 126.
[285] Ibid., p. 130.
[286] Ibid., p. 131.
[287] RCPSG, Glasgow Collection, Pamphlets Glasgow, 188 c 27, 10, 'Glasgow Apothecaries Company Centenary' (Glasgow, 1905), p. 6.

The history of the relationship between chemistry and medicine in the years when it became axiomatic for both educational expansion and commercial applications, particularly in Glasgow, is too rich to be fully explored here. But surely in previous histories on either subject the ties to practical medicine and its networks have been undervalued. One could cite as further examples the contributions laboratory work made to the specialist manufacture of glass instruments, and its industry, or its seminal influence on the study of a physiological phenomenon, animal heat,[289] derived from the latent heat discoveries of Black, Irvine and Robison. But the circle can provisionally be closed by reflecting on an epithet quoted from James Mill's *Essay on Education*. The namesake of the surgeon-apothocary Ninian Hill (a descendent of the family) who qualified as MD in 1836 had his dissertation printed 'when a candidate for admission' to the FPSG. He too must have believed in the benefits of a good scientific education. He quoted Mill on his title page:

> What is theory? It is the putting of the whole of the knowledge we possess upon any subject into that order and form in which it is most easy to draw from it good practical rules.[290]

---

[288] Ibid., p. 7.
[289] D. Denby, 'Physiology in the Enlightenment', *Glasgow Medicine*, 3, no. 6 (1986), pp. 10–11.
[290] Cited on title page, Ninian Hill, *Notes upon the Insufficiency of the Aortic or Semilunar Valves of the Heart* (Glasgow, 1836).

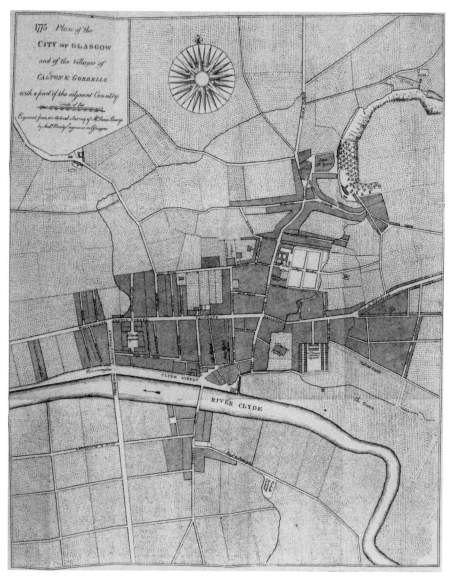

8. Plan of the city of Glasgow, 1775

7

Midwifery and General Practice

On 3 December 1739 the Faculty of Physicians and Surgeons of Glasgow proposed introducing examinations or 'trials' for midwives, treating them in the same manner as surgeons when granting a licence to practise. 'All midwifes', the FPSG decreed, 'after a certain time shall pass an examination and have a licence from the faculty before they be admitted to practise.'[1] Four months later a committee was appointed,[2] consisting of the Praeses, John Wodrow, whose teaching interests were in materia medica; George Montgomery, who was in clinical practice at the Town's Hospital and lectured on the institutes of medicine; Alexander Horsburgh, a leading city surgeon; and John Gordon, Lecturer in Anatomy and an advocate of better midwifery practice. The committee was charged with drawing up 'an form of an act' concerning the examination and licensing of midwives.

On 5 May, 2 June and again on 4 August 1740 the minutes of the Faculty show these proposals enacted and the examination committee reminded of their duty, the Faculty members 'having considered the many dismall effects of the Ignorance of Midwifes'.[3] The trials that were subsequently inaugurated proved durable, as the minute books record examinations of midwives until well into the nineteenth century. The Faculty's authority to regulate midwifery practice was based on the charter of 1599 and its ratification of 1672, which empowered the FPSG to 'Call before them and Examine all practisers in physick, surgery and pharmacy and if not qualified to Discharge them from practiseing under the foresaid penalty of fourty pounds'.[4] In this respect, the Faculty treated midwives no differently from other practitioners. They were subject to the same by-laws that provided for the prosecution of unlicensed practitioners, and were liable to the same penalty of £40. The one concession made to midwives was that the fee for their licence should be their only expense, the minutes explaining that 'As the Faculty have no other view but to prevent ignorant persons from practiseing midwifery, They appoint That Such [persons] as Shall voluntar[i]ly Submit to ane examination

[1] RCPSG, 1/1/1/2, Minutes of the FPSG, 1733–57, fol. 42r.
[2] RCPSG, 1/1/1/2, fol. 43r.
[3] RCPSG, 1/1/1/2, fos 44r and v, 46r.
[4] RCPSG, 1/1/1/2, fol. 46r and v.

towards their being Licensed Shall pay no freedome fyne nor be at any furder charge than two Shillings Six pence Sterling to be payed [to the] Clerk for each of their Licenses'.[5] The FPSG's decision was made public, the collector paying Dr George Montgomery 16s. 'for advertiseing about midwifes' on 23 February 1741.[6]

This was a bold departure, as the examination of midwives by a medical incorporation was without precedent. In Edinburgh the responsibility for licensing midwives rested with the magistrates and this was done by testimonials, not trials. Midwives were apparently not examined at all, but 'enter upon that difficult sphere at their own hand, without the least trial taken of their knowledge of the principles upon which they are to practise that art'.[7] They were required only to present to the magistrates 'a certificate under the hands of at least one doctor and one surgeon ... bearing that they have so much of the knowledge of the grounds and principles of this art, as warrants their entering upon the practice of it; whereupon a licence should be given them'.[8]

Any constraints on practice or redress for malpractice would therefore have been in the hands of the magistrates or the kirk session. The midwifery professorship in Edinburgh was not linked to licensing arrangements, and there were separate corporate bodies for the town's physicians and surgeons, which may have worked against a unified format. In fact, only one case of the Incorporation of Surgeons of Edinburgh licensing a midwife, on 19 February 1752, is on record. The candidate 'had been examined upon all the different sorts of birth, natural, laborious and preternatural, and on the methods of treating women after delivery and new born children [the committee finding] that the answers she gives to the severall questions we put to her are judicious and satisfieing and that she is in every respect extreamly well qualified to discharge the office of a midwife and well deserves the favour of a licence from the Corporation to practise the same'.[9]

The FPSG, on the other hand, was effective in its regulation of midwifery

[5] RCPSG, 1/1/1/2, fos 44r and v–46r and v.

[6] RCPSG, 1/1/1/2, fol. 63v.

[7] As quoted from the Town Council Records, 9 February 1726, in C. Hoolihan, 'Thomas Young, MD (1726?–1783) and Obstetrical Education at Edinburgh', *Journal of the History of Medicine and Allied Sciences*, 40 (1985), p. 331.

[8] Quoted ibid., pp. 331–32.

[9] We are obliged to Dr Helen Dingwall for this information. She adds, in a letter, that 'there were no other occasions of similar licences being given'. Midwives in Edinburgh would have been examined by the Incorporation of Surgeons, not the College of Physicians, and it is intriguing that only one candidate seems to have presented herself, particularly as Thomas Young lectured in midwifery there after 1756. Further research may clarify the issue, but this certainly implies a difference between the two Scottish cities.

qualifications. On 1 September 1740 Sarah Burmaster applied to be examined. Dr Montgomery, Mr Gordon and Mr Calder, Jr, were duly appointed 'To Examine the petitioner upon her skill or knowlege in Midwifery and to Report with their first Conveniency'.[10] On 1 December 1740 the Faculty approved the practice whereby the Praeses or Visitor called on any other two members of the Faculty to assist them in examining midwives, at any time, upon their applying to be examined. This same procedure was to be followed in future.[11] Soon after the introduction of examinations, four midwives were licensed on 1 December 1740. The number rose to fourteen in 1741, and in the following year was still as high as eleven. However, in 1743 only eight midwives were licensed – and for much of the century only about three women were licensed each year. The numbers increased again in the 1780s, reaching nine in 1786, and eight midwives were licensed on 2 May 1814, which is the last recorded entry. No fewer than fifteen candidates were rejected or fined and discharged between 1742 and 1743.[12] The FPSG records are instructive, revealing the candidates' marital status and the occupations of their spouses. Although the vast majority of the women were married or widowed, the use of the term 'daughter' in two particular instances suggests that the women in question were not married. Janet Wright, who was discharged from practising midwifery on 1 February 1742, is referred to in the records as 'Daughter of James Wright, Gardener in Glasgow', while Margaret Mitchell, who received her licence on 2 May 1757, is described as 'Daughter to Joseph Mitchell in Hamilton'.[13]

The fact that most candidates were the wives or widows of skilled tradesmen reflects Jean Donnison's comments about the social status of midwives in Britain more generally.[14] Shoemaking, weaving, tanning, watch-making and small-scale manufacturing were the most common occupations of the Glasgow midwives' husbands. Although a surgeon's widow from Hamilton was licensed, this was an exception. Most midwives were of lower social status than the surgeons, many of whom like James and Robert Houstoun came from educated, well-to-do families, or were making their fortunes through midwifery and obstetric practice, like William Smellie and William Hunter; or were in leading city practices, like John Gordon, John Moore, Thomas and William Hamilton, John Burns and James Towers. These social and educational differences between men and women widened further in

[10] RCPSG, 1/1/1/2, fol. 48r.
[11] RCPSG, 1/1/1/2, fol. 56r.
[12] RCPSG, 'List of Midwives Licensed'.
[13] Ibid.
[14] J. Donnison, *Midwives and Medical Men: A History of the Struggle for the Control of Childbirth* (New Barnet, Hertfordshire 1988), p. 39.

the second half of the eighteenth century, with women increasingly consigned to an educational and occupational ghetto.

The emphasis placed on professional standards was a distinctive feature of the FPSG's attitude towards midwifery. While female practitioners suffered because, other than being examined, they could not access the lucrative ladder to success available through FPSG patronage, Scottish surgeons moved decisively to exploit the advantages that better midwifery skills offered to general practitioners. Their success increased with every new discovery about the function of the female reproductive organs and the mechanism of normal and abnormal labour. Lectures rather than apprenticeships were the means of disseminating this new knowledge, and the preferred training ground for male midwives was at the bedsides of private patients rather than in hospital wards. In these respects, the west of Scotland medical network was pivotal. From the beginning, surgeons like Robert Houstoun, John Gordon and William Smellie took the initiative and made discoveries that culminated in the work of William Hunter, William Hamilton, James Towers and John Burns. In this way, Glasgow contributed more than most – on an axis that included London – to making midwifery an acknowledged medical discipline. Moreover, the rise of 'man-midwifery' had the decisive effect of blurring the obvious distinctions between physicians and surgeons. Active intervention in abnormal childbirth was part of surgical practice, yet by the end of the century physicians were attending pregnant women as 'accoucheurs'. The Glasgow story of midwifery is therefore both unique and indispensable in charting its development as a medical discipline.

When the FPSG drew up the curriculum requirements for its diploma in 1802, midwifery was mandatory: 'Mr Burns moved that an amendment should be added to the qualifications of country Practitioners viz. that they should have attended a course of midwifery'.[15] Similarly, when negotiations on medical qualifications reached a decisive phase in 1838, the FPSG insisted upon attendance at a six-month midwifery course consisting of 110 lectures.[16] Only the Society of Apothecaries in London required a lengthier period of midwifery training for its diploma, specifying that candidates must have attended two midwifery courses, each of six months' duration and comprising sixty lectures. In 1854 these requirements evened out across the country, with the universities of Glasgow, Edinburgh and Aberdeen, as well as the FPSG, the Royal College of Physicians of London, the Society of Apothecaries

[15] RCPSG, 1/1/1/4, Minutes of the FPSG, 1785–1807, 7 June 1802.
[16] RCPSG, 1/1/1/7, Minutes of the FPSG, 1835–47, pp. 150–51: indication of 1838 course requirements.

and the Navy Medical Board, all asking for attendance at a six-month course.[17]

As noted above, surgeons developed an early interest in improving midwifery. Not surprisingly, their enthusiasm was linked to the need to excel in a viable profession. The careers of the surgeons James Houstoun and Robert Houstoun exemplify this, as do the major contributions made to the development of midwifery by Smellie and Hunter and those influenced by them in both England and Scotland. Robert Houstoun, known today for his skills in surgical obstetrics, was also the first licensed city surgeon to be awarded the MD at Glasgow University. The option of receiving midwifery training abroad was open only to the elite, but this specialist teaching, which underscored the use of lectures and case studies rather than apprenticeships in training young practitioners, ultimately influenced the profession as a whole. Both Robert Houstoun and James Houstoun contributed significantly to the establishment of continental-style midwifery lectures as the preferred teaching vehicle in Scotland.

Robert Houstoun has achieved lasting fame as the first to perform a successful ovariotomy.[18] Significantly, he came from a traditional surgical background in Glasgow, where his father, Robert Houstoun senior, was a wealthy surgeon who entered the FPSG in 1669, and was twice appointed as Visitor in 1679 and 1691. Robert Houstoun junior, like the well-off James Houstoun, went to Glasgow University, but was then apprenticed (probably to his father).[19]

Robert's most famous case, which he was asked to present before the Royal Society in London in 1724, dates from 1701 and did not involve a birth but an ovarian tumour. His patient was Margaret Millar, the wife of a tenant farmer on the lands of the aristocratic Houstoun family, whom he treated as part of his practice in Renfrewshire. While in the country with another patient – Lady Anne Houstoun, the wife of Sir John Houstoun, to whom James was a full cousin – he was asked to attend Margaret Millar who was suffering from a greatly distorted abdomen. Robert Houstoun performed an operation to remove her growth and wrote up the case in detail. At this time

---

[17] *The Edinburgh Students' Guide to the University, Royal College of Surgeons* (4th edn, Edinburgh, 1854): comparative chart.

[18] For details see C. J. MacKinlay, 'Who is Houstoun? A Biography of Robert Houstoun, MD, FRS, 1678–1734', in *Journal of Obstetrics and Gynaecology of the British Commonwealth*, 80 (1973), pp. 193–200.

[19] The minute books of the FCPSG between 1688 and 1733 were burnt inadvertently, so the only record of Robert Houstoun being licensed is his signature in a roll 'of such worthy persons as have gifted books to the Surgeon's Library in Glasgow 1698', where he signed himself 'Master Robert Houstoune Chirurgeon Apothecary'. See RCPSG, 17/3, C. J. MacKinlay, 'A Biography of Robert Houstoun, MD, FRS, 1678–1734', p. 6.

he was a surgeon-apothecary, and obviously skilled enough to successfully remove a huge ovarian cyst from a woman then aged fifty-eight, who lived happily on for another thirteen years. This achievement has been hailed, albeit belatedly, as the first successful ovariotomy, predating that performed in 1809 by Ephraim McDowell in Kentucky by over one hundred years.[20]

By the time Robert Houstoun gave his report on the ovariotomy to the Royal Society in 1724, he had been awarded the MD. His interest in midwifery and obstetrics distinguished him throughout his career. Shortly before he made his reputation in Renfrewshire in 1701, Robert had attended the Paris maternity hospitals, and he obviously saw obstetrical cases as part of his surgical practice. A report of one such case of an extra-uterine pregnancy to the Royal Society in 1723 caused great upset, as these pregnancies were considered to be impossible at the time.[21]

Training in the wards of the Hôtel Dieu and the Charité in Paris distinguished those who aspired to the top of the profession. Robert Houstoun attended these institutions 'for two years together' in 1699 and 1700, while his contemporary and fellow Scot James Houstoun worked in both hospitals in 1714.[22] The latter wrote of his experience that:

> There I met with more laborious and fatiguing business than ever I had in the course of my life. Every twenty-four hours we brought about ten children into the world and there were only four midwives to go thro' all, night and day we were always employed. I continued in this hospital four months without setting my foot out of it four times. This is the only place for thorough instruction in this business; I brought nearly 300 women to bed in the time and everyone of the four midwives in proportion. We had cases of all sorts; we assisted one another and had a mistress midwife who directed the whole. In very extraordinary cases we called in the Master Surgeon of the Hospitals assistance.[23]

This intensive training was part of James Houstoun's ambitious strategy 'to prosecute my business that I must get my bread by or starve'.[24] Robert Houstoun, too, was an entrepreneur ready to sell his skills. As a licensed surgeon-apothecary with an upper-class clientele including the Cunninghames of Craigend and the aristocratic Houstouns of Houstoun House, Robert was able to invest in midwifery training in Paris. James Houstoun, on the other hand, decided that his midwifery skills would be best incorporated into practice as a physician. He followed the familiar route of obtaining

[20] H. Starke, *The Illustrated History of Surgery* (London, 1988), p. 193. Smellie also cited Houstoun's case.
[21] Ibid., pp. 198–99.
[22] MacKinlay, 'Who is Houstoun?', p. 194.
[23] As quoted ibid., p. 194.
[24] MacKinlay, 'A Biography of Robert Houstoun', p. 9.

an MD from Rheims, which was awarded on 12 August 1714, and subsequently becoming a licentiate of the Incorporation of Surgeons of Edinburgh on 5 March 1717.[25] Robert Houstoun pursued a more significant goal. The path which he took to obtain medical qualifications shows how much value was placed upon male midwifery skills in the west of Scotland, even at this early stage. From the status of a licensed surgeon, Robert eventually managed to persuade his alma mater, Glasgow University, to award him the MD, thereby elevating himself to a higher social and professional position.

He evidently invested a great deal of time and initiative in this pursuit: 'Mr Robert Houstoun, surgeon, who some time ago applied for the doctorate in medicine did still insist that he might be examined in order to his graduation'.[26] Robert eventually gained his MD on 3 January 1712, after undergoing 'tryals' at which his skills were assessed by Dr George Montgomery and Dr John Johnstoun, both practitioners in the city and active members of the FPSG.[27] Houstoun may even have had an eye on the medical professorship itself, as this was endowed shortly afterwards in 1714. It is of some significance that he insisted on undergoing a 'trial', as MD degrees in Scotland were then mostly awarded by testimonial. As one of its licentiates, Houstoun's initiative must surely have been sanctioned by the FPSG, which shows that the development of surgical skills, particularly those applicable to midwifery, was effectively narrowing the gulf between physicians and surgeons. Thomas Hamilton, too, later ambushed the divisions of the profession successfully, holding a surgeon's licence from the FPSG and managing to become Professor of Anatomy and Botany at Glasgow University, in spite of the fact that he did not possess an MD. He subsequently became the first person to lecture in midwifery at Glasgow.

Looking back on his life, James Houstoun wrote in 1753 that:

> The custom of the country is that the eldest son inherits the whole paternal estate and the rest of the children are thrown upon the world with little or nothing but education, which in my humble opinion is as good and more easily acquired in that country [Scotland] than in any other part of Great Britain. I have heard my mother say that I was the youngest of fifteen children so nothing fell to my share but education which I must say was as liberal as that country could afford.[28]

---

[25] R. W. Innes Smith, *English-Speaking Students of Medicine at the University of Leyden* (Edinburgh, 1932), p. 121.

[26] MacKinlay, 'Who is Houstoun?', p. 197.

[27] Ibid. This is not the same George Montgomery who was awarded the MD in 1732, but a different, earlier practitioner.

[28] J. Houstoun, 'The Memoirs of his Life', as contained in *The Works of James Houstoun, MD* (London, 1753). This quote is taken from MacKinlay, 'A Biography of Robert Houstoun', p. 8.

In fact this 'fifteenth son' whose 'only legacy' was his education was richly endowed, as the well educated were held in high esteem within Scottish society. James's medical education was lengthy and expensive: five years at Glasgow University, two years' study at Edinburgh, a further three years at Leiden and, thereafter, a period of clinical training in the Paris maternity wards.[29] During his time in Paris, he was reliant on financial support from his mother and was disingenuous at wheedling the necessary funds:

> Whilst I was in France, at my most expensive time, I drew too largely on my mother who was very loath to protest my bills but wrote to me of the straits and inconveniences I had put her to and I answered 'That I was in a popish country amongst some Scotch relations who were all papists and if she would allow me to turn papist I could live on one half the money I then spent'. The good honest old lady with the greatest sincerity and truth advised me for God's sake to continue in the faith for she would strip to the smock rather than that I should turn papist.[30]

A man of great ambition, the young James longed to enter those London medical circles where he knew midwifery, anatomy and botany were being studied, and where he might join a scientifically interested Scottish contingent.[31] Naturally, he sought out John and James Douglas, expatriate Scots who held considerable influence in London and who have been described as 'Whigs associated with the Court'.[32] Houstoun's attempts to 'insinuate himself into London man-midwifery' were largely unsuccessful.[33] However, it would be unfair to label his London career as a 'tragic farce'.[34] Although Houstoun may have been a minor player in the debate then raging over the merits of the forceps, it would be a great mistake to judge him on this criterion alone, as Scottish medical men were committed to a broader application of skills. Training in the Paris maternity wards was not primarily connected to learning about the forceps. Moreover, where no male professional standards yet existed, as in 'man-midwifery', Houstoun made a considered choice to translate and edit basic texts published on the Continent. This would have been a valuable contribution in Britain where obstetrical literature was scarce or not accessible in English. Conscious of the need for good midwifery textbooks, James Houstoun would have done everyone a service by making available French works on midwifery. He produced a

[29] Ibid.
[30] Ibid., p. 9.
[31] For details, see the chapter on 'Scottish Graduates in London', in J. Glaister, *Dr William Smellie and his Contemporaries* (Glasgow, 1894), pp. 279–97.
[32] A. Wilson, *The Making of Man-Midwifery: Childbirth in England, 1660–1770* (London, 1995), p. 86.
[33] Ibid., p. 87.
[34] Ibid., p. 86.

translation of Mauriceau's *Aphorisms*, which he passed to James Douglas, presumably seeking his approval. Unfortunately for Houstoun, his efforts were never published, although Douglas retained his manuscript.[35]

The book of Mauriceau's *Aphorisms* translated by Houstoun in 1719 may have been intended as a teaching text. Those interested in acquiring obstetrical techniques would surely have been grateful for such a volume, especially as London had no university and its medical incorporations had little or no interest in male midwifery practice. The Hippocratic aphorisms or those of the 'new Hippocrates', Boerhaave, were used as examination texts in Glasgow, both at the university and by the FPSG. In this context, Houstoun's translation of Mauriceau's *Aphorisms* makes sense, as does the fact that he also appended Deventer's observations on the obliquity of the womb to his translation.[36] Thirty years later in 1749, when midwifery was increasingly seen as part of general practice in Scotland, the Glasgow practitioner John Moore was training in Paris. In a letter to William Cullen, Moore noted that he was reading the works of Mauriceau and La Motte and attending as many deliveries as possible.[37] Mauriceau, then, was obviously not yet out of fashion. Indeed, his work was still recommended in 1768 when Thomas Young surveyed the textbooks available for the benefit of his midwifery students at Edinburgh University:

> The Books I would recomend are but few, and 1st. Mauriceau, Daventer, and La Motte are among the best authors, and the abridgement is the best, Dr Smyllie tho' [his work is] improper for one that knows nothing about midwifery, yet [it] is usefull afterwards.[38]

Like both of the Houstouns, Moore went to Paris to acquire a thorough knowledge of midwifery – 'a branch of my business which I freely own I have great need to study'.[39] At this time, the French were still regarded as the brokers of specialist training in midwifery and obstetrics, and it is significant that Moore mentions neither the Dutch nor the forceps as being of any importance.

Midwifery drew Glasgow close to London, establishing a network of support and communication that reached from the Faculty to the work of William Smellie. The Houstouns stood at the beginning of a long and fruitful

---

[35] Ibid., p. 87.
[36] Ibid.
[37] GUL, Special Collections, MS Cullen 91, Letter of John Moore to William Cullen (Paris, 1749).
[38] RCPSG, 20/1/6/1, Lectures by Dr Young, Professor of Midwifery in the College of Edinburgh, 22 November 1768, i, p. 25.
[39] GUL, Special Collections, MS Cullen 91, Letter of John Moore to William Cullen, Paris, 1749.

interchange. John Gordon and John Paisley, the two surgeons who did most to advance university anatomy lectures in Glasgow, influenced William Cullen, William Hunter and, not least, William Smellie, whose deep interest in midwifery was bolstered by the encouragement of both men. Similarly, Hunter's groundbreaking work on the gravid uterus was, in line with Paisley and Gordon's aspirations, never parochial, so as to connect with progressive medical science.

Indeed, this was a particularly intricate network. John Paisley was appointed as one of William Cullen's examiners when he applied to enter the FPSG in 1736.[40] At around the same time, William Hunter became surgical assistant to Cullen in Hamilton, from whom he received an introduction to the principles of midwifery. William Smellie was then in practice in nearby Lanark.[41] Hunter subsequently followed Smellie's example and moved to London, residing with him between November 1740 and August or September 1741.[42] Smellie, the oldest of this group, had already been working along innovative lines by making notes of the obstetrical cases he attended 'between 1722 and 1739 while I was practising in the country [Lanarkshire]'.[43] Cullen, who was Smellie's junior by thirteen years, gave him advice and lent him books, for Smellie was known as 'a colleague keen to borrow scarce books'.[44] Inglis, Smellie's partner in Lanark, showed him how to apply the noose in complicated births; Gordon instructed him, as Smellie himself testified, in the use of the blunt hook,[45] probably shortly after he completed his naval service and returned to Lanark in 1724. Although Smellie's trial and entry into the Faculty is not recorded (because of the loss of the second minute book), the Collector's accounts from October 1732 to October 1733 list the payment of £2 15s. 0d. Scots on 5 May 1733 as his freedom fine.[46] He seems to have fallen into arrears with his dues after that, because the next record of payment reads 'To Mr Smellie Surgeon his quarter accounts for eleven years 18s. 4d.'[47] However, his payments to the Faculty did not cease while he lived in England, as John Gordon paid Smellie's quarter accounts for 1746–49.[48]

[40] RCPSG, 1/1/1/2, fol. 21r.
[41] J. L. Thornton, 'William Hunter (1718–1783) and his Contributions to Obstetrics', *British Journal of Obstetrics and Gynaecology*, 90 (1983), p. 787.
[42] Ibid.
[43] A. H. McClintock (ed.), *Smellie's Treatise on the Theory and Practice of Midwifery: Edited with Annotations* (London, 1876), 'Memoir of Smellie', p. 2.
[44] Ibid., p. 3.
[45] J. R. Butterton, 'The Education, Naval Service and Early Career of William Smellie', *Bulletin of the History of Medicine*, 60 (1986), p. 17.
[46] RCSPG, 1/1/1/2, fol. 5r.
[47] RCSPG, 1/1/1/2, fol. 95r.
[48] RCSPG, 1/1/1/2, fol. 117v.

The assistance and encouragement of his surgical colleagues in Glasgow undoubtedly contributed to Smellie's status as one of the great masters of midwifery. Although it was in London that he made his reputation, his determination to find 'valuable jewels buried under the rubbish of ignorance and superstition' was nurtured north of the border.[49] In this respect, his career resembles that of William Hunter. Both men's desire to improve surgeons' education was in accordance with Scottish values and with the Faculty's educational and professional concerns in the 1720s and 1730s, and both recorded their disappointment with the midwifery training offered in England.[50]

Born in Lanark, Smellie returned there after completing a period of service in the navy and began an apprenticeship with a practitioner named Inglis in 1722. During his first year in practice he assisted at only two operative procedures on pregnant women, although he performed his first solo amputation in 1723.[51] In the 1720s he was probably unable to support himself from the fees he received. Even after taking over the practice on Inglis's death in 1729, Smellie was struggling financially, and by the second half of the 1730s he 'united the occupations of cloth merchant and practitioner'.[52] Success came gradually. Acquiring land piecemeal, Smellie was designated 'Apothecary' in one transaction and 'Chyrurgi' in another. As his reputation grew he began to be called 'some distance into the country' – to Hamilton, Wiston, and Carluke – and to forge links with other practitioners, particularly those interested in midwifery.[53] However, his apprenticeship training and his surgical experience in the navy provided only limited preparation for such challenges as having to cut the foetus in the womb and extract it. Smellie was well aware of his lack of obstetrical skills, and in his *Treatise on the Theory and Practice of Midwifery* he recalled his troubled feelings about his first intervention in difficult births.[54]

Smellie's use of the 'phantom' to demonstrate the principles outlined in his lectures was his most important contribution to midwifery, more significant even than the improvements he made to the forceps. 'Phantoms', 'manikins' or 'machines' quickly became indispensable in teaching men how

---

[49] W. Smellie, as quoted in McClintock (ed.), *Smellie's Treatise on the Theory and Practice of Midwifery*, p. 78.

[50] Ibid., p. 78.

[51] Butterton, 'The Education, Naval Service and Early Career of William Smellie', p. 13.

[52] J. Thomson, *An Account of the Life, Lectures and Writings of William Cullen*, 2 vols (Edinburgh, 1859), i, p. 18. As quoted in Butterton, 'The Education, Naval Service and Early Career of William Smellie', p. 14.

[53] As quoted in Butterton, 'The Education, Naval Service and Early Career of William Smellie', p. 16.

[54] Ibid., p. 13.

to deliver babies safely, and were therefore more revolutionary than the forceps. Their popularity was largely due to the fact that attendance at labours was still relatively rare for men, with even the Royal Accoucheur, William Hunter, making way for conventional midwives during normal deliveries.[55] Thomas Young, who was appointed as Professor of Midwifery at Edinburgh University in 1756, praised Smellie for bringing about a revolution in male midwifery practice by having his students replicate manual techniques on dummies until they had perfected their skills:

> It was Dr Smellie who made the greatest improvement in the method of teaching Midwifery by introducing the proper use of Machines in Teaching it, for it is impossible to make one understand the practical part without showing it upon the machines. He contrived machines in imitation of Women and Children, on which he was capable of showing all the Variety of Praeternatural Labours, as the method of turning the child, the obstacles which may occur, and the method of removing them, so that this way we deliver in all the variety of Praeternatural Births which could not otherwise have been done.[56]

Young made use of this form of practical instruction himself. Evidence for his hands-on approach abounds in the verbatim transcriptions of his lectures. What was later misconstrued as 'ungrammatical language' was in truth the record of his using teaching aids to effect in front of his students. The transcripts of his lectures contain frequent references to 'This here bone' or 'This Instrument, etc.'[57]

The first midwifery 'machine' was used by the prominent Parisian teacher and Accoucheur, Grégoire, as Thomas Young pointed out in his lectures of 1768. Young emphasised the value of practising deliveries on dummies, noting that in Paris 'they had only a wicker woman & a dead child before it began to spoil'.[58] Smellie's 'machines', however, were a great improvement: 'the schools were never compleat till Smyllies time, in which he did very much by inventing Machinery which is so very necessary to all mechanical operations, and also by making his students practise all Cases before they delivered women'.[59] During the ten years in which he lectured in London, Smellie taught more than 900 medical students.[60] He used dummies that were said

---

[55] J. Thornton, 'William Hunter', *British Journal of Obstetrics and Gynaecology*, 90 (1983), p. 787.
[56] Royal Medical Society Library, MS 302, 'The Lectures of Thomas Young', p. 6.
[57] See note on flyleaf in Royal Medical Society, MS 302, 'The Lectures of Thomas Young'.
[58] RCPSG, 20/1/6/1. *Lectures by Dr Young . . . 1768*, p. 9.
[59] Ibid., p. 9.
[60] W. Smellie, 'Author's Preface', reproduced in McClintock (ed.), *Smellie's Treatise on the Theory and Practice of Midwifery*, i, p. 27.

by a contemporary writer to have been composed of 'real human bones arm'd with fine smooth leather and stuff'd with an agreeable soft substance'.[61] These machines were very lifelike, even to the dummy being able to contract or dilate like a real child moving through the birth channel, and to replicate the 'motion of the joints'. The cranium of the 'phantom' was also praised by teachers of midwifery as it was 'elastick' and able to return to a natural form after being used in demonstrations.[62]

These 'machines' were clearly as essential and effective in encouraging students to develop the necessary visual and manual skills as were dissections and the making of preparations. Midwifery teaching had many useful affinities with the 'hands-on' approach advocated by Hunter in his anatomy lectures, and Smellie's innovation derived from the same educational insight. London opponents of Smellie periodically mocked his use of machines, but they soon became standard teaching aids. James Muir, for example, declared his intention to demonstrate obstetrical techniques upon the machine when he advertised a course of private midwifery lectures in the Glasgow press in 1757.[63] Mrs Nihell of London, who campaigned vigorously against the entry of men into midwifery practice, blamed the 'ingenious piece of machinery' for 'an innumerable and formidable swarm of men-midwives, spread over town and country'.[64]

The ability to learn and practise on dummies gave men the confidence to handle the real event successfully, and ineptitude became less of a problem. Young had commented that, prior to the establishment of regular midwifery teaching, surgeons 'frequently destroyed both Women and children' through 'their rash practice'.[65] The use of machines spread quickly: following the Glasgow practitioner James Muir's adoption of this teaching aid, Thomas Hamilton asked the university to pay for 'apparatus' to illustrate the midwifery course he began teaching in 1768. On 10 June of that year the Senate minutes record approval of a sum not to exceed £80 'to buy Machines for teaching midwifery as formerly proposed'.[66] The machines included a 'figure of a woman being part of the Midwifery apparatus lately bought by the College'. The safekeeping and repair of these teaching aids were items of finance in 1771 and 1789, showing that they were now a staple of midwifery

[61] 'Literaray Notes', *British Medical Journal*, 2 May 1914.
[62] Original quote without citation of source, ibid.
[63] *Glasgow Journal*, 25 April to 2 May 1757.
[64] Glaister, *Dr William Smellie and his Contemporaries*, p. 57.
[65] Thomas Young, 'The Lectures of Thomas Young', Royal Society of Medicine, MS 302, p. 3.
[66] GUABRC, UMM (Senate) 26643, 1763–1768, p. 324.

instruction at the university.⁶⁷ Having secured the necessary 'apparatus', Thomas Hamilton held 'a regular Course of Lectures upon Midwifery every session of the College'.⁶⁸

Almost as remarkable as his popularisation of simulated childbirth was Smellie's success in persuading women to allow the presence of his students in the delivery room. He presided over and lectured at the deliveries of 1150 'poor women'. As Thomas Young pointed out:

> Dr Smellie (in London) also gave an opportunity of attending Women in real labours, which was pretty much uncommon, and he had a Hospital where he took in poor women for this purpose so that he was the first who made any considerable improvement in the teaching of midwifery.⁶⁹

It is most important to understand that Smellie did not teach in hospital wards, but privately. In Paris, men trained in midwifery by attending the wards of the Hôtel Dieu or the Charité. Smellie's instruction took place outside hospitals, within the realms of private practice, and one could argue that this was safer, given the dangers of puerperal fever. Smellie's manner of teaching may have been very influential in Glasgow, where there was a marked reluctance to furnish maternity wards. As such, the Glasgow Royal Infirmary did not maintain one. Instead, midwifery teaching became part of the wider system supported by the FPSG in which the basic medical disciplines were taught at the university and in certified private lectures, feeding into the FPSG licensing examinations. Only in 1834 did the FPSG actively promote a lying-in hospital managed and staffed by its members.

Smellie's instruction was in line with teaching midwifery through private lectures and home attendance. He taught in London for almost twenty years between 1740 and 1759, 'at the bedside of women whom he supported financially during their lying-in on condition that he might be allowed to bring his pupils to watch and even take part in the delivery'.⁷⁰ This practical method of training was accompanied by systematic lectures, which, together with accounts of earlier deliveries he had attended, were ultimately incorporated into Smellie's three-volume masterpiece, the *Treatise on the Theory and Practice of Midwifery* (1752), *A Collection of Cases in Midwifery* (1754) and *Praeternatural Cases in Midwifery* (published posthumously in 1764). These books became classical reference works whose tone and style – precise,

---

⁶⁷ GUABRC, UMM (Sen.) 1/1/1 (1771–87), 5 July 1771, p. 36; SEN 1, Minutes: 1/1/2 (1787–1802), 10 June 1789, p. 15.

⁶⁸ GUABRC, UMM (Sen.) 26643, 1763–68, p. 324.

⁶⁹ Royal Medical Society Library, MS 302, 'The Lectures of Thomas Young', p. 7.

⁷⁰ Miles Phillips, 'William Smellie and the Maternal Mortality Problem', *Edinburgh Medical Journal*, 40 (1933), p. 133.

practical, observational and descriptive – made Smellie's clinical teaching a model for others. He commented too on cases described to him by his students, because he wanted 'to enable others to be as servicible as himself'.[71] In all, Smellie described 531 cases from his own practice, classifying them in 'collections'. The three volumes of his *Treatise* were also cross-referenced, rendering them of even greater service to both students and teachers of midwifery.[72]

Ten years after his arrival in London, Smellie went home to Lanark, aware of approaching death. After the cynicism and double-dealing that he had often encountered in the south, he was returning to an ethic of simplicity and straightforwardness – qualities that other reformers, too, believed to be the hallmarks of scientific integrity. In a letter to Dr Pitcairn, Smellie described himself as 'a man of learning and experience in practice', expressing the hope that he had 'gained respect and business by real merite'.[73]

William Smellie died in 1764, leaving his library of 300 volumes and the sum of £200 to the school in Lanark – a bequest that reflected the high value he placed on education and self-improvement.[74] However, his main legacy lay in the improvement of midwifery practice. Most of the men responsible for improving and institutionalising the teaching of that discipline in the second half of the eighteenth century were Smellie's disciples in both method and outlook.

Midwifery practice now began to favour men. The social and educational differences between men and women became ever more marked, and women were given no opportunity to develop a professional career beyond enrolling in classes taught by men. In 1772 the minute book of the FPSG records that Elizabeth Blair, widow of Alexander McKechny, late schoolmaster in Glasgow, and Elizabeth McNeil, spouse of James McNeilage, Wright in Glasgow, had attended Thomas Hamilton's lectures on midwifery and produced certificates. They were duly examined and passed.[75] This is the first mention of the use of teaching certificates in licensing midwives in Glasgow, demonstrating that women were complying with the 'trials' of the FPSG.

Thomas Young was appointed as Professor of Midwifery at Edinburgh University in 1756, one year before James Muir lectured in Glasgow.

[71] Ibid., p. 137. Miles Phillips takes this quote from Thomas Tonkyn's preface to a translation of La Motte's *General Treatise of Midwifery* of 1746, in which he discusses Smellie. Phillips, an eminent professor of midwifery and gynaecology, was in favour of using Smellie's case histories and cautions with regard to the forceps as teaching material for medical students even in the twentieth century.
[72] For a description of this method, see McClintock, *Smellie's Treatise on the Theory and Practice of Midwifery*, 'Memoir of Smellie', p. 19.
[73] J. Young, 'Dr Smellie and Dr W. Hunter: An Autobiographic Fragment', *British Medical Journal* (August 29, 1896), p. 2.

However, he is said to have held lectures in midwifery as early as 1750, and to have set up a ward in the Edinburgh Royal Infirmary in 1755 at his own expense.[76] Although Young's successors, Alexander and James Hamilton, were able to draw all midwifery teaching in Edinburgh to the university, Edinburgh's licensing system was less cohesive than Glasgow's. There were separate incorporations for physicians and surgeons, which was not conducive to the unification of medical qualifications, particularly in those areas where differences were emerging, such as midwifery. Men who wished to practise midwifery were licensed through the presentation of testimonials, one each by a physician and a surgeon, to the city magistrates.[77] Young's and the Edinburgh Hamiltons' testimonials were therefore the key to male midwifery qualifications, as Edinburgh University did not require course certificates in midwifery until 1833.[78]

Thomas Young's wards in the Edinburgh Infirmary were closed regularly due to high infection rates from puerperal fever.[79] This may help to explain why Glasgow did not have a specialist maternity hospital until 1834, although other important factors were also in play. The Glasgow medical school operated by uniting teaching from several sources with one point of entry to surgical practice, namely the FPSG licensing examinations. The FPSG maintained open access to these examinations, enabling recruits to the medical profession to receive their training at various institutions. In Edinburgh, a bias in favour of those who were educated at the university and in the hospital wards was evolving, effectively changing the power structures within the medical profession. In Glasgow, top city surgeons dominated a more diverse teaching enterprise.

James Muir qualified as a surgeon on 9 March 1736.[80] In 1757, he advertised lectures in midwifery for both men and women (who were to be taught separately) in the *Glasgow Journal*. That he taught women is born out by the fact that, in 1759, Muir specified that all prospective female students must be of a sober and discreet nature, 'vouched for by persons of character in the places where they reside'.[81] Asking for character references was not unusual, men too being required to give testimonials. Muir held another course 'for students' at the end of December or the beginning of

---

[74] McClintock (ed.), *Smellie's Treatise*, 'Memoir of Smellie', p. 6.
[75] RCPSG, 1/1/1/3, Minutes of the FPSG, 1757–85, p. 199.
[76] Hoolihan, 'Thomas Young, MD', p. 334.
[77] Ibid., pp. 331–32.
[78] Ibid., p. 337.
[79] Ibid., p. 340.
[80] RCPSG, 1/1/1/2, fol. 17v.
[81] *Glasgow Journal*, 15 October 1759.

January, all his courses using the machines so successfully introduced by Smellie.[82]

In 1763 John Anderson, the educational reformer and professor of natural philosophy, wrote to Baron Mure about establishing a chair of midwifery in the university. He proposed John Moore, the city surgeon, for the post, describing him as

> perfectly well qualified for the Office ... which will be similar to what was lately done for Dr Young in Edinburgh, will be approved of by the Publick, and be a real Advantage to the State.[83]

The crown, unfortunately, did not create a professorial chair, but in 1768 Thomas Hamilton began lecturing in midwifery at the University. A Regius Chair was eventually founded in 1815.

Male midwifery merged easily with general practice in Glasgow because there was little opposition to its practice. This was due in no small part to the identification at mid century of Glasgow's leading city practices with the FPSG. John Gordon, who was long intimately associated with the FPSG, had also nursed its interest in midwifery. In 1750 he was awarded the MD from Glasgow University and re entered the FPSG as a physician in 1755,[84] becoming President in the following year. In 1751 he established a partnership with John Moore, whose interest in the teaching and practice of midwifery has already been noted. Moore practised in Glasgow until 1777, and in 1772 entered into a new partnership with Alexander Dunlop. He, in turn, was related to John Anderson, the founder of Anderson's College, who had sought to promote Moore to a chair of midwifery at Glasgow University. Dunlop, whose son William has been mentioned as being among the first to give clinical lectures, was a colleague of John Burns at the Glasgow Royal Infirmary. Burns and Alexander Dunlop became partners after John Moore moved to London.[85] This tightly drawn phalanx of interests amongst leading practitioners ensured that midwifery and obstetrics were rapidly established in the city.

Three names stand out, however: William Hamilton, James Towers and John Burns. The oldest of the three, William Hamilton, became Professor of Anatomy and Botany at Glasgow University in 1781. Robert Cleghorn, the eminent clinician, wrote of him that:

> Besides anatomy, he taught botany and midwifery; which last he practised with

[82] *Glasgow Journal*, 25 April to 2 May 1757; Duncan, *Memorials*, p. 134.
[83] National Library of Scotland, MS 2524, holograph letters, no. 3, John Anderson to Baron Mure, 8 January 1763.
[84] RCPSG, 1/1/1/2, fol. 162v.
[85] 'The Late Dr Burns', *Glasgow Herald*, 24 June 1850, p. 4.

such success, that he was called to almost every difficult case near Glasgow. In October 1783 he married Miss Elizabeth Stirling, an accomplished lady, connected with several opulent families in Glasgow and its neighbourhood. From these connections his practice, already extensive, was very considerably increased.[86]

William Hamilton's reputation for excellence in attending women in childbirth emerged from his training in London with his father's friend and the natural successor to William Smellie's school, William Hunter. Hamilton began attending Hunter's anatomy lectures in 1777.[87] As a result of these lectures in which Hunter was assisted by William Cruikshanks, Hamilton became familiar with Cruikshanks' work on conception and generation, pregnancy, the causes of sterility and the anatomy of the gravid uterus. Cruickshanks' writings on generation, which were finally published in 1797, had their origins in these lectures.[88] Hamilton then attended courses at the London midwifery school of Thomas Denman and William Osborn. These were acclaimed as the best midwifery classes in London, and Hamilton made the most of them, coming away with several volumes of notes.[89] He subsequently returned to Glasgow to assist his ailing father, Thomas Hamilton, with his teaching duties at the university. 'I hope you go well at Glasgow altho' probably Edinburgh will keep the lead', wrote John Hunter from London. 'I mean some day to come and take a peep at you to see what you are doing'.[90]

His health growing progressively worse, Thomas Hamilton resigned his Chair in favour of his son in 1781. William took over the midwifery class, making judicious use of the knowledge he had acquired in London. He lectured in practical midwifery with the thoroughness of Osborn and Denman and on the diseases of women and children with as broad a sweep

---

[86] R. Cleghorn, 'A Biographical Account of Mr William Hamilton, Late Professor of Anatomy and Botany in the University of Glasgow', *Transactions of the Royal Society of Edinburgh*, 4 (1798), p. 39.

[87] GUL, Special Collections, MS Gen. 1356/32, Letter from William Hamilton to his father, 18 December 1777.

[88] W. Hamilton (no relation), *The History of Medicine, Surgery and Anatomy* (London, 1831), p. 305.

[89] GUL, Special Collections, MS Hamilton 120/1 and 120/2, Thomas Denman: Notes on lectures in Midwifery taken down by Mr William Hamilton; MSS Hamilton 120/2 and 120/3, William Osborn(e): Notes on lectures in Midwifery taken down by Mr William Hamilton. Osborn and Denman were heard concurrently, the lecture notes for Osborn's class beginning on 25 January 1780 (date inscribed on the front cover of MS Hamilton 120/1) and ending on 1 March 1780 (MS Hamilton 120/2). The lecture notes for Thomas Denman's class begin on 2 March 1780 (MS Hamilton 120/2).

[90] GUL, Special Collections, MS Gen. 1356/60, Letter from John Hunter to William Hamilton, 2 July 1789.

as Hunter. He evidently believed that obstetrics and gynaecology should be taught in combination, to meet the needs of the general practitioner rather than that of the accoucheur. He gave a three-part course covering the theory and practice of midwifery, the diseases of women, and children's diseases.[91] Starting, like Osborn and Denman, by giving an anatomical description of the pelvis, Hamilton proceeded to explain the diagnosis and treatment of various gynaecological disorders and reproductive problems. He then described the process of gestation before moving on to midwifery proper, his classes encompassing the different types of normal and difficult labour and giving advice on post-natal care. His lectures, like Hunter's, were delivered in conversational style, and he used simple, graphic language, aiming, it has been remarked, 'at perspicuity only and trusting for attention to the importance of the subjects he treated'.[92]

The advantage of machines, which his father had introduced to Glasgow University, had been impressed upon Hamilton still further during William Osborn's classes, where they were regularly used to demonstrate natural and preternatural labour and the correct way to apply instruments such as the crochet hook.[93] Osborn encouraged his students to exercise professional care and patience during complicated labours, and taught them to recognise dangerous situations in which assistance was required: 'We lay it down as an axiom that the shoulder or hand cases cannot be born by the efforts of nature without turning. And turning, if done with caution, is not so dangerous as leaving it to nature'.[94] This was one of the few instances when Osborn, a great believer in letting nature take its course, actively encouraged intervention. His pupils were well warned of the danger that could be done by hamfisted practitioners and were urged not to interfere unnecessarily, particularly when removing the placenta. They were told that if the placenta was retained in the contracted uterus they should wait for two hours before removing it manually, and that the 'after pains' should be made bearable, but not removed entirely since they were part of nature's healing process.[95]

---

[91] GUL, Special Collections, MS Hamilton 89, Heads of Lectures on Midwifery, 'Midwifery, Introductory Lecture'.

[92] Ibid., 'Midwifery, Plan of the Course and Books'. It is tragic that Hamilton did not live longer, as these headings for his course would most certainly have become a textbook. The MS contains only personal shorthand notes, an *aide-mémoire* for the lecture. He did not write out the full text.

[93] GUL, Special Collections, MS Hamilton 120/1, Dr Osborn's Lectures on Midwifery, 'Lecture 28'.

[94] Ibid., 'Lecture 29'.

[95] GUL, Special Collections, MS Hamilton 120/2, Dr Osborn's Lectures on Midwifery, 'Lecture 33'.

From Denman, too, Hamilton had learned caution, particularly when applying the forceps. Denman remarked to his pupils that 'a glorious variety of these [were] invented; & every one thought he was improving the art, better had they used the same pains in ascertaining the cases in which the use of instruments was necessary ... The avoiding [of] injury to mother & child will not depend on [the] instrument but on your own care and attention'.[96] Hamilton learned his lesson well. Patience, caution and the ethic of non-interference became the hallmarks of Glasgow practice.

The surviving volumes of William Hamilton's commonplace books contain brief descriptions of the twenty midwifery cases that he attended during his first few years in practice in Glasgow.[97] Only five of these were natural labours, the rest of the midwifery cases to which he was summoned being attended with various complications. Hamilton records that he was forced to use instruments on four occasions and his descriptions confirm that such assistance was absolutely necessary in every one of these cases. His actions are further justified in that he managed to save the lives of all four of the mothers and two of the babies involved – the two children who were lost had already been dead for some time before Hamilton applied the forceps to extract their bodies.

These books, in which Hamilton's list of the midwifery cases he attended is contained among a miscellany of other surgical descriptions, show him as a reformer eager to carry forward the work of Hunter and Cullen. Indeed, had he lived beyond the age of thirty-two his name might have ranked with theirs.[98] After Hamilton's untimely death in 1790 midwifery training at the university was continued by his partner in practice, James Towers, a first-rate practitioner who had entered the FPSG in 1787.[99] Towers, like Thomas Hamilton before him, had never taken the MD, and held the FPSG surgical licence while he taught at the university.[100] He subsequently supported its right to confer surgical degrees when the Chirurgiae Magister was conferred upon both himself and John Burns in 1817. In 1815 he matched Thomas Hamilton's achievement in managing to become a Professor of Midwifery

---

[96] GUL, Special Collections, MS Hamilton 120/2, Dr Denman's Lectures on Midwifery, 'Lecture 17'.

[97] GUL, Special Collections, MS Hamilton 83, Common Place Book, ii, fol. 174r and v; MS Hamilton 84, *Common Place Book 1782*, ii, fol. 96.

[98] T. Thomson, revised ed. of R. Chambers, *A Biographical Dictionary of Eminent Scotsmen* (London, 1875), iii, pp. 228–29.

[99] Duncan, *Memorials*, p. 264.

[100] James Towers is not listed in W. Innes Addison, *The Matriculation Albums of the University of Glasgow from 1728–1858* (Glasgow, 1913), although his two sons are. John and James Towers (junior) are both listed as sons of 'Jacobus, Medic[us] in urbe Glasgusi'.

without an MD. Towers' son John did not study for the MD either, instead taking the FPSG license in 1811 and graduating as CM in 1821. James Towers' application to teach midwifery at Glasgow University was approved by the Senate less than two weeks after Hamilton's death.[101] In this way, midwifery teaching was raised to a separate lectureship financed by the university on the same basis as those of chemistry and materia medica.[102] Just as William Hamilton had apparently 'inherited' the midwifery apparatus that had been used by his father, James Towers was given charge of the apparatus that had been used by William.[103]

Educated in Edinburgh and London, Towers had gained surgical experience at the Edinburgh Royal Infirmary before joining the staff of the Glasgow Royal Infirmary. In 1792 he established 'a lying-in ward for the more effectual instruction of his Pupils in the Practice of Midwifery' at the university.[104] Its primary purpose was to provide instruction to Towers' students, rather than a service for needy women. Like the infirmary, the lying-in ward catered for the respectable poor rather than the destitute. The university supported the hospital and the accompanying dispensary from the outset, and from 1794 Towers (who was married to the daughter of a bailie, James MacLehose) was also recompensed for admitting cases recommended by the town council.[105]

Undoubtedly, Towers' initiatives greatly strengthened midwifery teaching in Glasgow. However, his hospital was not the only place to offer such instruction. John Burns also taught midwifery along with surgery and anatomy at his private school in Virginia Street (later in College Street). Burns has been described as being 'through his works on the subject, the most popular expounder of midwifery, in his day'.[106] He entered the Faculty in 1796, and in 1800 was appointed as Lecturer in Surgery and Midwifery at Anderson's College in Glasgow,[107] where he continued to teach until he became the first incumbent of the Regius Chair of Surgery at Glasgow University in 1815.[108] Burns was a lecturer in the Hunterian mould, attaching great importance to a thorough knowledge of anatomy – and incurring public

[101] GUABRC, UMM (Sen.), 1/1/2 (1787–1802), pp. 51, 54, 57, 61.
[102] GUABRC, UMM (Sen.), 1/1/2 (1787–1802). The minutes for 1791 and for subsequent years show the annual renewal of these salaried lectureships.
[103] GUABRC, UMM (Sen.), 1/1/2 (1787–1802), pp. 51, 54, 57, 61.
[104] GUABRC, UMM (Sen.), 1/1/2 (1787–1802), 11 June 1792.
[105] R. Jardine, 'The Glasgow Maternity Hospitals: Past and Present', *Glasgow Medical Journal* (1901), p. 4.
[106] Duncan, *Memorials*, p. 173.
[107] SUA, B112, Minute Book of Anderson's Institution 1799–1810, p. 15; 21 June 1800.
[108] Duncan, *Memorials*, p. 265.

wrath by grave robbing, which, although not unusual at that time, was seen as 'violating the repose of the dead'.[109] In the preface to his *Anatomy of the Gravid Uterus*, published in 1799, Burns emphasised the importance of having a good medical education:

> No man will trust his own life, or the safety of those whom he holds dear, to any man, however powerful his recommendations may be, if he once detects him to be a blockhead .... The man who practices and is not on top of his profession is a murderer.[110]

His teaching certificates, like those of James Towers, were accepted by the Faculty when presented by candidates for examinations and both midwives and male midwifery practitioners attended his courses. When medical qualifications were under discussion in connection with the 1845 Parliamentary Bill, Burns pointed out that, unlike in England, midwifery had been taught in Scotland and seen as a preparation for general practice for nearly a century.[111]

As well as a highly competent lecturer, Burns was an authoritative writer on midwifery. He wrote his first book, *The Anatomy of the Gravid Uterus with Practical Inferences Relative to Pregnancy and Labour*, while he was a surgeon at the Glasgow Royal Infirmary and a lecturer at Anderson's College. Works on abortion and internal haemorrhage followed in 1806 and 1807,[112] and all three were published in one volume in New York in 1809. A volume of *Popular Directions for the Treatment of the Diseases of Women and Children* was published in London and New York in 1811.

The book that made his international reputation, however, was *The Principles of Midwifery: Including the Diseases of Women and Children*, which was published in London in 1809. Ten editions were printed over the next three-and-a-half decades, as well as two American editions, and translations were made into Dutch, French and German. Burns compiled his anatomical descriptions 'from dissections and preparations before me whilst writing',[113] and made revisions and additions to each successive edition. Professor

---

[109] T. Thomson, *Biographical Dictionary of Eminent Scotsmen* (London, 1875), i, 'John Burns', p. 252.

[110] J. Burns, *The Anatomy of the Gravid Uterus* (Glasgow, 1799), preface, pp. x, xx.

[111] Burns was intimately involved in setting curriculum requirements from 1802 to 1850. Single instances are given in Coutts, *A History of the University of Glasgow*, pp. 530, 562–66.

[112] J. Burns, *Observations on Abortion: Containing an Account of the Manner in Which it is Accomplished, the Causes Which Produced it and the Method of Preventing or Treating it* (London, 1806; 2nd edn, London, 1807; 2nd American edn, Springfield, Massachusetts, 1809). J. Burns, *Practical Observations on the Uterine Hemorrhage: With Remarks on the Management of the Placenta* (London, 1807).

[113] J. Burns, *The Principles of Midwifery* (London, 1811), preface, p. iii.

H. F. Killian, the director of the maternity hospital in Bonn,[114] who translated the sixth edition into German, required assistance when he came to the eighth edition because, as he explained in the preface, he considered this to be an entirely new work. Burns' works on midwifery and on women and children's diseases were the standard reference books of the time, and he was deservedly described as 'the most popular expounder of midwifery in his day'.[115]

This rapid acceptance of midwifery as a medical discipline in Glasgow had far-reaching implications for the medical profession as a whole. In Scotland, surgeons were not regarded as socially inferior to any other professional caste, which made it easier for midwifery to be accepted by both physicians and surgeons. A considerable number of Glasgow surgeons who were involved in midwifery practice became MDs. In Edinburgh, Alexander Hamilton – himself an MD – emphasised the unity of surgeons and physicians in their study of midwifery, despite viewing it as part of 'internal medicine', the classic terrain of the physician.[116] Physicians had long been charged with treating 'women's diseases', but the rise of male midwifery facilitated surgeons' entry into this field.

The ease with which midwifery was established as a medical discipline in Scotland contributed significantly to the early blurring of distinctions between physicians and surgeons there. Lectures in midwifery drew no distinction between its medical and surgical aspects, and common ground was established when either physicians or surgeons were called to normal or 'preternatural' births. The physicians had most to learn, as in 'turning' or other manual operations they needed to use techniques previously associated with surgery. Warnings with regard to overfrequent use of the forceps and other instruments, which were thematic in the teaching of Hunter, Hamilton, Denman and Osborn, helped to shape a 'British school' of midwifery which was seen to balance the French emphasis on surgical intervention (especially that of André Levret). The 'natural approach' that defined the emerging British school in turn influenced midwifery practice in other European countries, thereby neatly reversing the flow of knowledge from Europe to Britain that had occurred earlier in the century. German and Austrian medical practitioners, who were at the forefront of developments in European midwifery practice, weighed up the benefits of surgical

---

[114] J. Burns, *Handbuch der Geburtshülfe mit Inbegriff der Weiber und Kinderkrankheiten nach der achten, vollständig und gleichsam ein neues Werk bildenden Ausgabe*, ed. H. F. Killian, i (Bonn, 1834).
[115] Duncan, *Memorials*, p. 173.
[116] Alexander Hamilton, *Elements of the Practice of Midwifery* (London, 1775), p. vii.

intervention as favoured by the French against the British ethic of non-intervention. They sided with the British, or more accurately the Scottish, school.[117] As physicians ceased to be deterred by the thought of 'hands-on' midwifery, and surgeons in turn came to teach women's diseases, general practice in the proper sense of the term gradually became a reality.

The Glasgow Lying-In Hospitals

Hospitalised childbirth arrived with the development of specialist wards and therefore has its own distinct history. As soon as the new teaching hospitals emerged, practical instruction in the wards, at the bedsides of patients, became significant, but neither the Town's Hospital or the Royal Infirmary maintained specialist lying-in facilities. How childbirth and maternity care became part of mainstream medical agendas in the early nineteenth century was anything but simple. One strong influence were the teaching methods developed in the wake of Smellie and Hunter. Men were still having to prove what they had to offer. Only by exhibiting an excellent grasp of normal and difficult births would women call them to the bedside in labour. Thus the education of men remained a dominant vested interest that shaped attendance at childbirth. The lying-in hospitals never denied an overt mission to train male doctors.

This has a particular bearing on the period discussed here. As the good relations the Faculty and the University had achieved in educating the profession disintegrated, a story told in detail in the next chapters, separate paths to medical accreditation emerged. This and an expanding profession encouraged the creation of specialist hospitals. Pressures to set qualifying standards which in Scotland included midwifery placed this training centre stage. In 1834 the absurd situation arose that two rival lying-in hospitals were 'founded', both called the 'Glasgow Lying-In Hospital'. The Glasgow public was good enough to support both. But the University and the FPSG, with its symbiotic relationship to the extra-mural schools, had in effect created hospitals to serve their own ends. The overriding concern – and this drives home the point – was access to specialist training leading to degrees. Each accrediting body was convinced medical men needed to learn about childbirth at the bedside and each built affiliate institutions to do this.

This invariably led to debates over intervention and care at birth as efficacy

---

[117] P. Schneck, 'Die Anfänge der wissenschaftlich-klinischen Geburtshilfe an der Berliner Universität, 1810–1850', in P. Schneck and H. Lammel (eds), *Die Medizin an der Berliner Universität und an der Charité zwischen 1810 und 1850* (Husum, 1995).

was discussed. In turn these raised significant public health care issues. Deliveries in the lying-in hospitals involved women too poor or distressed to have comfortable surroundings or the support of families. Mortality figures became the measure of success for these institutions and were often compared to attendance at home births.

Teaching aims were thus soon embroiled with many other matters, not least the codes of conduct expected of voluntary charities. Moreover social and economic realities impinged heavily on the familial securities of earlier times. Poverty, in the wake of industrialisation, struck women caught in migrant labour patterns or the wage-earning fluctuations of providers, a problem made visible by the increase in births in lying-in hospitals. The first wave of acute female misery and dependence on charity seems to have gathered momentum between 1830 to 1850, but that is an impression only.

Other complications also ambushed what might still have been a simple coupling of medical teaching and charitable work. Lying-in wards were prone to puerperal fever. This and a high incident of stillborn births generated debates within the male medical fraternity, one basic question being whether home births or hospital deliveries were safer. Leading doctors in Glasgow took different sides on the issue, while some lying-in hospitals were forced to move premises at each virulent outbreak. Still, teaching midwifery remained the object, while the immediate context of hospital philanthropy and burgeoning public health worries complicated the plot.

Under William Hamilton and his partner, James Towers, midwifery instruction was taught in the manner of Smellie and Hunter, with an emphasis on anatomical instruction, the use of phantoms and hands-on experience, gained by attending patients in their own homes. John Burns, who, along with Towers, was instrumental in shaping midwifery teaching in Glasgow, was clearly part of this tradition.

There is no evidence that Burns actively campaigned for the establishment of a lying-in hospital in Glasgow, although the Rotunda in Dublin had been receiving patients since 1745 and there were several lying-in hospitals extant in London. In Glasgow a considerable hesitation is in evidence over the benefits of lying-in wards. Knowledge of the dangers of puerperal fever was probably at the root of this, and the practice of preferring home births bears out this cautious attitude. The reluctance to use lying-in wards was not a matter of mortality statistics alone, however. The university teachers in Glasgow preferred dispensary services and their teaching mission included the scientific scrutiny of intervention, as careful assessment was part of midwifery's new ethos.

Thomas Young's maternity ward at the Edinburgh Royal Infirmary was notorious for its closures due to outbreaks of puerperal fever, as were those of his successors, Alexander Hamilton and his son James.[118] In Glasgow William Hamilton had demonstrated an interest in the diagnosis and treatment of puerperal fever even before taking charge of his father's midwifery class. On 27 February 1780 Hamilton gave a paper on puerperal fever to the Cecil Street Medical Society in London.[119] He attempted to clarify scientifically the specific symptoms of the disease and to examine its causes and treatment, and his paper displays a graphic familiarity with puerperal fever both as observed in live patients and through post-mortem dissections. Hamilton clearly recognised the fever as an 'inflammation', but the modern theory of infection was not yet within his grasp:

> The Disease as it appears to us is not at all of a putrid or hectic kind, especially on its first attack, but on the contrary has all the hallmarks of Inflammation attending it. We likewise find that a quantity of putrid matter coming away from the uterus when the placenta or bits of the membranes have been retained has not produced the Disease, nor is it at all probable in the cases where the Disease has begun early (as immediately after labour) that the discharges could have acquired the least degree of putrisiency [sic].[120]

He was well aware of the low survival rates among sufferers, remarking that: 'the Prognosis must in this Disease be in general unfavourable, as we find it so frequently ending fatally'.[121] Hamilton retained an active interest in puerperal fever throughout his life, incorporating a special section on the disease into his midwifery lectures.[122]

Many years later, in 1838, John Burns thanked his colleague Dr Churchill in Dublin for a report on a lying-in hospital, adding that: 'We have of late had a good deal of puerperal fever here – and [the] very ill [are] as usual under my treatment'.[123] Burns, who would have had every opportunity to affiliate himself to such institutions, only once put his name down to support a lying-in ward, and this in order to lend his weight to the work of James

---

[118] C. Hoolihan, *Thomas Young, MD*, pp. 340–41, contains an extract from a letter by Young, describing an epidemic of puerperal fever in the lying-in ward in 1773. Alexander and James Hamilton founded their lying-in hospital in Park Place for the poor women of Edinburgh in 1793. See J. H. Young, 'James Hamilton (1767–1839) Obstetrician and Controversialist', *Medical History*, 7 (1963), pp. 62–73.

[119] GUL, Special Collections, MS Hamilton 83, Common Place Book, ii, 'On the Causes and Treatment of Puerperal Fever', pp. 124–56.

[120] Ibid., pp. 143–44.

[121] Ibid., p. 147.

[122] GUL, Special Collections, MS Hamilton 88, Heads of Lectures on Midwifery, 'Midwifery, Puerperal Fever'.

[123] Library of the Royal College of Physicians of Ireland, Letter from Dr J. Burns to Dr Churchill, dated Glasgow 28 March 1838.

and John Towers after James's death.[124] John Towers had succeeded his father, William Hamilton's former partner, in the Chair of Midwifery at Glasgow University in 1820, and had assumed control of the lying-in hospital. He remained in charge of this facility until he died in 1833. After a campaign to generate public support and subscriptions, the hospital was subsequently expanded and improved, and John's successor, William Cumin, acted as one of the superintendents there. Cumin's successor, John Pagan, discontinued the hospital's in-patient beds altogether in the 1850s, although the staff continued to attend women in their own homes, recording over 700 'out-patient' deliveries each year. In 1853 Pagan had calculated that the maternal mortality rate within the hospital wards was 1 in 77, while the mortality rate for women delivered by the hospital staff in their own homes was 1 in 325.5.[125] Accordingly, he used his influence to close the in-patient facilities because he felt that 'even the richest and best appointed' maternity wards were 'institutions of very questionable utility'.[126]

Dublin's lying-in hospital, the Rotunda, a forward-looking hospital with many links to Scotland was also troubled with puerperal fever. It accounted for approximately eighty-eight out of the 164 maternal deaths, more women dying because of it than any other cause.[127] Robert Collins, Master of the Rotunda from 1826 to 1833, graduated MD at Glasgow in 1822, two years before he took up the post. He was highly successful in reducing its ravages. In his book on midwifery of 1835, Collins recommended that patients be confined to their own homes. He also ordered a strict regimen of fumigation with concentrated chlorine gas and he had bedcovers washed and straw renewed 'should the most remote symptom of fever have been present'.[128]

Methods to reduce contact and a strict regime of cleanliness were already under intense discussion, recommended by the Glaswegian Alexander Hannay on the basis of his experience in Galloway. Hannay entered the FPSG in 1826 with a probationary essay on the pathology of puerperal fever. He wrote that his investigations were part of his obstetrical studies but were more immediately prompted by an outbreak of the disease in the autumn of 1823, in a village on the western shores of Galloway.[129] He was particularly

[124] GUL, Special Collections, Eph K/112, Report of a Public Meeting of the Subscribers and Others Friendly to the Establishment of the Glasgow Lying-In Hospital and Dispensary, Held in the Town Hall on 19 September 1834.
[125] D. A. Dow, *The Rottenrow* (Carnforth, 1984), pp. 40–41.
[126] As quoted ibid., p. 40.
[127] O. T. D. Browne, *The Rotunda Hospital, 1745–1946* (Edinburgh, 1947), p. 109.
[128] Ibid., quoted therein, p. 110.
[129] RCPSG, GCD 2.31, *Faculty Inaugural Essays, 1825–46*: A. J. Hannay, 'A Probationary Essay on Some Important Points Connected with the Pathology of Puerperal Fever' (Glasgow, 1825), p. 7.

interested in its spread by contagion, comparing it to typhus.[130] His description of the measures taken to prevent further infection invoke the sanitary precautions and isolation applied in outbreaks of fever but also the cleanliness later recommended by Ignaz Semmelweis. 'During her severe and fatal illness, she [a patient named Jacob] was visited by numerous acquaintances and friends, and particularly by a woman of the name of M'Gowan, then pregnant, who, on her [Jacob's] delivery, and previous to it, had been very assiduous in her visits and assistance. She was on the third day after delivery, attacked with the disease, which in her likewise proved fatal'.[131] He continued that soon after M'Gown's attack 'the midwife who gave her attentions in her confinement ... was called to a patient to whom I was also called'. He observed that 'the midwife attended in the same dress as that in which she attended M'Gown, and frequently went direct from the one patient to the other' resulting in the next patient being 'seized with the disease'.[132] Hannay remembered the midwife attending others wearing 'the same clothes'. He thus recommended 'the absolute exclusion of all who had visited those already affected from the chamber of the woman in child-bed'.[133] He was careful himself to 'change [his] own apparel' and to attend to 'frequent ablutions'.[134] In five cases these prophylactic measures obtained 'a success equal to my most sanguine wishes'.[135]

Hannay's treatise was reprinted in 1827. He taught medicine at the Portland School and was the Professor of Medicine at Anderson's College from 1828 to 1846. He worked with James Paterson, the midwifery professor at Anderson's College, in the fever outbreaks in Glasgow of 1847–48. They were instrumental in setting up the Barony Parish Fever Hospital and reported its statistics of treatment and expenditure.[136] These connections illustrate the broader awareness of infectious disease within which the prophylactic measures for puerperal fever Hannay recommended were discussed.

It could well be argued that maternity wards were primarily useful to men for teaching purposes and less so for women giving birth. As we have seen, the main concern of male midwives during this period was to gain experience of deliveries. James Towers solved this problem for the students of Glasgow University by establishing a small lying-in ward and a larger dispensary

[130] Ibid., p. 13.
[131] Ibid., p. 17.
[132] Ibid., p. 18.
[133] Ibid., p. 19.
[134] Ibid.
[135] Ibid.
[136] RCPSG, Pamphlets, A. J. Hannay and J. Paterson, *Report of the Committee of the Barony Parish Fever Hospital* (Glasgow, 1848).

service there. As Smellie had done before them, Towers and his students offered their medical services free in return for attending at the deliveries of needy women. The moral issues surrounding these 'deserving poor' arose only later, with appeals for subscriptions to fund charity lying-in hospitals. The University facilities, managed by James Towers and subsequently by his son John, were primarily concerned with dispensing medicines and advice, a point which was highlighted by the efforts after John's death in 1833 to turn them into a fixed institution with attending staff and hospital rules.

John Towers took over the lying-in hospital when his father died on 24 July 1820, and on 7 November 1820 the *Glasgow Courier* carried an advertisement announcing that his lecture course on the Theory and Practice of Midwifery would begin in two days' time, and that the New Lying-In Hospital in Rottenrow Street was now open for patients.[137] The advertisement stated that the cases treated there would be recorded and that it was Towers' intention to give occasional clinical lectures on the most interesting of these. The hospital was located at No. 85 Rottenrow. It was supervised by the Keeper, Mrs McDougald, who was listed in the Post Office directory for 1834–35 as Mrs McDougall, midwife, living in Balmano Place. Perhaps this was the same Mrs McDougall, the wife of a Glasgow weaver, who was licensed to practice midwifery by the FPSG in May 1808.[138]

Midwifery instruction was not completely dependent on lying-in wards because much of it centred on anatomy, on descriptions and explanations of 'difficult' births – those requiring surgical intervention – and on women and children's diseases. The Edinburgh Hamiltons, Alexander and James, as well as William Hamilton (no relation) in Glasgow structured their lectures to include sections on anatomy as well as 'internal' medicine. William Hamilton's case notes show that he was primarily called to 'difficult' births, only five of the twenty labours that he attended during his first five years in practice being 'natural'. The others were protracted or showed serious complications, such as pelvic deformities or breech presentations.[139] The emphasis in William Hamilton's lectures as in those of his teachers, Denman and Osborn, was on the difficulties caused by such complications. All three men taught that difficult births were not necessarily made so by ignorant practitioners, as was formerly assumed, but could be caused by abnormalities within the mother's body such as a narrowed pelvis. Accordingly, they taught

---

[137] Dow, *The Rottenrow*, p. 19.
[138] Ibid.
[139] Anne Cameron has described these cases in an unpublished M.Phil. thesis, 'Keeping it in the Family: The Contribution of Thomas and William Hamilton to the Development of Midwifery Teaching at Glasgow University, 1768–90' (Glasgow University, 1998), pp. 27–30.

how to detect these problems by internal examination.[140] For Denman, Osborn and Hamilton, it was of paramount importance that students should have a solid grounding in these medical procedures which were based, besides actual experience, on a firm grasp of anatomy. James Paterson, who lectured on the theory and practice of midwifery at Anderson's College, continued this tradition in his 1858 course, which began with lectures on the anatomy of the pelvis and proceeded to describe 'difficult' labours and complications at birth, only then describing the mechanism of normal labour.[141]

Midwifery instruction at Glasgow University had been continuous since Thomas Hamilton's first lectures for male and female students in 1768. In 1771 Thomas's sister Grace married William Irvine, who had lectured in chemistry at the university since 1769. During a visit to France that same year, Irvine procured several obstetrical instruments for his brother-in-law, which may quite possibly have been intended for his midwifery class. The university had given the considerable sum of £80 three years earlier in order to procure demonstration dummies and instruments.[142] Irvine was now getting the best from Paris:

> I have at last got your instruments – I wish they may be to your mind – I have spared no pains in Enquiring for proper workmen – there is not an instrument maker in Paris whom I have not visited ... In the case I have two iris knives of Wengal, two of de la Faye's invention which differ from the former in having a little bent, two couching needles, a Devil's Knife and a little Knife like a pallet for speeks, said to be new, and a little hook all mother of pearl handles, two pairs of forceps which I can't say much for and a pair of excellent bent scissors, all in a Case seven inches by four covered in green.[143]

In 1777 Thomas Hamilton's health deteriorated and he spent several months taking the water cure in Bath. During his absence from the university, James Monteath, a surgeon who was engaged in general practice and who had only recently entered the FPSG, advertised midwifery lectures in Glasgow.[144] This was not a rival enterprise, but the action of a colleague whose activities helped to ensure that midwifery instruction remained available. Thomas made a partial recovery and returned to Glasgow to resume his teaching duties, but he was conscious of his reduced capabilities. In 1780

---

[140] Ibid., p. 17.
[141] RCPSG, Glasgow Collection, Pamphlets, Glasgow, 11: 'Medical School, Anderson's University', p. 9.
[142] GUABRC, UMM (Sen.), 10 June 1768.
[143] GUL, Special Collections, MS Gen. 1356/3, Copy of letter from Dr W. Irvine to Professor Thomas Hamilton, 1 September 1771.
[144] Duncan, *Memorials*, pp. 134, 179, 261.

he formally requested that his son William be permitted to help teach his classes. The following year he resigned his Chair in William's favour.

James Towers, who continued William Hamilton's work after his death in 1790, had had a very similar training to Hamilton's in midwifery. Both men had attended the Medical School at Edinburgh University,[145] followed by a period of training in London. Towers, like Hamilton, may also have attended William Hunter's school. Towers summed up his avid pursuit of midwifery studies thus: 'I was led to study midwifery so fully, that against next winter I propose to teach in this place [the university]'.[146]

Ten years later, Towers' course at the University was augmented by that of John Burns at Anderson's College, and in 1826 the Portland Street School also began to offer midwifery lectures. The *Glasgow Courier* carried an advertisement for John Burns' 'class of midwifery for female practitioners' in his 'lecture room' in Virginia Street costing two guineas in August 1801.[147] Robert Watt's course on the theory and practice of physic, advertised in November of that year, gave notice that these would be 'elucidated by clinical cases in the Dispensary'. The dispensary mentioned was probably that of James Towers, as the newspaper entry notes: 'Poor women lying-in are now recommended to the Dispensary and attended at their own places with care, humanity and delicacy'.[148] Watt's extra-curricular teaching for the FPSG may therefore have included midwifery instruction. James Armour switched from the Portland Street School to Anderson's College as midwifery professor in 1828. In three decades the demand for midwifery teaching had mushroomed to support *three* separate courses, one at the university and two at extra-mural schools.

The generation of midwifery teachers who followed Hamilton, Burns and James and John Towers included James Armour, James Brown and James Paterson at Anderson's College, and James Wilson who lectured at the Portland Street School. The midwifery courses taught by these men also maintained the tradition of providing students with a thorough anatomical knowledge and explaining the use of those special delivery techniques developed by the school of William Smellie. Midwifery was taught in the extra-mural medical schools because it was now required knowledge for the FPSG examination. Moreover, it was a rapidly expanding enterprise for both teachers and practitioners in Glasgow.

Glasgow's lying-in hospitals were inseparable from these teaching enter-

---

[145] Ibid., p. 264.
[146] GUABRC, UMM (Sen.), 1/1/2, 1787–1802, 13 March 1790.
[147] *Glasgow Courier*, 8 August 1801.
[148] *Glasgow Courier*, 3 November 1801.

prises. James Paterson, lecturer in midwifery in Anderson's College, emphasised this point when he opened yet a third facility, the short-lived General Lying-In Hospital in 1843: 'An Obstetrical Institution, when properly conducted, should be open to all Students of the Medical profession, and should be *visited daily* [emphasis in the original] by the principal Accoucheurs'.[149] He noted the restrictions on teaching in the two extant lying-in hospitals. That connected with the University 'is exclusively confined to College students; and the one in St Andrew's Square, while open to the students of the Andersonian University and the Portland Street School, is, nevertheless, not frequently visited by the Accoucheurs, nor do students receive those clinical instructions from the Medical Officers, which it was the original intention that the Hospital should afford'.[150] In his own prospective institution he wanted not only a hospital 'conscientiously conducted' and offering 'a comfortable asylum and satisfactory assistance to the needy', but providing, above all, 'a valuable practical school for the acquisition of professional knowledge', which, as he reminded the public, gave 'the best guarantee for the fitness and obstetrical acquirements of those Students who are ultimately to become Medical Practitioners'.[151]

Paterson had become midwifery professor in 1841 and obviously wanted his own wards to instruct students. His student attendance in 1843 was almost fifty, having markedly increased. Average attendence hovered between thirty and forty students.[152] Paterson's student numbers were undoubtedly squeezed by John Pagan at the university and by clinical teaching at the Glasgow Lying-In Hospital, now run by James Brown and James Wilson, whose certificates fed the licensing requirements of the FPSG.

James Wilson was teaching at the Portland Street School while James Brown succeeded James Armour (who died in 1831) at Anderson's College. Wilson and Brown were the 'accoucheurs' responsible for the founding of the Glasgow Lying-In Hospital. James Wilson was originally a protégé of James Jeffray, Professor of Anatomy at Glasgow, 'who entertained a high respect for his abilities, and whose friendship he enjoyed during the remainder of the life of that venerable individual'.[153] His son James George became

---

[149] GUL, Special Collections, Eph K/119, 'Prospectus of the Glasgow General Lying-In Hospital (1843), p. 2.
[150] Ibid.
[151] Ibid.
[152] For details of the attendance at midwifery lectures between 1830 and 1860, see J. Butt, *John Anderson's Legacy: The University of Strathclyde and its Antecedents, 1796–1996* (East Lothian, 1996), p. 50.
[153] RCPSG, 1/13/8/29, Obituary of Dr James Wilson, senior from the *Lancet*, 3 October 1857, p. 381.

the lecturer for midwifery in Anderson's College in 1863. James Brown may be the same man who entered the FPSG in 1827 and gave a talk critical of educational standards in medicine before the Glasgow Medical Society, to which we refer below. He lectured in the Mechanics' Institute. In 1834 he became the Professor of Midwifery at Anderson's College and was intimately involved in the founding of the Lying-In Hospital whose chief physician he remained until his death in 1846.

The relationship between charity wards and teaching remained finely balanced. The purpose of electing surgeons to the Royal Infirmary, for example, was clearly to augment their professional education. The poor received free medical care, but they also had to accept the negative aspects of hospitalisation, as Richard Millar, the physician to the Royal Infirmary acknowledged:

> Our system of electing Surgeons has been often applauded as highly liberal, as containing nothing exclusive, and as ensuring quick succession of office. To the Surgeons, without doubt, it is highly liberal; to the sick poor, it is just the reverse of liberal. Should any man choose to be operated upon by a Practitioner not accustomed to the work, it is his own affair; but the inmates of our Hospital have no choice, they must either take the Surgeons provided for them, or entirely forgo the benefits of the charity.[154]

Lying-in hospitals offered medical assistance to pregnant women on the same basis. While their appeals for funds stressed the desperate plight of these poor, 'deserving' women, they placed just as much emphasis on the benefits that their presence afforded for the training of surgeons and physicians.

The first annual report of the Glasgow Lying-In Hospital, the rival institution to the University's teaching dispensary, described its patients as 'widows, whose husbands had died lately; others were wives deserted by their husbands in a state of extreme destitution'.[155] However, the report went on to say that:

> Some were brought from the streets under actual labour, others were sent to the Hospital by the officers of Police, perceiving their advanced pregnancy, and knowing the wretched condition of their ordinary abodes. In these circumstances it may well be imagined, that the Medical attendants were aware that some of these patients were unmarried: but it was thought to have sternly refused admission in such cases of emergency would have betrayed a degree of inhumanity to the unfortunate for which even the most rigid of the public would give them no credit. Two lives were in peril, one of which at least was an innocent one. In ordinary admissions the restriction of the house to confine charity to destitute married females only is rigidly adhered to. To have acted otherwise would have been in direct opposition to the moral views of the contributors, who disown

---

[154] R. Millar, *Medical and Surgical Establishments of the Infirmary* (Glasgow, 1828), p. 8.
[155] GUL, Special Collections, Eph K/116, 'Glasgow Lying-In Hospital and Dispensary'.

giving their aid to an Institution which tends in any degree to encourage vice and dissipation.[156]

The tone employed underlines both the dependent position of these female 'patients' and the thin line between respectability and Victorian condemnation. James Christie, the Professor of Physiology at Anderson's College Medical School who quoted the report in a later investigation, added (in 1888) that not until the Twenty-Fifth Annual Report (1860) did the Directors have the courage to announce that the Hospital was for 'poor and homeless lying-in women', at last dismissing moral distinctions.[157] In London, the practice of restricting entry to married women appears to have continued, as this factor was known to lower the death rates reported by the lying-in hospitals. Truly destitute women who had no familial support were more susceptible to infection and death because of 'exhaustion', a cause of death not entered in mortality statistics but familiar to all attending at labours. This may also have been the decisive factor in the high number of stillbirths over which John Pagan, Professor of Midwifery at Glasgow between 1840 and 1860, worried in the 1850s.[158]

John Pagan followed Cumin as Professor of Midwifery in 1840. He discontinued the in-patient facilities which it must be cautioned did not amount to a closure of the facility as such.[159] In Pagan's obituary a clear distinction was made between closing the wards and continuing attendance at deliveries in the home. 'It was on account of his decided views with regard to the origin and spread of puerperal fever that the wards ... have been for many years discontinued [from the date of writing in 1868]' but 'about 750 [labours] yearly, conducted by students and nurses' were carried out 'at the patients' own houses'.[160]

The *Medical Times and Gazette*, in an article on medical education in Glasgow of 1862, recorded the clinical and teaching activities of the Dispensary: 'The students of this Hospital also attend poor women at their own homes, and many witness the practice of the Dispensary which is open daily at 1 pm, when advice is given on the diseases of females and children, and the children of the poor are vaccinated'.[161] Pagan argued that if lying-in hospitals received patients at all, they should be separated in a system of

---

[156] Ibid.

[157] RCPSG, 1/13/8/29, James Christie, *The Medical Institutions of Glasgow: A Handbook* (Glasgow, 1888), p. 101.

[158] J. Pagan, 'Contributions to Midwifery Statistics and Practice', *Glasgow Medical Journal*, 1 July 1853, pp. 213–14.

[159] *Glasgow Medical Journal*, 1, new series (1869), 'Obituary of James M. Pagan', p. 130.

[160] Ibid..

[161] *Medical Times and Gazette*, 29 November 1862, p. 578.

small cottages.[162] This indicates that he was as aware of infection as the source of puerperal fever as Hannay and others had been. He gives the number of women 'delivered by the students attending my lectures' in the period when he taught midwifery (1840–53; the latter year marked the appearance of his article) as 8587, of whom thirty-six died and 8551 recovered.[163] These were mainly home deliveries and the appearance of 'puerperal convulsions' was low, being only five, three mothers dying and two recovering.[164] Of greater concern to him was the number of stillborn children. Pagan discounts the comments of a 'Dr Strang' that the causes for 'the enormous number of children reported stillborn in this city' were 'moral', a euphemism for infanticide.[165] There was never 'the least ground for suspecting foul play in any of the cases I saw'.[166] Mr Kirkwood, Pagan's assistant, assured him that 'physical explanations', amongst them the 'midwives tea', a drug not specified, given in the first labours were, to his mind, largely responsible for these deaths.[167]

The senior medical superintendent of the Glasgow Lying-In Hospital, James Wilson, in his report on the year 1851–52, was understandably nervous about the four deaths among the 395 women delivered in hospital considering the apprehension over 'contagion'.[168] 'The fact of four deaths having occurred among the women delivered in hospital, and none among those delivered at their own homes, and also a greater number of still-born children occurring in the former than in the latter, may appear to require some explanation'.[169] He maintained the fault did not lie with the hospital 'but solely in the character of those that resort to it'.[170] These are 'a more destitute and wretched class than the others, the greater proportion of them being without homes of any description'. He blames the deaths on the 'exhaustion' of women. This exhaustion had its roots in 'irregular living, starvation, and harsh treatment from their husbands, who have deserted them'.[171]

---

[162] 'Obituary. The Late Dr Pagan, Professor of Midwifery, Glasgow University', *Glasgow Medical Journal*, new series, 1 (1869), p. 130.
[163] Pagan, 'Contributions to Midwifery Statistics and Practice', *Glasgow Medical Journal*, 1, no. 2 (July 1853), p. 208.
[164] Ibid., p. 211.
[165] Ibid., p. 213.
[166] Ibid.
[167] Ibid., p. 214.
[168] J. Wilson, 'Report of the Glasgow Lying-In Hospital and Dispensary for the Year 1851–52', *Glasgow Medical Journal*, 1, no. 1 (April, 1853), p. 1.
[169] Ibid., p. 2.
[170] Ibid.
[171] Ibid., pp. 2–3.

Wilson, defending the lying-in hospital wards, aimed one final emotional shot at his opponents:

> If there have been no deaths in these out-door cases, and I have seen other reports of a similar kind alike favourable to life, then are we shut up to this conclusion, that our wretched wynds, vennels, and interminable closes – those places of filth and every kind of abomination – which have heretofore been considered the pest spots and sources of disease in our cities, must now be looked upon as the most favourable to health, and that they stand in no need of those sanatory measures which have so engaged the attention of our authorities and other friends of the poor.
>
> But what are we to do with those poor unfortunates, who have not even these insalubrious places of filth and squalor to live in? Are we to shut our hospitals, infirmaries, and fever-houses against them, because deaths occur in these resorts of the destitute? Who does not know that deaths are more frequent in hospitals then in some other places? and who does not know that hospitals are necessary evils, in consequence of the artificial state of our social system, where humane and neighbourly feelings are nearly extinguished, and where no man considers himself his brother's keeper? There may be reasons for shutting up hospitals, but not because deaths occur in them.[172]

Clearly Wilson was concerned with the issue of squalor and poverty and the place of hospitals responding to large-scale social pressures.

This playdoyer for the relief that lying-in hospitals could provide shifted the focus from teaching to care. In 1853 it was becoming clear that maternity hospitals needed to accommodate patients on a bigger scale whatever their ties to teaching institutions. Although the Glasgow Lying-In Hospital was born of the rivalry between the FPSG and the university it became a philanthropic institution much the same in character as the Royal Infirmary.

The Glasgow Lying-In Hospital was the showcase of the FPSG. It was the only maternity hospital recommended in the syllabus of the medical school of Anderson's University for those taking the licences of the FPSG. This was all very ironic, since James Towers and John Burns had been strong supporters of upgrading midwifery training by introducing the FPSG curriculum requirements in the subject in 1802. Nonetheless in 1834, on 19 September, subscriptions to the 'Glasgow Lying-In Hospital' were advertised, only a day after the renewed appeal to support the University Dispensary. Its relative size and number of births at first made it no bigger than the university one, earning it the suggested title of 'public' only through its support by the surgeon's incorporation.

---

[172] Ibid., p. 4.

The FPSG was no stranger to the politics of linking charitable medical care and training licentiates. Anderson's College and the Portland Street School lectures were primarily accredited through the FPSG. Indeed the university did not recognise certificates from the extra-mural schools after about 1837. The Glasgow Lying-in Hospital training in midwifery became acceptable to the surgeon's colleges, the Apothecaries' Hall and the Navy and Army Boards.[173] The hospital had twelve beds for in-patients. The number of women confined in its first year (1835) was 369 'and at their own houses during the same period, 635; making a total of 1004 women who received the benefits of this Institution'. Two physicians were to attend, while eight were to be employed as 'Out-door Accoucheurs for superintending the Cases of Women in their own houses'.[174] Only fifty-one women had been delivered in its first ten months, with forty-eight outpatient births.

Soon after opening, it was closed because of the deaths of two mothers 'both from inflammatory attacks incident to the puerperal state'.[175] The wards were shut down, thoroughly cleaned and fumigated, and the patients were dismissed. The hospital reopened ten days later and, not surprisingly, the medical attendants wanted 'as little public notice and discussion as possible'.[176] Unfortunately, this was not an isolated incident. In 1841, having relocated to St Andrew's Square, the hospital again needed to be closed down for cleansing and fumigation. Puerperal fever broke out once more in 1853 and in 1863.[177] When the third lying-in hospital, the 'General' was mooted, in 1843, a comparison with other major cities was made:

> In Edinburgh, which has a population of 164,451, there are *seven* Lying-in Hospitals and Dispensaries for Diseases of Women and Children; while in Glasgow, with a population of 273,147, and a far more numerous poor, can boast of only *two*, and these of the most limited description! In Dublin, with a population of 378,976, there are five or six of the kind of institutions referred to, and *one* of these contains no fewer than 125 beds, thus of *itself* furnishing to the poor of Dublin,

[173] RCPSG, Glasgow Collection, Pamphlets Glasgow, 11: 'Medical School, Anderson's University' (not dated, but probably 1858).
[174] GUL, Special Collections, Eph K/114, 'Glasgow Lying-In Hospital and Dispensary' and GUL, Special Collections, Eph K/116, 'Glasgow Lying-In Hospital and Dispensary', Report at the Annual General Court of Subscribers on 19 October 1835, p. 13.
[175] GUL, Special Collections, Eph K/116.
[176] Ibid.
[177] R. Jardine, 'The Glasgow Maternity Hospital Yesterday and Today', in Mrs Robert Jardine, *The Chapbook of the Rottenrow* (Glasgow 1913), p. 15. While it would be very interesting to gather more information about these outbreaks of puerperal fever in Glasgow, time constraints have prevented a specific local study. For a detailed treatment of this general topic, see Irvine Loudon, *Death in Childbirth: An International Study of Maternal Care and Maternal Mortality, 1800–1950* (Oxford, 1992).

nearly *four times* the accommodation which is as yet provided for the poor of Glasgow.[178]

Certainly Glasgow was lagging behind in provision for women so destitute they had to seek out these wards. In view of the size of their respective populations, none of the cities mentioned could be seen as generous in providing for destitute women. Lying-in hospitals were small. Those in Glasgow delivered between fifty and seventy women as in-patients and about the same as outpatients per year for each institution until the numbers rose dramatically in the late 1850s.

Towers' university lying-in ward reported in 1796 that during its four years in operation seventy women had been confined in the hospital, all of whom were deserving poor. Twelve of the women, moreover, had been taken into the hospital on the recommendation of the magistrates.[179] The town council consequently agreed to provide the hospital with an annual sum of ten guineas per year, which would be sufficient to cover the building's rent, on condition that only those who were recommended by a magistrate or clergyman of Glasgow would be admitted, except in cases of urgent necessity. They did not want poor women from adjacent parishes to be attracted into Glasgow by the medical attention offered at the hospital, only to remain there as burdens on the city.[180] The outpatient facilities were, of course, under no such restrictions.

In November 1826 John Towers applied to the town council for an increase in its subvention from ten to twenty pounds.[181] He explained the financial maintenance of the hospital to the university commissioners on 27 October 1827.[182] The hospital had been principally maintained by his father and subsequently by himself. Until 1826 the only additional capital had consisted of ten pounds per year from the town council and half a guinea paid by each student who attended. In 1826, however, he had received twenty pounds from the town council and ten pounds from the Trades' House together with the students' fees. John Towers claimed that the University had never contributed financially to the hospital, yet in 1792 the University had awarded James Towers an annual grant of twenty-five pounds. John's reply when asked whether the hospital would 'fall' should he stop teaching is very revealing: 'Were I to give it up it would fall at once'.[183]

[178] GUL, Special Collections, Eph K/119, 'Prospectus of the Glasgow General Lying-In Hospital' (1843).
[179] Dow, *The Rottenrow*, p. 17.
[180] Ibid., p. 17.
[181] Ibid., p. 20.
[182] Ibid.
[183] Ibid.

John Towers died on 14 September 1833. There was turmoil over the choice of his successor. In the interim John Burns and his son Allan continued the midwifery lectures. Finally William Cumin was officially appointed to the midwifery chair in October 1834. The following month an advertisement regarding the University Lying-In Hospital stated it was now open to patients. Dr Cumin was to be in attendance and medicines would be supplied free to all patients who required them. 'Persons applying for relief for themselves or children must bring a written recommendation from a Magistrate, Clergyman or the Elder of their proportion, or from a Contributor to the Dispensary.'[184] Provision for home visits was advertised. The dual purpose of a lying-in charity was once more clearly stated:

> The benevolent public are earnestly solicited to extend their support to this Institution, which promises to confer a great benefit upon the sick poor and their children at a very trifling expense, while it will afford a wide and interesting field of observation and improvement to Students of Medicine – in a department to which at present very few of them have any access.[185]

The renewed opening of the university hospital was promoted by the University and excluded the FPSG, which retaliated by campaigning for the rival Glasgow Lying-In Hospital. In 1834 the divisions over surgical degrees and the feuds over national recognition were still at their height, explaining why no one was sharing facilities.

The University Principal chaired the public meeting in the Royal Exchange Sale Rooms on 18 September 1834 to woo subscribers for the university teaching unit. Attention was drawn to the value and utility of this institution as no fewer than fifty poor women were confined there every year, students receiving practical training by attending the hospital. However, it was noted that the previous hospital could not cope with the number of impoverished women who required attention, nor did it prove adequate for rising student instruction.[186] A committee was appointed to oversee the expansion of the hospital including all the university medical teachers, John Burns, James Jeffray, Thomas Thomson and John Couper. Others had strong ties with medicine, among them Charles Tennant, Charles Stirling, and James Cleland.[187]

The university hospital was to be controlled by a board of management composed of the Lord Provost, the Sheriff Depute of Lanarkshire and the

---

[184] GUL, Special Collections, Eph K/92, 'The University Dispensary'.
[185] Ibid.
[186] GUL, Special Collections, Eph K/111, 'Prospectus of a Plan for the Extension and Improvement of the Lying-In Hospital in the City of Glasgow'.
[187] Ibid.

University Principal, as well as six directors elected on a yearly basis by those who contributed five pounds or more to the hospital on a single occasion or one guinea annually. Cumin was to be 'assisted by a Surgeon to be named by the Directors', who would hold office for three years and assume control of the hospital in the Physician's absence. A house large enough for six wards was found near the University.[188]

This can be compared with the prospectus for the Glasgow Lying-in Hospital. It envisioned that the 'six or eight practitioners of medicine ... shall be annually elected ... from medical men who have been at least ten years in practice, without reference to their being members of any Medical School or Corporation'.[189] The election of men who were not 'members' neatly circumvented the FPSG monopoly on accrediting teaching, which at the time required lecturers to be fully paid-up members of the FPSG. Nonetheless, both the founding committee and the committee for fundraising was exclusively recruited from the inner echelons of the FPSG. James Corkindale, appointed Treasurer, was particularly active at this time, steering the fortunes of the FPSG in its legal and political feud with the University. He had applied for the chemistry lectureship at Anderson's University and was later among its trustees. In 1834 he was President of the FPSG. James Wilson and James Brown were also of the FPSG camp.

By the first week in October 1834 the promoters of the Glasgow Lying-In Hospital advertised in the local newspapers for suitable premises, and the completed draft of the hospital's constitution and regulations was read before a meeting in the Religious Institution Rooms in George Square.[190] Twenty-one Directors were to be elected, including the Lord Provost, the President of the Faculty of Physicians and Surgeons, and Glasgow's two MPs. The hospital treasurer, secretary and matron were to be appointed on an annual basis – the matron was to be of exemplary character and qualified both by education and practice as a midwife. She was to be in charge of normal births, organise the attendance of students and pupil midwives, and 'take care that everything be done according to the rules of improved midwifery'. Trainee midwives were admitted and asked to pay fees of half a guinea for three months' instruction. They also trained in the ordinary tasks of nurses employed in general hospitals. The matron was the housekeeper, and it could be argued that this part of her job was the most important, as the first sentence of the regulation relating to her appointment referred to her duties in the domestic management of the institution.[191]

[188] Dow, *The Rottenrow*, p. 31.
[189] GUL, Special Collections, Eph K/114, 'Glasgow Lying-In Hospital and Dispensary'.
[190] Dow, *The Rottenrow*, p. 31.
[191] Ibid., p. 145.

The medical staff was composed of eight practitioners, two of whom would act as ordinary accoucheurs or hospital superintendents, two as consulting accoucheurs to serve in the dispensary, and four as out-door accoucheurs to supervise the attendance of women in their own homes (this was normally carried out by the advanced students). These initial proposals were revised several times before it was finally agreed that medical attendants would serve for up to four years at a time.[192] The first two ordinary accoucheurs were James Wilson, at the time lecturer in midwifery at the Portland Street Medical School, and James Brown, who had just been appointed Lecturer in Midwifery at Anderson's University. James Wilson has been regarded as the driving force behind the hospital, but D. A. Dow has pointed out he was not in fact involved in the initial debate about its establishment, and it was perhaps his extensive years of service there that created this impression. He views James Brown as the main force who exerted himself on the hospital's behalf until his death.[193] When James Wilson and James Brown offered their resignation in June 1837, according to the rules laid down in the constitution, the hospital directors asked them to stay on, both men continuing as ordinary accoucheurs until their deaths, in 1857 and 1846 respectively.[194]

Two buildings were thought suitable locations for the hospital, one in George Street and the other in the College Open, but both were rejected. On 22 October 1834 John Alston, one of the hospital directors, suggested that the Old Grammar School Building in Greyfriars Wynd, a short distance from the university, might serve the Hospital's needs.[195] The directors inspected the building and decided to rent the second floor and garrets for the annual sum of £30. The hospital received its first patient on 10 December. It had eighteen beds.

Nine years later the annual general meeting had to be postponed for a week because so few of the subscribers were in attendance. On 16 October 1843 it was necessary to halve the number of outdoor accoucheurs to four.[196] The problem lay in the large number of patients that drastically eroded its financial base. In 1842, its first full year after the move to St Andrew's Square, the hospital served 204 patients. In 1843 the women treated numbered 338. These figures do not include attendance at home deliveries.

By December 1856 the hospital was in desperate straits, having had to close down twice in the last twelve months due to outbreaks of puerperal fever, and having no room for a separate delivery area, so that operations were

[192] Ibid., p. 32.
[193] Ibid., pp. 42–43, for specific details.
[194] Ibid., p. 33.
[195] Ibid., p. 35.
[196] Ibid., p. 37.

carried out in front of the other patients. This situation was aggravated by the actions of the police, who continually pressured the hospital into accepting more patients.[197] The institution could clearly not continue in this way for much longer. Finally, in December 1858, the hospital's annual report revealed that a new permanent home had been acquired in a house at the corner of Rottenrow and Portland Street. The refurbished hospital opened in January 1860, boasting a kitchen, laundry, servants' room, lavatories, baths and closets, and a number of rooms containing between one and six beds.[198] This was the new beginning that lent itself to further expansion.

The history of childbirth in Glasgow has yet to be written, but even tracing a few of its constituent parts shows its vital importance to the profession's educational aims and its accreditation standards. The difficult interaction with poor relief and the plight of women at their most vulnerable could bear with much more research. It is inextricably linked with the rise of the hospital, and yet so much of childbirthing took place outside it. The home and the midwife were still significant alternatives for women expecting a child.

---

[197] Ibid., p. 46.
[198] Ibid., pp. 48–49.

8

Corporate Medicine and the Hospitals

'For their pains, attendance, charge and expenses, in recovering seventeen of the poor people that were wounded and shot by the military on the 25 June last, when the military was insulted by the mobb', the surgeons John Gordon, James Hamilton, Alexander Horsburgh, Thomas Buchanan, Alexander Porterfield, Thomas Dougal, William Stirling, and Thomas Hamilton, and the physician George Thomson were paid by the town.[1] Each was active in the FPSG and their treatment of the wounded in the Shawfield riots of 1725 showed their close ties with town affairs.

Early in the new century a powerful group of surgeons, the Marshalls, Gordon, Stirling and others, resolved to give medicine a new look. They undermined the closed shop of the craft, jettisoned the barbers, opened the doors to competitive new skills and expanded the FPSG's stake in medical management. They built up Glasgow's medical profession in diverse ways, from the integration with university teaching discussed above to the creation of new posts in hospital wards, under the corporate patronage of the FPSG. At the same time – and this was a stroke of genius – the Faculty bolstered its role as the local licensing agency, preserving its monopoly in training surgeons. After 1673 Glasgow's physicians were successfully reintegrated into the corporation, taking an evermore active part, particularly in the office of President until 1820. In this way the FPSG succeeded in uniting the whole profession under its roof while expanding its influence. The barbers were cast off to ply a less prestigious trade on their own, while the surgeons, anxious to acquire the aura of the educated rather than the merely apprenticed, identified themselves with the modern 'scientific' education that they were importing to Glasgow. This gave the Faculty's policy an ever greater leverage in the profession.

The story of the FPSG in the era of improvement began with a rebellion over restricted access to medical practice. This threatened to choke the life out of a profession on the brink of modernisation. The Incorporation of Surgeons and Barbers in 1679 passed a ruling to limit membership to the

---

[1] R. Renwick, *Extracts from the Records of the Burgh of Glasgow, 1718–38*, v (Glasgow, 1899), p. 245.

apprenticed or the sons or sons-in-law of freemen. Restrictive trade practices were not unusual, but this time the tensions proved dangerous to the town's expanding commercial interests. The town council took action against the Incorporation, accusing it of unfairly using its monopoly to secure its own advantage.[2]

At a time of political uncertainty – 1679 was the year of the 'western risings' culminating in the battle of Bothwell Bridge – the local craft guilds may have expediently closed ranks. However, apprenticeship rulings in general were already changing under the economic pressures of increasing commercial trade. Glasgow experienced an economic upturn before the Union of 1707 that is not generally acknowledged.[3] Unregulated trade and commerce were bolstering economic activity to the extent 'that some towns acknowledged its growth by adjusting existing legal restrictions'.[4] Economic growth eased the strictures of the apprenticeship system. Merchant's apprentices began to use the schools of the great Dutch mercantile centres of Amsterdam, Rotterdam and Dort 'where the arts of ciphering, accounting and languages were taught' between the 1680s and the 1720s.[5] This led to the setting up of similar schools in Scotland and to a slackening of the conditions of indenture. The same loosening of the rules for membership of the FPSG became apparent in the Henry Marshall case. It initiated a tense debate about protective guild practices which continued in the courts until 1691 and before the town council until 1709. It showed the widening split between the upwardly mobile surgical profession and the barbers.

The town council had been quick to see an advantage in recruiting skilled surgeons. Despite the Incorporation's 1679 rule to restrict membership they appointed a trained surgeon, Henry Marshall, an 'outsider' who had petitioned the town, one assumes with their full knowledge, to be allowed to practise. Henry Marshall was well connected and self-assured, and probably provoked the conflict directly to challenge the restriction. He was one of the new breed of surgeons trained elsewhere, perhaps in military campaigns.[6] His father was Patrick Marshall of Kilsyth; his brother John, of whom we have heard in connection with the physic garden, trained in Paris in 1677.[7]

---

[2] Duncan, *Memorials*, p. 80.

[3] T. M. Devine, 'The Merchant Class of the Larger Scottish Towns in the Later Seventeenth and Early Eighteenth Centuries', in G. Gordon and B. Dicks (eds), *Scottish Urban History* (Aberdeen, 1983), pp. 95–97.

[4] Ibid., p. 97.

[5] Ibid., p. 96.

[6] No evidence seems to hand about exactly where he trained or what his credentials were, but they must have been impressive enough to gain the confidence of the magistrates.

[7] Duncan, *Memorials*, p. 248.

John Marshall had an MA from Glasgow (1706), although his entry dates for the FPSG have been lost.

The litigation against his brother Henry ended with a decision by the Court of Session in July 1691. The Faculty won on the issue of the superiority of its royal charter over municipal authority, and thus, in law, was able to protect its restrictive practices. The town council, significantly, had argued that the restrictions of Faculty rules should not be used to limit qualified practitioners who were needed 'for the common weal of the burgh'.[8] While the FPSG won the action, it cost them representation on the Trades' House where they were no longer entered in the list of deacons.[9] These disputes severely damaged the identity of the Incorporation secured in 1656, which had seen it function as a guild of barber-surgeons. This traditional 'marriage of convenience' unravelled quickly as the surgeons gained in status through the post 1690 cooperation with the University on creating medical lectureships. Glasgow's ascendancy in trade,[10] and manufacture added impetus to the erosion of apprenticeship indenture. The rift between the barbers and the professional ambitions of the surgeons became permanent, a development also taking place in such leading medical centres as Paris and Leiden. Exploited by the modernising faction among the surgeons, loopholes were used to create a professional body sensitive to new medical objectives.

Traditional restrictions were ruptured again with the 1712 application for admission as a surgeon by William Stirling. An 'outsider' like Henry Marshall, he did not serve an apprenticeship. Stirling was of the modernising caste, not least in commerce where he brought linen manufacturing to Glasgow from Holland. He may indeed have been educated in Holland,[11] the Stirlings having strong medical and religious connections there. The first Walter Stirling married Helen, daughter of David Wemyss, the first Presbyterian minister of Glasgow, and the widow of Peter Lowe. His son John was imprisoned for three months in Edinburgh Tolbooth for hearing 'outed' ministers. John's second son was William, the surgeon; his brother, another

---

[8] An excellent summary of the case in Duncan, *Memorials*, pp. 79–81. Duncan seems only to have misjudged the expansive outlook of Glasgow after 1690.

[9] Duncan, *Memorials*, p. 81; Duncan reviews these conflicts pp. 79–81.

[10] Devine, 'The Merchant Class of the Larger Scottish Towns'.

[11] Walter, William's son, was born in 1723 and matriculated at Glasgow University. A Walter Stirling accompanied William Sinclair to Leiden and Paris in 1736–38, but the dates are too soon for the son of the surgeon. See K. van Strien, W. de Withlaan, 'A Medical Student at Leiden and Paris', part 1, *Proceedings of the Royal College of Physicians of Edinburgh*, 25 (1995), pp. 294–304. Walter is described as William Sinclair's roommate, and he and his friends went to Haarlem where 'they went to see the famous bleaching fields in the surroundings', p. 302. This Walter Stirling received his MD from Rheims on 28 November 1737, p. 644.

Walter, was a magistrate, and the third, John, was a bailie at the time of the Shawfield Riots. John and Walter 'were among the merchants named by McUre as undertaking the trade to Virginia, Caribby Islands, Barbados, New England, St Christopher's, Montserat and other colonies in America'.[12] The Stirlings were a leading commercial family who held sway among Glasgow's elite well into the nineteenth century.

The surgeons plainly wanted to admit William Stirling, and did so, asking as an entrance fee the aggregate of fines and expenses which would have accrued had he been apprenticed.[13] The already teetering relationship with the barbers went into another spasm of wrangling, this time on the 1000 merks fee, the barbers wanting their share. This further weakened the cause of restrictions on membership, with the surgeons acting increasingly like merchants, eager to loosen the rules and embrace the new. Indeed the kinship ties of these 'new' surgeons were both mercantile and familial. As one historian has remarked:

> This was an era of developing but still unreliable communications, high risks and unsophisticated commercial law. The business world was thus a tight nexus in which a merchant's reputation and that of his family was his most precious asset: to deal with kin and trusted acquaintances was not simply understandable but justifiable. Nepotism had a basic commercial rationale.[14]

Stirling joined forces with John Gordon, who broke the system of apprenticeship skills training by becoming a university lecturer on anatomy, and with Henry Marshall. These men became the 'Young Turks', holding sway in the governing body of the FPSG.

Henry Marshall, quite ironically, was called upon, when Visitor, in 1704, to answer the petition of the barbers to the town council against the surgeons because the latter had 'committed many unwarrantable encroachments upon the interest of the barbers contrary to the letter of deaconry'.[15] He answered that the letter of deaconry required revision as some of its provisions were 'distasteful' to the surgeons.[16] He signalled the barbers' incompatibility with the charter granted by James VI. Henry, as Visitor in 1704, was also the office-holder who must have been consulted by the university on the person most suitable to supervise the physic garden. His brother John was appointed Keeper. Indeed it was John who was 'specially recommended thairto by the

---

[12] G. Eyre-Todd, *History of Glasgow*, iii, *From the Revolution to the Passing of the Reform Acts, 1832–33* (Glasgow, 1934), pp. 299–300.
[13] Duncan, *Memorials*, p. 86.
[14] Devine, 'The Merchant Class of the Larger Scottish Towns', p. 100.
[15] Duncan, *Memorials*, p. 84.
[16] Ibid.

dean of facultie's letter: therefor the [university] faculty does nominate the said John Marshall for the said imployment'.[17] This overt nepotism shows how the elite closed ranks over the modernising of medical education, beginning with the 'physick and botanic garden'.[18]

The campaign to modernise the FPSG was further strengthened by the Marshalls in a campaign to obtain library books. Henry solicited books from his titled friends, and he and his brother John's own gifts added to the riches of the library newly housed in the Surgeon's Hall in Trongate after 1697. The elder Lady Barrowfield, for instance, donated books 'by the influence of Mr H. Marshall, bibliothecarius', as did Hugh Blair, 'Minister of the Gospell', and the earl of Wigton.[19]

William Stirling and John Gordon formed a close and powerful partnership. They formed a surgeon's 'business', adding Robert Wallace to their number. They also acted in partnership in the councils of the FPSG. One of their first actions was to widen Faculty membership. In 1718 they appealed to the town council against the Visitor to admit as an honorary member Andrew Vinniell, 'surgeon to generall Wrightman's regiment of foot', requesting that 'the faculty should meet and give him his diploma, quhich the present visitor refuses to execut, and therefore craving the councill to interpose their authority and ordain the visitor to expede and make out said diploma'.[20] The Visitor in 1716–18 was James Calder.[21] John Gordon succeeded him in October 1718, when his name first appears as present at a town council meeting. Under his presidency, in 1719, the letter of deaconry was finally rejected by the surgeons, although the legal battle dragged on until 1722.

As these developments indicate, a small group of close associates made up the medical profession's active leadership. The men who signed for separation from the barbers included Gordon, Stirling, Porterfield, Thomas Hamilton and Buchanan. Fourteen surgeons in all signed the document putting an end to the letter of deaconry 'to be in all tyme coming null and void', a process begun in 1719 and finalised in 1722.[22] In 1725 the same group of surgeons treated the wounded at the Shawfield riots. In 1733 the same names headed the list of medical men – seven physicians and eleven surgeons – who convened to steer the FPSG into the future.

---

[17] GUABRC, 26631, Faculty Meeting Minutes, 1702–20, p. 25 (Glasgow, 7 September 1704).
[18] GUABRC, 26631, Faculty Meeting Minutes, 1702–20, p. 30.
[19] Duncan, *Memorials*, p. 216.
[20] Renwick, *Records of the Burgh of Glasgow, AD 1718–1738*, p. 22.
[21] Duncan, *Memorials*, p. 247, under no. 162a.
[22] Ibid., p. 88.

The minute book from 1688 to 1733 having been accidentally burned while in the keeping of John Colquhoun, clerk to the Faculty, the active members had to define its regulations once more. 'Unfortunately', writes Alexander Duncan, 'no artist of the time had the opportunity of transferring to canvass the faces of these ancient fathers of medicine and surgery in Glasgow, in solemn conclave assembled, with rueful faces contemplating the charred remains of the book.'[23]

The faces of the 'ancient fathers' were in the early eighteenth century vigorous and youthful. The most active among them were always at the cutting edge of decision-making when medical concerns touched on education and caring for the sick poor. John Gordon was involved with the new anatomy courses, John Marshall with botany, while provision for the poor brought Gordon and Stirling into this core group. John Gordon entered the then Incorporation of Surgeons and Barbers in November 1713, passing his 'trial in anatomy' by dissecting the stomach and intestines of an animal.[24] He taught anatomy at the university a year later. John Marshall was Keeper of the Physic Garden, dying early, in 1719, while his brother, Henry, as we have seen, played centre stage in the drama which ended the partnership with the barbers. Henry Marshall had petitioned the town council to grant him 'such licence to set up as if he war reallie admitted with the said calling'.[25] The town did so, provoking legal action by the Faculty in 1691, which it won eventually in 1709.[26] In 1703, 1704 and 1705 Marshall was the Faculty Visitor,[27] and, characteristically, was not much inclined either to bow to the barbers or to keep the FPSG a closed shop. He must have enjoyed his role as a subversive moderniser from within.

William Stirling entered on 5 February 1712, and was examined under Alexander Porterfield, John Bogle and James Calder.[28] The Stirling family was eminent both commercially and in civic office. The early weaving manufacture was expanded by the nephew of the surgeon, also called William, who was the founder of the Stirling & Sons, which was engaged in producing turkey-red dye and in the bleaching industry.[29] This William was the son of John Stirling, brother to the surgeon, 'a bailie at the time of the Shawfield riot, and, though out of town at the time, [he] was arrested with Provost

---

[23] Ibid., p. 93.
[24] GCA, T-TH 14 5.2, Minutes of the Incorporation of Barbers, 1707–65, fol. 22v.
[25] RCPSG, 1/1/1/1b, pp. 273–74.
[26] J.D.Marwick (ed.) *Records of the Burgh of Glasgow, AD 1691–1717*, Scottish Burgh Records Society (Glasgow, 1908), pp. 16–19.
[27] Duncan, *Memorials*, p. 84.
[28] GCA, T-TH 14 5.2, fol. 13v.
[29] J. Irving, *The Book of Scotsmen* (Paisley, 1881), p. 500.

Miller, carried to Edinburgh by the dragoons and put on trial, and on the return home he shared in the demonstrations by the citizens, the jubilant shouts and ringing of bells'.[30]

Robert Wallace entered in December 1713, for his trial dissecting 'the liver of ane animall; he also held an discourse upon the parts containing and contained in the Louer parts of the Belly'.[31] Gordon and Stirling, and probably Wallace, described by a contemporary as being 'at the top of his profession',[32] formed one of the city's strongest surgical practices. Their interest in medical care and in health provision was intensified by their commercial stake in the city. Like Victorian civic philanthropy, eighteenth-century improvement and benevolence made common cause with successful capitalism. Gordon and Stirling's linen manufactory was known 'for weaving all sorts of holland-cloth, wonderful fine, performed by fine masters, expert in the curious art of weaving, as fine and well done as at Harlem in Holland. The masters of this improven manufactory are now united to such perfection, that noblemen, barons, gentlemen and citizens, and their ladies buys of them, and wears their linen'.[33]

William Stirling lived from 1682 to 1757, and his son Walter, born in 1723, became a magistrate of the city. Walter Stirling was a hunchback who dressed very plainly and his habits were characterised by 'great method and love of order'.[34] He was an only son and when he inherited wealth from his father he maintained the family's benevolent concern for the poor and for improvement in education. He gifted his house, one thousand pounds and certain shares in the Tontine Society to found Stirling's Library for the free use of everyone in the community.[35] On the board of directors of this educational charity three members of the Faculty, elected annually, were to sit, the first taking office in February 1791.[36] At this time the Faculty Hall stood on the east side of St Enoch's Square 'in its lower floor Stirling's Public Library was originally kept, and on either side of this were self-contained houses occupied by well-to-do citizens'.[37] Thus the threads of civic improvement criss-cross family firms, surgical business partnerships and the politics of the FPSG.

---

[30] Eyre-Todd, *History of Glasgow* (Glasgow, 1934), vol. iii, p. 300.
[31] GCA, T-TH 14 5.2, fol. 23v.
[32] Duncan, *Memorials*, p. 251.
[33] J. McUre, *History of Glasgow* (2nd edn, Glasgow, 1831), p. 257.
[34] W. Weir, typescript of FPSG History, quoting McUre, *History of Glasgow*, p. 227.
[35] Ibid., p. 226.
[36] Ibid.
[37] A. Aird, *Glimpses of Old Glasgow* (Glasgow, 1894), p. 83.

The Shawfield Riots

The Shawfield riots illuminate the Faculty's close relationship with the town. For Presbyterian Glasgow – although not for the Highlands – a Jacobite rebellion 'was a thing of the past' by 1725,[38] but the jittery nature of political control in Scotland did not accept any such surety. The Hanoverians' political supporters, under Sir Robert Walpole, cast a wary eye on anything developing into popular unrest or armed defiance. Of just such a threatening shape was the opposition to the malt tax, whose implementation in June 1725 was backed by Daniel Campbell of Shawfield, the MP for Glasgow. He blatantly exploited misconceptions that 'the present uneasiness of the People' would be inflamed to actions 'against the English' by the enemies of the government.[39] Campbell has been described as 'a parody of a hard-faced, grasping Hanoverian Whig merchant oligarch', but one who was also an enterprising capitalist at the centre of trade with North America and ventures in shipping to Spain.[40] His mansion, Shawfield, was an eye-catching demonstration of wealth in the town, and it was his furniture that was smashed as feelings ran high against the taxes for which he had voted.

The riot began as a show of popular displeasure, with the shouting of abuse and stones heaved at likely targets. It became political because the two companies of Lord Deloraine's regiment of foot under Captain Bushell were ultimately responsible to General George Wade, who was obsessed with Highland unrest, his suspicions not halting before Lowlands town magistrates.[41] The Provost of Glasgow, strictly in accordance with the law, had bidden Bushnell to hold fire. But the Captain, on the morning after the sacking of Shawfield House, when the 'mob' had reassembled and threatened his troops, fired indiscriminately into the crowd. The town records describe his actions:

> It has not hitherto appeared any of the troops were disabled or hurt by this insult, nor does it appear that the officer or souldiers under his command were at this time in any great danger. However, the commanding officer ordered them to fire, by which first fire two men were killed who had no way been concerned in the ryot and were then at a good distance from the guard house, and thereafter continued to fire by platoons towards the four principall streets tho in some of them there were no mob nor not so much as a single stone throun.[42]

---

[38] B. Lenman, *The Jacobite Risings in Britain, 1689–1746* (London, 1980), p. 205.
[39] Ibid., p. 207.
[40] Ibid.
[41] For the full story of these clashes and their political background see Lenman, *The Jacobite Risings*, pp. 205–15.
[42] Renwick, *Records of the Burgh of Glasgow, 1718–38* (Glasgow, 1899), p. 228.

The Provost again asked that the gunfire should cease, but pressed again 'by the inraged mob', who were incensed at the killings, the troops, in retreating, 'continued firing upon the streets, whereby in all there were nine persons killed, particularly one gentlewoman out of a window tuo stories high, some in the sides and others crossing the streets going about their lawful affairs, and sixteen dangerously wounded, quherby not above five or six at most, so far as can be knoun, were in any way concerned in the mobb'.[43] It was these wounded who were treated by the surgeons.

The aftermath of the riot produced an unprecedented incarceration of the Glasgow magistrates in Edinburgh instigated, in fact unlawfully, by the 'openly hostile and angry' Lord Advocate Duncan Forbes.[44] His action was nullified by the intervention of a legal action in their defence and they were never brought to trial. The debate on what had happened nevertheless remained a running sore because the question of the probity of Glasgow's magistrates had to be set against the Hanoverian identification of traitors in Scottish cities. One of the men who defended the city was the surgeon John Gordon. He and 'Andreu Martine and a third not named' formed a 'Secret Committy' and published a pamphlet, of which a 1000 copies were printed in 1725, because they

> found it necessary to draw up an account of the mobb at Glasgou, and the treatment of the Toun and Magistrates after it, in a better and shorter dress for the information of England, than the quarto paper published by W. Tennoch.[45]

Gordon's pamphlet was dedicated to Members of Parliament, and sent to every one of them, pillorying the Lord Advocate, Duncan Forbes. It was sent to London 'to Abram Henderson, Mr Brown or some other Scotsman there' to be distributed, 200 copies being seized.[46] Gordon was threatened with a trial, but again nothing was done. He had clearly not only acted to defend his city, but had overtly challenged the authority of the Lord Advocate. He appealed, as the dedication of his tract makes clear, to the powers and fair judgement of elected Members of Parliament, so lately charged with Scottish affairs, and not that of the reigning oligarch, showing an optimistic belief that representative government counted.

---

[43] Ibid., p. 229.

[44] Lenman, *The Jacobite Risings*, p. 212.

[45] R. Wodrow, *Analecta* (Edinburgh, 1842–43) iii, p. 248. The pamphlet may have been similar to 'A Letter from a Gentleman in Glasgow, to his Friend in the Country, Concerning the Late Tumults which Happened in that City, Printed in the Year 1725', SRO, RH2/4 (319) fol. 18, as cited in Lenman, *The Jacobite Risings*, chapter 9, p. 308n. We have not been able to trace Gordon's pamphlet.

[46] Wodrow, *Analecta*, iii, pp. 248–49.

John Gordon was not a newcomer to civic affairs. In 1718 he was put forward for the office of treasurer of the burgh, although John Auchincloss was elected.[47] In October 1718, in a meeting of councillors and the deacons of the crafts, John Gordon was listed as 'deacon of the surgeons'.[48] He was surely instrumental in the parting of the ways between barbers and surgeons, as he signed the document dividing their common assets on the side of the surgeons in 1722. He was at the centre of medical politics and deeply convinced of the rights of town government. Both he and his partner William Stirling staked their careers on medical improvement.

Medical Philanthropy and the Town's Hospital

This first took the form of provision for the poor, whom Gordon, Stirling and Robert Wallace offered to attend gratis if the town council would pay for medicine 'at ane easie rate'.[49] They were reimbursed by the town for these services in 1730;[50] and in 1736 Gordon and Stirling, 'surgeons in company', were appointed 'to be touns surgeon to Michaelmas next'. Their duties were 'to take care of the poor and apply medicines and drugs to them, and their salary for their drugs and medicaments and cures and services to be £10 sterling, and thereafter the touns surgeon to be annually chosen and elected'.[51] In 1738 the next 'touns surgeon' elected was James Hamilton,[52] who had been elected Visitor of the FPSG in 1734. Gordon and Stirling presumably held that office until Hamilton took it up.

From the late seventeenth century, the magistrates made payment for private cures of the poor injured in the town. For example, £30 Scots was paid to John Boyd, 'chirurgen', in 1700 for 'cureing ane fracture in the theigh of a poor boy' and for 'cureing of ane fracture in the arme of John Purden who had his arm broken at the last accidentall fire', as well as 'for cureing Thomas Rae of some bruises and spitting of blood gott be him at the said fire'.[53] Such specific payments continued, but paled besides the movements to institutionalise medical charity. The custom of paying different surgeons for medical treatments gave way to reliance on the town's surgeon, a development that led to what has been described as 'the first hospital in Scotland for the poor'.[54]

[47] Renwick, *Records of the Burgh of Glasgow, 1718–38*, pp. 37–38.
[48] Ibid., p. 36.
[49] Ibid., pp. 92–93.
[50] Ibid., p. 345.
[51] Ibid., p. 468.
[52] Ibid., p. 498.
[53] Ibid., p. 308.
[54] Aird, *Glimpses of Old Glasgow*, p. 83.

This Town's Hospital facing the Clyde, near Ropework Lane, was later described as substantial:

> While it existed it smoothed the path of many afflicted ones. The old fisher-market before its removal to the Briggate occupied the spot where the hospital stood. Behind the hospital, separated by a broad area, was a two-storey building. In the first flat were placed cells for lunatics and disorderly persons. On the second was an infirmary for the sick. The buildings accommodated 600 persons.[55]

Workhouse provision for the poor was quickly expanded to include medical provision. Plans were underfoot in 1730 and by 1731 £1300 had been subscribed and an annual income fixed at £820 by a tax on the citizens and by contributions.[56] The building was opened on 15 November 1733 and by December 1739 an infirmary with fifty beds which could accommodate sixty or seventy persons was completed.[57] It also had

> A large Chamber for Chirurgical Operations well illuminated; where it is intended that the Physician who attends the Hospital shall give Lectures to the Apprentices of Chirurgeons or other students upon the diseases which may prevail in the Infirmary.[58]

Hospital care had been in operation for at least two years previous to the opening of the special facilities, as the reports note, 'and many perfectly cured'.

John Gordon was the only surgeon elected to the board of directors in 1732 for 'the workhouse now in design' to be built by the town council,[59] a year before the 1733 meeting of the FPSG which sanctioned medical attendance by its physician and surgeon members. In November of the following year he was re elected.[60] There were twelve directors named, among them merchants, a wright, a maltman, and present and former bailies. They were to coordinate all proposals for the new town's hospital, and were 'empowered to do everything else that is requisite for beginning and carrying on this work'.[61] Medical provision for the sick poor, now entered a new phase, the

---

[55] Ibid.
[56] I. Brown, 'A Short Account of the Town's Hospital in Glasgow', *The Bibliotheck*, 1, no. 3 (1958), p. 37; For a comprehensive account of the Town's Hospital, see the article by Fiona Macdonald, 'The Infirmary of the Glasgow Town's Hospital, 1733 to 1800: A Case for Voluntarism?', *Bulletin of the History of Medicine*, 73 (1999), pp. 64–105, which arose from the work done for this book.
[57] Ibid., p. 40.
[58] Ibid., as quoted on p. 41.
[59] Renwick, *Records of the Burgh of Glasgow, 1718–38*, v, p. 368.
[60] Ibid., p. 407.
[61] Ibid., p. 370.

decisive one of institutionalising a controlling role by the FPSG in health management.

John Gordon's place on the board of directors coincided with the Faculty resolution of November 1733 to institute what, besides care for the poor, initiated hospital clinical teaching in Scotland. The director's board met officially on 15 November and adopted the Faculty resolution of a few days before which said:

> the subscribing physicians and surgeons being all of them much inclined to encourage and promote the good design of maintaining the poor in a workhouse already built at Glasgow for that purpose, do hereby in full faculty voluntarily condescend and agree among themselves that each of the six physicians subscribing will, according to their seniority as physicians admitted attend and visit for the space of a year from the poors' being first put in said workhouse the poor people to be kept in the Infirmary there and give their advice and prescriptions as to the sick and infirm from time to time as needful; and each of the eleven surgeons subscribing, according to their seniority as members of the Faculty will for the space of ane year commencing from the time of the poor being first put in said workhouse, visit and as surgeon attend the said House and do all the necessary business of a Surgeon to the poor in the Infirmary there, and furnish to them upon his own charge all drugs and medicaments necessary or to be prescribed by the physicians.[62]

The six physicians who signed were Peter Patoun, George Thomson, John Brisbane, John Johnstoun, John Wodrow and David Patoun. In July 1736 George Montgomery also signed what was in effect a contract for medical services. He would help inaugurate the hospital's clinical lectures. In 1739 the minutes record as additional signatories Robert Bogle, Hector McLean and David Corbett. The surgeons were Alexander Porterfield, Thomas Hamilton, James Calder, Sr, William Stirling, John Gordon, Robert Wallace, Thomas Buchanan, Alexander Horsburgh, James Hamilton, John Paisley and James Calder, Jr.

Glasgow matched Edinburgh in the development of hospital care, although the reliance on local ways of coping with the sick poor make true comparisons difficult. In this period decisions were municipally based with no overall national strategies, although medical men travelled both west and east to learn their skills. The Edinburgh hospital was founded by the College of Physicians in 1725 and opened with six beds in 1729.[63] John Gordon was one of its subscribers.[64] Thirty-five patients were treated in the first year. As a

---

[62] The original resolution as quoted in Duncan, *Memorials*, p. 92.

[63] See J. Comrie, *The History of Scottish Medicine* (London, 1932), p. 449.

[64] *An Account of the Rise and Establishment of the Infirmary ... Erected at Edinburgh* (Edinburgh, c. 1730), p. 24.

royal infirmary it received its charter in 1736 and opened on a grander scale in 1741. It was to house 288 sick with a separate bed for each patient and five cells for mad people.

Lectures at the bedside were given regularly by the physicians of the Glasgow Town's Hospital perhaps from 1739, but certainly from 1741 onwards, several years before John Rutherford began his lectures at the Royal Infirmary in Edinburgh in 1748. The decision taken at the meeting of the Faculty in November 1733 inaugurated regular medical attendance by the physicians and surgeons of Glasgow at an institution which combined care for the poor with treatment for the sick.

The traditional obligation of the FPSG, codified in its charter, changed its character at this juncture. The corporation's ties with city government forged to institutionalise the care of the sick poor were never to loosen. Poor relief triggered the creation of the teaching hospital. As teaching and improved medical observation were coveted objectives in medical eyes, under the umbrella of true charitable intent, medical men could now realise their ambitions in circumstances ideal for these purposes. The sick poor in charitable hospitals displayed a range of diseases and surgical cases confined to wards and therefore perfectly suited to instruction. Instead of the client relationship of the practitioner to the rich patient, these men and women were beholden to doctors.

The Town's Hospital was not just another civic foundation in the tradition of providing for the sick poor. It nurtured two vital and prototypical developments. Attendance in hospital wards was linked to the patronage of the FPSG and its corporate medical policies. The rota system established by the Faculty in 1733 provided all members of the FPSG with opportunities to teach and hone medical skills. A system of parities, it was enshrined in the medical provision of the Royal Infirmary when this larger charitable hospital was founded. Nearly a century later when Glasgow University wanted permanent clinical teaching in the GRI wards, this threatened the corporate hold of the Faculty, engendering a deeply divisive struggle. It took the heavy pressures of a national standardisation of clinical qualifications to induce the Managers of the GRI to take independent action. Even this solution favoured the Faculty, as we shall see. Finally, however, clinical teaching helped secure the independence of hospitals, transcending both university and FPSG agendas and leaving the hospital a potent third power base for the profession.

The management structures set up in 1733 reflected the wishes of the incorporation. This did not visibly alter with the founding of the GRI because the medical committees, and indeed the board of directors, maintained the same close relationship forged in the Town's Hospital. This medical network and the commercially and industrially based philanthropy upon which voluntary hospitals depended, served to define clinical teaching. Bedside observation gave medical men the opportunity to advance their science, an obvious priority in a century obsessed with improvement and classification. It also provided the means to excel in the educational stakes at a time when private and institutionally based teaching were competing in a free market.

Clinical teaching in hospital was different from private bedside instruction and teaching in 'private' courses, whether medical or surgical. It gave easy access to grateful and dependent (poor) patients. It showed the characteristics of disease and injury, helping to identify their diagnostic specificity and led to reliable statistics on morbidity and mortality.

Clinical Teaching

The beginnings of clinical teaching in hospital wards are to be found in the Town's Hospital, not in the Glasgow or Edinburgh Royal Infirmaries, even though these much larger institutions had the greater impact. The Glasgow Royal Infirmary was essentially only a grander extension of the ideas developed for the Town's Hospital. In both cases the FPSG determined the staff, and rotation of posts and controlled a significant part of the boards of directors, even though the university and the town were represented. Far from being separate entities, the two hospitals were ideologically of a piece.

In 1741 medical attendance was combined with instruction for apprentices and medical students. George Montgomery advertised in the *Glasgow Journal* a course on clinical bedside teaching fashioned after the 'Boerhaavian method', a label emphasising its clinical nature.[65] Montgomery, 'physician to the infirmary', was a member of the FPSG, and the Faculty clearly encouraged teaching. Already in 1739 and again in 1741 the report of the hospital stated that physicians should 'give lectures to the apprentices of the chirurgiens, or other students, upon the diseases which prevailed among the patients'.[66]

The often cited 'beginnings' of clinical lectures by John Rutherford at the new Edinburgh Royal Infirmary of 1748 actually followed these developments in Glasgow. Almost ten years after clinical teaching was inaugurated in the Town's Hospital, Rutherford applied to hold clinical lectures, like Montgomery, in 'the operation Room in the Infirmary'.[67] As he put it, 'The

[65] GUL, Special Collections, *Glasgow Journal*, 14, 26 October to 2 November 1741, p. 4.
[66] J. M. Cowan, *Some Yesterdays: With a Note upon the Development of Hospitals by Joshua Ferguson* (Glasgow, 1949), p. 35, for the 1741 report. For the 1739 wording, see above.
[67] EUL, Special Collections, LHB 1/1/2, Edinburgh Royal Infirmary Minute Book, 1742–49, 2, series 1, p. 169. The following quote is from: RCPSG, 20/1/2, MS of John Rutherford, Clinical Lectures (n.d.).

method I propose to pursue is to examine every Patient before you, least [lest] any Circumstance should be overlooked'. Then Rutherford gave his plan for the lectures 'the most usefull I know of', reviewing in succession the general history of disease, the causes of disease, his own opinion of the prognosis, and the treatment ('the indications of cure').

Montgomery continued his clinical lectures in 1742 and was still teaching in 1743–44.[68] He taught in the infirmary's operating room. His teaching was interrupted by the 1745 rebellion, but was resumed in 1746, this time at the University, where he taught the Theory and Practice of Medicine with William Cullen.[69] The *Glasgow Journal* carried the following announcement:

> *That* last year on Account of the Rebellion, the *colleges* on *anatomy*, the *theory* and *practice of medicine*, were interrupted. These *colleges* will open again, on the first Monday of November, at the University; where Gentlemen may depend on the utmost Care, and punctual Attendance, by *Robert Hamilton*, Professor of Anatomy, *William Cullen* and *George Montgomerie*, MDD.[70]

Cullen was chosen as 'physician to the infirmary' in 1747, together with Andrew Morris, and served the institution until 1755 when Dr Robert Dick offered his services.[71] Clinical teaching to students and apprentices can thus be said to have been continuous from 1741 onwards. These beginnings are auspicious in a number of ways. Clinical teaching united the university medical faculty, hospital care and the professional interests of the FPSG until the rivalries of the nineteenth century unfolded.

In 1740 the anatomy chair was being established, with Robert Hamilton occupying it from 1742 to 1756. Montgomery never held a university post, but probably taught university medical students, surgeons and apprentices together with Hamilton and Cullen in 1746, the year before Cullen's election as infirmary physician.

Montgomery was a Glasgow MD of 1732 examined under Johnstoune and Brisbane, William Cullen's MD following in 1740 and Robert Hamilton's in 1742.[72] Montgomery entered the FPSG in 1735, Hamilton in 1743 and Cullen in 1744, as soon as he came to Glasgow. They combined to form a powerful coalition across all the important medical institutions. They succeeded one

---

[68] GUL, Special Collections, *Glasgow Journal*, 20 September to 27 September 1742, p. 4; *Glasgow Journal*, 24 October to 31 October 1743, p. 4.

[69] Mitchell Library, Glasgow Room, Stack 5ASH. 75, *Glasgow Journal*, 268, 8 September to 15 September 1746, p. 4.

[70] *Glasgow Journal*, 8 September to 15 September 1746, p. 4.

[71] MLRBM 641982, Glasgow Town's Hospital, Minutes of Directors' Meetings, 1732–64, p. 174.

[72] Duncan, *Memorials*, pp. 253–54.

another as Presidents of the Faculty for six years (Montgomery in 1743–45; Hamilton in 1745–47; Cullen in 1747–49). Moreover, Robert Hamilton succeeded Cullen in the chair of medicine when the latter moved to Edinburgh in 1756, and when Hamilton died in the following year his brother Thomas, the partner of Cullen, succeeded him in the professorship.

In the 1730s Cullen's practice in Hamilton attracted the medical men who shaped Glasgow medicine. When Cullen was veering between surgery and medicine, he approached John Gordon with thoughts of an apprenticeship, but instead devoted himself to meeting the challenges of 'internal' medicine. Cullen's ties with Glasgow's medical men underpinned their clinical ambitions. Thomas Hamilton and his son William Hamilton carried forward clinical teaching, as did William's close friend Robert Cleghorn. William Hamilton and Robert Cleghorn gave clinical instruction in the Town's Hospital and the Infirmary in the 1780s, demonstrating the continuity of teaching as the transition was made from one to the other. Thus the Glaswegian model of tripartite interests – corporate investment in good surgical practice, the raising of the Glasgow University profile in educating medical men, and the skilful use of the hospitals for these aims – became established.

Cullen was one of the key players. He became President of the FPSG just as his obvious teaching strengths came to the fore. He was prominent in the core group teaching in the Town's Hospital and fashioned the link between clinical teaching and theoretical knowledge through his lectures. Above all, he invested more effort than anyone in Scotland before him in acquiring a sound and thorough knowledge of medical theory and medical case notes, including the clinical observations now so important in the medical advances on the Continent. His lectures and books show him thoroughly acquainted with contemporary European medicine.

An Edinburgh student with west of Scotland connections, the son of Captain Coventry of Fairhill Park, close to Cullen's old practice in Hamilton, described him in 1785 when he had become the doyen of medical professors in Scotland:

> I was struck by the venerable, fine intelligent countenance of Dr Cullen. He had a noble, gentlemanly appearance, a fine, bold Roman nose, a good forehead. His age and his large wig descending in ample curls, and his rather tall person, with fine figure, made him appear venerable ... Cullen had the liberality of a prince, and never was so happy as when entertaining his friends or befriending young men. In this manner he spent his large income, while Monro hoarded his to spoil his son, thus removing the necessity of exertion, and ruining the once celebrated medical school of Edinburgh. You scarcely now hear of Monro, who

certainly was a first-rate anatomist, but Cullen's fame, like that of Hippocrates, will never die.[73]

Cullen showed the same spirit of conviviality in the art of making friends and supporting students while he was in Glasgow. This goes some way towards explaining the fortuitous introduction of modern medical thinking in Scotland. Cullen used English when he first taught medicine in 1746 at the university, breaking with the Latin tradition.[74] Because he used case studies in teaching, the vernacular was appropriate and made his observations accessible especially to apprentices, who were not matriculated and had no need to know classical languages, as well as any others who paid to attend and earned certificates rather than degrees.

Cullen's teaching combined clinical instruction with the systematised structure of a lecture. This was exactly the model adopted in the progressive continental medical schools. Cullen was to join the ranks of Herman Boerhaave, Gerhard van Swieten, Friedrich Hoffmann and Georg Ernst Stahl in the estimation of contemporaries. They all pioneered the linking of observational case histories with scientific systematisation. This combination of raw observation and its interpretation, in the theory and practice of medicine, became the paradigm in which teaching operated. As Robert Whyte remarked when he was teaching with Cullen in the 1760s in the Edinburgh infirmary, bedside teaching had a strictly defined role of hands-on practical instruction:

> The chief intention of the Clinicall Lectures is to illustrate the cause and Symptoms of all the parts treated here [in the infirmary] and not to delve into the Subtile Reasoning which I suppose to be discussed elsewhere.[75]

This almost classic definition was echoed in later decades. J. A. Easton, in 1849, in an introductory lecture to this course of clinical medicine, published at the request of the students, lauded the role of the hospital 'as a school of Practical Medicine'.[76] 'While for the knowledge of the doctrines and principles of Science you resort to the Academic Hall', he wrote, 'it is to the wards of the Hospital that you repair to learn how to observe carefully, to

---

[73] L. M. Liggett (ed.), 'Extracts from the Journal of a Scotch Medical Student of the Eighteenth Century', *Medical Library and Historical Journal*, 2 (1904), p. 106.

[74] Coutts, *A History of the University of Glasgow*, p. 488.

[75] RCPSG, 20/1/3, MS of R. Whytt Clinical Lectures of 1760. On the flyleaf are notes by Dr James Finlayson of May, 1881, to the effect that 'Whytt was known to have been associated with Monro and Cullen in the work of Clinical Lecturing', and on p. 255 of the MS there is proof of 'the lectures being delivered in an Infirmary'.

[76] J. A. Easton, *Lecture Introductory to a Course of Clinical Medicine Delivered in the Glasgow Royal Infirmary* (Glasgow, 1849), p. 6.

discriminate accurately, deduce logically, and prescribe skillfully.'[77] The rest of Easton's lecture is a lesson on taking a case history while reviewing the knowledge learned in abstract form, essential to placing the case into context.

Cullen was well versed in the European literature of observational case histories, citing, for example, Georg Ernst Stahl as a main contributor. Teaching from cases also had an important place in private practice, so that bedside teaching was a variant of an accepted form of observational knowledge. In 1772, this time in clinical teaching in Edinburgh, Cullen was still driving home the central role of observational case histories to his students. 'We must be conscious of the difficulty of observation', he insisted,[78] warning them that 'whenever you see a Case narrated with few circumstances you are to be suspicious of it'.[79] Scientifically 'a principal fallacy arises from the general credulity of mankind'.[80] Cullen immediately addressed the meaning for clinical pathology, whose office it was 'to explain the diseases of the body and to explain the operation of the remedies and thereby to lead to a more certain means of curing Diseases'.[81]

Thomas Hamilton did not have an MD, having entered the medical world through apprenticeship. He was apprenticed to John Crawford, one of the contenders for the anatomical chair in 1740, and then became a partner of Cullen's in the Duke of Hamilton's country seat of Hamilton in Lanarkshire. Thomas received his licence from the FPSG on completing his 'trial' in September 1743 and entered the Faculty in 1751,[82] the year in which he moved to Glasgow to join the 'business' of John Gordon. Gordon had, in the meantime, received his MD (1750) and no longer practised surgery, having asked John Moore and Thomas Hamilton to take his place. The Stirling-Gordon-Moore-Hamilton practice, spanning two generations, was probably the most powerful one in Glasgow at the time, connecting most of the important medical men of the city. Both Thomas Hamilton and his brother Robert had city practices, as did Cullen and Black, and, of course, all had close ties with the Hunters before and after they moved to London.[83] Thomas

---

[77] Ibid., p. 9.

[78] Clinical Lectures by Dr Cullen, 1772/73, Royal Society of Medicine, MS 96, Thomas Dale, M. D., vii, p. 560.

[79] Ibid., p. 558.

[80] Ibid., p. 559.

[81] Lectures on the Institutions of Medicine by W. Cullen MD, Royal Society of Medicine, MSS Cullen 101–104, 1772–73, vi, Pathology, p. 2.

[82] RCSPG, 1/1/1/2, fol. 75r; Duncan, *Memorials*, p. 254.

[83] This information must be pieced together but is apparent, respectively, under R. and T. Hamilton, G. Montgomery, W. Cullen and John Moore, in the Roll of Members in Duncan's *Memorials*.

Hamilton had the distinction of being made a professor with only the licence of the FPSG as a qualification. He became the professor of botany and anatomy in 1757, and held the post until his death in 1781.

In 1780/81 his son William Hamilton took up his father's post. He was appointed on no less a recommendation than William Hunter's. The Hamiltons were connected with William and John Hunter 'by early friendship and a constant intercourse of good offices'.[84] William Hunter was apparently happy with William Hamilton's professional zeal, and 'particularly pleased with them in the son of his old friend, that, after the first season, he invited Mr Hamilton to live in his house, and committed the dissecting room to his care'.[85]

This acknowledgement of William Hamilton's qualities was due as much to Hunter's medical principles as to family ties. William Hunter attests to this in writing to Thomas Hamilton in 1780:

> Your son has been doing everything you could wish, and from his own behaviour, has profited more for the time than any young man I ever knew. From being a favourite with every body, he has commanded every opportunity for improvement that this great town afforded during his stay here; for every body has been eager to oblige and encourage him. I can depend so much on him, in every way, that if any opportunity should offer for serving him, whatever may be in my power I shall consider as doing a real pleasure to myself.[86]

This opportunity to be of service soon presented itself. William Hunter was instrumental in helping William Hamilton's appointment to his father's chair in 1781. He wrote to the Duke of Montrose: 'That from an intimate knowledge of Mr Hamilton, as a man, and as an anatomist, he thought him every thing that could be wished for in a successor to his father, and that it was in the interest of Glasgow *to give him*, rather than his to solicit the appointment'.[87]

---

[84] R. Cleghorn, 'A Biographical Account of Mr William Hamilton, Late Professor of Anatomy and Botany in the University of Glasgow', *Transactions of the Royal Society of Edinburgh*, 4 (1798), pp. 35–63.
[85] Ibid., pp. 36–37.
[86] Ibid., p. 37 (obviously quoted from letters that Cleghorn had to hand in 1792).
[87] Ibid., p. 38. Emphasis in the original.

Physicians were using a post-mortem autopsy to understand not anatomy but clinical medicine.[88] The Royal Medical Society in Edinburgh had its origins in the mid 1730s when six keen students assembled to learn from 'an offer ... for pecuniary gratification, of the body of a young woman, a stranger, just then dead by a fever of ten days standing'.[89] These students were not surgeons, indeed they all afterwards had distinguished careers as physicians. But they made common cause with surgeons to borrow the anatomical amphitheatre of Alexander Monro primus for the purposes of the dissection. This took over a month to complete, and resulted in the foundation of the society, whose purpose was the scientific discussion of its medical findings. Fifty years on, William Hamilton and Robert Cleghorn shared this search for diagnostic certainty in their clinical case notes. The culture of clinical science spanned private practice and public institutions. Robert Cleghorn's record of his cases includes autopsy findings; indeed case notes formed the basis of his clinical teaching in the period before his work at the Royal Infirmary. Cleghorn gave his first clinical lectures in 1787 and 1788 'on patients under his care in the Town's Hospital'.[90] In the papers of William Hamilton the same concern with clinical diagnosis, corroborated through autopsies on his patients, was strongly in evidence. Although no name is entered for a consulting physician to the Town's Hospital from 1779 to 1786, these years corresponded to William Hamilton's active teaching career in Glasgow.[91] His recorded autopsies point to the use of institutional settings, and thus to the Town's Hospital.[92] Autopsies were not rare in hospitals either in London or Paris. They augmented dissection in connection with anatomical teaching, honing in on a specific purpose: better diagnostic skills.

[88] This was particularly true of William Hamilton and Robert Cleghorn, as their case studies attest, in MS Hamilton 84, and RCPSG, 1/10/2, Adversaria Cleghorn.

[89] J. Gray, *History of the Royal Medical Society, 1737–1937* (Edinburgh, 1952), p. 15. This quote is taken from a letter by Dr William Cuming to Dr John Coakley Lettsom, of 14 October 1782, when the latter was compiling a Life of John Fothergill.

[90] Cowan, *Some Yesterdays* (Glasgow, 1949), p. 35.

[91] The physicians attending the Town's Hospital are listed in Macdonald, 'The Infirmary of the Glasgow Town's Hospital', appendix, pp. 104–5.

[92] GUL, Special Collections, MS Hamilton 84, e.g. a dissection on 30 December 1783, pp. 81–83.

A vital link between the pursuit of medical knowledge in public institutions was being established. Medical experience often gained in private practice was transferred to hospital wards and supported by public charitable funds. Scientific inquiry was accessible on the wards and in public teaching, while the ideology of civic improvement was served.

Robert Cleghorn was deeply committed to the scientific mission made possible by case observation:

> This case is another proof of the necessity of attending to local symptoms, and to the feelings of the patient, however bizarre these may appear and however contrary to nosology; Fancy or Whim or a nosological name is sufficient to check enquiry or exertion in the herd of practitioners ...

These remarks were part of his case notes on the illness of his friend, John Oswald, who died in 1800:

> This case occasioned much anxiety to me during its progress and now gives rise to many questions. What was the origin of the disease? Many years ago having been very agile he [Oswald] leaped several times and felt a pain in his side from which he was not free for a long time. Could that lay the foundation of the suppuration under which he ultimately sank? By injuring the part it might predispose, but it is hardly conceivable how it could do more, considering the interval of time and the goodness of his health in many particulars. His colour was good but too florid, each cheek having a red circumscribed circle, and his eye-lids being very red. His appetite too was good especially for breakfast (in winter he ate three rolls), but his strength decayed and his sleep was unrefreshing.[93]

Finally, Cleghorn reached his diagnostic conclusion:

> I judged right and prognosticated too truely, but I saw no symptoms of danger at all in this last attack. The only error consisted in not introducing a Bougie into the Bladder (for we did into the Urethra) because thus we should have discovered the state of the prostate, but unfortunately we could not have removed it. Mercury might have been tried, and a regimen (which J. mentioned indeed) adhered to; for he was injured by some late indulgences, and by the treatment, e.g., travelling, emeticks, Bark, etc.[94]

Robert Cleghorn agonised over the fate of his patients and sought scientific means to alleviate disease, recording precisely the emergence of each symptom and reviewing medical advice and treatment. The next generation of medical lecturers in Glasgow, while poking fun at their elders, knew well the strength and validity of clinical experience. 'You must respect him', wrote Adam Boyd to William Mackenzie in 1814 of Robert Cleghorn, 'he is the

---

[93] RCPSG, 1/10/2, Adversaria Cleghorn, Case of John Oswald, pp. 77–78.
[94] Ibid., p. 79.

very essence of honesty – and though he cuts two baskets of strawberries per day – with a sack or two of gooseberries – and comes forth of a morning with a beard like a field of barley, and exhibits at night, stretched asleep on four chairs, a bum like the Infirmary Door – nonetheless he has a most medical brain.'[95]

Clinical teaching moved from the Town's Hospital to the Royal Infirmary in 1794, giving a further boost to both bedside instruction in hospital wards and medical management. The effect of these related developments was to create a powerful new institution that would rival the medical faculties and the incorporation.

Medical philanthropy was irrevocably changed by the focus on clinical teaching. To provide free care for the sick poor had been one of the original obligations in the charter of 1599. Now care of the sick poor came to serve the needs of the medical profession as a provider of jobs and skills. It was placed high on the agenda of curriculum reform and became a prerequisite to the granting of diplomas and degrees. But even more importantly hospital wards gave the scientific investigation of disease a new base. From within its networks came the studies and methods that operate under the name of public health. Epidemiology, the development of health statistics and managing the diseases associated with industrialisation got their start in the hospital and academic careers of infirmary medical staff. The cholera epidemic of 1832 provoked intense medical activity while cases of typhus and other fevers showed that infectious diseases were associated with bad housing and overcrowding. Industrial accidents also increased dramatically and accounted for a majority of cases in the surgery wards of the Infirmary.

Private practice, on which the apprenticeship system was based, could not cope with disease and injury on the scale produced by the Industrial Revolution. The social dislocation caused by mass immigration, together with the spread of factory work, made hospitals rather than families the base for treatment. Moses Buchanan gives a revealing insight into the extent of disease and injury treated in the Glasgow Royal Infirmary by mid century. His tables show the importance of industrial accidents and health hazards, as well as the many diseases suffered by the poor in Glasgow. The injuries treated in 1845 were largely the results of accidents to railway workers, mechanics and others with occupational risks.[96] Between 1795 and 1846 110,199 cases of disease were admitted – others would be dealt with as outpatients or in other hospitals. The problems of large-scale ill health in an industrialising society

---

[95] RCPSG, 24/2/22, Papers of Dr William Mackenzie, Adam Boyd to Mackenzie, 10 August 1814.

[96] M. Buchanan, *Statistical Table*, Medical Reform, 1846, p. 28.

were now overshadowing the individual clinical case history which mutated to generic descriptions of disease. The nosology, or classification of disease, that had interested Cullen now served to boost interest in public health.

A talented generation of Glasgow medical men explored this mass phenomenon of disease. James Adair Lawrie, Robert Perry, Richard Millar, Andrew Buchanan, Robert Watt and Robert Freer between them tackled early public health problems in the form of epidemiological and statistical studies. Lawrie was the son of a long line of ministers in the Church of Scotland. He had 'a much coveted appointment as surgeon in the East India Company's service' and returned to Scotland in 1828.[97] He was an MD of Glasgow University (1822) and country licentiate of the FPSG (1822). After his return he was appointed professor of surgery at Anderson's College (1829) and admitted as a member of the FPSG on 1 March 1830 paying the requisite fee to the Widow's Fund. Lawrie was very active in the FPSG in committees on the Royal Botanic Institution, as manager of the Lock Hospital (1839–40), manager of the GRI (1842–44), member of the Library Committee, trustee of the Widow's Fund (1843–50) and seal keeper (1846–47). He was also a founder member of the Medico-Chirurgical Society (1844). In this last capacity he encouraged the early discussion of ether as an anaesthetic. He gave a paper on the use of ether in surgical operations in July 1847 and one on the use of chloroform in midwifery in April 1851. But much of his standing derived from his study of cholera which he investigated first in India and then in Sunderland, Newcastle and Gateshead. In the introduction to his *Essay on Cholera* of 1832, Lawrie wrote: 'I met with the disease in Calcutta, in 1823, while attached to the General Hospital in that capital; it appeared at Purtabgurh Malwa, while I had medical charge of the Rampoorah Local Battalion, and was the only medical officer at the station ... After Cholera broke out in the North of England, I was one of the crowd of medical men whom the disease attracted to that quarter from all parts of Great Britain and many parts of the Continent'.[98]

Both Lawrie and Robert Perry, who also pursued links between poverty and disease, dedicated their works to the Lord Provost in the hope 'of alleviating ... suffering under disease'.[99] They also claimed that 'the influence of your exhalted station ... will be best calculated to insure the attention to the wants and circumstances of the poor'.[100]

[97] J. MacLehose (ed.), *Memoirs and Portraits of One Hundred Glasgow Men*, ii (Glasgow, 1886), p. 171.
[98] J. Lawrie, *Essay on Cholera* (Glasgow, 1832), preface, p. v.
[99] Ibid., Dedication ... to the Lord Provost.
[100] R. Perry, *Facts and Observations on the Sanitary State of Glasgow* (Glasgow, 1844), Dedication ... to the Lord Provost.

Lawrie's and Andrew Buchanan's work in the cholera epidemic in Glasgow won glowing tribute:

> Night and day he laboured with his life-long friend, Professor Andrew Buchanan, both in the hospital established for the reception of cholera patients and in private practice, to discover the cause and the remedy for the dire malady. His 'post-mortem' examinations were numerous and his personal labour unremitting, and though no great addition was made to the existing knowledge, yet all the methods of investigation then known were fully and diligently employed. That terrible epidemic gave Lawrie a hold of practice which he never lost.[101]

As with most medical men of that generation, Lawrie never retired: 'He left his work only to die'.[102]

Lawrie's friend Andrew Buchanan, Professor of Physiology (1839–76) and surgeon to the GRI (1835–62), was the son of a manufacturer, and began practise in 1824, two years after graduating as MD in Glasgow. His papers on the medical management of the poor 'got him into such trouble as to cause him to resign his connection with the journal [the *Glasgow Medical Journal*] and also his post as a district surgeon'.[103] He was in charge of the poor in one of the worst districts of the city 'and there he contracted the first of three attacks of typhus fever from which he suffered during his career. He nearly lost his life, as did his four pupils, all of whom at one time were laid up with the same dire scourge ... which for so long annually claimed numerous victims from among the members of the profession in Glasgow'.[104]

Robert Perry was a surgeon and later physician to the Royal Infirmary for thirty years and physician to the Fever Hospital. He graduated MD in Glasgow in 1808 and entered the Faculty in 1812. He was a member of the Glasgow Medical Society and a founder of the Western Medical Club (1845). Perry distinguished typhus from typhoid fever and published his findings in the *Edinburgh Medical Journal* of 1836, having first read them to the Glasgow Medical Society. His work was based on 4000 cases he examined on the wards and in post-mortem examinations.[105] In 1835 under the auspices of the Glasgow Medical Society, his observations on typhus were investigated in situ by Drs William Weir, Young, John Pagan, John Macfarlane and George Watt. They remained sceptical, but Perry has been credited ever since for separating typhus from typhoid fever. He was President of the FPSG in

---

[101] *Memoirs and Portraits of One Hundred Glasgow Men*, ii (Glasgow, 1886), p. 172.
[102] Ibid., p. 171.
[103] Ibid., p. 45.
[104] Ibid.
[105] T. Gibson, *The Royal College of Physicians and Surgeons of Glasgow: A Short History Based on the Portraits and Other Memorabilia* (Edinburgh, 1983), p. 125.

1843–45. At that time he published his findings on the connection between disease poverty and crime in his *Facts and Observations on the Sanitary State of Glasgow* (1844).[106] This work was based on the reports of district surgeons and brought before the public by 'one connected with a large public hospital, where ample opportunities are presented of tracing the influences which favour its spreading or increase its virulence'.[107] Its focus was an epidemic in 1843 in which 'the number of coffins given out to the poor in these districts was 1378 ... and the death rate was probably between fifteen and twenty percent'.[108] The description of poverty was as hair-raising as in the later descriptions of James Burns Russell, the first Glasgow Medical Officer of Health: 'a man, his wife and four children all of whom had fever ... occupied the whole space [where they lived] as a bed ... and [I] had to supply them with what was necessary through the window'.[109] Despite the unrelenting confrontation with human misery and suffering, Perry held popular lectures to educate the public 'to detect the changes which accompany the approach of disease',[110] showing a touching faith in medical advice with his introductory quote: 'To die is the fate of man, but to die with lingering illness is generally his folly'.[111]

Much remains to be done for the historical appreciation of these early medical epidemiologists and advocates of public health.[112] The scale of the problem was immense. Richard Millar, the Lecturer in Materia Medica at Glasgow University (and later its professor in 1831) was another man with a public conscience. He published *Clinical Lectures on the Contagious Typhus* in 1833 and in 1828 he tabulated the medical cases (as opposed to surgical cases) treated in the Royal Infirmary. In the first thirty years of the institution the medical cases numbered 23,451 while surgical cases treated were approximately half of this number (12,334). Millar wrote 'The comparative importance of the two sets of diseases as met with in our Infirmary therefore may be fairly said to be no less decided than as we have already found it to be in extra-hospital practice'.[113] In his clinical lectures on contagious typhus

---

[106] Duncan, *Memorials*, p. 272.
[107] Perry, *Facts and Observation*, p. 1.
[108] Gibson, *A Short History*, p. 131.
[109] Ibid.
[110] R. Perry, *Prospectus of a Course of Popular Lectures on the Animal Economy* (Glasgow, 1815), p. 4. 'Animal Economy' meant physiology.
[111] Ibid., p. 1.
[112] One gap has been most happily closed with a book on Glasgow's first Medical Officer of Health James Burn Russell, E. Robertson, *Glasgow's Doctor James Burn Russell, 1837–1904* (East Linton, 1998).
[113] R. Millar, *Medical and Surgical Establishments of the Infirmary* (Glasgow, 1828), p. 6.

the last section is devoted to reports from hospital, mortality rates and the specification of fatal cases with dissections.

Moses Buchanan concurred with the view that hospital wards in Glasgow were a mirror of the health of the general public:

> Hospitals ... are, in my opinion, of all charitable institutions, the most laudable; and whether publicly or privately endowed, if properly managed, do more to the relief of suffering humanity, and are less liable to abuse, than any other species of eleemosynary endowments. How often do we find the most industrious, in the labouring classes of society, arrested in their laudable career by severe accidents, by fever, or by dangerous internal disease? and what situation, at such a moment, can be so fitted for their treatment as an hospital? where, I hesitate not to say, they enjoy advantages from which those in a higher rank of life are precluded.[114]

Buchanan then returned to the scientific and educational gains to be made from patients so neatly grouped for medical scrutiny. This, indeed, was the inner spring on which the clockwork now turned: 'Surgery, to be well taught, ought, I think, to be less a treatise, than a demonstration of disease ... the student, who is well directed, will learn more in one week in the wards of an hospital, such as this [the GRI], than during a whole session of attendance on surgical lectures, however ably written or amusingly, or eloquently, enforced'.[115]

The Glasgow Royal Infirmary

On 15 May 1788 Alexander Stevenson wrote to the FPSG to inform them of the proposed 'situations' for the Infirmary. He asked that as many members as found it convenient should look at the sites to 'examine which of them will be most eligible'.[116] Stevenson entered the FPSG as a physician on 12 July 1749.[117] He succeeded Joseph Black as Professor of Medicine in the University in 1766. The son of a notable Edinburgh physician, he was a Glasgow graduate with a good practice.[118] John Moore quipped that he had 'a vast deal of physic contain'd in his wig!'.[119] He was President of the FPSG in 1757–58 and 1773–75.

---

[114] M. Buchanan, *Lecture, Introductory to a Course of Clinical Surgery, Delivered to the Students of the Glasgow Royal Infirmary* (Glasgow, 1831), p. 6.

[115] Ibid., p. 17.

[116] RCPSG, 1/1/1/4, Minutes of FPSG, 1785–1807, 15 May 1788.

[117] RCPSG, 1/1/1/2, fol. 173r. There is a discrepancy here with W. Innes Addison, *The Matriculation Albums of the University of Glasgow from 1728 to 1858* (Glasgow, 1913), p. 81, which gives Stevenson as a *member* of the FPSG in 1756.

[118] J. Mackie, *The University of Glasgow, 1451–1951: A Short History* (Glasgow, 1954), p. 226.

[119] Coutts, *History of the University*, p. 496.

The Glasgow Royal Infirmary was erected in the centre of the city. At this time 'the High Street and the adjoining localities, less densely populated than now, were the places of business, and even of residence, of many of the principle merchants and shopkeepers; and from the College to the Cross Tower the ancient thoroughfare looked quaint and picturesque, with crow-stepped gables abutting on the line of vision, such as may still be seen along the quays of Antwerp and the streets of Ghent'.[120] Beyond the university and the cathedral 'on a bold ridge of hills' rising 'on a gentle eminence of the table land' stood the site of the Archbishop's Castle with abundant good water at hand and the impossibility of any nuisance being erected in the vicinity.[121] Here the Infirmary was opened in 1794 and completed in 1832, 'where, after the hour of visit, the sick poor receive advice gratis from the physicians and surgeons of the establishment, and where those patients who have lines of admission from subscribers, must wait an examination by the above-mentioned office bearers, before being sent to their respective wards'.[122]

In November 1786, a year before the site visit, in response to another letter of Dr Stevenson's, the FPSG set up a committee to consider the plans for the Infirmary. It was convened by the Praeses as chairman (Dr Peter Wright) and included Stevenson, Robert Marshall, Robert Cleghorn, William Hamilton, Alexander Dunlop, Archibald Young and the Visitor (James Monteath). It reported in 1787:

> The Committee are of opinion that the Establishing [of] an Infirmary in Glasgow will not only be a great benefit to the distressed poor, but tend highly toward the improvement of Physic and Surgery, and therefore they Recommend it to the Faculty to be assisting in so good a work and to subscribe towards its support as much as the state of their funds will permit.[123]

This report was signed little over a month after Stevenson's letter, one of the quickest responses ever to be found in the Faculty minutes, surely a sign of the ground being well prepared. The report was approved by Faculty on New Year's Day 1787.

Officially the GRI was launched through the exchanges of these letters, including the crucial ones between Professors George Jardine and Alexander Stevenson as recorded in the university history.[124] Unofficially the foundation

---

[120] J. Veitch, *Memoir of Sir William Hamilton, Bart* (Edinburgh, 1869), p. 3.
[121] M. S. Buchanan, *History of the Glasgow Royal Infirmary* (Glasgow, 1832), p. 4.
[122] Ibid., p. 7.
[123] RCPSG, 1/1/1/4, Faculty Minutes, 1 January 1787, fol. 44.
[124] Coutts, *A History of the University of Glasgow*, pp. 496–97. See also J. Jenkinson, M. Moss and J. Russell, *The Royal: The History of the Glasgow Royal Infirmary, 1794–1994* (Glasgow, 1994).

was driven forward by alliances between mercantile wealth and the powerful medical elite, uniting important town institutions like the FPSG and the University as never before, an alliance that was to crack apart resoundingly some few years later.

Stevenson became seriously ill and resigned his professorship in 1789, dying in May 1791. His nephew, Thomas Charles Hope, taught his classes for him from then on. Hope was in an advantageous position to succeed to the professorial chair and duly did so, staying in the post until 1795 when he moved to the Chair of Chemistry in Edinburgh.[125] Hope had been the Lecturer in Chemistry at Glasgow from 1787 to 1791, as the Professorship in Medicine became his only on Stevenson's death.[126] Robert Cleghorn succeeded Hope in the chemistry lectureship in 1791. In the critical years before the GRI opened, Cleghorn was President of the FPSG (1788–91), then its Librarian (1792).[127] He had an insider's knowledge of the medical faculty and of corporate strategy within the FPSG, as well as a commanding role as manager and physician to the GRI. Cleghorn was physician to the Infirmary from 1795 to 1810, alternating in the post every two years with his colleague Dr Richard Millar. Cleghorn married into the Glasgow banking elite: his wife, Margaret Johnston, was the widowed eldest daughter of Andrew Thomson.[128] Thomson was the partner of Robert Carrick in Glasgow's famous Ship Bank. Robert Carrick was one of the bailies of the city in 1796.

Cleghorn was close to David Dale the industrialist and philanthropist who founded New Lanark. They were linked through the Literary Society and the Humane Society, both of them being office bearers in each society. They were also together on the first Board of Managers of the Infirmary in 1795, Cleghorn elected as the representative of the FPSG. Dale built his first mill in 1783, and by 1795 he had 'four mills at work, driven by the Clyde, and giving employment to 1334 persons'.[129] The labour for his textile works was first recruited from the Town's Hospital. It's non-medical charities had a programme of instruction for the children of the poor including spinning (the 'mechanick employments') taught along with basic schooling.[130] Dale was involved financially and personally with the Town's Hospital until 1798 after which he supported the new Infirmary. As a Director of the Town's

[125] On Hope, see J. Kendall, 'Thomas Charles Hope, MD', in A. Kent (ed.), *An Eighteenth-Century Lectureship in Chemistry* (Glasgow, 1950), pp. 157–63.
[126] Coutts, *A History of the University of Glasgow*, p. 498.
[127] For Cleghorn, see G. Thomson, 'Robert Cleghorn, MD', in A. Kent (ed.), *An Eighteenth-Century Lectureship*, pp. 164–75.
[128] Ibid., p. 171.
[129] L. Stephen (ed.), 'David Dale', *Dictionary of National Biography*, xiii (London, 1888), p. 384.
[130] D. McLaren, *David Dale of New Lanark* (Milngavie, 1983), p. 82.

Hospital his name first appears in November 1787. He was also on its manufacturing committee (1791 and 1792–94). This brought him into intimate contact with the educational, manufacturing and medical side of the charity.[131] By a quirk of fate Dale's custom-built house on Charlotte Street later became the Eye-Infirmary with Dale's octagonal study in use as the operating theatre.[132] His public service and experience of office involved him in the founding of the Royal Infirmary. He was a manager in its first year of active service, in 1795 and gave money every year until 1806.[133] His initial contribution in 1795 was £200, the largest donation given by any one single person.[134] His political leanings were clear. Dale founded a new Protestant communion on congregational principles, known as the 'Old Independents' and served a personal ministry among the convicts of Bridewell prison.[135] Of Dale's son-in-law, Robert Owen, an 1842 biographical entry has it that he 'acquired an unenviable reputation by being the founder of the new sect called "Socialists" '.[136]

Dale's philanthropic spirit was in tune with Cleghorn's liberal and humane caste of mind. Cleghorn was hoping to become Professor of Medicine in 1795 when Hope went to Edinburgh as he taught Hope's classes in the interim. Coutts writes that Jeffray, the Professor of Anatomy and Botany, offered to teach the class of medicine, but a similar offer by Dr Cleghorn was preferred 'which drew forth dissent from Jeffray and some others'.[137] Cleghorn puts a different gloss on this, writing to Professor John McLean of Princeton University in April 1796:

> You have now probably heard that Dr Hope is gone to Edinburgh, the Assistant and Successor of Dr Black. I have taught his class this season by appointment of the Faculty, but Dr [Robert] Freer of Edinburgh has now got the class because I am a Democrat. The manoeuvres against me would have surprised you or any honest man even while you breathed this servile air, but now they would appear to you utter ... [here a portion of the letter is destroyed] ... remember me most kindly to the Millars or to any of my other acquaintances who have had the wisdom and good Fortune to cross the Atlantic.[138]

The Millars mentioned may or may not have been relatives of Richard

---

[131] Ibid., pp. 76–79.
[132] Ibid., p. 52.
[133] Ibid., p. 84.
[134] Ibid., p. 87.
[135] L. Stephen, *DNB* (London, 1888), p. 385.
[136] W. Anderson, *The Popular Scottish Biography* (Edinburgh, 1842), p. 237.
[137] Coutts, *A History of the University of Glasgow*, p. 498.
[138] Letter as quoted in p. 175, n. 35, in Thomson, 'Robert Cleghorn, MD', in Kent, *An Eighteenth-Century Lectureship*.

Millar, the lecturer (1791–1831), and later Professor of Materia Medica (1831–33). He was President of the Faculty many times (1800–2, 1806–8, 1818–20, 1826–28). In the lectureship on materia medica he was Cleghorn's successor. Richard Millar also had a conscience, labouring fervently in the severe 'fever epidemic' of 1818–19.[139] He wrote on contagious typhus, as we have seen, and on the history of medicine. Like Cleghorn he was an original manager of the GRI and an adamant critic of its bias toward surgeons. Millar's 1828 pamphlet on *Medical and Surgical Establishments of the Infirmary* was clearly meant to jolt his colleagues and the public into an awareness of a fiefdom maintained by the Faculty which he felt no longer reflected modern practice. Millar objected to the rotation system in the Royal Infirmary being primarily of use to surgeons. He felt it unfair that a physician's clerks had to move to the next surgeon's clerkship 'whether he deserves it or not, and to the exclusion, possibly, of more meritorious applicants, thus monopolising two offices that are better exercised separately'.[140] He adds 'as if in our Infirmary here, Surgery were every thing, Medicine nothing'.[141]

His description of the meaning of the rotation system for the two branches in medical practice is revealing:

> In common life, this individual [the surgeon] is at once Surgeon and Physician, the last much more frequently than the first, since for once that he appears in the former capacity, he acts, at least twenty times, in the latter; but the moment he sets foot within the hospital, the scene alters, he disengages himself entirely from his medical attributes, and is converted into the pure, and genuine Surgeon.[142]

This marked discrepancy between posts in the Infirmary and private practice was upheld through the rotation system preferred by the FPSG. Surgeons benefited from it, as Millar noted: 'When a Glasgow Surgeon sets out in business, he has plenty of time for the Infirmary; he is even anxious to officiate for the sake of operating, as operations abound there much more than in private practice. But as life advances his views alter. The fee of the Surgeon is small, and to subsist he must have numerous patients. When his practice increases, therefore, he has not time for the Hospital'.[143] The physicians had higher fees and fewer patients giving them more time to attend the Infirmary. This conflict over the powerful place of surgeons maintained through the agency of the FPSG coloured the medical life of the city. The

---

[139] Duncan, *Memorials*, p. 267.
[140] R. Millar, *Medical and Surgical Establishments of the Infirmary* (Glasgow, 1828), p. 3.
[141] Ibid.
[142] Ibid., p. 4.
[143] Ibid., p. 8.

power struggle was centred on the Glasgow Royal Infirmary and to follow it explains how hospital medicine finally brokered its own path between conflicting demands.

Medical Infighting

Hospital ward attendance reflected the system decided by the FPSG. The Royal Infirmary was managed by a twenty member board of managers of which ten members were permanent, four appointed by the FPSG, three by Glasgow University and one each by the Trades' House, the Merchants' House and the Church of Scotland. The other ten members of the board were elected by the General Court, who, in turn were selected by the subscribers.[144] The rotation system favoured the surgeons, they having a shorter term and more posts available, thus weighing the influence on the board of managers towards the policies of the FPSG.

In August 1794 the managers of the Royal Infirmary considered the report of a committee on medical management. The report was signed by Thomas Charles Hope, Robert Wallace, Alexander Dunlop and James Towers. Two physicians were to be chosen annually, at least one of whom was to leave office every year. Four surgeons were to be chosen annually as well, at least two of whom were to rotate and would not be eligible again for at least two years.[145] Both physicians and surgeons were to attend the infirmary at a stated hour to examine patients and instruct students. These students, and surgeons' apprentices, paid five guineas or two guineas annually to attend.[146]

These medical staffing arrangements came under siege in January 1796. The point at issue was clearly about skills training in the profession. A representation by the FPSG was made to the General Court. It was signed by the surgeons James Monteith, John Riddell and William Anderson, who had been elected by the Faculty to the three positions on the board of directors which were theirs to fill.[147] The FPSG surgeons wanted changes in the plan of attendance. The objective was 'that as many of the medical people as possible shall have an opportunity not only of acquiring that additional knowledge and experience afforded by the observation and treatment of the variety of diseases an Infirmary brings under review ... but [also] of seeing

---

[144] J. Jenkinson, M. Moss, J. Russell, *The Royal: The History of the Glasgow Royal Infirmary, 1794–1994* (Glasgow, 1994), p. 23.
[145] HB, 14/1/1, Glasgow Royal Infirmary Managers' Minutes, 1787–1802, pp. 111–12.
[146] HB, 14/1/1, p. 112.
[147] HB, 14/1/1, pp. 178–79; RCPSG, 1/1/1/4, fos 113r,v.

the many and varied operations that are performed there and of acquiring that dexterity in operating which practice alone can give'.[148]

The aim was access for as many surgeons as possible to gain practical training in surgical operations. The surgeons wanted a quick rotation in posts and suggested a system whereby two physicians would serve for six months and six surgeons for two months each. The Faculty would lay before the managers a list of physicians and 'privileged surgeons' resident in Glasgow from which they could elect the staff. This system was to ensure 'against the abuses of election' observed in other hospitals which were seen in the document as leading to 'canvassing and cabal'.[149] It was also to guard against 'the justly dreaded evils of indiscriminate Rotation'.[150] Most importantly their plan hoped to ensure that: 'It excludes no member of Faculty for any length of time whose services the Infirmary would wish to have' while also giving 'to the attending students the most ample opportunity of seeing varied practice'.[151]

This plan was obviously to the advantage of the FPSG surgeons and did not go unopposed. The dissenting party was led by a former President, the physician Robert Cleghorn. He was backed by others engaged in teaching at the University, Thomas Charles Hope, James Towers and Robert Cowan, who all eventually held professorships. They were supported by surgeons from prominent Glasgow families: Robert Wallace, Alexander Dunlop, William Couper, Charles Wilson and Archibald Young. This group argued that 'the Faculty never made an offer of gratuitous services, nor does its subscription entitle it to any peculiar [particular] controul over the Managers, far less the power of dictating to them'.[152] Cleghorn and his supporters backed the right of the Infirmary Managers to regulate medical attendance independently. 'On the present plan a sufficient number of Surgeons will most probably be employed, and those who do not actually attend the house are at all times welcome to examine any particular patient, or to witness any operation that may appear interesting.'[153]

The managers, asserting their independence, rejected the plan of the FPSG to change the rotation system and scolded its 'misapprehension of the present plan of medical arrangement approved of and established by the General Court'.[154] The managers wished to pay due attention to the FPSG, but did

[148] RCPSG, 1/1/1/4, fol. 115v.
[149] RCPSG, 1/1/1/4, fol. 115r.
[150] RCPSG, 1/1/1/4, fol. 116r.
[151] RCPSG, 1/1/1/4, fol. 116.
[152] RCPSG, 1/1/1/4, fol. 116v.
[153] RCPSG, 1/1/1/4, fol. 117r.
[154] HB, 14/1/1, p. 179. This extract from the records of the Royal Infirmary was also recorded in the Faculty minute book, RCPSG, 1/1/1/4, fos 126r, 126v.

not like the interference with 'the clear and undoubted right which the General Court have by their Charter to establish such rules and arrangements as they shall judge proper without the interference of any other persons whatever'.[155]

Further interference was not slow to follow. In the same year nine members of the FPSG refused to attend the Infirmary until 'a plan of medical attendance more liberal than the present be adopted'.[156] The next salvo also came from this group, Alexander Nimmo, then President of the FPSG, protesting at the appointment of Dr McDougall, a nephew of the anatomy professor James Jeffray. Nimmo protested 'against the appointment of any person not having a licence from the Faculty of Physicians and Surgeons of Glasgow to the office of Surgeon'.[157] The Faculty's grip on the medical management of the Infirmary was only weakened by the General Court's right to make house rules, but it made its presence felt through its hold on surgical licensing and by its ability to withhold services. These two measures were deployed again.

In February 1810 the managers tried to free themselves from the FPSG's monopoly. They refused 'to enact a law … which would … prevent the Hospital from having the benefit at any time of medical or surgical assistance however eminent unless the Practitioner was previously enrolled as a Member of the Faculty of Physicians and Surgeons of Glasgow'.[158] In 1812 the managers were petitioned by the University to appoint one of their number to give clinical lectures that would satisfy its curriculum requirements. In this case, too, threats were made and an impossible situation ensued which was resolved by ad hoc decisions on the part of the Infirmary management. In juggling the often irreconcilable demands of different factors of the medical community, the Infirmary emerged as a 'third force'.

The climate was by now turbulent. Around 1810 the Glasgow Medical School, as a city-based consortium of institutions, began to raise standards, enacting curriculum requirements which were practically indistinguishable from one another.[159] But in other sectors dissent was rife. The FPSG tightened its monopoly powers on licensing surgeons by direct legal prosecutions, no longer only issuing 'letters of horning'. In regard to the Infirmary it was behaving as if it had monopoly rights as well. When the managers refused to enact a law restricting their staff to members of the FPSG, it threatened reprisals at law. The FPSG paid the Professor of Law for a considered opinion

---

[155] HB, 14/1/1, p. 179.
[156] HB, 14/1/1, p. 189.
[157] HB, 14/1/1, pp. 191–92.
[158] HB, 14/1/2, Glasgow Royal Infirmary Managers' Minutes, 1803–1812, p. 229.
[159] See below, Chapter 9, 'Time of Crisis'.

which was 'that as the Infirmary is situated within the Royalty [city of Glasgow] the managers do not possess the power of naming any surgeons to practice there who is not previously entered with the Faculty, and it also appears to one that if a Physician practice surgery, he also must enter with and submit to the trial of the Faculty'.[160] That was precisely the warning the FPSG had wished for to maintain its grip on Infirmary medical staff.

The battle over rotation became enmeshed with educational reform, in particular the need to regularise clinical teaching in the new Royal Infirmary. The archaic splitting of competencies between surgeons and physicians was exacerbated by the FPSG and the university operating separate systems of accreditation. Neither of these bodies either wanted or were capable of maintaining hospitals. But both began to set requirements in hospital attendance for candidates wanting degrees. Clinical teaching was done in the mode familiar from the Town's Hospital. Students were simply collected for a class run by the current incumbent of a post. Their accreditation was good for whatever system of qualification they then pursued. Physicians like Cleghorn must have taught as they did in the Town's Hospital, expounding on separate cases, teaching diagnosis and treatment. The surgeon John Burns, then himself just qualified, used just this opening when he proposed to the Board of Management that he teach a clinical course. Bedside teaching was an accepted part of the routine in an eighteenth century hospital, but it was not formalised as part of a degree programme.

The status of hospital work for medical practitioners was changing rapidly. It enhanced perceptions of competency. This combined with demands to school potential practitioners before licensing them in the broadest possible range of cases and to expose them to the ethos of the profession. By now most medical men were conscious of the fact that the skills that would recommend them to private patients were gained in hospital practice. Concerns for proper conduct were part of this new profile. The surgeon's clerk was to instruct dressers, the lowest on the medical totem pole, 'to study neatness and elegance in dressing',[161] and to make sure, in the case of dissections, that 'the body [is] decently sewed up, and dressed before being delivered to the dead person's friends for interment'.[162]

Clinical teaching changed from an auxiliary subject whose worth was understood, but whose function in medical education was not settled, to a requirement. This development was not straightforward, even though its merits were never questioned since its incorporation in the teaching of most

---

[160] HB, 14/1/2, pp. 206–7.
[161] HB, 14/1/1, p. 126.
[162] HB, 14/1/1, p. 123.

Scottish lecturers. The standardising of Scottish medical education produced pressures that forced institutions into realignment. As hospitals became centres of instruction, tipping the balance on that side rather than on their charitable work, their educational value was recognised. The work that a clerk did was undeniably the best possible preparation for medical practice. John Burns exemplary advance up the ladder of Royal Infirmary posts more than illustrated their strategic importance. John Burns was appointed to the office of physician's clerk in May 1795.[163] His duties (as laid down in the regulations for physician's clerks in November 1794) were to take an accurate record of the cases under the care of the physicians, in order to brief them before they visited the patients; to attend the patients with the physician and write down their symptoms and the medicines prescribed, as dictated by him; to visit the patients in-between times – and more frequently according to the danger they were in – in order to give account to the physician on his next visit; and 'to transcribe such of the cases of the medical patients into the ledger as shall be pointed out to him'.[164] He was to keep an account of and receive the money paid by patients admitted into the hospital; to lay before the managers, at their monthly meetings, an account of the patients then in the hospital – divided into medical, ordinary and surgical cases, servants and supernumeraries – and to give a particular account of the servants employed in the hospital.[165]

Burns soon rotated into a more prestigious position. In July 1795 a letter was laid before the meeting of the managers from Mr Burns, 'Student of Medicine', requesting that he be appointed Apothecary and House Surgeon in place of Mr McArthur. The 'Medical Gentlemen having recommended him as a person properly Qualified for that office', the meeting unanimously chose him as apothecary and house-surgeon.[166] Burns resigned the joint post in March 1796, 'having intimated that it will not be convenient for him to execute that office after the first of May'.[167] He was subsequently elected (by signed lists), in January 1797, one of four surgeons for the year, to continue in office until the first Monday of February 1798.[168]

By May 1797 Burns had sent a letter which was laid before the quarterly meeting, stating:

In the Edinburgh and many of the English Hospitals lectures are read on Surgery,

---

[163] HB, 14/1/1, pp. 158–59.
[164] HB, 14/1/1, p. 125.
[165] HB, 14/1/1. p. 125.
[166] HB, 14/1/1, pp. 160–61.
[167] HB, 14/1/1, p. 183.
[168] HB, 14/1/1, p. 203.

but as yet nothing of the kind has been attempted in this Infirmary – it is evident however that such a course of Lectures if properly conducted, would not only be beneficial to the students, but also to the Medical School established here, and particularly to the Infirmary, as every additional means of improvement provided here must increase the number of the students, and consequently augment the fees of the Hospital. This course does not interfere with the Lectures of any Professor, but on the contrary has been approved by many of those who fill the medical chairs at Edinburgh, the only place in Scotland where this plan has as yet been carried into execution. In order to ascertain how far the undertaking will succeed in this place I request permission to deliver a Course of Clinical Lectures on Surgery this summer during the months of my attendance in the Infirmary.[169]

The managers duly approved Mr Burns's proposal and agreed to take up his offer of lecturing.[170] Burns was re elected as surgeon in January 1798.[171] He was still being elected surgeon in 1808.[172] Not content with surgical lectures he also lectured on midwifery.[173]

The 1797 initiative on clinical lectures in the Infirmary set a long chain of events in motion. Alexander Dunlop proposed in 1801 to give clinical lectures on surgical cases, which was approved.[174] In February 1810 students moved to have a clinical lecturer appointed. They argued:

The rising reputation of the medical school in Glasgow is sufficiently evident from the numbers who now attend it, and it must certainly be the desire of all those who wish well to its interests, as well as those of humanity to see that every deficiency be supplied and every source of information be laid open. The advantages arising from a course of Clinical lectures many of the students have already experienced, and they cannot allow this opportunity to escape of stating their sentiments of gratitude to the Gentlemen to whose exertions they are indebted for the lectures delivering this season in another department. It is too little to say they have much more than supplied their own want of information and sufficiently compensated for their own defect of Observation.

A course of lectures on Clinical Surgery is still wanting and the anxiety of the

[169] HB, 14/1/1, p. 209.
[170] HB, 14/1/1, p. 209.
[171] HB, 14/1/1, p. 222.
[172] HB, 14/1/2, p. 166.
[173] HB, 14/1/2, p. 205.
[174] HB, 14/1/1, pp. 290–91. This request is sometimes credited to *William* Dunlop his son. 'William Dunlop was bred a Surgeon, and was the son of Mr Alexander Dunlop, who, in his day, was the most eminent Surgeon in Glasgow. William was a remarkably able and ingenious Man, and was the first to deliver Clinical Lectures in the Glasgow Infirmary; but he had no great love for his Profession, and this, perhaps, is the reason why we find him a Partner with, and Assisting, Samuel Hunter in The Herald. He fell into bad health, and retiring to the Island of Tenerife in the hope of amendment, Died, and was Buried there in 1811'. Compare Duncan, *Memorials*, p. 268, who writes that he died in 1809.

students on the subject has been fully manifested by the attention and approbation they have bestowed upon the Gentlemen who has generously and gratuitously and in their opinion with the highest professional ability attempted to supply the defect.[175]

This 1810 memorial by students raised the usual hackles. A permanent clinical lecturer was not desirable in view of the rotation principle. But the management committee allowed:

> The Committee are unanimously of opinion that Clinical lectures on cases of Patients in the surgical wards of the Infirmary would be a useful addition to the Medical School, but at the same time they do not see how such an appointment can be made while the surgical Attendance on the Hospitals is carried on according to the existing Regulations. Were a permanent lecturer to be appointed, he must either give lectures on Patients treated by other Surgeons, or he must be appointed a permanent Surgeon to the Hospital, and have in his power to make choice of such surgical patients as he may think proper to treat and lecture upon. To the first plan no attending Surgeon we conceive, will submit, or if they did, do we see with what propriety or advantage lectures could be given on cases where the lecturer could not possibly be acquainted with the Surgeons motives and reasons for his practice, to the appointment of a permanent Surgeon there are many objections with which the Managers are well acquainted.
>
> The Committee are of the opinion that if a lecturer were appointed, a sufficient number of Students would not be found to render the lectures an object of Emolument to the Lecturer, without which they would soon be given up. It is well known how difficult it is to collect Students sufficient to induce the Physicians to give regular Clinical lectures on their patients, and both in London and Edinburgh the Medical lectures have been uniformly better attended than the Surgical.
>
> The Committee recommend to the managers of the Infirmary to give every Encouragement to the Surgeons in attendance on the Hospital who may choose to give Clinical lectures on Patients under their care.[176]

On the same day a report by the committee appointed to consider the proposal for augmenting the fees or salaries to the medical staff gave their opinion that 'upon the event of the Honorariums or fees paid by the Students attending the Infirmary, shall exceed the sum of £80 mentioned in the Report then the managers shall take into their consideration what addition should

---

[175] HB, 14/1/2, pp. 230–31.
[176] HB, 14/1/2, pp. 238–40.
[177] HB, 14/1/2, pp. 240–41. Salaries of the two attending physicians were then £30 p. a. each and to the four surgeons, £10 p. a. each, subject to a regulation by which no physician or surgeon could receive any salary till he had attended at least two years without any salary, ibid., p. 240.

be made to the salaries of the physicians and surgeons from that fund'.[177] The committee found that the fees paid by these students now exceeded £200 per annum and that 'there is still the prospect of the increase of that fund', so they proposed raising the salaries of the two physicians to £50 each and those of the four surgeons to £20 each, 'the whole salaries of the Medical Establishment amounting to £180 – a sum below the average of the fees paid by Students'. But according to regulations, no physician or surgeon was to receive a salary until he had served two years gratis, 'by which regulation the annual amount of the expense of this establishment will seldom or ever amount to £180 per Annum'.[178]

After this discussion, Dr Cleghorn moved,

> that thereafter any Physician in actual attendance on the Infirmary may give a course of Clinical lectures on his Patients to the students attending the Infirmary – the course to consist of two lectures each week for three months. And if both Physicians shall chuse to give a course at the same time they shall take different days, the eldest Physician having the right of naming his days first, and when both shall lecture in the operation room each shall fix an hour that is convenient for the house.

Secondly, 'If no Clinical course be given by the attending Physicians before 20th of November, yearly the Managers shall appoint a Clinical lecturer for the season if they find one & if they find any call for Clinical lecturers'.[179] On 5 November 1810, the managers deferred the consideration of Dr Cleghorn's motion because it had not been given in signed by him according to form.[180] Cleghorn subsequently withdrew the motion, but in February 1811, Jardine proposed the following motion which Cleghorn seconded:

> Whereas some inconvenience has arisen from the electing the Physicians and Surgeons, particularly with respect to Clinical Courses of Lectures in the end of January, it is moved That the Physicians and Surgeons shall be elected annually on the first Monday of August and shall enter upon their respective offices on the first of November thereafter.
>
> That the attending Physicians shall have the power of giving two courses of Clinical lectures viz one from the first November to the first of February, and the other from the first of February to the first of May at the rate of two or three lectures every week, and that the senior Physician shall have the option of taking the first or second course – But if the same Physicians shall be re-elected a second year – the junior Physician shall have his Option of taking the first or second course.[181]

[178] HB, 14/1/2, p. 241.
[179] HB, 14/1/2, p. 242.
[180] HB, 14/1/2, p. 244.
[181] HB, 14/1/2, p. 263.

The physicians elected on the first Monday of August were, on or before the first Monday of November, to signify their intention of giving the clinical lecture courses. If they did not agree to do so, the managers were to appoint one or two suitable persons. But if one of the elected physicians agreed to give the clinical lectures, he was to have his option of the courses, or with the consent of the managers was to give both courses.[182] The quarterly meeting of the managers held in May 1811 agreed to these points.[183]

In 1812 and 1824 the issue of a permanent clinical lecturer appointed by the university to comply with their curriculum requirements was raised again. Before going into detail on the continuing fracas this raised, it might be well to reflect on what was *actually* provided. The Royal Infirmary records show an *actual* provision of clinical lectures throughout the period leading up to the Medical Act. It secured clinical instruction in Glasgow by granting physicians and surgeons in hospital posts the right to lecture. Admittedly this was an ad hoc system, in their view the one most compatible with acceptable behaviour to patients. No doctor would be happy to have someone not in charge of the treatment explain the case, a point of view the hospital management supported. Thus *several* physicians and surgeons continued to hold clinical lectures. In 1812 a proposal for a permanent lecturer was turned down, but Richard Millar and Robert Cleghorn were elected to give clinical lectures.[184] On 7 November 1814 Robert Freer and Robert Watt were elected 'to give clinical lectures on the patients in the Infirmary'.[185] In 1815 Watt was unanimously re elected and another physician, Robert Graham, who became the university professor of botany in 1818, was elected by a majority of votes. They too gave clinical lectures.[186]

Clinical lectures were thus a regular feature of Infirmary life, held by those approved by the managers. The managers also sanctioned occasional lectures. Granville Pattison 'expressed his desire to deliver occasional Lectures on the cases of Patients under his care to the students attending the Infirmary'. This was allowed 'provided they shall not be given in the Infirmary nor interfere with the hours of the Clinical Lectures which are delivered under their appointment'.[187] At a meeting of 1 November 1816 Meikleham moved 'that a Committee be appointed to inquire into the course of clinical teaching that may be thought proper for the Infirmary'.[188] There was no formal report

---

[182] HB, 14/1/2, pp. 263–64.
[183] HB, 14/1/2, pp. 271–72.
[184] HB, 14/1/3, Glasgow Royal Infirmary Managers' Minutes, 1812–1819, p. 34.
[185] HB, 14 /1/3, p. 77.
[186] HB, 14/1/3, p. 144.
[187] HB, 14/1/3, p. 172.
[188] HB, 14/1/3, p. 175.

from this committee. But in 1818 William Cumin requested leave to hold a series of surgical lectures. The Infirmary issued tickets, but only those students 'attending the house' could enter for them.[189] The income from student tickets was not negligible, being £382 13s. 0d. in 1812 for 'tickets, shop fees and diplomas'.[190]

In 1812 the FPSG had threatened boycotting medical services if the university introduced a clinical lecturer. In 1824 the issue arose again and they again called on their members to boycott. These actions finally drove clinical teaching into the hands of hospital management. The hospital soon arbitrated the conditions for the teaching required by the universities and the incorporations. Until the 1830s however the development of clinical teaching attached to the university degree courses was severely hampered by FPSG objections. This point was not lost on the Professor of Surgery, John Burns, who complained to the Royal Commission visiting the universities of Scotland in 1827 that he had been 'anxious to illustrate the doctrines I deliver in the College [University] by reference to patients and cases ... in other words, I wish to accompany it with clinical instruction'.[191]

He continued: 'There is an hospital in this city, or Royal Infirmary, which was established about thirty years ago, but there is no benefit derived in the way of University instruction from that Hospital. Students are allowed to attend upon paying fees, but there is no clinical instruction given by any Professor in the University, nor is there any connection between the Royal Infirmary and the Medical School within the University'.[192] He said that the University and he, personally, had made repeated applications 'to give lectures upon a certain number of surgical patients, and to connect that instruction with the lectures to my own class ... the request has been declined'.[193] He continued: 'The heads of public bodies [who were ex officio on the managing board of the GRI] have given every countenance, and the Lord Provost, Convenor, and Dean of Guild, have certainly given their support to the measure, but still the surgeons have had influence to procure a majority of managers on their side'.[194] He suggested to the commission that legislation be introduced for cities containing universities with medical faculties, to induce hospitals to serve the teaching needs of the profession.

---

[189] HB, 14/1/3, p. 261.
[190] HB, 14/1/3, p. 18.
[191] J. Burns, evidence to Royal Commission on 8 January 1827, *Evidence, Oral and Documentary, Taken and Received by the Commissioners, July 23 1826 to October 12 1830*, ii (London, 1837), p. 127.
[192] Ibid.
[193] Ibid.
[194] Ibid.

He then squarely laid the blame for the situation in Glasgow on 'the influence of the surgeons in the town who attend the hospital'.[195]

In the 1820s the Infirmary had a regular income from students attending the hospital and was considering paying a salary for services on the basis of this income: 'the opinion of the Managers [was] that the fees of Students form a legitimate source of pecuniary remuneration to the Medical and Surgical Attendants, since from their exertions alone such fund was originally created and at present subsists'.[196] The income varied, of course, but was found to have been at a maximum in 1814, with £439 14s. 0d., and a minimum in 1819 of £214. The income could reasonably be calculated at between £230 to £360. Surgeons were paid £80 per annum and physicians £50 per annum. In 1820 there was no decision to increase salaries, but the managers abolished half-yearly tickets in favour of yearly tickets. 'The truth is a half yearly attendance at an hospital is a term too limited either for the instruction of the Pupil or for his ulterior views in life. Thus it is quite useless for obtaining a Diploma in Surgery from any of the three Royal Surgical Colleges of the empire, or from the Faculty of Glasgow, while it is no less inadequate for an examination before the Army or Navy Board, or for the service of the East India Company'.[197] All of this indicates the very important consolidation of hospital teaching. The university was fighting to gain access and the FPSG was keen to keep control of posts to qualify surgeons. Teaching was also available to the many students who would qualify elsewhere, through the surgeon's colleges, the Apothecaries' Hall, the Navy and Army Boards, or the East India Company. The income from student tickets, as recorded in 1826, had risen steadily, from a little over £80 in 1796 to almost £500 in 1825, showing the benefit of keeping student fees in-house.[198] Indeed the Medical Committee now referred to its teaching in terms of 'the Medical School of the Infirmary'.[199]

Clinical lectures were an obvious prize. The University wanted the Professor of Surgery to be allowed 'to illustrate the principles taught in the university by cases occurring in the Hospital, and thus enabling him to lead the Student practically to apply those lessons, which he receives in the general course'.[200] John Burns was teaching 222 students in 1826,[201] a figure consistent with the over 200 students in the classes of the Professor of Anatomy. The

[195] Ibid.
[196] HB, 14/1/4, Glasgow Royal Infirmary Managers' Minutes, 1819–1825, p. 63.
[197] HB, 14/1/4, p. 55.
[198] HB, 14/1/4, 'Annual Report: Student and Shop Fees', p. 37.
[199] HB, 14/1/4, p. 62.
[200] HB, 14/1/4, p. 249.
[201] Burns, *Royal Commission Evidence*, p. 130.

volume of students attending classes was much higher than those qualifying in Glasgow. But the university interest was considerable and it meant bolstering their curriculum requirements. They wanted control over the qualifications of the teacher, in effect saying that they rejected the staff surgeons at the Infirmary. In 1825 a compromise was reached

> that the clinical course in place of three months should consist of six – that it should be divided into two periods with two different Lecturers each to participate equally in the emoluments; – that the Directors should appoint one Lecturer for three months and that the University should recommend another for three months more – that the period during which the College Lecturer should attend should be that of the junior Surgeon – and the Senior Surgeon nominated by the Directors should lecture another three months, the priority of time to be a matter of arrangement. Such a plan should appear to your Committee pretty much calculated to meet the difficulties of the case ... it divides the patronage ... it confers the same pecuniary emoluments ... and the Junior Surgeon will thus enjoy the benefit of the skilled science and experience of an advanced practitioner.[202]

The same plan was to be followed in medical clinical lectures.

This rather reasonable compromise was not to the liking of the FPSG, whose war with the University seems to have clouded any sense of the benefit for the whole rather than the parts. It damned attempts at compromise and then effectively delayed them out of existence, for which James Corkindale was responsible. His tactics were not straightforward as he seemingly supported the principle of compromise, only to postpone it for so long that the internal appointment of clinical lecturers was never changed.[203] The FPSG selfishly guarded its monopoly right to qualify surgeons despite the pressures to deregulate local corporate hegemony.

In answer to the proposals of the University, the red herring of the university having little say in appointing regius professors was put forward. Notwithstanding the University's own representations at the time against the Crown on this issue the Faculty maintained that 'neither the Directors [of the GRI], the University or the Faculty can have any controul' if the Infirmary allowed regius professors to teach![204] The FPSG gave the game away when it wrote: 'it should however be kept in mind that the Faculty too have a curriculum and that they have for more than two hundred years been giving diplomas in Surgery, while the University about nine years ago either usurped the power or at least adopted the practice'.[205] The CM degree was the obvious

---

[202] HB, 14/1/5, Glasgow Royal Infirmary Managers' Minutes, 1825–1829, p. 16.
[203] HB, 14/1/5, pp. 159 and 162.
[204] HB, 14/1/5, p. 111.
[205] Ibid.

thorn in the FPSG's side, this prompting a remark that the University 'derives from their popish bull' (the founding charter) its powers to grant degrees, [while] the FPSG had 'Royal and Parliamentary authority'.[206] The Faculty said it would 'never consent' to university lecturers as part of Infirmary teaching because this would deny 'full and free competition in lecturing to their own members when acting as Infirmary Surgeon'.[207] This argument was tantamount to retaining a closed shop for FPSG members. 'Free and full competition in lecturing' was not the whole truth either, since the 'private' teaching which fed the qualifying requirements of the FPSG was restricted to members of the FPSG only. Membership dues, unlike licence fees which were not equivalent to membership, were prohibitive at the time.

John Burns was fair enough not to criticise this 'private' teaching before the Royal Commission on Education. When asked whether 'any lectures upon Medicine [are] given in this city by gentlemen who are not members of the University?', he replied 'Yes, several' and that they were 'countenanced by the Faculty of Physicians and Surgeons' and that they were 'all members of that body'.[208] Asked how damaging this was to the University, Burns replied: 'Why, they cannot interfere with the prosperity of the University, if no undue advantages are taken. So long as the power of licensing surgeons is possessed by the University, and so long as their licentiates have privileges equal to those possessed by the licentiates of the College of Surgeons and Faculty of Physicians here, they cannot do any injury. I am rather of the opinion that private teaching is useful'.[209] That was a generous remark, considering the determined opposition Burns met at every turn from the FPSG for his having elevated licences to university degrees in surgery.

The FPSG was pursuing its own narrow agenda, but its actions had wide-ranging consequences outside its immediate quarrels. By the FPSG hindering the university from gaining a foothold in clinical teaching it strengthened the independent position of the hospitals. Hospital physicians and surgeons gained their own power base in the medical committees and began to wield an ever increasing influence over the medical profession.

A curious definition of bedside teaching was put forward in the document opposing the University's bid for clinical teaching. In it the FPSG jealously guarded the practitioner's complete control over medical treatment, suggesting that medical practice is idiosyncratic to the man in charge. The FPSG claimed that clinical lectures are:

[206] HB, 14/1/5, pp. 111–12.
[207] Ibid.
[208] Burns, *Royal Commission Evidence*, p. 132.
[209] Ibid.

not general dissertations on surgical subjects but a detail of the particular thoughts which pass thro' the mind of the practitioner during his management of the individual diseases under his care. A Surgeon to the Infirmary must first determine the nature of the disease by contrasting it with others, to which it is similar, next he must deliberate on the different means of cure which his science and experience suggest and must weigh in his own mind the reasons why he has recourse to others, when the first curative measures have not proved successful. Now clinical lectures are nothing more than the practitioner detailing to the students in a plain and easy way what passes in his own mind during these reflections and every well-educated surgeon who can practise judiciously is surely able to tell what he is doing, and this in the opinion of those who can judge of the matter is the only useful clinical instruction.[210]

This view echoed the FPSG's mind-set of accrediting surgical skills in the manner of a craft guild, a practice it had given up when it supported the teaching role of lecturers. It *now* disclaimed 'general doctrines' for clinical lectures because 'not the general nature of Fractures, dislocations, ulcers, but the particular circumstances and management of the individual broken, dislocated, or ulcerated limbs that are at the time before the young gentlemen who witness the practice' were central to the corporation's claim to license effectively.[211]

But even as the FPSG emphasised the practical work of the hospital doctor, he was emerging as a hospital specialist. The qualifications needed to become a clerk or dresser were acquired more selectively. In 1807 the candidates for surgeon's or physician's clerk needed to have only basic requirements. Candidates 'for either of these offices must produce certificates of their having attended, either here or in some other Medical School, the Classes of Anatomy, Practice, Chemistry, and Materia Medica; it being further necessary that Candidates for the office of Surgeon's Clerk should have previously acted as Dressers, for six months, in some Public Hospital'.[212] Even this ruling was met with dissent by some members of the FPSG, who wanted even less in the way of formal qualifications. In 1819 the medical committee saw dressers as on the bottom rung of successful careers and indicated that candidates 'should possess a requisite Medical Education'. This was defined as 'attending at least the following classes, during the time specified, viz., two six months courses of Anatomy, one six months course of the practice of Physic; one six months course of the Institutes of Physic; one six months course of the materia medica; one six months course of chemistry', with six

---

[210] HB, 14/1/5, p. 113.
[211] HB, 14/1/5, p. 114.
[212] HB, 14/1/2, p. 133.

months experience in an apothecary's shop.[213] This was equivalent to the surgeon's licence before the 1824 revision of the curriculum.

Standards had come full circle, the education of surgeons and physicians crossing all the boundaries set up by the Faculty, the University and the Hospital. The education of a practitioner promised to remain intact in Scotland, whatever the tribal rivalries. The *Medical Times and Gazette* of 1862 even paints a pleasurable picture of medical studies in Glasgow:

> The social aspect of this city will be a matter of interest to gentlemen who intend to spend the greater part of four years within its limits. Inferior, as far as Medical education is concerned, to no city in the empire, we believe it to be superior to some as respects the economy of living and social 'surroundings'. Glasgow landladies, as our experience went, are always civil and obliging. A student, desirous of a room for himself, may be accommodated for 5s. or 6s. a week, in addition to his commissariat requirements, a note of the expense of which is usually presented fortnightly. Students commonly live in pairs, and engage a sitting-room and bed-room from 8s. to 11s. per week. The favourite *habitats* of Medical students are Albion-street and George-street, and the short streets leading from the latter. On a Saturday in winter, the student fond of skating will find plenty of ice in the vicinity of Glasgow; and during his summer residence may transport himself in a few minutes to the green fields of the country, or, for a sixpenny steam-boat fare, to the lochs and scenery of the far-famed Firth of Clyde. And on Sundays he may, with great advantage to his mental and bodily health, preface the Medical lectures of the week with the sermons on still more exalted subjects, delivered by talented clergymen, in whom, at the present time, Glasgow is very abundant.[214]

---

[213] HB, 14/1/4, p. 32.
[214] *Medical Times and Gazette*, 29 November 1862, p. 579.

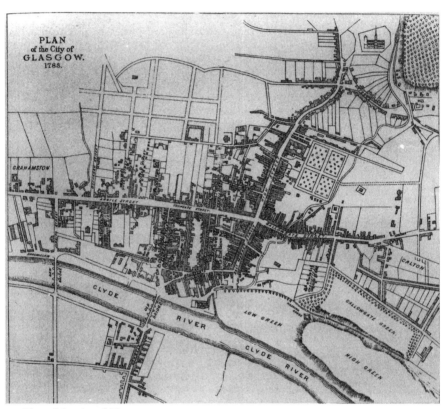

9. Plan of the city of Glasgow, 1783.

9

Time of Crisis

'In every one of its ranks, medical, surgical or obstetrical, numerous and melancholy examples of the utter incompetency of licenced charlatans present themselves daily'.[1] This damning assessment of licensing practices caught the temper of the times. Criticism from within the profession was rife, as the worth of diplomas and degrees was scrutinised. The men who argued for and against regulating the medical profession and, especially, about who was to be in control of it, were themselves products of a mosaic of learning where licences and degrees meant different things in a competitive world. Medicine needed a recognisable standard, and yet national and local loyalties, as well as overarching rivalries, made this a campaign as intricate and as far-flung as defeating Napoleon.

The man who coined the phrase 'licenced charlatans' was educated at Glasgow University and Anderson's Institution.[2] He was speaking to the Glasgow Medical Society which met in the Faculty Hall in Glasgow from 1814 onwards. George Watt was an adamant reformer in 1831, arguing that 'in the current age' public opinion was 'strongly set against the existence of all Corporations and monopolies' and wondered if they should not be 'done away with'.[3] This opinion was at the radical end of the spectrum, but one which touched a common vein of support. Dr Watt had just been appointed Professor of Medical Jurisprudence at Anderson's College and the reform debate did him no harm, probably adding to his grasp of the issues. He later became the Faculty representative on the General Medical Council from 1860 to 1863. As Watt weighed in on the debate over standards, a number of developments were impacting on the role the FPSG had chosen for itself. While it wanted to improve qualifying standards, it was also assuming monopoly powers, both in licensing surgeons and in accrediting teachers, an authoritarian mantle to which, as we shall see, many in the profession objected.

Improving standards in medicine through educational means was never

---

[1] RCPSG, 2/1/20, *Glasgow Medical Essays*, 18, no. 5, p. 3.
[2] Duncan, *Memorials*, p. 284.
[3] RCPSG, 2/1/20, *Glasgow Medical Essays*, 18, no. 5, p. 1.

in question in Scotland. Glasgow took the initiative in making curriculum requirements mandatory. The University and the FPSG acted in tandem in 1802 as new requirements were coordinated and increased from a minimum course to a full-fledged four-year curriculum fifty years later. This was one of the unsung successes of early nineteenth-century Scottish medical reform: devising a common medical curriculum that assured a basic level of knowledge for *both* surgeons and physicians. It was the cornerstone of the early emergence of the general practitioner in Scotland. One medical reformer remarked in 1830 that the strict division of physicians and surgeons survived only in England, the graduates of the Scottish universities having the ability to work in both areas.[4] Henry Robertson, MD, of Edinburgh, pointed, in 1827, to the legal standing of Scottish degrees because they were awarded on the basis of both public and private examinations, and not, as in England, 'with a prescribed matriculation alone'.[5]

In the west of Scotland the consensus on standards was built around unified, or very similar, curriculum requirements, a monumental achievement given the high mobility usual in those seeking accreditation. Glasgow University compiled a record of the places of study of its degree candidates in 1820, showing how common was the practice of changing from one university or lecture course to another. For instance, 'Mr Kelly studied in Dublin two years and last year at this university'; or 'Mr William Henry Booly studied at Edinburgh and London and two years at Glasgow'; or 'Mr Archibald Smith, three years Edinburgh and two years at Glasgow'.[6] Dublin was one of the main feeding grounds for Glasgow MDs, but Edinburgh and London feature in numerous biographies of eminent Glaswegian lecturers as well as in this snapshot of students, telling the tale that a 'free trade' in medical education was still common.

The cause of higher standards in medical education was argued for early on by one of its key proponents, John Burns, in 1799. In the preface to his teaching textbook, *The Anatomy of the Gravid Uterus*, he urged a rigorous attention to medical studies, admonishing students that:

No man will trust his own life, or the safety of those whom he holds dear, to

---

[4] T. Foster, *Observations of the Union ... Necessary between the Hitherto Separated Branches of the Medical Profession ... Addressed to W. Lawrence* (Chelmsford, 1830), p. 5.

[5] H. Robertson, *A Letter Communicated to the Monthly Gazette of Health Addressed to the Graduates of the Scotch Universities and General Practitioners on the Illegal and Unwholesome By-Laws of the College of Physicians of London, Establishing a Monopoly in Favour of the Graduates of Oxford and Cambridge, in Opposition to the Articles of Union between England and Wales* (London, 1827), p. 8.

[6] GUABRC, Senate Minutes, 1/1/4, 1 April 1820.

any man, however powerful his recommendations may be, if he one detects him to be a blockhead.⁷

He called a doctor who was ill-prepared 'a murderer' saying it was 'unwarrantable and criminal' to undertake practice when not sufficiently qualified.⁸ His sentiments served a practical purpose. Just three years later Glasgow University and the FPSG made it compulsory to complete similar basic courses of study as prerequisites to their degree and licensing examinations. The more haphazard system by which candidates were examined only on certain aspects of their knowledge was on its way out in both institutions. The 'trial' or examination before a board, sometimes accompanied by a thesis, was retained, but now certificates showing knowledge in certain basic areas were needed. As the University MD requirements and the licensing requirements for surgeons became indistinguishable, a basic standard of qualifications agreed across institutional boundaries for both surgeons and physicians was established.

In 1802 the FPSG tightened requirements for surgeons, while the University revised requirements for the MD degree. Glasgow University was pushing for higher standards as early as March of that year,⁹ and these activities were mirrored in FPSG activity and reports a few months later. At the University three years of study with one year's attendance at courses given by the medical faculty were now considered a minimum. A list of courses was stipulated and this can be said to have become the core curriculum of the medical school, even though certificates of study from other teachers outwith Glasgow were valid. Thus the curriculum requirements for an MD comprised anatomy, chemistry, surgery, pharmacy, the theory of physic, the practice of physic, materia medica, and botany. The examinations, to be taken after course requirements were fulfilled, were held in Latin and covered, first, anatomy and surgery; secondly, the theory and practice of medicine; and, thirdly, chemistry, materia medica and botany. The old practice of having to defend an aphorism from Hippocrates and comment on a case history was retained. Only the writing of a thesis was a matter of choice. It was to be defended before a 'comitia', a group which could include members of Senate, students and others who were interested. Andrew Napier was the first to receive his MD under these new requirements on 14 September 1802.

---

⁷ J. Burns, *The Anatomy of the Gravid Uterus with Practical References Relative to Pregnancy and Labour* (Glasgow, 1799), p. x.

⁸ Ibid., p. xx.

⁹ On these changes see Coutts, *A History of the University of Glasgow*, pp. 541–43; and GUABRC, Senate Minutes, 1/1/2 to 1802 and 1/1/3, 12 August 1802 to 8 January 1819 for developments in that period.

In June 1802 the report of the Committee on 'country licentiates' was laid before the FPSG, to be enacted into a by-law on 5 July of the same year. It stipulated that 'no candidate for a diploma should be admitted to examination without adducing proof of their having received previously an adequate professional education, viz., that they have attended two courses of Lectures in anatomy, one in the Practice of Physic, one in Chemistry, one in Materia Medica, together with a six months course at some Hospital, and that they have seen Practical Pharmacy in some Surgeon's or Apothecary's Shop at least during a period of three months'. John Burns moved that 'country licentiates' should also have attended a course of midwifery, which was accepted.[10]

In January 1810 John Burns added a course in surgery to the curriculum requirements for a FPSG licence. This completed the minimum systematic instruction the Faculty felt was essential for practitioners. It further proposed that curriculum requirements should be made compulsory. Obviously the profession was tightening accreditation procedures. Failing the production of specified certificates the candidates would not now be eligible for the diploma.[11] Burns's motion was entered among the by-laws of the Faculty. The motion also proposed that certificates accepted for licences would have to be signed either by university teachers or by members of the College of Physicians or Surgeons of London, Edinburgh or Dublin; or, indeed, members of the Glasgow Faculty. While this step fitted the logic of tighter standards at the time, it was to lead to grievous criticisms of the Faculty's monopoly powers in the not too distant future. In a prescient move Dr Peter Wright, prominent in Glasgow practice and five times President of the FPSG, realised the significance of regulating private teaching and asked the vital question whether the by-laws of 1802 could be honoured by certificates 'from a Licensed and regular Practitioner' whose lectures the candidate attended.[12] The Faculty were of the opinion that this was sufficient, validating previous practice.

In 1804 the University of Edinburgh reached an agreement with Glasgow that it, too, would set a minimum curriculum. This was less strict than Glasgow's, omitting, significantly, surgery and pharmacy, the main qualification areas of the surgical practitioner. The three-year requirement concurred with that initiated by Glasgow. Setting residency and curriculum requirements was meant to sort the sheep from the goats. Medical schools like Glasgow and Edinburgh drew a line between their degrees and places

[10] RCPSG, 1/1/1/4, Faculty Minutes for 7 June and 5 July 1802.
[11] RCPSG, 1/1/1/5, 1 January 1810.
[12] RCPSG, 1/1/1/4, Faculty Minutes of 22 February 1805.

like St Andrews and Aberdeen, or Cambridge and Oxford, which gave paper degrees (requiring no personal examinations) and did not stipulate a standard curriculum.

Glasgow University was the first to require both midwifery and surgery as MD qualifications. In 1812, following the 1802 and 1810 requirements enacted in the FPSG by-laws, Glasgow University made midwifery obligatory. The level of competence demanded of licencees remained high. In 1810 the FPSG examined in anatomy, surgery, physiology and pharmacy. Candidates wrote an essay and gave practical discourses (or demonstrations) of their skills in the treatment of cataract, hernia, or surgical procedures such as amputation of the thigh, lithotomy, aneurism or extirpating the breast.[13] In 1817 the CM required two courses in anatomy and surgery (one year each), and one course (of six months) in materia medica, pharmacy, chemistry, institutes of medicine, practice of medicine, midwifery, and in the principles and practice of surgery, together with twelve months attendance in a hospital ward. The difference between Scottish and London requirements could still be cited in 1836 by Alexander Hannay in his introductory lecture to the medical classes of Anderson's University: 'the London courses ... are only of three months' duration, and in most instances only of three lectures a week; whereas the courses given by the teachers of Anderson's University and others in Glasgow ... consist of five Lectures and examinations every week for six months'.[14]

The Faculty raised its requirements in three successive waves, the first from 1802 to 1810, the second in the 1830s and the third in the 1850s, just before the Medical Act. Between 1830 and 1834 the curriculum specified 'additions' in practical anatomy (meaning dissection), practical chemistry (laboratory chemistry), clinical medicine and clinical surgery. Dissection of legally obtained bodies was made possible by the passing of the Anatomy Act of 1832, making corpses increasingly available for medical education. Clinical medicine and surgery were added as requirements after the battle over clinical teaching at the GRI was resolved.[15] Botany and medical jurisprudence were added as new requirements in 1831 and 1834 respectively. Eighteen months attendance on the wards was made mandatory while the six months to be spent in an apothecary's shop was reduced from a previous requirement of twelve months. These regulations were printed in the 1845 booklet *Chartered Rights and Privileges of the Faculty of Physicians and*

---

[13] Compilation from RCPSG, 1/1/1/5, Minutes of the FPSG, 1807–20.

[14] A. J. Hannay, *Remarks on the Origin, Nature, and Importance of Medical Science ... An Introductory Lecture to the Study of Medical Science* (Glasgow, 1837), p. 13.

[15] See the previous chapter, 'Corporate Medicine and the Hospitals', pp. 293ff.

Surgeons of Glasgow. Already the course of study was impossible to complete under three years, equal to the terms of an MD qualification. Certificates needed to be signed by professors or lecturers at a university or members of the RCP or RCS, respectively, of London, Edinburgh or Dublin or by members of the FPSG.

The 1830s requirements for licensing as a surgeon met university standards by stipulating a written essay accepted and passed by the FPSG board of examiners. Latin, however, was needed by surgeons only in small doses. Two MDs in 1825 and 1827 published their 'probationary essay' as part of their 'Admission ... in Conformity' to regulations on becoming members of the FPSG.[16] This probationary essay was a far cry from the elucidation of a case or surgical operation hitherto used to examine candidates. The best of these essays could now be counted as contributions to medical science. George Watt's essay, for instance, was on the pathology of pulmonary consumption. This fitted well with the scientific papers given at medical society meetings with their close links to the FPSG, for example the Glasgow Medical Society.[17]

The 1852 curriculum was even more demanding. Compulsory courses covered the usual subjects, anatomy, practical anatomy (dissection), surgery or surgery and military surgery, chemistry, practical chemistry, theory or institutes of medicine (physiology), practice of medicine, materia medica, midwifery and diseases of women and children, clinical medicine, clinical surgery, medical jurisprudence and botany. One course of six months was seen as sufficient in each of these subjects with only anatomy, practical anatomy and surgery requiring more time. The attendance in a 'general hospital' was raised to twenty-four months, and the six months apprenticeship in an apothecary's shop was now entitled 'practical pharmacy' and could be taken in the laboratory of a surgeon, an apothecary's shop, in a hospital dispensary, or a chemist's or druggist's shop. The 1852 curriculum recommended that candidates also attend a three months course in an eye hospital, attend a lying-in hospital and a syphilitic hospital. Pathological and comparative anatomy, a course in natural history and languages (Greek, French, German or Italian) were specifically recommended.[18] These even stricter curriculum requirements ensured that the men who graduated between 1800 and 1850 became the first doctors to have a standardised professional profile.

---

[16] A. J. Hannay, MD, *On the Pathology of Puerperal Fever* (Glasgow, 1825); J. Spittal, MD, *A Summary View of the Practical Utility of the Stethoscope in Diseases of the Chest* (Glasgow, 1827).

[17] For information on the medical societies, see J. Jenkinson, *Scottish Medical Societies, 1731–1939* (Edinburgh, 1993).

[18] RCPSG, 1/1/1/8, Minutes of the FPSG, 1848–59, pp. 170–209.

As the requirements in Glasgow became interchangeable, the discrepancies of locally idiosyncratic accreditation seemed more of an obstacle then ever. As critics of the incorporations put it: 'should a practitioner, regularly licensed by incorporation A, or even a member of that incorporation, attempt to practice within the limits of B, his conviction and its consequences would be as certain as corporate illiberality and corporate jealousy could make them'.[19] The body of the profession wanted better standards applied and divisions between surgeons and physicians abolished: 'we conceive that the tripartite constitution of the profession may, with great propriety, be abandoned'.[20] Impartial boards organised regionally should recognise attainments 'acquired alone by industry and talent'.[21] The medical corporations 'though connected with one and the same profession', were – according to grassroots critics – 'coming into continued and disgraceful collision'.[22] Radicalised medical reform issued from ever greater numbers dissatisfied with the exclusive internal reforms undertaken by the incorporations. Membership of the incorporations became 'the root of those dark and time-honoured abuses which like unhealthful parasites, have clung round every tendril of our polity'.[23] This 'polity' was opposed to an elitist self-interest steering the reform debate. As curriculum reform continued – with which the profession had few quarrels – the issues of universality and openness crystallised. In Glasgow the key issue became 'monopoly'.

Graduate numbers rose, both in terms of university degrees and in licensing surgeons. After the University of Glasgow began to confer surgical degrees, choices on where to take the qualification grew, while differences in training were becoming increasingly negligible in Scotland. With its requirements very similar to an MD or CM university degree, the FPSG entered a prolific phase of awarding over 2000 licences to surgeons between 1801 and 1859. The exact number of recorded licences comes to 2070 with a breakdown for each ten-year period of 72, 302, 424, 457, 339 and 476 awarded respectively. On average this meant about thirty-five licences granted per annum but fluctuation rather than a regular pattern was usual.[24] From 1827 to 1837/8 Glasgow University awarded 202 CM degrees and 146 MDs. From 1837/8 to 1848 the CM was down slightly to 197 while the MD made considerable inroads, with 525 awarded. More than half of these (258) were examined

---

[19] *Glasgow Medical Examiner*, 1 (1831–32), 'The Combat Deepens', August 1832, p. 115.
[20] Ibid., p. 117.
[21] *Glasgow Medical Examiner*, 1 (1831–32), 'May 1831', p. 34.
[22] Ibid., p. 33.
[23] Ibid.
[24] Figures as compiled in RCPSG, 1/13/2/2B, W. Weir, Faculty Memoranda, p. 406.

between 1845 to 1848.²⁵ Overall this averages out to around twenty CM and a little over thirty MD degrees conferred each year between 1827–48. In the period after 1860 the CM degree taken at University flared into significance once more, doubling in numbers over MD degrees.²⁶ That means only ten fewer surgeons being accredited, while together the figures amount to fifty-five surgeons entering practice per year from Glasgow. MDs entering practice were slightly more numerous, and on the increase. Their areas of practice were now also shading into one another. No wonder the English Apothecaries Act, in its frequent revivications, was attacked ever more vehemently by the Scottish corporations and universities.

The breakdown of the income gained between 1786 and 1821 from 'county' and 'town' licentiates – with 'county' licences predominating by far – shows the success of the FPSG in regularising surgeon's accreditation. The revenue received over this thirty-five year period came to £296 18s. 0d. with licence fees at £2 2s. 0d. in 1786, £4 4s. 0d. in 1799, and £5 5s. 0d. from 1803 to 1821. Town licentiates' fees were higher, costing either £15 15s. 0d. or £21.²⁷ The licences awarded between 1786 and 1799 were only about two per year. The surge in licences corresponded to the introduction of the curriculum requirements, eleven country licences being awarded in 1802, with five to seven awarded until 1808; another jump is then visible, to around twenty per annum between 1810 and 1812. In 1813 eighteen licences were awarded plus nine diplomas for midwives and one apothecary's licence (other individual diplomas are sprinkled in with the tabulation of licentiates). From 1814 to 1816 the licences leapt significantly from fifty-two to more than seventy, only to fall again to an average of about twenty per year between 1818 and 1821.²⁸

In comparison Oxford and Cambridge awarded only fifteen medical degrees annually in a ten-year period, while Edinburgh and Glasgow had an annual rate of about 160 MD degrees awarded.²⁹ These figures are all hard to pin down, given the itinerant style of medical learning, but one thing is eminently clear: Scottish licentiates and MDs were both more numerous and better educated than in England. However, the Royal Colleges in London, close to Westminster, and the four MPs allotted to Oxford and Cambridge, gave power and influence to medical institutions south of the

[25] Coutts, *A History of the University of Glasgow*, p. 546.
[26] Ibid., p. 577. The figures given here are: from 1862–91 the MD was conferred on 687 men while the CM was awarded to 1794.
[27] RCPSG, 1/1/1/6, Faculty Minutes, 2 July 1821.
[28] Ibid.
[29] I. Loudon, *Medical Care and the General Practitioner, 1750–1850* (Oxford, 1986), p. 271. The period covered is 1826–35.

border disproportionate to their educational achievements. Andrew Buchanan, appointed Professor of Materia Medica at the extra-mural Anderson's College in 1833 (and no friend of the 'ancient' universities), saw the situation with unadorned practicality:

> Now there is no mode of being legally constituted a Physician but by obtaining the degree of Doctor in Medicine, either from an English or from a Scottish University, and, to do the Scotch Universities justice, the system of education which they prescribe, irrational though it be, is nevertheless much better, and cheaper, than the system of Oxford and Cambridge. The Universities of Edinburgh and Glasgow are preferred, not because they are good, but because there are none better to be had.[30]

Andrew Buchanan's own institution possessed teachers of distinction, but it, in turn depended on diplomas and licences being granted by the incorporations, St Andrew's University or the Army and Navy Boards. Edinburgh and Glasgow universities rejected 'private' lecturers in either city – Anderson's faculty being considered 'private' – while recognising lecturers in Dublin and London. The parting shot of Moses Buchanan, yet another 'private' lecturer at the Portland Street School, was: 'they fear such rivalry, knowing well that if their baneful monopoly were abolished, many among them might lecture to empty benches'.[31] As these remarks attest the competitive market in medical education was as fierce at home as elsewhere.

Glasgow University was the first to make the MD degree dependent on curriculum and residency requirements. The Glasgow curriculum in medicine and surgery (including the innovative CM degree) was in place before a draft medical reform Bill went to Parliament in 1816. This legislation only mentioned the Colleges of Surgeons in London, Dublin and Edinburgh. It did not mention any Glasgow-based institutions. National pressures to standardise qualifying procedures tended to ignore Scottish achievements and were based squarely on an English model and *English* politics.

In 1856, as another wave of legislative proposals was in the offing, James Lawrie, Burns's successor as Professor of Surgery in the University of Glasgow, not only stressed the stricter requirements of Scottish qualifications but, more importantly, that a double qualification was becoming standard in Scotland. He drew attention to the fact 'that a majority of the medical men in Scotland hold both a University degree and a surgical diploma'.[32]

---

[30] A. Buchanan, *Of Monopolies in Learning: With Remarks on the Present State of Medical Education* (Glasgow, 1834), pp. 7–8. He became Professor of Physiology in Glasgow University in 1839.

[31] M. S. Buchanan, *Remarks on Medical Reform* (Glasgow, 1846), p. 22.

[32] J. A. Lawrie, *Letters on the Charters of the Scotch Universities* (Glasgow, 1856), p. 22.

The difference between Scottish and English medical practitioners, he wrote, consisted mainly in the larger proportion of Scottish men being graduates of a university, two-thirds of whom now held MD degrees.[33] He made clear that there were no 'pure' physicians left in Scotland: 'The wealthy city of Glasgow, with its 400,000 inhabitants, has not *one*, we are *all* general practitioners'.[34] He added that 'fully one-half of the resident Fellows of the Royal College of Physicians of Edinburgh hold surgical diplomas; while two-thirds of the Fellows of the Royal College of Surgeons of Edinburgh, and of the Glasgow Faculty, are Doctors of Medicine'.[35] That most practitioners in Scotland were members of an incorporation, but also held university degrees meant that a large portion of the medical profession had 'voluntarily submitted themselves to a double qualification'.[36]

This situation was reflected in the 1823 and 1831 rules regulating attendance in the Glasgow Royal Infirmary. The Medical Committee's report on the proper qualifications of a Physician to the Infirmary stated:

> That medical practitioners in Scotland consist at present of three classes. The first comprehends those who possess a degree of Doctor of Medicine, who confine their practice exclusively to the treatment of internal diseases, who dispense no medicine to their Patients and who from their expecting higher fees than the rest of their Brethren are for the most part employed only in consultation with other members of the profession. The second class consists of those who also possess a medical degree, but practice both internal and external diseases and dispense medicines to their Patients. A third class comprehends those who without possessing a degree of Doctor of Medicine practise in the treatment both of internal and external diseases and dispense medicine to their patients – Practitioners of the first class are usually called Physician and are few in number. Those of the other two classes are generally named Surgeons and constitute the great body of the profession. The education of the two first classes are precisely the same and their degrees, with the exception of those procured from the College of Aberdeen and St Andrews, may be considered as proof that they have received a complete medical education. Of the third and by far the most numerous class, a large proportion have received as complete an education as the other two classes and enjoy most deservedly as large a share of the public confidence as those who have acquired degrees.[37]

This clear depiction of Scottish medical practice emphasised the equal education of all its 'classes'. In May 1831 the FPSG wanted to know about 'in future allowing Doctors in Medicine, tho' general practitioners, to be

[33] Ibid., p. 23.
[34] Ibid.
[35] Ibid.
[36] Ibid., p. 24.
[37] HB, 14/1/4, GRI Managers' Minutes, iv, pp. 193–94.

eligible to the situation of Physician to the Hospital'.[38] The FPSG thought all MDs should be eligible because:

> The resident members of Faculty are nearly eighty in number, of whom one half are Doctors in Medicine, and the other half Surgeons, but the education of both classes are precisely the same, and in point of acquirement and practice they are of equal footing. Of the Doctors in Medicine, the great majority are in active practice as General Practitioners, a few seldom practice Medicine at all, and two act chiefly as consulting Physicians.[39]

Throughout all the bitter fighting over control of the profession and the ways and means of regulating it, the fact should not be lost sight of that Glasgow University and the FPSG redefined accreditation requirements successfully. After 1816 they had to do this with a wary eye on possible legislation. It was the many Bills before Parliament that left many of their achievements hanging or turned them into dogfights over jurisdiction. However, neither the FPSG or Glasgow University could be judged selfless, nor willing to jettison their local or national political advantage. Educational improvements were often relegated to the back benches as ancient medical fiefdoms were defended, and destructive infighting crippled more democratic and progressive solutions.

Medical reform in the 'national' arena turned on the English problem of no accreditation or regulation of the general practitioner, so different from the Scottish situation. The control of the London Royal Colleges did not extend to the provinces. This vacuum of provision in England, there being no provincial colleges on a sufficient scale to cater to rising numbers, swept the Scottish reforms before it, mowing down the progressive with the lacklustre. These English-based moves (often confused with 'national') to regulate parts, or all, of the profession impacted heavily on the Scottish institutions. They were not going to use Scottish medical education, or the higher status of its surgeons and surgeon-apothecaries (i.e. general practitioners) as a starting point. Scottish reformers opposed to the Faculty were particularly bitter about the English Colleges and the Apothecaries Society:

> We find first a Royal College of Physicians in possession of a charter, granted by Henry the Eighth ... with which they would lord it over their brethren in the nineteenth – a charter whose basis is the paramount absurdity that Physicians can alone be educated at the Universities of Oxford and Cambridge – schools of no note in the annals of Medicine ... We find next a Royal College of Surgeons governed by a self-perpetuating council, distinguished in all their acts by a selfish anxiety for their own interests ... and we come lastly to an Apothecary's Company,

---

[38] HB, 14/1/6, GRI Managers' Minutes, vi, p. 192.
[39] HB, 14/1/6, GRI Managers' Minutes, vi, p. 191.

the essence of whose existence is the astounding fact that a person called upon to treat disease must not use the means of doing so – must not administer those remedies in the study of which his life may have been spent.[40]

Since at least the seventeenth century the surgeon-apothecary in Scotland had operated freely across the wide spectrum of medical practice.[41] The President of the FPSG was a physician, and the Visitor, always a surgeon, acted as the head of the Faculty's surgeons, so that there was in effect a joint or dual presidency. 'During the whole period in which this dual presidency existed there is no evidence of the slightest jar or want of harmony between the two parties', as the Faculty historian A. Duncan observed.[42] The distinction between physician and surgeon in highest office of the FPSG was lifted in the 1820s when the by-laws were changed to allow surgeons to become President.[43] Outside the internal leadership of the FPSG, however, and especially as relations with the university broke down, things were not as smooth.

The licence of the Faculty was in reality a qualification for a surgeon in general practice. Even before the curriculum was codified (and certainly after) the skills examined were broad enough to be suitable for general practice. Prosecutions of surgeons within the jurisdiction of the Edinburgh Surgeons in the east, and the Faculty in the west, tended to be for lack of qualifications rather than for overreaching boundaries, such as compounding or administering drugs which became a central issue in London. Concern centred on local jurisdiction, not with the range of medical practice. In 1811 MDs from various universities were called to account before the courts by the FPSG for not having a surgical licence, pointing strongly to their engaging in 'general practice'. This proved a watershed because the Faculty was prosecuting not quacks but 'general practitioners' holding legitimate patents. As the opposition put it: 'In 1811 they charged those *who had been examined* [our emphasis] and were practising in this city, with the sum of £16 15s. in addition to £5 5s. paid for their certificate of qualification; and in 1824 they commenced prosecutions against several surgeons who refused payment, which prosecutions are still pending. In 1821 they assumed to themselves the sole privilege of lecturing in Glasgow and refused to receive the tickets of any lecturer in the city if not a member of their body'.[44] Hackles were raised

---

[40] *Glasgow Medical Examiner*, 1 (1831–32), column heading: 'May 1831', p. 33.
[41] See D. Hamilton, *The Healers: A History of Medicine in Scotland* (Edinburgh, 1981), p. 162; H. Dingwall, ' "General Practice" in Seventeenth-Century Edinburgh: Evidence from the Burgh Court', *Social History of Medicine*, 6 (1993), pp. 125–42.
[42] Duncan, *Memorials*, p. 93.
[43] Ibid.
[44] *Glasgow Medical Examiner*, 1 (1831–32), column heading 'May 1831', p. 37.

– while reform was applauded by the general practitioner he now rose against the narrow rule of corporate interest. The body of the profession deplored that they remained unrepresented, 'remained without an advocate attached to [their] interests in the councils of the nation',[45] because only corporate interests bartered the spoils of reform amongst themselves.

It seems clear that in the first half of the nineteenth century in Glasgow the tripartite division of the profession did not translate into the same power structure as in England. It was not the overriding concern of the Scottish medical profession to enforce divisions along ancient fault lines. They wished to perpetuate what the FPSG had stood for in the Enlightenment, attention to educational and professional needs in a mobile and commercially linked profession. Liberal views were strong in Glasgow and battle was rife within the key networks of Glasgow medicine: the Faculty, hospital posts, private practice, and, of course, professional or lectureship posts in the university.

In England most general practitioners held the licence of the Society of Apothecaries in addition to their surgical licence, yet those who practised pharmacy or midwifery (essential elements of general practice) were barred from becoming council members of the Royal College of Surgeons. The same prohibition applied in the Royal College of Physicians, and thus general practitioners were denied any controlling influence on the policy of these bodies, despite accounting for the majority of the rank-and-file membership. Furthermore, neither of the English colleges would take on the education and examination of general practitioners, for fear of losing the social status of pure physicians or surgeons.[46]

General practitioners in England attempted to gain professional recognition and a political voice within the College of Surgeons or via a new college of general practitioners, or through new national organisations like the Provincial Medical and Surgical Association (later the British Medical Association). The medical reform debate was initiated as a campaign to establish the independent professional identity of the general practitioner. At the beginning of the nineteenth century, the general practitioner in England was indistinguishable professionally from the army of irregular practitioners (unlicensed and alternative practitioners) and was denied recognition in his own right by the English Colleges. Against this background general practitioners

[45] Ibid., p. 1: 'Introductory Address' setting out the issues of medical reform for the *Examiner*.
[46] See I. Waddington, 'General Practitioners and Consultants in Early Nineteenth-Century England: The Sociology of an Intra-Professional Conflict', in J. Woodward and D. Richards (eds), *Health Care and Popular Medicine in Nineteenth-Century England* (London, 1977), pp. 164–88; and idem, *The Medical Profession in the Industrial Revolution* (Dublin, 1984) pp. 33–49.

as a group began to develop a sense of autonomous professional identity. The founder of the Lincolnshire Benevolent Medical Society, Dr Edward Harrison, initiated a campaign for conservative reform in 1804–9 to codify the training of the three classes of practitioners. The plans were blocked by the Royal College of Physicians of London. During the next thirty-five or so years general practitioners conducted a continuing campaign for professional recognition, which was only spurred on by the reactionary Apothecaries Act of 1815. The government's attempt to regularise the status of the apothecary followed the line of the English colleges in upholding the old professional divisions and firmly associating the general practitioner with the humble dispenser of drugs. The underlying problem of the emergence of the new category of the general practitioner was pressing but not tackled immediately.

The Provincial Medical and Surgical Association (PMSA) was founded in 1832 by the Edinburgh-educated Worcester physician, Charles Hastings. This body finally became the British Medical Association in 1855. Largely composed of general practitioners from outside London, this body aimed to represent a unified profession, was open to all branches and allowed all types of practitioner to hold its offices.[47] By the late 1840s and early 1850s three basic demands had crystallised. The first was political representation for general practitioners (either within existing corporations by admission to the higher offices, or through a new college of general practitioners). The second was provision of a broad-based medical education and examination. The third was recognition of their equal status as practitioners by inclusion in a common register.[48]

Medical journals were important in facilitating and emphasising the coherence of the general practitioners as a separate professional group. The *British Medical Journal* had evolved from a number of antecedents by 1857. The *Lancet* was also of critical importance, particularly under the editorship of Thomas Wakley, as a campaign journal for the reform movement. Wakley used the journal to criticise the abuses of the corporations and he engineered a House of Commons Select Committee on Medical Education in 1832. This was the Select Committee before which the Glasgow medical fraternity emphasised the quality of Scottish medical education. The de facto general practitioner had been working hard for some two hundred years, in the guise of the surgeon-apothecary and was, in Scotland, extant by the expedient of

[47] The Association became the British Medical Association in 1855, taking this name from a rival London-based body which had folded in 1845, ostensibly because its plans for medical reform were not gradualist enough. See Loudon, *Medical Care and the General Practitioner*, pp. 280–82; N. Parry and J. Parry, *The Rise of the Medical Profession* (London, 1976), pp. 123–30.
[48] See, I. Waddington, *The Medical Profession*, pp. 87–91.

taking dual qualifications. Long before the Medical Act of 1858, surgeon's licences and MD degrees from Glasgow and Edinburgh, the cities that fielded most medical qualifications had carved a niche for the GP. Eventually the Medical Act provided a de jure common register of all practitioners and allowed a *single* qualification as the requirement to practice. This brought a common denominator into play by distinguishing only between registered practitioners and those beyond the pale.

Trouble Ahead

Licensing by the FPSG took an upward turn between 1802 and 1821, but membership of the Faculty was declining rapidly between 1800 and 1851. Only two members joined in 1800, and fewer than five in the years 1801–11. Only the years 1813 and 1819 brought double figures. All the remaining years saw on average three surgeons or doctors becoming members.[49] Dr Alexander Dunlop Anderson pointed openly to the possible extinction of the Faculty in 1849, as the result of seriously dwindling membership. At the core of the problem was the contribution to the Widow's Fund.

The Widow's Fund, perhaps inadvertently, contributed to the most significant development of the period between 1800 and 1858, the financial ring-fencing of a governing medical elite, making FPSG membership exclusive and expensive. The Faculty forfeited its role as the representative of the majority of practitioners, while the FPSG engaged in medical politics in terms of its own interests. It was the deliberations of the few that structured the responses to national legislation, to medical regulation and to licensing procedures. The governing council of the FPSG took many important decisions, while excluding a swelling number of medical professionals who were in practice but *ante portas*.

In the 1830s the FPSG was heavily criticised, as were the corporations in England, for excluding the voice of the run-of-the-mill general practitioner. The 1811 restructuring of licensing fees was the starting point for this undemocratic development. With new tariffs set for country and town practitioners, at five guineas and twenty guineas respectively, the ruling council thought there would be no further excuse for not getting a licence. Accordingly, measures to prosecute practitioners unlicensed by the FPSG were set in motion.

But the 'cheap rate' for licences corresponded with an increase in the fees for the Widow's Fund, now a requirement for full membership. The motion to reduce fees stipulated that licences could be granted to practitioners who

---

[49] Duncan, *Memorials*, pp. 268–92: taken from the Roll of Members.

'should have no right to be members or have any claim on the funds of the Faculty Library or Widow's Fund'.[50] In 1816 the contribution to the Widow's Fund went up to £150, a truly punishing sum at the time.[51] This fund was an assurance scheme for its members in the event of disability or the death of the principal breadwinner. The rise in entry money had jumped from £50 in 1792,[52] when the fund began (on 8 June) as a regularisation of the earlier practice of annual charitable outlay (amounting to about £100 per year) made to the poor or struggling members of the medical profession, their widows, children and other relations.[53] The core funding came from a sum transferred from the Ordinary Faculty Fund (which was used for instance for legal expenses). This process was repeated three times and the total transfer by 1847 was £6957 2s. 9d.

The Widow's Fund brought the Faculty into disrepute because it made membership unobtainable for most practitioners. Criticism of the Faculty was scathing: 'Their entry-money in 1777 was for those who had been apprentices of different standings from £3 3s. to £10 10s. and to a stranger £21. In 1792 they increased their entry-money to £50, for entrants of all denominations and formed a Widow's Fund; in 1811 it was raised to £100 and in 1816 the entry money became £150 with compound interest upon this sum for every year the entrant was above 25'.[54] Licences, although feed at the lower rate, were compulsory for any who were not 'pure physicians'; in reality this meant almost all medical practitioners, as there were few 'pure' MDs. Moreover, the monopoly in teaching accreditation which the FPSG imposed meant anyone teaching for profit outside the university was subject to their membership terms. This was the issue that raised the wrath of the profession: 'In 1821 they [the FPSG] assumed to themselves the sole privilege of lecturing in Glasgow and refused to receive the tickets of any lecturer within the city, if not a member of their body. In 1829 a Memorial on this subject ... signed by seventy Physicians and Surgeons was submitted to the Faculty to which no reply has been received'.[55]

Legal advice was taken on the fund's inauguration from John Millar, the then Professor of Civil Law at the University of Glasgow. He supported the action of the FPSG and a charter of erection and seal of cause were duly obtained from the Glasgow magistrates and council, sanctioning regulations

---

[50] RCPSG, 1/1/1/5, 5 November 1811.
[51] Duncan, *Memorials*, pp. 159, 209.
[52] Ibid., p. 209.
[53] Ibid., p. 158.
[54] *Glasgow Medical Examiner*, 1 (1831–32), p. 36.
[55] Ibid., p. 33.

prescribed by the Acts of Parliament regulating the operations of Friendly Societies.[56] Thus after 1792 all new Fellows were obliged to pay not only the entry freedom fine, but also a sum to the fund according to age and scale of future benefit contemplated for dependants. There is some suggestion that a portion of Licentiates' fees were also diverted into the fund. The entrance and fund charges were again increased from Whitsunday 1841, but a new lower rate of contribution and benefit was also offered. However, a delay in having the new regulations approved by the Lord Advocate, as was necessary under the Friendly Society Acts, led to a threatened legal challenge by three Fellows in April 1843.[57] This action focused Faculty minds on the issue, and legal advice was again sought in 1847 on the whole problem of the fund. The opinion of counsel was that the disputants had no case but more generally that

> the Faculty being a public Corporation established by public authority for public purposes could not refuse to admit to it any party qualified according to law, and willing to pay the usual Freedom fine, because such party would not consent to become a Contributor to the Widows' Fund.[58]

The issue was resolved by the 1850 Act for regulating the privileges of the FPSG which abolished the fund and created both the Fellowship and the Council.

While it may have benefited destitute dependants, the Widow's Fund divided the FPSG from its historic roots in full membership for all surgeons and physicians practising in its chartered territories. Instead it created an administrative and political elite. As the full membership dwindled, and those the FPSG licensed were excluded from decision-making, the governing elite pursued its own politics and became less accountable. It became harder for young practitioners to enter or influence the FPSG. Although its wealthy elite was comparatively liberal and highly-regarded professionally, it was nonetheless a tightly-knit interest group.

Membership of the FPSG developed into a significant financial hurdle for aspiring practitioners, while at the same time accreditation as a teacher recognised by the FPSG became more difficult. These two developments were inextricably linked. Only a member, not a licentiate, was recognised as

---

[56] RCPSG, 1/1/1/8, 3 January 1848 meeting, paper extracted in minutes entitled 'Memorial for the Faculty of Physicians and Surgeons of Glasgow. For the opinion of Counsel', pp. 9–10. This legal opinion came from Andrew Rutherford and Alexander Dunlop, and is dated 18 September 1847.

[57] RCPSG, 1/1/1/8, MC, p. 13. The three were Alexander Fisher, Alexander King and John Crawford, all of Glasgow.

[58] Ibid., p. 17.

a valid signatory of a certificate for students taking a course. The curriculum requirements made certificates mandatory and these were restricted to FPSG members or bodies outside Glasgow as listed in the 1810 enactments (members of other colleges and the universities). In 1818 James Armour, the future Professor of Midwifery in Anderson's College, was clearly struggling as he was too poor to pay the membership fees: 'I mean the borrowing of money to pay the Faculty. I had once pleased myself with the idea of submitting to any kind of deprivation or hardship rather than load myself with borrowing ... But these notions have been obliged to yield to necessity'.[59]

While the representatives of the profession argued the ins and outs of medical qualification, those hoping to practise evaluated their chances. They discussed career prospects, particularly the benefits of teaching, and how to make ends meet. 'It will take some time before you can relish the common place tittle tattle of a Town filled with weavers and mechanics', wrote the Glaswegian Harry Rainy, later Professor of Forensic Medicine at Glasgow University, to his fellow surgical licentiate William Mackenzie, son of the muslin manufacturer James Mackenzie and founder of the Glasgow Eye Infirmary, while the latter was abroad. Rainy teased Mackenzie, whose future medical career he helped fashion as 'a person whose mind [is] refined by the classic scenery of Italy'.[60] But what about less fortunate friends, such as James Armour, Rainy asked, who had no inherited wealth. Rainy wanted to know 'whether he [Armour] should take a shop'.[61] Armour himself asked Mackenzie about the comparative securities and advantages of maintaining the traditional surgeon's 'shop' while evaluating the advantages of 'private' teaching:

> The medical school [this refers to both university and extra-mural teaching] here this session has by no means fallen off in the way I imagined it would. Pattison raised his fee to 3 guineas – has between 90 and 100 students – lectures every day an hour on Anatomy and another on Surgery – Jeffray has 150 students at the least. Were you to attempt lecturing here you would undoubtedly get some of these: for some years you must expect few, but that any where. You would have an opportunity of being elected one of the Infirmary Surgeons in a few years – a thing I once heard you say you would very much wish. Another formidable objection is paying £150 to the faculty but you might Lecture I believe and have your tickets received as a Licentiate or if that would not do, could you not take a Diploma just now whilst you are in London as Allan Burns did, and

---

[59] RCPSG, 24/2/50, Dr William Mackenzie's Letters, James Armour to W. Mackenzie, 7 April 1816.

[60] RCPSG, 24/2/124, Dr William Mackenzie's Letters, Harry Rainy, to W. Mackenzie, 4 December 1816.

[61] Ibid.

then Lecture as one of the Licentiates of the faculty of Glasgow – this however I merely throw out as a hint to be considered by you – you can ascertain the expence and advantage of the two plans ... I have just now asked Brown what he thinks of your returning to Glasgow and he agrees with me in thinking that provided you can be content with being a *moderately great* man it is from many circumstances the best place after all —he thinks you may even Lecture. When I talk to you of joining me in my shop it is with this view. Were you coming to Glasgow with a view to Lecture much of your time must necessarily be employed in getting ready your rooms for that purpose – were [you] to take a shop you would find it would require a good deal of trouble and some little expence to fit it up – but more than this had you a shop you must attend to it, at the beginning at least, otherwise you would never get any practice. Now were you with me, the shop is ready – you might go about your lectures whilst I would be in the shop & and we could easily send for you for any one who wanted you ... If you come here & think we shall be better by ourselves I shall give you all the assistance I can in procuring & fitting up another shop. I am not quite sure of your opinion that in order to get practise one must lecture – on the contrary one is much more likely to get one who is always to be found when a patient calls. I have been seven months in practice I have made about £20, have got £3 – I pay £20 for my shop, will not sell more than £20 medicines yearly.[62]

The surgeon James Armour was barely breaking even although he was among the group of medical men who were making their way professionally. He and Harry Rainy kept a very close eye on the possibilities of entering the ranks of 'private' medical teaching, a recognised element in establishing a lucrative practice. Armour believed the combination of the traditional 'shop', the business premises at which surgeons dispensed, combined with teaching was a formula for success. Unfortunately, teaching accreditation was a minefield, both, as he mentions, in attracting sufficient numbers of paying students and in paying the membership fees of the FPSG.

The power to restrict teaching to those approved by the incorporations and the universities was seen in terms of a monopoly. Teaching became a closed shop. The extra-mural schools, like Anderson's College and the Portland Street School staffed their medical lectures only with men who were members of the FPSG because they were dependent on its licensing. Indeed textbooks on medicine from Glasgow all carried the imprimatur of 'Member of the Faculty of Physicians and Surgeons'.

Andrew Buchanan's medical talk before the Glasgow Literary and Commercial Society in 1833 on monopolies in learning took up the issue. He pointed out that mercantile communities are sharply aware of the dangers

---

[62] RCPSG, 24/2/116, Dr William Mackenzie's Letters, James Armour to W. Mackenzie, 9 March 1818.

of monopoly. He castigated the universities for granting a monopoly to a *single* person in each subject so that to become a physician a man 'must be trained by this individual', it being of no consequence 'that he may have been originally unfit ... or that he since may have become lazy, or dissipated, or superannuated'.[63] The monopoly of the incorporations he saw in a somewhat better light because the competition was greater, the Edinburgh College of Surgeons having 103 and the FPSG having ninety-nine members, the monopoly thus being vested in a hundred individuals.[64] Buchanan proposed a national system of education that would dispose of both these monopolies and introduce six licensing colleges for both teachers and 'practitioners in various branches of medicine'.[65] These would be spread throughout the United Kingdom (London, Liverpool, Glasgow, Edinburgh, Dublin and Belfast). He warned: 'Knowledge is in this country an interdicted commodity having no marketable value but when exposed for sale by a monopolist'.[66]

The Society of Regular Physicians and Surgeons within the City of Glasgow was the first breakaway organisation to set themselves against the FPSG. In a meeting of 14 April 1812 the FPSG Committee on unlicensed practitioners warned 'that a very daring attack has been made upon the powers and privileges of the Faculty by a number of irregular practitioners who the Committee are informed have made application to the Magistrates of the city for a Seal of Cause appointing them a Society of Regular Physicians and Surgeons'.[67] This may or may not have been the same group who sent a signed protest to the FPSG in 1819.

This latter memorial raised the question of what the fees of entry meant in terms of entitlement. It noted that the Faculty 'styled' some as members and 'we, the Licentiates entertain considerable apprehension whether the Faculty really does or does not intend to concede to us these privileges'.[68] By 'these privileges' the memorialists meant the normal rights given to members, especially the exemption from burdens such as quartering soldiers, military ballotting, watchings and tolls. They deplored the division from full members of 'a distinct species of practitioners' namely the Licensees and demanded clarification of their relationship to the corporate body that was meant to represent them. They stated, in anger, that the twenty guineas of the examination fee only 'purchased for us immunity from prosecution by

---

[63] Buchanan, *Of Monopolies in Learning*, p. 7.
[64] Ibid., p. 9.
[65] Ibid., p. 18.
[66] Ibid., p. 21.
[67] RCPSG, 1/1/1/5, Faculty Minutes, 14 April 1812. We have found no further traces of this group to date.
[68] RCPSG, 1/1/1/5, Faculty Minutes, 6 December 1819, p. 236 (not signed).

the Faculty'. Twenty-eight licentiates signed, including William Hood and James Brown, who may or may not be the same Brown who supported the Glasgow Lying-In Hospital.

Immunity from prosecution was not to be taken lightly. 'How disgraceful to the age we live in', commented the *Glasgow Medical Examiner*, 'that the Physician or Surgeon who may legally and creditably exercise his abilities in one portion of the island, dare not do the same in another district without incurring pains and penalties and subjecting himself to a ruinous and vexatious law suit?'[69]

The overriding concern of these splinter organisations was democratic because, as the memorialists rightly noted, the exclusion of the licentiates from membership transformed the executive of the FPSG into an elitist and unaccountable semi-public body. In 1824 further signs of an unabated opposition by some to the managerial elitism now practised by the successful and wealthy 'full members' appeared. The Glasgow Faculty of Medicine was founded as another mutation of corporate displeasure at exclusive practices. Meeting in the so-called Cowpox Hall, they drew up a constitution which invited *all* 'Fellows ... having Medical Degrees from the universities, or Surgical Diplomas' from all the Colleges, or 'any respectable Foreign College or School of Medicine or Surgery, or who have passed their examinations at the Military or Naval Medical Boards', to be eligible for membership.[70] Twenty-seven medical practitioners signed as members. In 1833 they sent a petition to Parliament to protest 'the restrictions imposed by the unconnected and independent Medical Corporations of this country [which] ought to be entirely removed'. Anyone who was a licentiate of 'any of the existing examining bodies' should be deemed competent to practise in any part of the British Dominions. Fees ought to be fixed by legislation and not the 'caprice' of the licensing boards. All licensed practitioners should be free to teach and no lecturer 'should be considered eligible to the office of examinator'.[71] This bill of rights for British licensed practitioners was echoed, surprisingly, in meetings held by the Glasgow Medical Society. William Hood, who had protested earlier, became the fifty-third member of the Glasgow Medical Society in 1828, but died in 1830. Attacks on the Faculty did not mince words, especially since the *Glasgow Medical Examiner*, an organ proud to be an opponent of monopoly, stated its mission as dragging 'the illiberals of this body and their measures before the public'.[72]

---

[69] *Glasgow Medical Examiner*, 1 (1831–32), p. 57.

[70] RCPSG, 40/1, Minutes of the Meetings of the Glasgow Faculty of Medicine, 1 October 1824.

[71] RCPSG, 40/1, Minutes of the Glasgow Faculty of Medicine, 24 June 1833.

[72] *Glasgow Medical Examiner*, 1 (1831–32), p. 131, under the title 'September 1831'.

The governing elite of the FPSG was not kind in rejecting pleas for either reform or greater openness. The 1819 Memorial of the Town Licentiates was repudiated with the following words:

> You have practised unmolested and we never heard of any other expectation being entertained till we perused your memorial ... We can hardly believe that any of our members would be so ignorant of laws recently enacted as to promise you other privileges. If they have done so, they are unauthorised and we as a Body are not answerable ... *Licentiate* is the appropriate term to express the condition in which your diplomas have placed you. You have a name and that name conveys an exact idea of your character. The Laws we have adopted in your case are comfortable to the practice of other medical bodies like ourselves who make a distinction betwixt licentiates and members.[73]

The subsequent Memorial of 1829 to the Faculty questioned fee levels, privileges and monopoly rights, and was signed by seventy medical men 'Doctors of Medicine, Army and Navy Surgeons, Surgeons holding diplomas from the London and Edinburgh Royal Colleges and from the University of Glasgow' and licensees of the FPSG, styling themselves in protest as 'a species of member who have nothing to do with the Faculty Laws but to obey them, nor with the funds but to pay the sums ... a species of serf'.[74] No reply was forthcoming by the Faculty a failure repeatedly admonished in the *Glasgow Medical Examiner*. The political conclusion of the reform groupings could be no other than 'to submit our grievances to the legislature'.[75]

Medical reform debates touched all echelons of the profession, regularly raising the ghosts of division. The talented and ambitious argued solidly for higher educational hurdles. In a paper read before colleagues in the Glasgow Medical Society in 1824, the licentiate James Brown from Paisley voiced his concern about lowering standards. He was keen on restrictions when others were adamant on opening the profession. Brown wrote:

> Ask the public in general and after having it explained that such and such are the qualifications required of students by the faculty [the Faculty of Physicians and Surgeons of Glasgow]. The common sense feeling on the subject will be that certainly the faculty believe such a course to be sufficient for all the purposes of medical education. It is from a belief in this as a first principle that many a poor youth both in town and country is drawn from his humble occupation as a tailor or a shoemaker to enter upon this fascinating course which promises to make a doctor of him at a cheaper rate both with regard to money and time than his quondam [past] master agreed to teach him the mysteries of his calling. The

---

[73] RCPSG, 1/1/1/4, Minutes of the FPSG, 1807–21, fol. 238r–238v (20 December 1819).
[74] *Glasgow Medical Examiner*, 1 (1831–32), p. 37.
[75] Ibid., 'On the Question at Issue betwixt Licentiates and the Faculty', p. 90.

thing is plain and palpable – reducible to a calculation in pounds shillings and pence.[76]

Brown wished to draw blood with his allegation that lowly tailors and shoemakers would gain a quick entry to practice, sanctioned by the Faculty. After a century of ensuring that surgeons were drawing even with physicians it was ironic to argue against opening the medical profession. On the other hand, the actions on internal reform had assured that the FPSG made decisions on a restricted membership basis. It was logical, perhaps, to attract increasing numbers of licensees while excluding them from the inner sanctum. But the body of the profession wanted something very different: universal access, affordable fees and representation of their views on reform, demands on which the FPSG was not forthcoming.

'Lowering standards', the problem Brown addressed, may have endangered the status of surgeons as 'gentlemen', mainly an English problem; on the other hand it could accommodate a healthy intake of bright, industrious young men who would pay fees. By the 1820s Glasgow educational opportunities began to incorporate men from social strata other than the 'middle' class. John Garnett and George Birckbeck at Anderson's College were expanding 'science' classes to include 'working tradesmen' by 1800.[77] As 'mechanics' institutes' were established throughout Britain 'in the shadow of Glasgow',[78] a debate ensued as to their ability to attract the working classes with an educational system based on lectures. Significantly, it was argued that the 'mechanics' institutes had 'failed' because those attending them were 'not of the class of mechanics, but were connected with the higher branches of handicraft trades, or are clerks in offices, and in may instances young men connected with liberal professions'.[79] In other words institutions ranging from the 'extra-mural' to further education were in all reality catering to a 'middle class' which mainly included skilled craftsmen. Those whose access to privilege was curtailed by meagre income faced significantly better prospects by education expanding outside the expensive and time-consuming requirements of the 'ancient' university.

The FPSG was in effect the greatest beneficiary of these developments, as it licensed increasing numbers, but also kept membership in its councils to the few who could afford it. It could rely on candidates from the extra-mural

---

[76] RCPSG, 2/1/13, vol. x, no. 1, pp. 5–6.

[77] J. Butt, *John Anderson's Legacy: The University of Strathclyde and its Antecedents* (East Linton, 1996), p. 31.

[78] E. Royle, 'Mechanics' Institutes and the Working Classes, 1840–1860', *Historical Journal*, 14 (1971), p. 308.

[79] Ibid., quoted there, p. 309.

schools taking its licensing examinations, and thus shift its allegiance from the university to other medical schools. As Anderson's College and the Portland Street School of Medicine attracted more students in the 1820s, the FPSG provided an essential service: medical accreditation. The extra-mural schools taught the medicinal curriculum but they could neither license nor grant an MD. As the ties between the FPSG and Anderson's College became more intimate, men like Dr James Corkindale became trusted insiders at both organisations. The FPSG soon realised it possessed a power-base independent of the university. This was an important factor as relations with Glasgow University came under severe strain over jurisdictional issues and national recognition.

Legal Prosecutions

The decisions made in Faculty to rationalise its membership and licensing structures had the effect of setting the medical practitioners' teeth on edge. The liberal practice of periods of grace in which to apply to the Faculty for examination, often of a year, fell victim to the 1811 decision to lower licentiate fees. Because leave to practise became affordable, the FPSG felt it could now move forward a campaign to prosecute in law those who practised irregularly.[80] 'Irregular' came to mean any one not complying with the rules of the incorporation.

The first indication of a more rigorous pursuit of 'irregulars' came in 1805 when the Committee on Unlawful Practitioners was raised.[81] In 1811 the committee reported the magistrates were to be called upon to act because the Faculty had neither police powers nor the money to pursue cases.[82] The Faculty deplored that 'since the period when the prospect of ruin to their funds obliged them to desist from the prosecution of delinquents, the number of irregular practitioners had increased very much indeed'. It was on this account the Faculty sought the assistance of the magistrates 'as guardians of the public welfare' to prosecute on their behalf. The procurator fiscal was to act against anyone practising surgery or pharmacy in Glasgow who was not a licentiate or member of the FPSG.[83]

The magistrates declined but suggested a memorial to the criminal court, the Lords of Judiciary. This memorial stressed that those prohibited from practising by the Faculty should be prosecuted as public delinquents. Mr

---

[80] See below, especially the case of John Cross MD.
[81] Duncan, *Memorials*, p. 155.
[82] RCPSG, 1/1/1/5, Faculty Minutes, 6 January 1812.
[83] RCPSG, 1/1/1/5, Faculty Minutes, 13 December 1811.

Hugh Miller for the FPSG attended the court in Edinburgh in January 1812 and the court issued an act of adjournal on 14 March 1812. An act of adjournal requires one person to give satisfaction to another within a specified period. This enabled the Faculty to ask the procurator fiscal to prosecute all persons illegally practising medicine and surgery within their jurisdiction. But in an entirely novel move men with MD degrees were called to account.

In February 1813 four practitioners were asked to appear before the Faculty and produce their certificates. These men were John Steel, with an MD from Aberdeen University, who refused to be examined by the Faculty; Andrew Reid, with an MD from St Andrews University, who also refused to submit to examination; John Maxwell, who held a licence from the Royal College of Surgeons of Edinburgh and refused examination; and John Todd, whose diploma is not recorded but who also refused to be examined.[84] All were fined £40 Scots and prohibited from practice. Later in the same month a further six were summoned, three asking for examinations in May. Roderick Gray and William Hood refused examination, as did John Campbell. At the end of February, when they were also summoned, William Herron and Francis Cavenagh refused. All together, twelve medical practitioners asked for postponement of examinations until May and this was granted, while five refused examination and were to be fined. The sub-committee of examiners (Moses Gardener, John Nimmo, John Gibson and Hugh Miller) were 'to consider what may be the most advisable plan for prosecuting those persons who encroach upon the privileges of the Faculty within the City of Glasgow'.[85]

Those who refused examination were obviously intending to make a political point. The first summoned were holders of university MD degrees from Aberdeen and St Andrews. The main issues were twofold: firstly, whether an MD would enable the holder to practise surgery or pharmacy as distinct from medicine within the jurisdiction of the FPSG. In effect, and contrary to its charter, the FPSG challenged the jurisdiction of a university degree. Secondly, the territorial jurisdiction of the FPSG was tightened, in the sense that it would prosecute all practitioners, whatever their degrees or diplomas, and force them to be examined under both pain of fines and interdict in medical practice. This second point was obviously as contentious as the first because it called into question the validity of leave to practice issued by such important bodies as the Naval and Army Boards and the Royal Colleges in Edinburgh, London and Dublin. The 1812 act of adjournal thus had much to answer for: it reopened fundamental questions of jurisdiction just when many agreed these must be widened.

[84] RCPSG, 1/1/1/5, Faculty Minutes, 1 February 1813.
[85] RCPSG, 1/1/1/5, Faculty Minutes, 1 March 1813, p. 108.

The legal action of the FPSG attacked graduates of other institutions, not the unlicensed. In 1815 three of the men who had refused examination in 1813 (John Steel, Andrew Reid and Roderick Gray), and three others: (James Watt, Peter McDougal, and John Cross), were summoned before the criminal court. The court decided in favour of upholding the FPSG's right to local jurisdiction in 1826, but did not decide the issue whether being awarded an MD degree legitimately included practising surgery.[86] Thus the distinction between a licence and an MD degree was maintained, but the boundaries of practice remained a grey area.

The whole episode infuriated not only the Faculty's previous ally, the university, but legitimate practitioners at large. The 'ruinous and vexatious lawsuit' was seen as intimidation.[87] The *Medical Examiner* attacked by printing the letter of a 'gulled licentiate', intimating that fees were not paid:

> from a conviction that it was a just or legal demand; but rather for the purpose of avoiding an expensive and, perhaps, ruinous law suit. In my own case, I refused payment, although repeatedly urged and threatened by the official personage whose office it was to collect the fines; and it was not until an action at law was instituted against me for the amount, that I paid it with expenses, – not from any new light which broke in upon me as to its justice, but partly to allay the clamour, and satisfy the wishes of my friends, who trembled at the idea of a young man setting himself up in open hostility to the Faculty; and partly to escape from the vexations, uncertainties, and expenses of a legal investigation which might have involved me in ruin, by incurring an expense which I was totally unable to defray.[88]

University degrees had never been confined to a territorial jurisdiction. John Burns, no fool, turned this fact to advantage. The 1812 prosecutions may have substantially contributed to Burns changing sides and initiating the CM degree. The submissions by the university in the legal action insisted medicine was not only a practical field, but an art and a science. As surgery had in the recent past proved itself of an exacting and scientific nature, a degree in it, so it was argued, sat very well with the university brief to teach and examine in the sciences and the arts. Burns had become Professor of Surgery in 1815 and countered the onslaught of the FPSG on MD degrees and its interdiction on practising surgery illegally, with the new degree of Chirurgiae Magister (CM) approved by the University in 1816. In 1817 he took this degree himself and consistently displayed it on the title pages of his many subsequent publications. James Towers, a surgeon licensed by the

---

[86] Duncan, *Memorials*, p. 162.
[87] *Glasgow Medical Examiner*, 1 (1831–32), p. 57.
[88] Ibid., p. 87.

FPSG and made professor of midwifery in 1815, never took the MD but had the CM conferred together with Burns.[89]

Burns had been quick to advise the FPSG on improving curriculum requirements, now he moved the accreditation of surgeons to an academic level. Burns' logic was that of the improver, gained with his autodidactic climb to the first ranks of educationalist and moderniser, not least as champion of the new discipline of male midwifery. Glasgow University was the first in the country to introduce a separate degree in surgery, the CM, which for ten years proved more popular than the MD. It was timely on two accounts – because of the lawsuit with the FPSG and to counter new legislative moves on regulating surgery. Edinburgh University CM degrees were only instituted after the Medical Act in 1859–60 and confirmed in 1866.[90]

The lengthy legal actions between the Faculty and the University of Glasgow lasted thirty years. In legal terms it was a protracted and complex case argued before several divisions of the judiciary. In political terms its impact had nothing to do with the narrow point of law: whether the FPSG possessed the right to examine all practitioners of surgery within the boundaries of its jurisdiction as set out in its original charter. This last point, in the final decision before the House of Lords, given on 7 August 1840, was upheld.[91]

The political impact went beyond the narrow legal brief. When the prosecutions started, the FPSG only wanted to ensure that licentiates *within* Glasgow could easily be examined by creating a two-tier system of membership. It stirred up a hornet's nest because men licenced by degrees or diplomas from other bodies were summoned. In effect the FPSG challenged the validity of medical patents generally accepted when they were not in direct collision with the old incorporations' means of challenging them. For instance the Army and Navy Boards recognised a plurality of licences and degrees, creating a universality of its own making. While the judicial arguments on the chartered rights of two ancient institutions were pursued into the fine print, the surgical degree of the University of Glasgow, the CM, remained attractive. The CM was specifically mentioned as valid in the Passenger Vessels Acts from 1819 on,[92] while the FPSG licence was omitted. John Burns, on his appointment as Regius Professor of Surgery, was encouraged by the crown to support the education of military and naval surgeons.[93]

---

[89] GUABRC, 1/1/3, Senate Minutes, 6 May 1817, p. 317.
[90] W. S. Craig, *History of the Royal College of Physicians of Edinburgh* (London, 1976), pp. 551–52.
[91] Duncan, *Memorials*, p. 170.
[92] 1819, 1819, 1822.
[93] GUABRC, 1/1/3, Senate Minutes, 1 November 1815, p. 271r.

The legal prosecutions did not alter the popularity of the CM degree, around twenty being awarded per year from the 1820s to the late 1840s. Only from late 1830s did the MD surpass it at around fifteen per annum by 1838, rising to fifty per annum by 1848.[94] Nor were the number licensed by the FPSG affected. These remained at approximately fifty per year.[95] In 1822 official recognition of the CM in the Passenger Vessels Act still smarted. Dr James Corkindale, troubleshooter for the FPSG, circulated a statement to all Scottish MPs that while the University had been awarding diplomas in surgery for the past five years 'no other University of the United Kingdom has interfered with the function [of the four corporations of surgeons] tho' it is believed that the Constitution and rights of all the Universities are in this respect precisely the same'.[96] Dr Corkindale and Mr Duncan Blair of the FPSG were of course not well liked for their role as enforcers of the Faculty prosecutions, as the *Medical Examiner* noted, 'from the very active part these gentlemen took in tracing out unfreemen and in collecting fines'.[97]

Criticising the constitution of the university smacked of legalese, as the FPSG was using the rights awarded in 1599 to argue against the CM. Nor had it hesitated in 1812 to act against holders of the MD degree practising in Glasgow, a contravention of its charter's original policy to unite physicians and surgeons in one body. The 1812 Act of Adjournal allowed them to use the sheriffs and magistrates to prosecute for the Faculty.[98] The FPSG did not baulk at bringing before the Procurator Fiscal 'all Medical Practitioners not duly qualified',[99] where 'not qualified' meant simply not examined by them. The medical practitioners prosecuted under this action were in possession of MD degrees, two of them from the University of Glasgow.

The flavour of the FPSG prosecution and its disregard of prior usage was caught in the letters of John Cross, MD:

> Altho' I compeared before a Committee of the Faculty on January last, shewed them my degree from the College of Glasgow and announced my resolution to repair from the practice of surgery and pharmacy until I should be able to join the Faculty conveniently, yet I have received a libelled summons to the Court of Session.
>
> I have been credibly informed that the Faculty have always hitherto had the liberality whether from custom or by a positive enactment to allow any person to practise surgery and pharmacy for one year unmolested. It is not yet four

[94] Compiled from Coutts, *History of the University*, pp. 546, 577.
[95] RCPSG, 1/1/1/6, Faculty Minutes, 1 July 1822.
[96] RCPSG, 1/1/1/6, Faculty Minutes, 1 July 1822.
[97] *Glasgow Medical Examiner*, 1 (1831–32), p. 87.
[98] RCPSG, 1/1/1/5, Faculty Minutes, 6 April 1812.
[99] RCPSG, 1/1/1/5, Faculty Minutes, 6 January 1812 and 6 April 1812.

Months since I came to town. Upon this ground I am hopeful I will have the goodness not only to withdraw the prosecution but to grant me liberty to practise all three branches of the healing acts for the other eight months'.[100] He then states if this customary indulgence is denied he will 'practice physic exclusively'.

On 1 March 1813 a number of practitioners were summoned, most of whom held MD degrees but admitted to practising surgery and pharmacy.[101] The defence stated that all had 'regularly studied medicine at the Scottish Universities' for which they were granted MDs, possession of which entitled them to practise as surgeons as well as physicians without the necessity of trial before the corporation. The FPSG charter automatically admitted those in possession of a university degree to practise physic, but the Faculty contested whether this gave them 'the right of practising all the subordinate branches of the science', where their Charter expressly excluded 'all persons whatever' from practising surgery within their bounds unless admitted by the Faculty.[102]

On 14 November 1815, the Lord Ordinary (the judge of the Court of Session) having considered the proceedings and documentation thus far, found the Faculty entitled to pursue their action. He found that John Steel, James Watt, Peter McDougal, John Cross and Andrew Reid, the defenders, were authorised on the basis of their diplomas and testimonials, to practise medicine within the Faculty's jurisdiction; but he also found that no one could practise surgery or as an apothecary or druggist without an examination before the Faculty. Indeed, one defender, Roderick Gray, was found insufficiently qualified to practise medicine, surgery or pharmacy![103]

The Lord Ordinary's interlocutor (or decree) was reviewed by the Lords of the Second Division who agreed with his findings. They were unanimous in their support of the Faculty's entitlement, and their only point of doubt concerned the rights granted through university degrees. Ultimately, all but one of the judges (Lord Robertson wanted more information) decided that a university MD gave no title to practise as surgeons within the bounds of the FPSG. As Lord Bannatyne put it in simple terms: 'A person may be qualified to act as a physician, but he may not be skilled to practise as a surgeon'.[104] Even the Lord Justice Clerk who favoured the case of the defendants, recognised the claim to monopoly in the charter that obliged the

---

[100] RCPSG, 1/1/1/5, Faculty Minutes, 7 March 1814.
[101] RCPSG, 1/1/1/5, Faculty Minutes, 1 March 1813.
[102] SRO, EXCS 233/57/IOD, Revised Case for the FPSG, 1833.
[103] Ibid.
[104] Ibid.

law to support it 'no matter whether odious or not'. The charter, he agreed, distinguished between three divisions of the healing arts – physicians, surgeons and apothecaries – and the Faculty were entitled to ensure that no person practised surgery without undergoing examination. The law could not uphold parts of the charter and not the rest. The Lord Justice Clerk remarked that: 'It is said that it would be a degradation for these MDs to submit to an examination on surgery. Why, it may be said to be a degradation for a MD to practise surgery at all'. The judgement read: 'Their [the FPSG's] possession of the privilege of debarring MDs, though possessed of the highest University honour in medicine, from practising within their bounds, without being admitted by the corporation, is solemnly declared'.[105]

These judgements at law upheld territorial rights, a judgement essential to a later crucial fight for survival when the Faculty negotiated with its sister incorporations. It did little to enhance the estimation of the FPSG in the eyes of its opponents. These were in fact hoping the decision would go against the Faculty. The *Medical Examiner*, reporting a rumour about the University winning the case, raised the hopes of the profession: 'If so we will have the pleasure of seeing one other blow dealt to the absurd existing system ... no one will, for a moment, weigh betwixt the University degree and the Faculty diploma, with the present small different of price'.[106]

In 1840, after the appeal to the House of Lords, the Faculty once more succeeded in defending their territorial jurisdiction. But it was a pyrrhic victory, representing the last throes of local jurisdiction. The law upheld the past while in reality the 'three branches of medicine' were merging in Scotland.

Parliamentary Bills

Local balances of power – or overt rivalries – became embroiled in national initiatives to standardise medicine. For an ancient chartered body whose rights were territorially defined and remote from Westminster, nationalisation was an obvious hazard. The restrictive Apothecary's Act of 1815 made medical careers difficult if not impossible for Scottish graduates wanting to practise in its London jurisdiction. The Passenger Vessels Act of 1803 increased employment opportunities for surgeons because it stipulated that no vessel with fifty or more on board could clear a port without a surgeon. The surgeon was required to have passed an examination at London's Surgeon's Hall or at the Royal Colleges of Surgeons in Edinburgh or Dublin.

[105] Ibid.
[106] *Glasgow Medical Examiner*, 1 (1831–32), p. 76.

TIME OF CRISIS 369

There was no mention of the FPSG. As irritating as a dripping tap, outside pressures forced local jurisdiction over medical qualifications to engage with supraregional developments. The inauguration of a surgical degree by Glasgow University in 1815 increased the threat. University degrees, by their very nature, were not bound by territorial restrictions, having a universal validity.

In Parliament on 25 June 1816 a Bill for Regulating the Practice of Surgery throughout the United Kingdom of Great Britain and Ireland was introduced.[107] Its proposals stipulated that no one could practise surgery without being examined by one of the three Royal Colleges, London, Edinburgh or Ireland (Dublin). Secondly, the Bill proposed that whoever received the testimonials of these Royal Colleges 'shall be entitled and shall have the right to practise Surgery in any and every part of His Majesty's Dominions, any law or custom to the contrary notwithstanding'. Thirdly, any surgeon granted a diploma by the Army or Navy for active service was entitled to practise surgery 'in any and every part of His Majesty's Dominions'. Fourthly, any male person who wanted to practise midwifery had to obtain the licence of a Royal College. The Bill, however, *omitted* mention of the Faculty of Physicians and Surgeons of Glasgow or university degrees.

The passage of the Bill through Parliament was called to the attention of the Faculty by Kirkman Finlay, MP for Glasgow. Subsequently Dr Richard Millar read a draft to the Faculty warning that it would 'materially injure the rights and privileges of this Faculty'. Dr Millar and the surgeon Hugh Miller, who had helped shepherd the Act of Adjournal of 1812, important for prosecuting 'illegal' practitioners, through its legal hurdles, were appointed to watch the progress of the Bill and give advice to Kirkman Finlay. Finlay, as the MP for the burgh of Glasgow, took a deep interest in medical matters. He was on the board of managers of the Royal Infirmary and a successful Glasgow businessman. In October 1816 it was decided that jurisdictional restrictions and (pernicious) oversights were to be resolved by applying for a Royal Charter. This would bring the Glasgow Faculty in line with other licensing bodies and automatically extend their jurisdiction by accreditation through the Army and Navy Boards.

On 7 October the committee appointed to confer with Kirkman Finlay reported. This set in motion the attempt of the FPSG to gain 'the name of a Royal College of Surgeons etc.' a strategy by which, had it succeeded, much subsequent infighting amongst the medical profession might have been avoided. As rivalries over jurisdiction amongst the incorporations were rife this move – logical in its own way – disturbed the FPSG's most cunning

[107] A Bill for Regulating the Practice of Surgery throughout the United Kingdom of Great Britain and Ireland, 25 June 1816 *Parliamentary Papers* (531), ii, 807.

adversaries closer to home: the Edinburgh Royal Colleges who were involved in plans to create regulation on the basis of a college for each section of the kingdom. Westminster reform was adopting a tripartite strategy, based on strengthening the medical incorporations in the three 'capitals' of Edinburgh, London and Dublin. Pamphlet attention in consequence focused almost exclusively on Edinburgh institutions when reviewing the Scottish elements of medical educational provision.

In this Edinburgh usually came out well and Glasgow was either neglected or passed over. In 1845 'Lucius', writing his *Remarks on Medical Reform and on Sir James Graham's Bill*, quoted Dr Thomas Wakley that Dublin was 'in gradual decline' and the course at Glasgow University was 'very deficient',[108] which was hardly the case. In 1833 Glasgow University had approved and implemented a major report on improving the curriculum for medical and surgical degrees;[109] and in 1834 had passed a motion that medical examinations would now be written ones.[110] The FPSG was also raising its curriculum demands in the 1830s.

Edinburgh's standards were certainly not above criticism, a medical student from Ireland noting, for example, in 1824, that it was 'by no means such as I was led to suppose ... being on the whole inferior to Dublin'.[111] Not even the fire engines 'bear comparison'.[112] Alexander Monro (tertius) and other lecturers in Edinburgh fared no better in these observations: 'Dr Duncan's lecture before breakfast, Hope after from 10 to 11, rather a pleasing lecturer, suited to the tastes of *amateurs* and the capacity and information of beginners, but not very scientific ... At one, heard Munro on Anatomy. Think him by no means a bad lecturer tho' he does not stick close enough to his subject. He was on the formation of the bone. Alas! how poor did his specimens and preparations appear in *my* eyes, tho' he seemed perfectly well pleased with them'.[113]

Jealous questions arose as the Royal College of Physicians of Edinburgh caught wind of the FPSG's petition for a royal charter. The Faculty petitioned the crown in 1817 for a royal charter setting out 'their rights as equal to the other Royal Colleges' immediately after the 1816 legislation that intended to regulate the practice of surgery. The FPSG suspected that the RCPE needed

[108] Lucius, *Remarks on Medical Reform and on Sir James Graham's Bill* (London, 1845), p. 8.
[109] GUABRC, sen. 1/1/5, pp. 91–92.
[110] GUABRC, sen. 1/1/5, p. 115.
[111] PRONI (Belfast), Papers of Dr John C. Ferguson (1818–63), D. 1918/2/5, Letter to his father, Thomas Ferguson, 20 November 1824.
[112] PRONI, D. 1918/2/4 Diary, Wednesday, 17 November 1824.
[113] Ibid., entry for Monday, 15 November 1824.

no reminding that the Bill introduced in Parliament by Attorney General Garrow would extend the powers of all the Royal Colleges in the kingdom and 'the present charter wanted by them is merely to obviate some alleged difficulties from their not having per expressum the Title of Royal College altho' long ago incorporated by a Royal Charter corroborated by acts of Parliament and frequently sustained in the Courts of Judicature'.[114]

The Edinburgh College was up in arms – although not revealing its true aims to the FPSG – because it was manoeuvring behind the scenes to become the sole Scottish college of physicians. The RCPEd inquired, rather too innocently, if the proposed charter would give the FPSG jurisdiction in 'different parts of the kingdom to practise'.[115] To which the FPSG replied that, whilst they had the right to challenge unlicensed practitioners only within their own bounds, they had the right to grant diplomas to anyone who had completed the medical curriculum, was examined and found qualified.

In December 1818 the Faculty took up arms against the progress of the proposed Passenger Vessels Act, c. 124, eventually passed in 1819, where Glasgow University was first recognised as granting surgical degrees valid in a national context. The FPSG was not mentioned even though it, rather than the University, was the traditional port of entry for surgeons. In a petition of 1820 to the Lords Commissioners of His Majesty's Treasury printed at the expense of the Faculty, it called attention to the high standards its own licensing maintained and to the quality of the Glasgow School of Medicine:

> Previously to any Licences being granted by Your Petitioners' Faculty, they invariably require a complete and regular curriculum of medical education, in all its departments, and they subject the Candidate to a rigid examination as to his skill and knowledge. There has been long established in this City a School of Medicine, which has been largely completed by the liberal care of His Royal Highness the Prince Regent. The Medical Teachers of the University are Members of Your Petitioner's Body, and generally form a part of their Committee for examination. They have, besides, many private Teachers, who have been highly useful in their own department. Your Petitioners and their Licentiates serve an extensive Infirmary, and other Hospitals, belonging to this city, where the Students have continued access to see the practice of the Healing Art exemplified in every one of its branches. From these circumstances, Your Petitioners' diplomas are held in estimation, and are sought after, not only by those to whom they are absolutely necessary, but also by others who practise in districts over which Your Petitioners have no control, but where they have ever been regarded as sufficient testimonials of surgical skills and erudition.

[114] RCPSG, 1/1/1/5, Faculty Minutes, 18 October 1817.
[115] Ibid.

Questions on jurisdiction were forcing an agenda of political manoeuvring which often obscured the more important issue of the educational campaign to implement better standards. Jurisdictional disharmony became particularly fraught as the Royal Colleges pursued aims of their own, and this drowned out solid achievements in bettering medical standards from many other quarters. The 'school of medicine' the FPSG so proudly mentioned was an accomplishment achieved *across* institutional boundaries. It represented a civic coalition between several bodies involved in educating medical men. Now the FPSG, the other Royal Colleges and the universities were plunged, by their own actions as well as outside pressures, into the divisive question of how they would accept each other's patents.

This meant a heightened and often acrimonious debate over the different approaches to medical education. This was accelerated by the Royal Commissions. The Royal Commissions on Scottish universities, appointed in 1826 and reappointed in 1830, investigated various issues about the administration and affairs of the universities. But Scottish medical education was different to that in England and this was drawn to the attention of the Commissioners. As John Burns remarked, when questioned before the Select Committee on Anatomy in 1828,[116] in England anatomy and surgery teaching are generally attached to a hospital. In Scotland these subjects were taught at university. He added that private teachers were also active in Glasgow and that students from the University attended their classes as well as his own. He saw no threat to the University in this and thought competition beneficial to the quality of teaching.[117]

The private teaching mentioned by Burns in his evidence before the Royal Commission of 1826 included the courses on offer in extra-mural schools such as Anderson's College. The medical students at Anderson's took the qualifying examinations at the University and the FPSG. One of Glasgow's medical high-flyers educated in Glasgow commented on the clear difference between Glasgow and London in educating medical men: 'You may tell him [a London physician] that Armour is very different from an English booby that thinks himself qualified to practise medicine by an attendance of two or three winters on classes in hospitals. Armour possesses good abilities improved by a liberal education'.[118] This liberal education was the direct result of schooling in Glasgow and attending the medical facilities in the city. Edinburgh operated a similar shared institutional training of medical

[116] Report from the Select Committee on Anatomy, appendix 5, pp. 129–30, Communication of Dr Burns, Professor of Surgery at the University of Glasgow.
[117] Evidence to the Universities Commission, ii, Evidence by Dr John Burns, pp. 126–32.
[118] RCPSG, 24/2/83, Dr William Mackenzie's Letters, James Armour to W. Mackenzie, 18 June 1818.

men, although Edinburgh University's medical faculty was impatient at being under the thumb of the town's council. In 1827 Andrew Duncan, Jr, of Edinburgh, stated that 'a fluctuating body, such as the town council, is not qualified to regulate those details [the medical curriculum] of university discipline'.[119] But he too saw the Royal College of Surgeons of Edinburgh not as a 'hostile school' but thought that competition was 'advantageous to both'. His 'free trade' leanings in education even condemned the idea of a single medical curriculum.

Major moves toward a 'new curriculum' were once more in progress in 1834, with the FPSG meeting to discuss this question on 29 January, 3 February, 6 February, and 12 March. Their requirements were not far removed from those stipulated by a university committee headed by John Burns and Charles Badham, the Professors of Surgery and of Medicine, respectively.

The deliberations of the University came to fruition in 1839 when medical studies were made compulsory for four years. In addition a certificate of good moral character by two respectable men was needed, with evidence that the candidate was over twenty-one years of age. Every candidate was to undergo full examinations on all subjects in the curriculum. The requirements for the surgical degree were little different from those specified for medicine.[120] On 30 January 1834 the FPSG pressed for a uniform system for all those studying for medical degrees.[121] On 6 February 1834 the Faculty resolved that a petition be presented to Parliament on the benefits of a uniform curriculum for candidates for medical degrees. Certificates of instruction from the Royal Colleges of Physicians and Surgeons of London, Edinburgh or Dublin and the FPSG were all to be recognised.[122] The incessant criticism from organs such as the *Glasgow Medical Examiner* mounted in the years from 1829 to 1832 had had an effect. These were the issues liberals had campaigned for.

Dr James Corkindale of the FPSG lobbied Mr Warburton MP on the desirability of a uniform curriculum for 'all candidates for medical degrees' in 1839.[123] He wanted qualifications to be recognised across the whole of the United Kingdom. But the FPSG, for all its level-headed reform proposals scuppered its chance of success by including proposals not to the taste of

---

[119] A. Duncan, Jr, 'On Medical Education', *Edinburgh Medical and Surgical Journal*, 92 (1827), p. 2.
[120] GUABRC, 26704 Senate Minutes, 1/1/5: 10 October 1829, 15 November 1833, 12 April 1839, 15 April 1839 and 29 April 1839.
[121] RCPSG, 1/1/1/6, Faculty Minutes, 30 January 1834.
[122] RCPSG, 1/1/1/6, Faculty Minutes, 6 February 1834.
[123] RCPSG, 1/1/1/6, Faculty Minutes, 6 February 1834, p. 564

other players in the educational field. Corkindale – who could not let the point go – argued before the Select Committee of the House of Commons on Medical Affairs that university professors be examined on their suitability as teachers and that these 'teaching certificates' have an equal standing with private teachers. This undermined the requirement of in-house teaching the universities saw as essential for establishing the worth of their degrees. The universities were also to be excluded from giving diplomas in general practice or to 'pharmacians'. This would of course rehabilitate the Faculty as the sole agent awarding diplomas to the 'surgeons-apothecary', by now indistinguishable from the general practitioner. Lastly, the surgeon's education was to be the same as the physicians, a demand which, in real terms, existed in Scotland already and served only to remind England of how much remained to be done there.[124]

Conflict in Regulating the Profession

On 7 July 1817 the Faculty minutes note that Kirkman Finlay had met with Sir William Blisard and Mr Norris on the progress of the proposed Bill on regulating the practice of surgery. It was suggested that the FPSG apply to become a Royal College of surgeons and this would cost no more than £300.[125] 'This secured all you have at present and you get everything when you are made a College of Surgeons.' In September 1817 the warrant of the Prince Regent on the petition for the royal charter was laid before the Lord Advocate.

A month later representatives from the University of Glasgow met with a delegation from the FPSG. The University had placed a caveat (objection) with the Lord Advocate against the petition for a royal charter. The meeting discussed the proposal of the university to have a common medical curriculum and that the FPSG accept that the degrees of the university were sufficient evidence of medical competency.[126] This of course went to the heart of the matter as the litigation against MDs practising surgery and pharmacy within the bounds of the FPSG was in full flow. The university meant to keep its own accreditation of surgical competency through its new CM degree. The meetings, there were two, could have led to an amicable solution centred on the willingness of both parties to agree a standard curriculum. In those terms Glasgow could have found a common medical voice with a mutually agreed curriculum whose standards were exemplary for the time. This, in

[124] RCPSG, 1/1/1/6, Faculty Minutes, 12 March 1834.
[125] RCPSG, 1/1/1/5, Faculty Minutes, 7 July 1817.
[126] RCPSG, 1/1/1/5, Faculty Minutes, 20 October 1817, pp. 199–202.

effect, did happen, as the FPSG and the University raised curriculum standards in tandem throughout the first half of the nineteenth century. But they did this separately and not by common resolution. In the meeting between the FPSG and the university delegation the FPSG did not think their position was derogatory to the University.[127] But on all other questions dissent reigned.

The litigation over the rights to license interfered heavily with the petition for a royal charter. The Lord Advocate could not act while important issues of entitlement were at law. In 1822 the Passenger Vessels Act before Parliament again omitted mention of the validity of FPSG diplomas. This was traced to the hostility of the Lord Advocate, who thought the Royal College of Physicians of Edinburgh had the superior authority over medical affairs in Scotland. The Lord Advocate was lobbied to correct this slanted impression, but James Lawrie was still setting the record straight in 1856 when he wrote: 'The College [of Physicians of Edinburgh] is prohibited from erecting a school of medicine, or conferring medical degrees ... The right and privileges of the four Scotch Universities are expressly reserved, and their graduates are to have the "power" of practicing in Edinburgh without fine'.[128] Lawrie cited the Latin of the charter to make it absolutely clear that the validity of the MD degree is derived from university privileges, not from an examination by the Edinburgh College of Physicians. Indeed the College was prohibited from examining doctors.

On 13 April 1825 the FPSG petitioned Parliament yet again because another Bill omitted the FPSG diploma which 'has the effect of injustice to our Body, is injurious to the interest of our Pupils, and is a discouragement to the Students attending our Medical School'.[129] In 1827 the FPSG and the Royal College of Surgeons of Edinburgh petitioned against the newest version of the Apothecaries Bill because it disqualified persons with Scottish diploma from practice in England and Wales.[130]

In the meantime the court case on the validity of the CM degree dragged on. The FPSG were prosecuting CM degree holders practising within their jurisdiction, while the University had raised an interlocutor with the Lords' Ordinary in 1833. This set the fundamental question of whether the university degree could grant the holder power to practise the 'art of surgery' and that they did not need to be licensed by the FPSG or examined by them. The FPSG defended its chartered rights to examine practitioners through thick

---

[127] RCPSG, 1/1/1/5, 20 October 1817, fol. 201.
[128] J. A. Lawrie, *Letters on the Charters of the Scotch Universities and Medical Corporations and on Medical Reform in Scotland* (Glasgow, 1856), p. 10.
[129] RCPSG, 1/1/1/6, Faculty Minutes, 13 April 1825.
[130] RCPSG, 1/1/1/6, Faculty Minutes, 4 June 1827.

and thin. The issue of territorial jurisdiction was, of course, integral to the ancient chartered rights and thus the rock upon which the FPSG justified its existence. If it could not be upheld in law, the whole edifice would vanish. But all around, like the flood waters rising around the ark, the prevailing trend was to abolish local jurisdiction. However, the backwardness of the legal decision, raising from the dead the already mouldering division between the surgeon and the physician, resulted in something more significant than reinstating territoriality. The case at law made it impossible that the other royal colleges would succeed in their subterfuge of ignoring the Glasgow incorporation through not mentioning it in proposed legislation. The legal fracas in the goldfish bowl of Glasgow made the FPSG visible in the larger British context. The Lord Advocate could not claim he was unfamiliar with the long history of licensing surgeons in Glasgow, even if his views were at first hostile. He had not been a great supporter in the campaign for the royal charter.[131] Nonetheless, with the chartered rights of the FPSG argued before the courts from 1812 to 1840, no one could deny its existence.

In 1833 the Scottish Colleges were fighting the Apothecaries' Bill with a memorial to the Home Secretary and a petition to Parliament. Writing on this very subject Andrew Buchanan clearly laid out the Scottish point of view:

> The system of medical education in England and Scotland is to be brought under the consideration of Parliament very soon, in consequence of a quarrel among the monopolists themselves, as to the extent of their respective privileges. The worshipful Company of Apothecaries of London claims all England for its own, denying the right of Scotch Licentiates to practise medicine in any part of that country and treating as ignorant interlopers all who attempt it. There can be no doubt that this act of intolerance is prompted by a lust for gain, and not by any zeal for the interests of medicine, since it is well known that the Scotch Licentiates are better educated, and therefore generally speaking better qualified to practise the various branches of the art, than the Licentiates of Apothecaries' Hall.[132]

The Edinburgh Royal College of Surgeons requested that the FPSG give up their right to examine other colleges' licentiates if the Scottish diplomas were recognised in the legislation. The FPSG agreed, but pointedly stated that they would exempt from this the Glasgow CM.[133] On 12 March 1834 the Select Committee of the House of Commons on Medical Affairs asked a member of the FPSG to go to London to give evidence.[134] This invitation acknowledged the FPSG as a partner in consulting on legislation.

---

[131] RCPSG, 1355/4, 'Report by the Lord Advocate to his Majesty, 7 January 1821'.
[132] Buchanan, *Of Monopolies in Learning*, p. 19.
[133] RCPSG, 1/1/1/6, Faculty Minutes, 13 February 1833.
[134] RCPSG, 1/1/1/6, Faculty Minutes, 12 March 1834.

Treacheries abounded, however, as the Irish Medical Charities Bill appeared in Parliament in 1838 with the FPSG's name once more omitted. This would have destroyed the usefulness of its diplomas in a country with which the west of Scotland had close ties. Dr James Lawrie and Dr Alexander Panton had gone to London for the FPSG to have the Faculty name put on 'an equal footing with those of the Royal Colleges of Ireland, London and Edinburgh'. They arrived in London two days before the Bill was to be voted for commitment to the House of Commons. There they were entertained by Mr French, MP for Roscommon, who was in charge of the Bill. The Bill had been delayed and the opposition of the Irish to the name of the FPSG being put forward was shared, according to Mr French, by the 'most influential Members on Scotch questions in the House of Commons'. Objections were overcome, yet the Bill did not proceed, as indeed it was opposed by the leading Irish Members on both sides. The delegates from the FPSG said that its failure had nothing to do with their arrival and their success in having the name and rights of the FPSG entered into it. The only 'efficient' support for the Bill had been from the Lord Advocate and the Attorney General. All the Scottish Members that the delegates had called on had been in favour of the rights of the FPSG. They 'received efficient assistance and advice from the Faculty's agent W. Grahame'.[135]

In 1839 a committee was formed to deal with reconciling legislative issues. It suggested that the degree of MD entitle the holder to be a general practitioner. This was opposed by the body of FPSG members, who thought it would weaken the legal case against the University still before the Court of Session.[136] In the 1840s a plethora of bills to regulate the medical profession (1840, 1842, 1844, 1845, 1846, 1847, 1848, 1854, 1856, 1857) were introduced, none of which succeeded. As these legislative tides washed through the House, the FPSG was only saved from disaster by its vigilance regarding its sister colleges.

Between 22 April and 6 May 1844 a Bill was floated proposing the FPSG become the Royal College of Physicians and Surgeons of Glasgow. This new charter was to be granted if they abandoned their claim on absolute local jurisdiction.[137] They were prepared to do so, but other events were in the offing.

Two years later the full plot was revealed. It unfolded in a report of the President to the Faculty about his trip to London. He had been invited to a meeting of delegates from the London College of Physicians, the London

---

[135] RCPSG, 1/1/1/7, Faculty Minutes, 6 August 1838.
[136] RCPSG, 1/1/1/7, Faculty Minutes, 4 February 1839.
[137] RCPSG, 1/1/1/7, Faculty Minutes, 22 April 1844 to 6 May 1844, pp. 466–67.

College of Surgeons, the Apothecaries' Society, the National Institute of Medicine and Surgery, the Edinburgh College of Physicians and Edinburgh College of Surgeons. It appeared that they had been working on the heads of a Bill for some time while omitting to inform the FPSG. The heads of the Bill did not mention the FPSG once and would have consigned it to oblivion to make room for an Edinburgh Royal College of Surgeons. This body was to become the sole Royal College of Surgeons of Scotland. As the only Scottish College of Surgeons, it would examine and license in Scotland. No one would be able to register from Scotland unless they went through this new RCSS.

The President appeared before the Committee of the House and gave evidence to them on the proposed change of the Royal College of Surgeons of Edinburgh to the Royal College of Surgeons of Scotland and its manifest injustice to the FPSG. The proceedings of the committee were stopped immediately to assess the position of the FPSG and to allow them to meet the English bodies. Mr Wakely and other members of the Committee ordered the Edinburgh and Glasgow representatives to draw out a plan to amalgamate the two bodies. Initially the RCSE representatives agreed to the proposal but it was later refused by the College authorities in Edinburgh.

The FPSG President Dr William Weir reported that he had met the English representatives on 15 May 1848. They had already been contacted, and the Edinburgh Colleges had been in discussions about new charters giving them the privilege to become the Royal Colleges of Physicians and Surgeons, respectively, of Scotland.[138] Had this Bill gone through, as the President said: 'all the Members and Licentiates of the Faculty in Glasgow would lose their status and be degraded into irregular practitioners, and that not one of them could register until he was enrolled as a member or Fellow of one of the new colleges, be again examined and pay whatever fees might be fixed upon'. This predicament he said would affect all the practitioners in the west of Scotland: 'I stated this fully and in strong language to the meeting. I objected to the document on these grounds, and I concluded by claiming for the Glasgow Faculty and the Members thereof all the rights and privileges, individual and corporate which the Edinburgh Colleges might obtain'.[139]

There was a certain irony in the 'predicament' of the FPSG, it having so vehemently pursued the exclusion of all other licensees and medical degrees from its own jurisdiction. Nor were the Colleges sincere about working together. The President had challenged the Edinburgh College of Surgeons representatives at the meeting in London 'if I was not correct in saying that

---

[138] RCPSG, 1/1/1/7, Faculty Minutes, p. 63.
[139] RCPSG, 1/1/1/7, Faculty Minutes, 27 May 1848, p. 66.

they in Edinburgh expected to become a Royal College of Scotland and he answered 'most certainly'.[140]

A year later on 16 November 1849 the Royal College of Surgeons of Edinburgh gave notice of their intention to change its name to the Royal College of Surgeons of Scotland. On 19 February 1850, the FPSG agreed to drop the pursuit of a new charter as a Royal College in exchange for the Surgeons of Edinburgh desisting in its aim of designating themselves the Royal College of Surgeons of Scotland.[141] This opened the way for the all-important Act for Regulating the Privileges of the Faculty of Physicians and Surgeons of Glasgow which was given the royal assent on 10 June 1850.

This Act was crucial in finally resolving the many upheavals of the past forty years. The members were now designated Fellows of the Faculty. These Fellows and the licentiates had the same privileges as those of other colleges, while as an incorporation retaining the right to pass by-laws relevant to the work of the Faculty. Territorial jurisdiction, so long and so jealously guarded, was given up. The Fellows and licentiates of other Colleges and the graduates of the universities had the right to practise within the bounds of the FPSG. This right became reciprocal in the United Kingdom. Membership of the Widow's Fund was no longer compulsory or a requirement for becoming a Fellow.[142]

On 3 November 1856 the FPSG noted a minute of agreement between the Faculty, the Royal Colleges of Physicians and Surgeons of London, the Royal College of Physicians and Surgeons of Ireland, the Royal College of Physicians of Edinburgh, the King's and Queen's College of Physicians of Ireland, and the Society of Apothecaries of London, 'to regulate and equalise the education, examination and privileges' of all of them.[143] In June 1857 the FPSG gained representation on the proposed General Medical Council, this having been secured earlier in the year and reported to the FPSG on 2 March 1857.[144] Finally, on 2 August 1858, Dr James Watson minuted that the Medical Bill of Lord Elcho had passed both the House of Commons and the House of Lords: 'The Faculty had obtained all the privileges, and were confirmed in all the rights claimed by them, the same as enjoyed by any Royal College in the Kingdom'.[145]

---

[140] RCPSG, 1/1/1/7, Faculty Minutes, p. 67.
[141] RCPSG, 1/1/1/7, Faculty Minutes, 19 February 1850, pp. 145–46.
[142] Duncan, *Memorials*, p. 230.
[143] RCPSG, 1/1/1/7, Faculty Minutes, 3 November 1856, p. 453.
[144] RCPSG, 1/1/1/7, Faculty Minutes, 1 June 1857, p. 489.
[145] RCPSG, 1/1/1/7, Faculty Minutes, 2 August 1858, p. 555.

The Final Act

The Royal Colleges in London were so worried by the prospect of imminent government legislation on medical reform in 1844–45 that they assembled a grand alliance of United Kingdom medical corporations to draw up an alternative Bill. The hope was that any measure which demonstrably had the support of a majority of the chartered representatives of the profession would be likely to appeal to the government. The Faculty was to play a continuous role in the joint negotiations between the corporations through its delegates to London conferences, and at meetings of the Scottish Branch of the General Conference of Corporations in Edinburgh.

In February 1856 Thomas Emerson Headlam, a barrister and Liberal MP for Newcastle, introduced into the House of Commons on behalf of Hastings and the Provincial Medical and Surgical Association (PMSA, later the BMA) a revised Bill, 'To alter and amend the Laws regulating the Medical Profession'. In April Lord Elcho (later Earl of Wemyss) introduced a rival Bill. Headlam's Bill was more sympathetic to the claims of the Royal Colleges of London for the maintenance of existing status divisions within the profession. It recommended a register graded into classes of practitioner, and that all practitioners should be examined by and enrolled in the appropriate corporation before registration. Elcho's Bill harked back to the claims of the general practitioners with its idea of a common register. It also sought to protect the position of the universities, particularly the Scottish institutions, rather than the corporations. Both Bills were referred by the President of the Board of Health, W. F. Cowper, to a Select Committee. This committee, itself chaired by Cowper, produced its own Bill which was unlike the other two and reflected a different agenda.[146] Stokes has characterised this new Bill as proposing 'a Medical Council which was to be a department of State'.[147] The council was to be headed by the President of the Board of Health and twelve Crown nominees. Registration was to be on a common, simply alphabetical register, and graduates in medicine would not have to join a corporation. In short the provinces of the corporations were to be annexed by the state, before which all practitioners were to be equal.

---

[146] See Waddington, *The Medical Profession*, p. 97. Chapter 6 gives a detailed account of developments in Parliament and between the Royal Colleges in London, between 1856–58, upon which our narrative relies heavily. On the progress of medical reform in this period see also Sir G. Clark, *A History of the Royal College of Physicians of London*, 3 vols (Oxford, 1966), ii, chapter 35.

[147] T. N. Stokes, 'A Colerigdean against the Medical Profession: John Simon and the Parliamentary Campaign for the Reform of the Medical Profession 1854–58', *Medical History*, 33 (1989), pp. 343–59.

The Faculty had already been busy in following developments in London. Acting on the urgent advice of their then London parliamentary solicitor, Mr Alexander Grahame,[148] who carefully watched and reported on the progress of measures of reform, the Faculty had sent a deputation consisting of the then President, Dr Robert Hunter, and Treasurer, Mr George Watt, to London to conduct any necessary negotiations.[149] On 1 April Grahame arranged a meeting with Headlam in the lobby of the House of Commons at 5 p.m., and a conference was held in one of the committee rooms. The deputation put their case: that the Faculty approved of the general principles of the Bill but wanted Glasgow to be named as one of the examination centres for the preliminary entrance exam to professional education, and some powers over the licensing of physicians (namely 'the privilege of granting after examination Letters testimonial to Physicians').[150] Headlam agreed to the two points, but suggested that one of the Glasgow MPs (rather than he) should present the Faculty's petition to the House and move the amendment for placing the Faculty 'in their proper position in the Bill'.[151] The deputation then saw Mr MacGregor (a Glasgow member) and he consented to place the amendments on the day's orders. He received the petition on Wednesday 2 April and immediately presented it to the House, although the deputation were disappointed to find that in the following morning's newspapers it was 'not properly or specially referred to'.[152] The deputation then observed the Commons debate:

> By the kindness of Sir James Anderson we were admitted to the Speaker's Gallery and heard the debate on the Bill ... in rather a large House for a morning sitting it seemed impressed with the interest and importance of the question and of the necessity for legislation on the subject of medical polity.[153]

Consideration of the Bill was however postponed and Headlam's Bill was referred to a Select Committee along with Elcho's.

The Faculty very soon became involved in the joint action of the corpor-

---

[148] Of the firm of Grahame, Weems and Grahame, 30 Great George Street, Westminster. Some of the very detailed firm's bills to the Faculty are preserved in the RCPSG Archives. See RCPSG, 1/5/3, 'Receipts, Bills and Papers relating to Lawyers Fees and other business, 1844–67', which contains copies of bills for February to July 1856. This makes it clear that Grahame did much of the work of composing petitions and briefing the delegates on the complicated details of the progress of the various bills. Interestingly, Grahame was also employed from early May 1856 by the RCSEd (RCSEd, CM, 8 May 1856, p. 219).
[149] RCPSG, 1/1/1/8, Faculty Minutes, 8 April 1856, p. 430.
[150] Ibid., pp. 430–31.
[151] Ibid.
[152] Ibid.
[153] Ibid., p. 432.

ations. On the 18 June, Watson and Dr Alexander Dunlop Anderson, on behalf of the Faculty, proceeded to London to oppose the new Bill which had been produced by the Select Committee under Cowper and was considering Headlam's and Elcho's Bills. This Select Committee Bill, produced in secret,[154] with no opportunity for consultation, greatly worried the corporations. The crux of the problem was the composition and powers of the proposed Medical Council. The RCSE issued a statement warning that:

> The Medical Bill, as reported to the House of Commons by the select committee, would in effect destroy the College by transferring its powers to an untried, uncertain, and fluctuating body, of purely experimental character, called a Medical Council.[155]

The Bill would give the Medical Council wide powers to define medical education and set a standardised curriculum. This was an object which all the corporations had desired for some time, but the Council was also to be entirely nominated by the State, and was to retain the examination fees. The corporations saw this as the State taking over their functions and leaving them without any purpose or means of financial survival. The Faculty delegates (Watson and Anderson) objected to the Bill because 'The Council, the Registrars and all the subordinate officers were to be Crown nominees'.[156] A contemporary report of the RCSEd was more forthcoming on the corporations' fears:

> The almost unlimited powers given by the present Bill to the Medical Council [are] highly objectionable. The College ... in their Report on Mr Headlam's original Bill gave their assent to the principle of Crown nomination, but this was under the idea that the only duties of the Medical Council would be to fix a minimum course of study and to exercise superintendence over the examining

---

[154] There was however a leak of some of the Select Committee's conclusions, which further intensified the climate of concern. Andrew Wood, President of the RCSEd, reported that he had been offered a 'sketch of the Bill' by the MP Adam Black. He had declined because the condition was that he could not discuss the information with either his council or college. But he had obviously learnt enough to be worried that the Bill would be 'most ruinous to this College', and further reported to the College that 'the course which measures threatened to assume in the hands of the Committee, might be utterly subversive of the rights and interests of this body'. Quotations from RCSEd College Minutes (CM), xiv, 1854–59, pp. 217–19.

[155] 'Statement issued by the Royal College of Surgeons of England in reply to the 1856 Select Committee Report' *Lancet*, (1856), i, p. 699. Cited here from Stokes, *A Coleridgean*, p. 355. See also RCS Eng. Minutes of Council, 10 July 1856, pp. 108–9, for the College's petition against the Select Committee Bill.

[156] RCPSG, 1/1/1/8, 1 December 1856, pp. 458–59, 'Report of the Delegates from 18 June till 1 December'.

boards; they ought not however to adhere to this principle of nomination for a Medical Council on which it is proposed to bestow such arbitrary and indefinite powers as those contemplated by this Act [sic]. The Council under this Bill, equally with that under Mr Headlam's original measure, is open to the objection that it will come entirely to supersede those bodies which have hitherto regulated the Medical Profession. The Medical Council under the Bill will be entrusted with powers which no medical corporation ever possessed, and there is nothing in the Bill which provides any security that the Council may not be composed mainly or entirely of persons who neither represent the profession nor possess its confidence, or who have never given much attention to the subject of Medical Education, and who have no experience in framing regulations for the qualifications of teachers, and the instructions of students.[157]

The report concluded that:

Besides conferring on a Medical Council powers of control over the Medical profession unknown to any of the other learned professions in this Country, it introduces an entirely new principle – that, namely, of superseding the functions of examination, and licence, the whole of those bodies by whom these powers have hitherto been exercised with so much advantage to the public, and conferring them upon a mixed body, now to be called into existence, the advantageous working of which it is impossible to predict.[158]

The corporations' position was then that a state council should not be given wide powers which would threaten their independent existence, but should be only 'for the limited purposes of fixing a minimum course of study, and of exercising such a superintendence over the various licensing Boards as should secure the faithful discharge of their duties as examiners'.[159] Only a council dominated by the nominees of the corporations should be entrusted with wider powers. The corporations clearly perceived the situation now as a battle with the state for the maintenance of their professional autonomy.

The Faculty also for its own part objected specifically to the following provisions of what was usually known as 'Cowper's Bill': Faculty examiners were to be outnumbered on the proposed Scottish licensing board for general practitioners by examiners from Edinburgh and the universities; the council was to control the appointment of all examiners and the resultant exam fees; and holders of the Faculty Licence (Surgery) were to be forced to take a

---

[157] RCSEd, CM, 'Report of President's Council on Bill, as amended by Select Committee', 1854–59, pp. 226–27. Report presented to College, 16 June 1856.
[158] Ibid., p. 229.
[159] RCSEd, CM, pp. 231–32, 16 June 1856, Petition to House of Commons opposing Select Committee Bill.

further qualification in medicine, in spite of the fact that this was already covered in the Diploma.[160] On account of such objections ('chiefly by the Scotch and Irish delegates'), Cowper's Bill was withdrawn on 8 July.[161]

Out of these negotiations between the corporations on opposition to Cowper's Select Committee Bill emerged the idea for joint action. Most accounts of this joint action are partial and omit its origin, relying as they do on evidence from the records of the London Colleges.[162] Faculty Minutes record that, 'it was suggested that the corporations should join and frame a form of Government for themselves'.[163] The minutes of the RCSEd, which are generally more full than their Faculty equivalent, note that it was some MPs who opposed the Select Committee Bill who told the delegates of the corporations that 'it would strengthen their hands in arguing against the Bill could [the delegates] be prepared to state that there was any chance of agreement amongst the corporate bodies as to some plan of medical reform'.[164] The delegates of the three Colleges of Surgeons and the Faculty then met Cowper himself, voiced their objections and asked him to withdraw the Bill. Cowper replied in terms that must have pushed the corporations to finally work together:

> Mr Cowper had little to say to all this except that as the medical bodies had hitherto found it unprofitable to come to any agreement amongst themselves it seemed expedient that legislation should proceed from a Select Committee of the House and that he had heard nothing to induce him to withdraw the Bill, and that it was his intention to go on with it.[165]

Accordingly on 27 June, the Faculty's representatives, again Watson and Anderson attended a meeting of the Royal Colleges of Surgeons of England

---

[160] 'Report of the Delegates from 18 June till 1 December', RCPSG, 1/1/1/8, Faculty Minutes, 1 December 1856, pp. 458–59.

[161] Ibid., p. 459. See also RCSE, Minutes of Council, 10 July 1856, p. 109.

[162] See Waddington, *The Medical Profession*, p. 98 (he relies mainly on the archives of the RCS Eng. with some input from those of the RCP Eng.); Stokes, *A Coleridgean*; Clark also notes that the London Colleges are now stung into action, but he neglects the role of the other United Kingdom corporations even more than the previous two commentators. See Sir G. Clark, *A History of the Royal College of Physicians of London*, p. 725. Conversely, the Scottish literature also often omits the national picture: see C. H. Cresswell, *The Royal College of Surgeons of Edinburgh* (Edinburgh, 1926), pp. 293–94. The best account of these developments is in W. S. Craig, *History of the Royal College of Physicians of Edinburgh* (Oxford, 1976). But again here the title of the relevant chapter 12, 'College Participation in Medical Reform', indicates the partiality of the treatment.

[163] RCPSG, 1/1/1/8, 'Report of the Delegates', p. 459.

[164] RCSEd, CM, volume for 1854–59, 'Report by Council of the RCSEd of Steps Taken by Them in Reference to the Medical Profession Bill', p. 237.

[165] Ibid.

(RCS Eng), Edinburgh (RCSEd) and Ireland (RCSI). The bodies all now agreed on a document laying down a plan of action which was known as the 'Treaty of London'.[166] This was specifically intended to be a jointly-agreed basis for unanimous action. It would pledge the corporations:

> to active concert for the purpose of taking such steps in the way of reform by the corporate bodies themselves, as will tend to obviate the necessity for such destructive legislation as has been recently proposed, and of removing the grounds for the very noxious and unpleasant agitation, which has recently, and for years, distracted the attention of medical men, and relaxed rather than fostered legitimate practical reform.[167]

The treaty laid down:

> 1. That a Council be established to consist of representatives chosen equally by and out of each body respectively, to meet annually at such time and place as may be agreed upon.
>
> 2. That the Medical Council shall at their annual meeting prepare a Register in such form as they may agree upon, of the several Fellows and Licentiates of those Colleges represented upon the Council, to be printed and published under their joint sanction.
>
> 3. That the Medical Council shall consult respecting all matters relating to preliminary and professional education and examination with a view of regulating Medical and Surgical Education, and leading to uniformity and reciprocity of privileges of the members of each division of the Profession in the United Kingdom.
>
> 4. That these articles be submitted to the consideration of the several Colleges of Physicians, with the expression of an anxious desire that they should accede to them.[168]

It was also agreed that a preliminary meeting of the Medical Council should be held at an early date in London to get things underway and to: 'draw up proposals for such measures as seemed suitable to the wants of the

---

[166] RCPSG, 1/1/1/8, 'Report of the Delegates', p. 459.

[167] RCSEd, CM, ibid., p. 238.

[168] 'Articles of Agreement between the Several Colleges of Surgeons of the United Kingdom and the Faculty of Physicians and Surgeons of Glasgow', signed in London on 27 June 1856. RCPSG, 1/1/1/8, Faculty Minutes, 7 July 1856, pp. 439–40. As well as the Faculty representatives, also present are these other signatories: Mr Lawrence, President RCS Eng; Dr Andrew Wood, President RCSEd; Dr R. C. Williams, President RCSI; Dr James Stannus Hughes, Council Member RCSI; and H. Maunsell, Secretary of the Council RCSI. The treaty is recorded in the minutes of all of the UK corporations as an important step forward. See also, for instance, RCS Eng, Minutes of Council, 7 August 1856, pp. 121–22.

profession particularly in regard to uniformity in education, examination, and privilege of the general practitioner'.[169]

In the meantime the Council of the RCS Eng, who seem to have initiated the plan for joint action, now took the lead again. The President, Benjamin Travers, and six other eminent members of the council had met and settled on 'what in their opinion should form the basis of a Medical Bill'.[170] They then met twice with the President, Registrar and four other representatives from the Royal College of Physicians of London (RCPL) at which there was general agreement on the principles for framing a Bill.[171] By 16 October there had been three further conferences, two of which had been attended by representatives of the Society of Apothecaries. It had also been decided that the general conference of the corporations of the United Kingdom should be held at the end of October. This took place over the 21, 22, 23 and 24 October 1856, with high-ranking representatives from the London and Edinburgh Surgeons and Physicians, the Irish Surgeons, the King's and Queen's College of Physicians of Ireland, and the Society of Apothecaries, as well as Watson and Anderson from the Faculty.[172]

The Scottish and Irish delegates were under the impression that discussions were to be based on the Treaty of London,[173] but, again the agenda was dictated by London and the conference debated the detailed proposals for a Bill emanating from the RCS Eng, which had been agreed with the other London corporations in July.[174] Watson and Anderson again attended for the Faculty and they reported that:

> The result of these conferences was the adoption of various heads of agreement, having for their objects to regulate and equalise the education, examination and privileges of the Licentiates of the several Royal Colleges and the Faculty; to institute an authentic registration of duly qualified practitioners; and to construct a general Council under the superintendence of which these necessary objects should be carried out.[175]

A printed document was circulated by the RCS Eng as a basis for a Bill and this was amended then approved by conference and was subsequently

---

[169] 'Report of the Delegates', 1856, p. 460.
[170] RCSEng., Minutes of Council, 7 August 1856, pp. 124–25.
[171] Ibid. See also RCPL, Annals, 29 October 1856.
[172] See RCSEng., Minutes of Council, 7 August 1865: pp. 124–25, 127–28.
[173] Both the Faculty and the RCSEd had prepared areas of discussion based on the treaty: see RCPSG, 1/1/1/8, Faculty Minutes, 6 October, pp. 450–51; RCSEd, CM, 7 July, pp. 242–43.
[174] See RCSEd, CM, 2 November, p. 298; and RCSE, Minutes of Council, 7 August 1856, pp. 124–25.
[175] RCPSG, 1/1/1/8, Faculty Minutes, 3 November 1856, p. 453.

known as 'Amended Proposals for a Medical Bill'.[176] Watson and Anderson described the document as the basis 'for immediate united voluntary action, and secondly for a future legislative measure to be in due time procured from Parliament'.[177] The Faculty delegates also reported that 'the measures for reform in contemplation by the united medical bodies would prove most satisfactory to the Faculty and very beneficial to our interests generally'.[178] And this in spite of the omission of the Faculty's name from all sections listing the corporations in the draft Bill.[179]

The Corporation Bill clearly asserted the right of the professional elite to control the profession independent even of any state-regulated framework. It began with the words: 'As far as possible, all existing rights and privileges shall remain untouched'.[180] There could be no more succinct an expression of the attitude of the English corporations, and indeed the Glasgow Faculty itself, towards medical reform. However, concessions had by now been made to the general practitioners. In England, they were to be examined in medicine and pharmacy by a board drawn from the RCPL and the Society of Apothecaries; in surgery by the RCS Eng; and, importantly, in midwifery by a joint board of the three corporations. They were then to be enrolled as in the RCSEng. as 'members of that College, and Practitioners in Medicine and Midwifery'. In Scotland (and Ireland) all these examinations would be handled by the colleges. No longer was a degree enough for practice, now holders would also have to enrol as members of a college of physicians. Together these provisions increased the power of the colleges generally, since a registered practitioner now had to be a member of a college of physicians or surgeons.

The Medical Council was to consist of seventeen representatives of the corporations and universities with only a minority of six crown nominees. As Stokes has written, this version of the Medical Council thus 'first and foremost represented the medical corporations: it could not claim to be the medical department of the State'.[181] The Medical Council also had a limited role. It was established:

> for the purpose of effecting uniformity of education in the United Kingdom; for regulating the several subjects on which candidates for the diplomas or letters

---

[176] 'Report of Delegates', p. 466.
[177] RCPSG, 1/1/1/8, Faculty Minutes, 3 November 1856, p. 454.
[178] Ibid., p. 455.
[179] See 'Report of Delegates', p. 466. An amended reprint was demanded and hastily arranged.
[180] 'The Medical Reform Bill of the Corporations', a complete reproduction of the Bill appearing under this title *Lancet* (1856), ii, 8 November, p. 522.
[181] Stokes, *A Coleridgean*, p. 355.

testimonial of the respective Colleges shall be examined; for determining as to the fitness and efficiency of medical schools seeking recognition; and the mode of the annual publication of the Register of legally qualified practitioners.[182]

Furthermore there were to be separate registers for physicians and surgeons.[183]

The Faculty wholeheartedly and officially endorsed these proposals at their meeting of 1 December 1856, the motion noting that they

> embody most of the principles of Medical reform which have long been entertained and advocated by this Faculty, and ... do cordially approve of the same as the basis of a new legislative enactment for regulating Medical and Surgical education and practice throughout Her Majesty's Dominions.[184]

The new procedures were uniform, so the Scottish corporations now entered into a period of discreet negotiation to work out the arrangements for the conjoint examination of the general practitioner in Scotland involving Physicians' and Surgeons' colleges. The point was that uniform education and examination under the Bill would confer uniformity of rights to practise throughout the Empire. Under this Bill the validity of the Faculty's qualifications would be universally assured.

Watson and Anderson attended the first meeting of what was called the Scottish Branch of the General Conference of the Corporations of the United Kingdom, in Edinburgh on 21 November 1856. On the table were two plans for the administration of the proposed new conjoint Scottish board for licensing general practitioners. The first proposed an amalgamation of the Faculty with the Edinburgh surgeons to form a new body – the Royal College of Surgeons of Scotland – with two divisions: one in Glasgow and one in Edinburgh. The Faculty title was to be given up and the new Glasgow body would be called the Royal College of Surgeons of Scotland in Glasgow. Importantly, a new charter bearing this name was to ensure its relative autonomy: it was to be able to hold property in this title 'and enjoy the immunities and discharge the duties at present enjoyed or discharged by the Faculty'.[185] Each city division was to appoint equal numbers of examiners to the board with diets being held in each city. Fellows of Faculty who wished to join the Edinburgh Physicians as a result of this arrangement were to be

---

[182] 'The Medical Reform Bill of the Corporations', a complete reproduction of the Bill appearing under the title *The Lancet* (1856), ii, 8 November, p. 522.

[183] See Waddington, *The Medical Profession*, pp. 100–1; Stokes, *A Coleridgean*, p. 355.

[184] RCPSG, 1/1/1/8, Faculty Minutes, 1 December 1856, pp. 467–68, resolution proposed by George Watt and seconded by Watson and unanimously passed. Copies of the resolution, signaling the Faculty's agreement to the Proposals were sent to all UK corporations.

[185] 'Report of Delegates', p. 462.

admitted *ad eundem* to the RCPEd, and the Physicians were to supply assessors to the conjoint exam board to make up one third of its complement.

The second plan involved no change to the current names or organisation of the three Scottish corporations, but merely suggested that they combine to constitute a 'General Licensing Board for Scotland', each body contributing an equal number of examiners and diets being held in both cities.[186]

The branch conference, having elected a president and secretary (Drs Wood and Maclagan respectively) and secretaries for each city (Maclagan, and Watson for Glasgow), discussed the two plans. The Edinburgh Surgeons objected to both, and Wood suggested a third possibility: no title changes, no conjoint exam board but uniform exams by each city's surgeons with assessors attending from the Edinburgh Physicians. The Faculty objected to this plan, and the question was referred to full meetings of each body to come up with a decision.[187] The Faculty's final opinion was to reject Wood's plan but to support the operation of either of the first two.[188]

At a further meeting of the Scottish branch in early February 1857 it was agreed that for the present the Faculty and the RCSEd should continue to operate as separate licensing authorities for general practitioners, with the involvement of the RCPEd which would provide two assessors to each board and attest to examination in medicine on the reverse of the diploma certificate. The curriculum for the diploma was to be the same for both boards, and practical midwifery, natural history and comparative anatomy were to be added. There would also be a common preliminary examination before entering professional education. This would include English composition, Latin, mathematics, and natural philosophy, and candidates would also be encouraged to study German, French and Greek.[189] These negotiations would begin again after the passing of the Medical Act, and lead to the establishment of the Double Qualification in 1859.

Watson then attended a large meeting of the General Conference in London on the evening of 9 February in the hall of the Royal College of Physicians.[190] The whole draft Bill was considered clause by clause over four

---

[186] Ibid., pp. 463–64.

[187] Ibid., pp. 464–65.

[188] RCPSG, 1/1/1/8, Faculty Minutes, 1 December 1856, pp. 468–72, resolution proposed by George Watt and seconded by Dr Tannahill.

[189] RCPSG, 1/1/1/8, Faculty Minutes, 2 February 1857, report of delegates (Watson and Anderson) from Scottish branch meeting, Edinburgh, 21 January 1857, pp. 478–84.

[190] Reflecting the importance of this meeting there was a broad and high-ranking attendance: for the RCP Eng, the President, Dr Mayo and the Registrar, Dr Francis Hawkins (who was Secretary of the English Branch Conference and acted as Secretary for this meeting), and Dr Burrows, Dr Nairne and Dr Alderton; for the RCS Eng, Mr Travers, Mr Green, Mr C. Hawkins,

successive days. Eventually everything was agreed (including the Scottish arrangements) and the Bill was adopted. Watson noted that the Faculty name appeared everywhere 'where a Royal College or Surgical Licensing body was referred to'.[191] However Watson had to battle on one important point: Faculty representation on the Medical Council.

The English and Irish Branch Conferences had both put forward schemes which both proposed representation for the various corporations and universities of the United Kingdom and government nominees who were not to be office bearers or councillors of the corporations. This last point was a concession to the demands for political representation of general practitioners. However, no provision was made in either scheme for a Faculty representative. Watson protested that this broke the Treaty of London, the basis for all these negotiations, which had stipulated equal treatment for all bodies. His protest was met with arguments about Glasgow's provincial status, the small number of diplomas it issued annually when compared with the 450 conferred by the RCS of London, and 'the feeling of the profession out of doors which was frankly declared to be unfavourable to the Faculty'.[192] Watson replied with a strong defence of the Faculty's claim:

> Your Delegate pled your Charter – its antiquity – its ratifications by two special Acts of Parliament, the carefulness shewn by the Faculty since its institution to be fully abreast of other licensing boards in its requirements – and the fullness of its examinations both in regard of ordinary literature and professional knowledge. He pled the rapidly growing population and mercantile importance of Glasgow, soon to become a second London – the extent and excellence of its Hospitals and its two important Medical Schools – the University and Anderson's Institution – but especially he dwelt on the great fact that according to the terms of the Treaty of 27 June last, the seat of the Faculty was secured – this was a fait accompli not to be reasoned on – that the delegates of the Faculty had assisted at every branch and general Meeting of the Conference having been regularly summoned to attend as an integral part of the arrangements – and that the very presence of the Delegate of the Faculty at the present Meetings was a demonstration of her conceded right.[193]

Mr Stanley and Mr Lawrence; for the Society of Apothecaries, Dr De Grave, Mr Taggart, Mr Brunt and Mr Simeons; for the University of Oxford, its MP, Dr Acland; for the RCPEd, the College Secretary, Dr William Seller; for the RCSEd, Dr Wood; Watson for the Faculty; for the Royal College of Physicians of Dublin, Drs Miligan and Duncan; for the Royal College of Surgeons of Ireland, Drs Williams, Maunsell and Hughes; and for the University of Dublin, Dr Harrison. See 'Report of Dr Watson to Council Read 24 February 1857', RCPSG, 1/1/1/8, Faculty Minutes, 2 March 1857, p. 486.

[191] Ibid., p. 487.
[192] Ibid., pp. 488–89.
[193] Ibid., p. 489.

In spite of this robust defence, the debate continued for three nights with no resolution of the Faculty's position. Finally the Oxford MP and University Delegate, Dr Acland, suggested referring the point to the arbitration of the three MPs sponsoring the Corporations' Bill (Sir William Heathcote, MP for Oxford University; Joseph Napier, MP for Dublin University and Headlam). After consulting with the Faculty's Council by telegram, Watson reluctantly agreed to this as the Faculty's best chance. He felt he had made no impact on the English Delegates and that even a Bill without a Faculty representative was better than dividing the meeting and losing the rest of the agreed Bill. Even the Corporations' Bill as it stood was better than a government Bill 'in which the Faculty probably would not have been recognised at all.'[194] Before leaving London Watson paid visits to each of the three 'umpires' and arranged for Grahame, the Faculty's parliamentary solicitor, to draw up a document detailing the Faculty's claim.

In concluding his report on these negotiations Watson felt moved to express his thanks to the Edinburgh delegates, especially to Dr Wood, President of the RCSEd, who had seconded all of Watson's motions. Without this support, Watson felt that 'he would not have obtained the favourable concessions which were made to the Faculty at the Conference'.[195]

At the beginning of June, Watson was able to report to the Faculty that the arbitration had been successful and that the Faculty 'was now recognised as a constituent part' of the proposed Medical Council.[196] Meanwhile, the Corporations' Bill had been introduced into the House of Commons by Headlam [197] and was given its first reading on 27 February 1857.[198] The second reading was scheduled for 18 March, but this was delayed by the end of the Parliamentary session. The corporations used this time to attempt to smooth the passage of their Bill. A deputation with representatives from 'nearly all the medical corporations', as well as Headlam and about twenty other sympathetic MPs, met with the Prime Minister, Lord Palmerston, on 9 June.[199] The deputation informed him of the imminent introduction of the Corporations' Bill and asked him either to support it or not to block it by the introduction of a government Bill. Watson was part of this deputation and reported Palmerston's comments to the Faculty:

---

[194] Ibid., p. 490.
[195] Ibid., p. 497.
[196] RCPSG, 1/1/1/8, Faculty Minutes, 1 June 1857, p. 501.
[197] It is unclear what part he played (if any) in the development of the Bill, but he was the natural choice, particularly as seven of the clauses of his Bill of February 1856 were incorporated in the Corporations' Bill. See Waddington, *The Medical Profession*, pp. 99 and 214 n. 9.
[198] RCS Eng, Minutes of Council, 11 March 1857, p. 156.
[199] RCS Eng, Minutes of Council, 10 June 1857, pp. 174–75.

although he did not promise Government support to the Bill as he had not read it, he said that if it really were as it was said to be, calculated to settle this vexed matter in a manner apparently for the good of all concerned he and his colleagues would be but too happy to countenance what seemed to be agreeable to the representatives of such an unusually large [number?] of Medical Corporations.[200]

But parliamentary acceptance of the corporations' Bill was by no means assured and the debate over professional or state control continued. In the debate on the first reading of the Corporations' Bill Lord Elcho's point that, 'It was possible ... that there might be a unanimous feeling among the medical corporations in favour of this Bill, without a corresponding unanimity among the great body of the profession',[201] made it clear that one possible defence of state control of the profession was to enable the general practitioners (and not just the elite) to have some say in the running of the medical profession. Again this point hinged on the Composition (and powers) of the central Medical Council. As was noted above, the corporations' Bill envisaged a council of elite representatives. The Select Committee Bill, a version of which Elcho had reintroduced on 15 May, proposed a council of crown nominees. Some of these might have been representatives of the wider profession (general practitioners) as well as representatives of the elite (corporations).

In June the corporations put out a joint statement backing Headlam and their own Bill and opposing Elcho's Select Committee Bill. This repeated the old complaints, stating that no other profession was subject to the same centralised control by the state, and adding that Elcho's Bill was only supported by certain universities.[202] Marischal College, Aberdeen, and the universities of Edinburgh and Glasgow had refused to be persuaded into agreement with the Corporations' Bill,[203] although strenuous attempts had

---

[200] RCSEd, CM, volume for 1854–59, 'Report of Dr Watson to Council' 'Report by Council of the RCSEd of Steps taken by them in reference to the Medical Profession Bill', p. 237. The word in square brackets is obscured by water damage in the original and has been inferred from the context here.

[201] *Hansard*, Third Series, 145, 13 May 1857, col. 245. Cited from Waddington, *The Medical Profession*, p. 103.

[202] 'Reasons, on Behalf of the Medical Incorporations, in Favour of Mr Headlam's Medical Bill, and against that of Lord Elcho', June 1857, RCPSG 1/5/3.

[203] See 'Statement of the Universities of Marischal College at Aberdeen, Edinburgh, and Glasgow, Relative to Degrees in Medicine', 24 February 1857, and the reply, 'Remarks by the Scottish Medical Incorporations on the Statement of the Universities of Marischal College Aberdeen, Edinburgh, and Glasgow', both in RCPSG, 1/5/3. The corporations had however obtained the consent of Oxford, Dublin, Queen's University, Ireland, and St Andrews to their Bill.

been made from March of 1857,[204] largely because it did not recognise medical degrees as sufficient for registration. Candidates for registration had rather to be licensed by a corporation. Glasgow, Edinburgh, and Aberdeen supported Elcho because he held out the possibility that degrees might be recognised for registration. The corporations held to the line repeated by the Faculty on 19 May stating that they:

> decidedly object to any scheme of compromise of which the basis is the recognition of the holders of the University honours as being in virtue thereof legally entitled to practise in any department of the profession without having previously after examination obtained the licence of one of the Medical or Surgical Corporations.[205]

The Corporations' Bill finally came up for its second reading in the House of Commons on 1 July, and was criticised by many members for being old-fashionedly elitist, and protective of outmoded monopolistic privileges.[206] Nevertheless there were some supporters of what one member called 'professional aristocracies',[207] and the Bill passed by a majority of 147 (225 for and 78 against). However, Headlam had failed to get Palmerston to take up the Bill as a Government measure and it thus finally succumbed to the usual fate of the Private Member's Bill, unsupported by the government, and ran out of Parliamentary time. Headlam was forced to withdraw the Bill in July.[208]

The developments which now led to the passing of the Medical Act on 2 August 1858, were conducted largely without the involvement of the corporations, although their General Conference and Scottish Branch continued to meet and review developments.[209] This is in itself significant of the fact that Cowper, the framer of the Act, was largely unsympathetic to the corporations. However, the corporations did have one final, telling say in the content of the Act.

Cowper introduced his new Bill on 23 March 1858. Framed in tandem with John Simon, his right-hand man as Medical Officer at the Board of Health, the Bill aimed to secure a single, uniform, and very general qualification as

---

[204] See, RCPSG, 1/1/1/8, Faculty Minutes, 19 May 1857, which details negotiations with the Universities. See also Craig, *History of the RCPE*, pp. 273–77.

[205] RCPSG, 1/1/1/8, Faculty Minutes, 19 May 1857, Minute of Faculty Council of 19 May extracted in minutes of Faculty, p. 509.

[206] See Waddington, *The Medical Profession*, pp. 104–5.

[207] Ibid., p. 105.

[208] See RCSEng, Minutes of Council, 9 July 1857, p. 183.

[209] See RCSEng, Minutes of Council, 15 March 1858 8 and 14 April 1858. The General Conference had met on 10 March, 13 April and 20 April to organise its opposition to Cowper and Elcho's new Bill and to continue to urge the government to adopt the Corporations' Bill or to support its private introduction by Headlam.

requirement for registration and practice (which would be neither solely existing degrees or licences). This would be run by an extremely powerful General Council of Medical Education and Registration. The Council would define the qualification and have power to force existing bodies to cooperate in examining in order to create a sufficiently broad-based curriculum. Such a Bill was, predictably, welcomed by the British Medical Association,[210] on behalf of the bulk of the profession, and opposed by the corporations.

In this opposition there seems to have been no direct contact with the hostile Cowper. The corporations (often now solely the London colleges acting independently) utilised direct channels to the government, notably to the Home Secretary, Spencer Walpole.[211] The corporations met Walpole on 17 April, and a full deputation of conference, including Scottish representatives, met with him on 12 June. Walpole refused to support the reintroduction of the Corporations' Bill,[212] but such pressure may have influenced his decision to intervene to force certain amendments which reflected the corporations' line on Cowper even before his Bill went into its committee stage. Cowper was required to emasculate the Medical Council: its authority to force corporations to unite for exams was removed and it was left only with powers to request details of curricula.

However, the extent of the corporations' influence here should not be overstated. Walpole himself was a staunch Whig politician, who believed in progress through the moderate reform of existing institutions, and not wholesale social restructuring.[213] As such, he was naturally predisposed to the maintenance of the rights and privileges of existing institutions, and viewed the corporations as the traditional (and therefore best) mode of regulating medicine. Such a man would not have seen the necessity of introducing the state into an area where existing institutions could regulate themselves.

On the floor of the House, the corporations' other ally, Headlam, checked any backdoor attempts to reintroduce into the Bill the need for those registering in the future to be qualified in both medicine and surgery, thus protecting the existing licensing rights of the corporations.[214] Furthermore, the Bill as finally passed contained no stipulation that the majority of representatives on the Medical Council were to be anything other than the elite of the corporations.

[210] The PMSA had changed its name in late 1855 to the BMA. See E. M. Little, *History of the British Medical Association, 1832–1932* (London, 1932), pp. 76–77.
[211] Ibid., pp. 106–32 passim; Stokes, *A Coleridgean*, pp. 356–59.
[212] See brief report on 17 April meeting in RCSEng, Minutes of Council, 22 April 1858, p. 251.
[213] See Stokes, *A Coleridgean*, p. 357.
[214] Ibid., p. 122.

TIME OF CRISIS 395

Thus the Medical Act was a victory for the corporations, in as much as they retained control of their profession, albeit through a new body created by the state. The corporations had nine representatives on the new council, the universities eight, and the remaining six were crown nominees, and the council had limited powers. As Stokes has argued, the attempt by Cowper and Simon to create a new state-controlled profession, possibly informed by Coleridgean ideas about the creation of a clerisy or national policy-making intelligentsia, had been thwarted by the traditional interests of the corporations mediated through a reactionary Whig Home Secretary.[215] That the attitude of the corporations was identical to that of Walpole is nicely demonstrated by the comments of Dr Wood, President of the RCSEd, on the Corporations' Bill during the deputation to Palmerston in early June 1857. The reform it accomplished, he noted approvingly, was achieved,

> by having the present machinery so modified as to meet the requirements of the times; in fact, it reforms without destroying. Lord Elcho's Bill, no doubt, carries out many of the objects of medical reform, but ... in doing so, sacrifices all existing bodies. It should not be forgotten that those institutions are peculiarly British; they do not exist abroad; they were established in strict conformity with the principles of the British constitution, and with the view of giving to the medical profession the power of self-government. To disturb them, therefore, would be retrograde, and not progress in legislation. To hand over the profession to be regulated solely by a council appointed exclusively by the Crown, as the Bill of Lord Elcho [the Select Committee Bill] proposes, would be to establish an un-English despotism.[216]

It is indicative of the Whig nature of the British state in this period that this kind of moderate reform, geared towards the preservation of the power of peculiarly British institutions through new mechanisms, is also reflected in the reform of the Home Civil Service during the 1850s.[217] The voice of liberals and radicals in the profession however was ignored, neither their passionate belief in 'no distinctions ... but those of talent',[218] nor their public campaign to abolish 'petty tyrannies',[219] nor hopes to sweep aside 'forever ... away from the land ... monopolies of every description' seemed to matter.[220] Even the vocabulary of 'despotism' applied to the incorporations

---

[215] Stokes, *A Coleridgean*, pp. 357–59. See also Sir G. Clark, *A History of the Royal College of Physicians of London*, p. 728.
[216] 'Medical Reform: Deputation to Lord Palmerston', *Lancet*, 13 June 1857, p. 614.
[217] P. Gowan, 'The Origins of the Administrative Elite', *New Left Review*, 162 (1987), pp. 4–3.
[218] *Glasgow Medical Examiner*, 1 (1831–32), p. 2.
[219] Ibid., p. 131.
[220] Ibid., p. 56.

in the era of reform by the great body of general practitioners was now seen simply as foreign, or 'un-English'.

The Faculty, like all the other corporations, expressed satisfaction at the final passing of what it perceived as a satisfactory measure of medical reform. The Faculty minutes on the passing of the Act describe it as the outcome of 'the great struggle amongst the medical corporations'.[221] This is indicative of the Faculty's perspective as a small corporation fighting for, and gaining, full recognition. The broader picture of reform was however also one of a great struggle with different elements of the governing elite about the role of the medical profession.

---

[221] RCPSG, 1/1/1/8, Faculty Minutes, 1 August 1858, p. 555.

10

The Best Place to Study Medicine

Conflicting claims about Glasgow as a place to train in medicine were made in the press and medical community in the years leading up to the Medical Act of 1858. Medical journals discussed the new opportunities offered by private schools of medicine and showed how practical demands, such as experience of dissection, were now met more widely. Others criticised what was already in place, hoping to achieve further reform. These debates established what a 'general practitioner' was and they created and defined the image of the general practitioner in the public eye.

The public perception of the general practitioner was intimately associated with the changes in medical teaching. The rise of the private medical schools and the shifting emphasis in the university diplomas toward surgical degrees meant an increased public awareness of anatomical studies. The other crucial change centred on the extension of the medical curriculum. By its expansion and standardisation it made medical studies more broadly available. The course of studies required by the FPSG for the successful completion of their diploma became mandatory, even though it did not undertake to teach any of these requirements; secondly, the university's development of the course of studies required for its degrees expanded. The private medical schools, thirdly, awarded course certificates that were accepted by the Royal Colleges of Surgeons, the FPSG, and Naval or Military Boards.

From the 1820s onward, as discussions over reform blossomed, curriculum requirements were scrutinised. In general, reformers argued for the inclusion of a broader base of foundation courses and recommended attention to ancillary subjects. This push for a better curriculum had repercussions on licensing practices. The reformers wanted the licensing bodies, in particular the FPSG, to take the lead in raising the standards. This was to be done by extending the curriculum requirements, resulting, the reformers hoped, in a more uniform and better-qualified practitioner, backed up by the diploma of the FPSG. Thus the developments in anatomical and surgical teaching, the extension of curriculum requirements as argued for by reformers and the response of the FPSG to these pressures, shaped the transition from the Enlightenment to nineteenth-century medical practice. The ability of Glasgow to respond to these demands and changes, as it became one of the leading industrial and exporting cities of the empire, set the scene for the claim, often reiterated by its medical teachers, that it was, assuredly, 'the best place to study medicine'.

By 1856 James Adair Lawrie was linking the need for better regulation with the reforms in the teaching curriculum already on the agenda of the reformers. Lawrie had taught in Anderson's University and was appointed Professor of Surgery in Glasgow University in 1850. He was elected in 1858 as the representative of this university and St Andrews on the General Medical Council.[1] He wrote as an advocate of compromise, urging all the institutions involved in accreditation (the universities and royal colleges) to accept a universal standard rather than regional fragmentation. Reform along these lines meant persuading medical colleagues, the government and the public that a 'general practitioner' was a wise option and that it was possible to agree on the education and skills such a creation would demand. Of course this entailed an agreed curriculum. Lawrie's suggestions were forward-looking and timely pieces of advice. As he explained in 1856: 'the appellation "general practitioner" is of modern origin, and finds no place in any of our Scotch Universities or corporation charters'.[2]

This 'modern origin' of the general practitioner, as Lawrie quite rightly emphasised, was rooted in the changing medical curriculum, specifically in the major expansion of anatomy teaching, but also in formative subjects like chemistry, in which the extra-mural schools in Glasgow were strong and innovative. Chemistry had always been an important subject attached to medicine, as William Cullen and Joseph Black had demonstrated.

The driving force in changing to modern teaching methods after 1740 had been anatomical and surgical skills training which the Hunter brothers had popularised in London and which had become so important through the teaching of the Hamiltons for the west of Scotland. As has been shown, eminent Scots surgeons travelling back and forth between Glasgow and London (and indeed the Continent) expanded the skills and knowledge available to Glasgow. Thus it is not surprising that the University of Glasgow continued to modernise its anatomical teaching substantially in the first decade of the nineteenth century. Its competitors, the private medical schools, were even quicker in establishing anatomical and surgical instruction, modernising their teaching with all the visual and practical aids the Hunters had made exemplary. Primarily that meant the provision of teaching

---

[1] J. MacLehose (ed.), *Memoirs and Portraits of One Hundred Glasgow Men*, 2 vols (Glasgow, 1886), ii, pp. 171–72.

[2] J. A. Lawrie, *Letters on the Charters of the Scotch Universities and Medical Corporations and on Medical Reform in Scotland* (Glasgow, 1856), p. 5.

materials, in the form of specimens, preparations and, most crucially, of corpses.

The procurement of corpses, before the Anatomy Act of 1832, was hugely damaging to the reputation of surgeons, and yet, at the same time, anatomy lessons and training in surgery were popular as never before. The intake of students was high and reputations were made on the basis of good lecturing and clear demonstrations. One source assesses the number of medical students at Scottish universities at about 10,000 between 1751 and 1850.[3] From the turn of the century the private schools prospered, Anderson's College reaching an apex of popularity in medical studies after 1828 and the Portland Street School in operation from 1826 to 1844. The private schools, the revamped curriculum and the more stringent requirements of the University brought pressures to bear, especially on the FPSG. Its corporate reactions to governmental bargaining over national licensing have just been described. In contrast the FPSG was largely successful in localised campaigns to reform qualification standards. It controlled the teaching at the private medical schools. There was also considerable cross-over, both in students and teaching methods with University anatomical courses.

The 'general practitioner' gradually gained a distinct profile through these various developments. The campaign for broader curriculum requirements eventually brought with it a perception of the distinct skills a 'good doctor' possessed. The revolution in anatomical teaching and surgical skills training via course work was a victory for 'modernism'. The medical profession was at last identified with an 'external' set of criteria, namely an agreed curriculum that could not fall below a minimum standard. These standards were pushed ever higher by reformers, who would not agree to the traditionally minimalist and private examinations by the FPSG. Finally the regional basis of accreditation was removed in favour of a national register (codifying standardisation and professional self-regulation). The 'surgeon-apothecary' had replaced the 'learned physician' and both metamorphosed into the new hierarchies of general practitioner and consultant. Glasgow became – arguably – 'the best place to study medicine', a phrase coined by the first historian of the Royal Infirmary and the Professor of Anatomy at the Andersonian, Moses Buchanan. But this vote of confidence flew in the face of public mistrust and the bickering of the profession at large.

Glasgow's 'Resurrectionists', or graverobbers, coloured the public view of

---

[3] L. R. C. Agnew, 'Scottish Medical Education', in C. D. O'Malley (ed.), *The History of Medical Education*, UCLA Forum in Medical Sciences, 12 (London, 1970), p. 260.

a profession that risked early death from contact with infection. The public view of the surgeon was possibly as a ghoulish villain, as depicted in stories about the 'hue and cry' of the 'perfect mob' that 'beset the approaches to the College and High Street vowing vengeance on the body snatchers'.[4] Its metamorphosis to a more professional image was a step of deep significance, embodied to this day in the architecture of the city:

> What a change, too, in the outward appearance of the newer churchyards! Compare the graveyard at St David's or beside the Cathedral, with their damp, airless vaults, and great menagerie-looking cages surrounding the tombstones, with the elegantly laid out and tastefully finished sepulchres of the Necropolis and Sighthill. What an amount of money must have been laid out in iron, to render these resting places secure, grim, forbidding! How much more pleasing to the feeling of friends, when they go to indulge their natural grief, to see the tomb unfettered, even though it be only covered with the simple turf of flowers, or marked by the plainest stone! In this way, too, they can be better arranged, with spaces for flowers, shrubs, and trees, and admit of considerable ornamented decoration, which has been taken advantage of to render the burying-places just mentioned, and many others, tasteful as gardens, as well as useful places of internment.[5]

But even as the city's merchants and business men, in the 1840s, buried their dead in the new, stately Necropolis, within a stone's throw of the Royal Infirmary, the earning of a living as a surgeon was a precarious enterprise. For the aspiring practitioner, the skills that were attained in the medical schools and the consequent career opportunities were crucial. The margin between success and failure was slim. Training was at a premium and the certificates gained in the private schools or the degrees conferred by the universities, were the key to hospital appointments, teaching posts, partnerships in practice, service in the navy or on merchant ships. One reformer, George Watt, in 1831, stated that it was an almost universal complaint among medical men at the time that the profession was overstocked. In the 1830s and 1840s Scottish medical schools, particularly in Edinburgh, produced more practitioners than could be accommodated in practice in Scotland or in military or naval service abroad. The Napoleonic wars had encouraged some 300 practitioners a year in the services, many of whom had come onto the medical job market in the 1820s.[6]

Just how precarious this adventure of medical training could be is well exemplified by one of the stories told about the controversial anatomical

---

[4] G. Buchanan, 'On the Effects of Mr Warburton's Anatomy Bill, and the Facilities for the Study of Practical Anatomy on Glasgow', *Glasgow Medical Journal*, 2 (1855), p. 438.

[5] Ibid., p. 442.

[6] Loudon, *Medical Care and the General Practitioner*, pp. 208–9.

teacher, Robert Knox. He was well-known in his time as a lecturer, but also as one who tainted his reputation as a doctor by his involvement with the Burke and Hare murders. His biographer, Henry Lonsdale, who knew him personally, writing not long after his death, tells a story about Knox retrieving a crust of bread from where it fell on the ground. Knox and his colleague, Dr Andrew Adams of Glasgow, were walking together and deep in conversation. Knox spied a large crust of bread, picked it up and placed it where it could be seen, high on a wall. Knox told his friend he was never able to pass by a wasted piece of food without being aware of the wretched look on the face of those who had to go hungry, so he never left food to rot on the street.[7]

What this story conveys is the nature of a medical career not far from the brink of poverty, trading only on talents and skills. If added to this suspicion of unsavoury practices abounded, the reputation of the medical man might sink towards the ambiguous spheres of figures portrayed in tales such as Frankenstein.[8] Mary Shelley's novel was published in 1818 and the fiction of a man resurrected through science was part of popular credulity. The confidence of the public in medical men had sustained a serious assault through the clandestine activities surrounding the procurement of corpses for anatomical dissection. What may have redeemed the 'general practitioner' was the merit of 'science' broadly conceived: if the anatomical schools were linked to the bogeyman image of resurrecting the dead, they also traded on the image of enlightening the public about the merits of sound sciences, those of chemistry, metallurgy, botany and zoology.[9] The extra-mural schools inaugurated the great movement toward democratic expansion in education bringing it within the grasp of the working man. Thus innovation mixed with suspicion was an almost natural product of the process of expanding the medical curriculum.

Knox can be taken as an exemplary figure of his time. His involvement with the horror attached to the murderous venture of men who not only 'resurrected' dead bodies but also killed to supply the wants of anatomical teaching showed how a tainted image of the inquisitive and experimental man of science evolved. The profession of surgery was touched by the distaste and violent reactions created by disinterring corpses for dissection. The irony was that the experience gained by these means resulted in greater public

---

[7] This description is contained in an anonymous review of Lonsdale's book on Knox: 'A Sketch of the Life and Writings of Robert Knox, the Anatomist: By his Pupil and Colleague, Henry Lonsdale', *Glasgow Medical Examiner*, 2 (1869–71), pp. 356–58.

[8] M. Shelley, *Frankenstein or the Modern Prometheus* (first published in 1818; Penguin edn, 1992).

[9] D. E. Allen, *The Naturalist in Britain* (2nd edn, Princeton, 1996).

confidence when used in practice. Knox personifies that paradox. Lionised in his day as an eloquent and exacting lecturer, he became unemployable in universities because of his association with Burke and Hare. 'As the star of his greatness was declining, [he] lectured at the Portland School of Medicine', wrote his biographer Henry Londsdale, who cited William Weir, President of the Faculty from 1847–49 and for many years its Treasurer,[10] and a mainstay of the Portland and Andersonian Medical Schools. Weir described listening to Knox 'as a fascination beyond reading 'Robinson Crusoe' in his youth; that of all the lecturers he ever heard, Knox was *facile princeps* in richness and abundance of illustration'.[11] But Knox's ambivalence reflects that of Glasgow. As Lonsdale puts it:

> In Glasgow he gained some friends and admirers, but no bread and butter. So small was his class, that he returned his fees to his pupils before November was out. The city of big chimneys, bigger piety, and biggest of all in the use of alcoholic liquors in Scotland, was not for Robert Knox, the heterodox man of science and the claimant of individual rights.[12]

The city of 'big chimneys' and a big drink was also one of liberal leanings and interest in innovation and getting ahead. The city tolerated both the tales of daring connected with the Resurrectionists and the costs of making – as well as publicising – Glasgow as the best place to study medicine. The ambition and accomplishments of the men involved in engineering the network needed for good medical studies show through in the often divergent assessments given of the city.

The surgeon and poet Thomas Lyle, who was a medical student between 1812 and 1815, and joined the register of the FPSG in 1817,[13] disparaged the state of all of the anatomical schools of Scotland 'because it is now quite impossible to procure more than two or three bodies in the course of a year by exhumation'. On Glasgow he observed that 'bodies are so scarce that they are salted in summer and hung up to dry like Yarmouth herrings'.[14] The same student dented the reputation of the University Dissector, Jardine, by calling his sessions nothing 'more or less than a den of tumult and uproar, of wrangling and bawling and every species of schoolboy mischief'.[15] However, this same student, for the same period, praised the professor of anatomy,

---

[10] Duncan, *Memorials*, p. 275.

[11] See Henry Lonsdale's review of Knox's life, *Glasgow Medical Examiner*, 2 (1869–71), p. 357.

[12] H. Lonsdale, *A Sketch of the Life and Writings of Robert Knox, the Anatomist* (London, 1871), p. 259.

[13] Duncan, *Memorials*, p. 100; the *Dictionary of National Biography* in its edition of 1893 gives this date wrongly as 1816, see under Thomas Lyle.

[14] Thomas Lyle, as quoted in L. R. C. Agnew, *Scottish Medical Education*, p. 256.

[15] Ibid.

James Jeffray, for his 'melody of eloquence' and how spellbound his listeners were 'whenever he raises himself up from his tripod, and emerges from behind the revolving mahogany dissecting-table to the middle of the arena fronting the students ...'[16]

The revolving mahogany dissecting table gives a special clue that all may not be as troubled or 'schoolboyish' as Lyle would have the reader think. A revolving dissection table was neither cheap nor antiquated and indeed pointed to the updating and modernising of the anatomy facilities in the university. Under Jeffray, in 1811, a year before Lyle's student attendance, the Anatomy Department was rebuilt as the Hamilton Building. The accommodation consisted of the dissecting room, the professor's study, the lecture theatre, preparation and anterooms and a reading or oseology room generally known as 'the Bone Room'. Three good skylights had been installed and spaciousness characterised the dissecting room.[17] The crowding that Lyle describes must have come from the many students, not the inadequacy of the facilities.

As for the claim that bodies were so inaccessible, even before the Anatomy Act of 1832, that they were 'dried and hung like Yarmouth herring', others write of just the opposite. A correspondent in the *Lancet* stated that in 1816 and 1817 Dr John Barclay's dissecting rooms in Edinburgh were supplied from the College Street dissecting rooms in Glasgow and that despite the numbers, 'new students had to wait only three to four days to be provided with material for dissection, while in London it was common to wait a month'.[18]

College Street had a particular history of its own and it serves to illustrate the modernisation of medical teaching. Moreover, it shows how the non-university sector of instruction in the practical skills necessary for a medical career tallied well with an ongoing commitment to educational reform. College Street was set up as an anatomical school by the two Burns brothers, John and Allan. There they taught anatomy, surgery and midwifery and the sixteen-year-old Allan Burns was put in charge of the school's dissecting rooms. Allan Burns had outstanding skills in specimen preservation, but was not qualified for surgical practice in Glasgow.[19] In 1804 Allan Burns was invited to go to Russia to help found a hospital on the Scottish model and to be physician to the Imperial Court. He returned in 1805 to find that his

[16] As cited in F. L. M. Pattison, *Granville Sharp Pattison: Anatomist and Antagonist, 1791–1851* (Edinburgh, 1987), p. 16.
[17] S. W. McDonald, 'The Life and Times of James Jeffray, Regius Professor of Anatomy, University of Glasgow 1790–1858', *Scottish Medical Journal*, 40 (1995), p. 119.
[18] Ibid., p. 119.
[19] Pattison, *Granville Sharp Pattison*, p. 22.

brother John had been debarred from teaching anatomy because he was implicated in a case of grave robbing. Allan Burns then gave the lectures on anatomy and the principles and operations of surgery.[20] His reputation was so great that he figures in one of the attacks against the FPSG. In the *Glasgow Medical Examiner* of 1831, 'Zeno' writes:

> We cannot forget that we once had such a lecturer among us as Allan Burns; a man, we are proud to say, who despised the Faculty as much as we do – a man whose talent they have never matched – yet one whose tickets the illiberal fools, to their eternal disgrace, never recognised, though, in spite of this drawback, such was the commanding nature of his abilities, that his lectures were always overflowingly attended.[21]

Both men had a major impact on the teaching of anatomy and surgery in Glasgow, their influence based on the private medical schools. Allan Burns strongly impressed, among others, Granville Sharp Pattison and Robert Cleghorn, with his first-rate knowledge of anatomy and, together with his senior associate, Andrew Russell, built up his extensive museum.[22] Thus the private medical schools were supplied both with well-trained teachers and with the teaching materials that were a sign of the quality of the enterprise. It was this collection that featured prominently in the advertisement for Granville Sharp Pattison's teaching when he began his own series on anatomy, physiology and the principles of operations of surgery in 1814 in College Street.[23] Allan Burns's collection went to Granville Sharp Pattison when he replaced John Burns as the lecturer in anatomy and surgery at Anderson's College in 1818. Burns only officially resigned this position in 1817, despite having been appointed professor of surgery two years earlier.

Granville Sharp Pattison edited Allan Burns's major publication, *Observations on the Surgical Anatomy of the Head and Neck*, in a second edition after Burns's death (Glasgow and London, 1824). Pattison characterised the *Observations* as 'a most valuable standard work ... It contains no hypothesis nor theories, but consists entirely of pathological inferences, drawn from the most acute and accurate observations on the anatomical structure of the parts'.[24] Pattison indicated that 'my late dear friend ... published the present

---

[20] Ibid., p. 23.
[21] 'Zeno', 'The Medical Schools of Glasgow', *Glasgow Medical Examiner*, 1 (1831–32), pp. 185–86.
[22] Pattison, *Granville Sharp Pattison*, p. 23.
[23] Ibid., p. 61.
[24] A. Burns, *Observations on the Surgical Anatomy of the Head and Neck: Illustrated by Cases and Engravings, with a Life of the Author and Additional Cases and Observations by Granville Sharp Pattison* (2nd edn, Glasgow and London, 1824), p. xv. The first American edition was published in Baltimore, 1823.

volume as the commencement of a series, it having been his intention to have proceeded upon the same plan, and described all the other parts of the body, so as to have formed a *complete system of Surgical Anatomy*.[25] The first edition of the book appeared in 1811 in Edinburgh,[26] and a German translation, with a preface by Professor Johann Friedrich Meckel, was published by the University of Halle printer, Renger, in 1821.

Burns was known as a specialist in vascular preparations, as distinct from those in spirits, and here he seems to have distinguished himself above and beyond what the Hunters and Monros had accomplished.[27] Apparently Burns was very adept at enriching pathology through close observation of alterations in the morbid parts, and thus fully complemented the scientific research of John Hunter, who, above all, sought to vindicate practical anatomy by its usefulness as a corrective diagnostic tool. Burns's primary interest was in cardiac disease and here, according to contemporary opinion, he was a leading investigator.

John Burns, as much as his brother, seems to have stirred deep interest across the Atlantic and in Germany. His publications raised him into the first ranks of writers on medical subjects, especially in midwifery. John Burns's first book appeared in 1799 in Glasgow, *The Anatomy of the Gravid Uterus with Practical Inferences Relative to Pregnancy and Labour*. Works on abortion (1806), and on uterine haemorrhage (1807) followed.[28] These were published in an American edition in New York in 1811 in one volume (covering the anatomy of the gravid uterus, abortion and uterine haemorrhage). A second New York publication appeared in 1811, *Popular Directions for the Treatment of the Diseases of Women and Children*. The *Practical Observations on the Uterine Haemorrhage with Remarks on the Management of the Placenta* of 1807 is dedicated to Alexander Monro, Jr, 'in testimony of esteem for his professional abilities and for his private friendship'. The author described himself on the title pages as 'Lecturer in Midwifery and Member of the FPSG'.

The book that made John Burns's international reputation was entitled *The Principles of Midwifery: Including the Diseases of Women and Children*, first published in London in 1809. The second edition, published in London in 1811 was 'greatly enlarged' and states in the preface that Burns takes the anatomical descriptions 'from dissections and preparations before me whilst

---

[25] Ibid., Introduction, p. ii. Emphasis ours.

[26] Illustrated with ten places and numbering 415 pages.

[27] G. S. Pattison, 'Life of Allan Burns', introduction, in A. Burns, *Surgical Anatomy of the Head and Neck* (Baltimore, 1823), pp. viii–ix.

[28] John Burns, *Observations on Abortion* (London, 1806); idem, *Practical Observations on the Uterine Haemorrhage with Remarks on the Management of the Placenta* (London, 1807).

writing'. Subsequent editions were enlarged and revised, appearing in 1814, 1817, 1820, 1824, 1828, 1832, 1837 and in 1843 (10th edn). Two German translations were made, the first a partial one of the 6th edition (in 1827) and a complete one in 1834, of the 8th edition, by Professor H. F. Killian, the director of the maternity hospital in Bonn.[29]

This quite remarkable publication record of both of the Burns brothers, and particularly their authority in the anatomical sciences (all the works were recognised by contemporaries as originating in direct observation from dissections or from original preparations, or indeed from observational case histories), shows them to have been leading lights in the investigative scientific tradition of the Hunters. Their influence went far beyond their local reputation as surgeons, anatomists and specialists (foremost, in John's case, as an obstetrician and gynaecologist) and demonstrates the high quality of teaching and specialist medical expertise available in Glasgow through the extra-mural schools. Like the majority of the teachers at Anderson's College, John Burns later transferred to a professorship at the University.

One of the most remarkable characteristics of Glasgow was its intimacy. For all the sharp feuds between John Burns and the FPSG leadership, for example, the social life of the city was claustrophobic. The men gathered together in taverns, or sat together on committees charged with helping Glasgow's poor: not least of these the charitable trust of impoverished sons of ministers to help the sons of the manse. Three of the most powerful surgeons in Glasgow in the early nineteenth century, Burns, William Couper and James Monteath, were all Presidents.

Glasgow's surgeons and physicians belonged to the inner circles of commerce and bolstered the Faculty's influence. The surgical 'businesses' that dominated the town also dominated the FPSG. John Gordon took care to ease Cullen into Glasgow medical life – while himself choosing to practise with John Moore. The best men were ambitious and knowledgeable, but were often cut down in their prime, like William Hamilton or Allan Burns, John's son. The Burns brothers, the Hamiltons, the Hunters, the Moores, the Hills, the Coupers, the Stirlings and others gained influence and repute from their family backgrounds – and significant civic and financial support.

The Burns brothers were sons of the longest-serving Church of Scotland minister in the Barony Church, the Rev. John Burns. He had been assistant

---

[29] J. Burns, *Handbuch der Geburtshülfe mit Inbegriff der Weiber-und Kinderkrankheiten: nach der achten, vollständig und gleichsam ein neues Werk bildenden Ausgabe*, ed. H. F. Kilian, i (Bonn, 1834).

to the Rev. Lawrence Hill, the father of the surgeon Ninian Hill.[30] Two other brothers, James and George, were very successful Glasgow shipping entrepreneurs, establishing the company that was to begin with the Liverpool trade and end up with joint Admiralty contracts with Cunard for the American steam routes across the Atlantic to Halifax and Boston. The Burns shipping empire also supplied the steamers for the day passage through the Crinan Canal or the night passage 'round the Mull'; or, as it was put poetically, the boats went:

> gliding along canals or battling with the Atlantic, meeting at Oban, crossing and recrossing, plunging into the lochs, winding along the sounds, threading their way among the islands, fine pleasure boats for the flock of summer swallows, stout trading boats summer and winter serving the whole archipelago, linking with the world the lonely bay or the outer islet, freighted out with supplies of all sorts, and shapes, freighted in wool and sheep, Highland boats and Highland bodies: surely the liveliest service in the world![31]

The Burns brothers come to life in contemporary descriptions of them. John is portrayed as 'a little beneath the middle height, and rather slight in figure; his countenance is rather indicative of a steady power of application than expressive of that rapid and brilliant thought which constitutes the more sparkling ... leaning back in his chair, with one leg laid over the other, and with finger on chin, in the most easy colloquial manner; we have heard the Professor proceed into the minute practical details, respecting the disease under consideration, without having recourse twice to his notes during the lecture of an hour'.[32] Of Allan Burns, too, a cameo portrait exists: 'He was small, delicate and boyish looking, and the first time we saw him enter the theatre to deliver his introductory lecture surrounded by several of the private party, who were stout robust fellows, we mistook him for one of the apprentices, till he emerged from the crowd, advised to the table and commenced reading his lecture'.[33] The description of Pattison is also complementary, he being seen 'as little inferior to ... his late master [Allan Burns]; while he was idolised by the students for the great care and attention he bestowed in grounding them in the principles of anatomy'.[34] Allan Burns

---

[30] See above: J. MacLehose, *Memoirs and Portraits*, i (Glasgow, 1886), 'James Burns' (brother to John and Allan), p. 59.
[31] Ibid., p. 68.
[32] 'Sketches of the Medical Lecturers of Glasgow, No. 1: Dr Burns, Professor of Surgery in the University', *Glasgow Medical Examiner*, 1 (1831–32), p. 74.
[33] Thomas Lyle, as quoted in F. L. M. Pattison, *Granville Sharp Pattison*, p. 22.
[34] Ibid., p. 24.

died in 1813, 'from a puncture got in dissection',[35] and it was left to the others to embark on extensive careers.

Granville Sharp Pattison's career was marked, as John Burn's had been, by the troubles attendant on the violent dislike of the public for secret disinterment. For nearly a century, from 1744, when the first riots in Glasgow took place, to 1832/34, when the Anatomy Act was passed, the tension between those seeking skills and those who abhorred and sought redress for the desecration of graves was palpable. It was a conflict pursued at law and surgeons and students paid the price, even if, in later, less troubled times, the risks were clothed as adventures.

It was in 1809 that the Virginia Street medical school of the Burns brothers found new premises in College Street. These were on the north side and close to the east end and, apparently, students lived in the upper flats on each side of the street, stretching double lines of cord from window to window to transfer small articles across to each other. The school was close to Inkle Factory Lane, giving access to Ramshorn churchyard. This geographical advantage gave the school a dubious but exciting role in the procurement of corpses for dissection.

The problem of procuring corpses was a delicate one. It was a criminal offence to disinter the dead. It raised strong emotions over the sanctity of the dead and vivid fears connected with the afterlife, in which the powers of the deceased were seen as still potent. The elaborate rituals of burial were not just respectful of those who died, but also insured that the dead did not harm the living. The very title of 'Resurrectionists' given to the medical students who illegally obtained corpses bore witness to the fears attached to the dead rising from the grave. The deep distaste of the citizens of the city for 'graverobbers' can be measured by how they tried to discourage their activities: iron cages over graves before 1832 and the watchmen and use of pistols in cemeteries. Even allowing for the kind of exaggeration that can sometimes creep into the stories told by medical men some time after the event, the truth was grim. Private dissections depended on paying for a corpse, not an activity within the boundaries of the law, although, as in Paris, the law often turned a blind eye, it is said, because the need for surgical training was felt to be an honourable objective.[36]

---

[35] 'James Burns', in MacLehose, *Memoirs and Portraits*, p. 59.
[36] T. Gelfand, 'The "Paris Manner" of Dissection: Student Anatomical Dissection in Early Eighteenth-Century Paris', *Bulletin of the History of Medicine*, 46, p. 129.

George Buchanan commented on the troubled days before the Anatomy Act by recalling how students 'had to repair to continental schools for what was denied to them in this country, or at least was obtained with the penalties of the law hanging over their heads'.[37] In 1855 George Buchanan, who was the son of Moses Buchanan,[38] extolled the virtues of practical anatomy (dissection) for the teaching of medicine in Glasgow.[39]

Buchanan's testimony to the early days of anatomical instruction for surgeons contrasts the practices in London and Glasgow for procuring corpses, in order to demonstrate the superiority of Glasgow's teaching of practical anatomy, both past and present. In London surgeons and medical students paid 'a set of men of great bodily strength, and of usually low and depraved habits, as well as ungovernable temper, who were styled 'resurrection men' [to undertake] the supply of the dissecting rooms'.[40] The rivalries among the gangs and their unsavoury methods meant that the price of a corpse in London amounted to £12 'and even at that rate it was difficult to get the men to work'.[41]

In Glasgow, on the other hand, 'there were no men that made it a business to obtain bodies'.[42] The supply of bodies to the dissecting rooms had to be undertaken by the anatomical teachers and their students themselves. Buchanan then relates the experiences of 'one not infrequently present' amongst the private resurrectionists of Glasgow:

> The grave of the recently buried body was carefully observed and marked during the day, and a band of sufficient number – usually four or six – was made up. The party, provided with a dark lantern, an old carpet, a sack, and shovels and pickaxe, took advantage of the first dark, cloudy, perhaps windy night, in order that their proceedings might be the better concealed. Sentries were posted to give notice of any alarm which might get up, and the principals entered the graveyard by climbing the wall, at a distance from the gate. The grave was then opened to about half its extent, and by dint of hard labour, about one third of the coffin lid was exposed. The strongest of the party now entered with the lantern to perform the most difficult part of the whole, and over the open grave was thrown the carpet above mentioned, for the double purpose of concealing the light and

---

[37] Buchanan, 'Warburton's Anatomy Act', *Glasgow Medical Journal*, pp. 431–33.
[38] There are a confusing number of Buchanans. There was George the son of Andrew Buchanan, who was Professor of Physiology at Glasgow University (1839–76); There was also George the son of Moses Steven Buchanan, the chronicler of the GRI and Professor of Anatomy in Anderson's University. He was the son of George Buchanan who became the distinguished Professor of Clinical Surgery at Glasgow University.
[39] Buchanan, 'Warburton's Anatomy Act', p. 434.
[40] Ibid., p. 434.
[41] Ibid.

deadening the noise of the working. The coffin-lid was wrenched open by a short crowbar, and, by sheer force, was broken off where it remained covered with earth. The noose of a strong rope was now put round the neck of the body and handed up to the others outside, who soon pulled up the corpse to the surface. It was then wrapped up in the sack and carried off to a convenient place, and the grave was filled up and covered, great care being taken to leave the surface as near as possible in the same condition in which it was found.[43]

This graphic and detailed account of medical students graverobbing seems mainly to have been connected with the private medical school in College Street. One episode recounted by Buchanan was that of the famous case of Mrs MacAllister's body, involving Granville Sharp Pattison, who was tried before the Court of Session in Edinburgh. The verdict was 'not proven', but 'the amount of popular excitement produced ... the more general adoption of mortises and iron stanchions around the tombs'.[44] The contrast to the period after the Anatomy Act was vivid, as Buchanan is at pains to emphasise. Clearly to be noticed:

is the altered feelings with which the students are regarded by the citizens [today]. In [former] days the name of a medical student was always associated with everything that is bad; and, in not a few instances, the pursuits in which they had to engage were not calculated to elevate their sentiments; but now they are, with propriety, accounted among the peaceful and regular inhabitants of the city.[45]

From the 1840s the city of Glasgow could truly boast a renaissance of modern medical teaching. The most prominent teachers in the private medical schools did not hesitate to laud the assets of 'their' city in print. Nor were they unsupported by the statistics that they could quote. These showed that Glasgow was indeed the place in which instruction was easily available and affordable for men of modest means.

'Prices ... such as the most indigent can afford'. This statement of Moses Buchanan's refers specifically to the ability to pay for a corpse for dissection. His *Remarks on Medical Reform* of 1846, however, expressed his more general sentiments that Glasgow was the best place for an affordable and accessible education. Moses Buchanan's essay was typical of a general trend in all the writings on the Glasgow schools of medicine. They portrayed medical learning as easy to acquire in the city on the Clyde because the institutional provision was comprehensive. The teaching hospitals were interlinked with specialist infirmaries which again connected with the system of course instruction in the private medical schools as well as those offered in the

[42] Ibid., p. 435.
[43] Ibid., pp. 435–36.
[44] Ibid., p. 438.

University. Despite the fierce feuds over licensing, this constituted a viable 'school of medicine' made of excellent component parts. These parts may have been separate, but the choices made by students linked them to effect. Instruction was plentiful and, as was often pointed out, students could go to two universities at once: the established one granting its own degrees, and Anderson's University (officially so called from 1828 to 1877), whose certificates were honoured by the diploma of the FPSG. Where one finally graduated was a matter of choice or financial resources.

In regard to Anderson's, as well as the University, it was said that 'English and Irish flocked to them'.[46] There can be no doubt that the school of medicine worked as a city-wide enterprise because of the plurality of institutions offering instruction within it. This very 'plurality', however, offered a unique 'consumer choice' for those wishing to take advantage of different types of education.

Glasgow became attractive to medical students because of financial incentives. By 1862 the licence granted by the FPSG only cost £10 (surgery); and £16 for a double diploma of Physician and Surgeon. The fees for attendance at the Royal Infirmary, one correspondent wrote, 'contrast favourably with the high charges of similar institutions'.[47]

A measure of how well classes were equipped with material aids can be deduced through the availability and price of corpses. George Buchanan used the statistics of the Inspector of Anatomy for Scotland to tabulate the number of corpses available in the city in the nine years from 1846 to 1854. The unclaimed bodies available for purposes of dissection numbered 1490, of which 567 were used for the schools while 923 were left over to be buried at the city's expense rather than that of the anatomical teacher's own.[48] Buchanan applauds the fact that corpses were available for practising surgical operations. Surgery performed on dead bodies he wrote 'is one of the most important parts of practical anatomy'.[49] He cites the cases where several naval surgeons were in port and asked if they could refresh their knowledge of surgery by practice on corpses. Buchanan was able to provide these in the Anderson's anatomical theatre at very short notice (within days).[50]

Instruction in the anatomical courses was intense. At the University James Jeffray instructed for one hour each weekday with a review examination every Saturday. In the afternoons students spent four hours in the dissecting

---

[45] Ibid., p. 442.
[46] *Medical Times and Gazette*, 29 November 1862, 'Medical Education in Glasgow', p. 578.
[47] Ibid., p. 578.
[48] Buchanan, *Warburton's Anatomy Act*, pp. 439–40.
[49] Ibid., p. 440.

room with a second daily set of lectures by the demonstrator.[51] Already in 1801 Jeffray was praised because under his auspices:

> Glasgow bids fair to become one of the first and best medical schools in Europe. The hospital is happily situated, liberally supported, and well attended. The College has lately received a most valuable acquisition to its treasures in the museum of the late celebrated Dr William Hunter, for whose arrangement apartments are now building on a most judicious and enlarged scale, under the immediate inspection and superintendence of Dr Jeffray.[52]

These praises for anatomical instruction glide over easily into naming the other factors responsible for Glasgow's strengths in medical education.

In Anderson's University the full range of medical lectures was offered: anatomy and surgery, botany, chemistry, midwifery, and the theory and practice of medicine. Like Jeffray at the University, Moses Buchanan was careful to explain the terms on which he taught his course in anatomy and surgery to prospective students. His classes too were well supervised and individual study opportunities were available in libraries and collections. Buchanan praised the anatomical museum, the library and reading room; he noted the models in plaster, papier-maché and wax, the medical periodicals and the 'library of bones' that could be studied.[53] He explained the weekly review he made a part of his teaching. A question was to be solved by a team of four students for which a prize was given. He emphasised the special value he placed on surgical operations performed on corpses, a practice also introduced by John Burns in the university course in 1839.[54] For this 'modern' practice and its usefulness Buchanan cited the 'recent regulation' enacted by the medical department of the British Navy that 'no one can be promoted from Assistant to Senior Surgeon before going through a complete course of all the operations of Surgery on the dead body'.[55] But above all Moses Buchanan was at great pains to laud the opportunities offered in clinical instruction at the Glasgow Royal Infirmary. He tabulated a number of statistics at the end of his essay on anatomy and surgery in Glasgow to prove his point. In his 'candid opinion, after a most careful examination of every hospital at home or on the Continent of Europe ... the Glasgow Royal Infirmary presents to the eye of the medical and surgical pupil the best

---

[50] Ibid., p.
[51] Pattison, *Granville Sharp Pattison*, p. 15.
[52] J. Birsted, as quoted in MacDonald, *Life and Times of James Jeffray*, p. 121.
[53] M. Buchanan, *Lecture Introductory to a Course of Anatomy: Delivered to the Students of Anderson's University, Glasgow* (Glasgow, 1842), p. 28.
[54] Coutts, *A History of the University of Glasgow*, p. 528.
[55] M. Buchanan, *Remarks on Medical Reform: Being the Substance of a Lecture Introductory*

microcosm of disease which is anywhere to be met with'.[56] Obviously the GRI had put aside the factionalism of FPSG and University disputes to promote in-house clinical teaching. The 1830s power struggles were now transferred to parliamentary lobbying.

Buchanan compared this microcosm of clinical teaching with London and Edinburgh. The latter, he claimed, was hampered by monopoly. Only certain teachers presented clinical cases, while in Glasgow 'all the cases in hospital are, or may be clinical; every operation is brought under review, and every inspection properly explained'.[57] In the statistical tables that compared England, London, Ireland and Scotland and their hospitals for the year 1845, Glasgow's Royal Infirmary rated lowest of all in costs to students and highest of all single institutions for its number of beds. The hospital that came nearest to it in size was St Bartholomew's, London, with 440 beds, the GRI having 450. Its average number of surgical operations was 140, while St Bartholomew's had 120. But costs counted most and made the real difference, with all the London hospitals charging considerably above the costs to students in Glasgow. Guy's hospital in London charged £50 as a fee for two year's attendance, while Glasgow asked only £7 7s. 0d. Edinburgh's fees were also above Glasgow's (at £12 12s. 0d.) with only ninety operations per year. Equally instructive are Buchanan's tables on the types of operations done in Glasgow: he gives diseases treated and operations performed from 1795 to 1846. These details bear out the wide diversity of diseases treated and the regular performance of the most common operations.[58]

Others writing in the medical press verify this picture of Glasgow as a good place to study medicine. Articles written in 1862 draw together the type of education available and, especially, the clinical study and specialist hospital experience to be obtained.[59] An anonymous contributor to the *Medical Times and Gazette* of 1862 lists three principle constituents needed for a successful course of medical education. The courses of study given in the University and leading to its three medical degrees, the MB, the CM (at the time the only university degree in surgery) and the MD all provided these. Anderson's University offered medical teaching equal in quality to that of the University. Its certificates were accepted wherever a student wished to take a degree, but 'a large portion of them ... take the licence of the Faculty of Physicians and Surgeons, Glasgow, either for the diploma of Surgeon (£10), or for the double

to a *Course of Anatomy Delivered to the Students of Anderson's University* (Glasgow, 1846), p. 14.
[56] Ibid., p. 14.
[57] Ibid., p. 15.

diploma of Physicians and Surgeon (£16) granted by the Glasgow Faculty and the Edinburgh College of Physicians conjointly'.[60]

The strong position of Anderson's derived from its excellent lecturers. Arguably this was the positive side of the restrictive practice by the FPSG to monitor teachers by regulating whose certificates were acceptable. Until the Widow's Fund was finally abolished completely (in 1850), teaching was tied to membership in the FPSG. The 'cheap' rate fixed for courses at Anderson's allowed it to compete for students. As every teacher at Anderson's was advertised as being a member of the FPSG, it was surely a policy sanctioned by its councils. Thus qualification by way of Anderson's University medical school was a good choice at a financial saving, as the courses in Anderson's were all £1 1s. 0d. cheaper than at the University.

Clinical teaching and hospital ward experience were now readily available in Glasgow. In 1862 the GRI had 600 beds and the fees for attendance were lower than at all other comparable institutions. Instruction given at the bedside was augmented by clinical lectures on both surgical and medical cases. These lectures met the requirements of the examining boards. All operations and post-mortem examinations were performed in the presence of students. The GRI had become an exemplary teaching hospital by mid century. Other clinical experience was available for midwifery and obstetrics in the Rottenrow Lying-In Hospital and in the University Lying-In Hospital and Dispensary. The first named listed attendance at about 600 cases per year and the numbers at the University hospital are not given, but were also in that league (750 cases were mentioned by John Pagan when he opted to discontinue in-patient facilities). Cases in the Eye Infirmary numbered around 3000 out-patients annually. The Lock Hospital (venereal diseases) and the new Dispensary for Skin Diseases were other treatment centres open to medical students.

Clinical lectures in hospital wards were now part of the requirements needed either for a university degree or the licence of the FPSG or its conjoint diploma. The third medical examination in the university included clinical surgery and clinical medicine as well as midwifery. The certificates required were for general hospital attendance for two years, attendance at six midwifery cases and six months spent in outpatient practice either at a hospital or dispensary or with a registered practitioner.

The University programme for medical study seems exemplary for its time. It was a four-year course beginning with anatomy (including dissection with

[58] Ibid., pp. 25–28.

the plentiful supply of corpses for which Glasgow had a reputation) and chemistry in the first year. Botany followed in the first summer and in the second and third winter surgery, physiology, and the practice of medicine (internal disease and its treatment) followed. Materia medica, midwifery and forensic medicine were added in the fourth year. Examinations were both written and *viva voce* covering all subjects taught and held at the end of the second, third and fourth year.

Medical education had shifted significantly from the early eighteenth century to the years just after the Medical Act of 1858. Teaching had been revolutionised since Gordon's 1714 anatomical and surgical lectures which changed the closed-shop apprenticeship system to the open access of the private lecture system. This led in turn to the academic ascendancy of the surgeon. The former craftsman climbed to equality with the physician in the eighteenth century. Glasgow's story shows how crucial were merit and skills. In a commercial city the competitive ethos developed in textile manufacture, the chemistry of dyeing and other ways of producing better goods did not halt before medicine. Integrated with this commercial elite, the surgeons of Glasgow applied its methods. This encouraged importing what was useful from abroad and making those at home innovative. Botany, anatomy, surgery, midwifery and chemistry were improved in turn and, like yeast to the dough, created two systems of medical education: the university's and that of the FPSG's consortium of extra-mural schools. The same city surgeons assured the private schools their success. They were also responsible for medical care in the new hospitals, specialist and voluntary alike. In effect they engineered the creation of the teaching hospital from its charitable workhouse beginnings to the educational success of the GRI in the 1860s.

The medical curriculum evolved concurrently. The two main schools, Anderson's and the University medical school, offered comprehensive programmes of medical teaching conforming to medical reform. The clinical side of teaching was integrated with this system through curriculum requirements needed for both university degrees and the licensing bodies. What had been, in mid eighteenth century, a system of free trade medicine, in which travel to Paris, London and the Continental countries provided the knowledge and skills for the elite, had now developed into a system of education for the many men of few means. The system of education and its accreditation through regulatory bodies was standardised through the 1858 Medical Act. In an unbroken succession the variety and quality of medical teaching available in Glasgow made it advantageous to go and learn there.

Learning in medicine now centred on the curricula established by institutions of learning. These became the canonical reference points of medical knowledge and formed a system which spread, octopus like, through medical institutions. More uniformly teachable, this standard fare needed clinical knowledge as an observational and empirical corrective.

The chance to observe a wide range of cases gave the theoretical side of curriculum requirements substance. Thus the teaching hospital together with systematised requirements became the fulcrum on which modern medicine turned. The teaching hospital made available the clinical cases which the universities and the private medical schools could not provide. The large teaching hospital also, obviously, became the institution in which accidents, diseases and operations could be dealt with when the population of a city had become both transient and poor. The poor were the sleeping partners of the medical enterprise. They, in turn, influenced the increasingly anonymous relationship between doctor and patient to which we have become habituated. In the voluntary hospitals, as in the military hospitals in which surgeons also trained, 'scientific objectivity' was cultivated, isolating disease and treatment in a search for 'origins' and 'natural histories'. Case histories were key tools in training the doctor. The patient became anonymous. Modern medicine had come a long way from the narrative voice making the patient's perceptions palpable in the consultation letter.

Reform in medicine between 1800 and the 1860s turned on all of these issues. Glasgow's reform debates sprang from a liberal tradition sympathetic to improvement, commerce as sensitive as politics to the French and American cries for change. Even if trading ceased, the debates did not. Medical reform became a topic of great interest against this background and perhaps really began in earnest after the 1780s. The medical debate itself was quite specific. At its core was the need for young practitioners to detach the profession from piecemeal regulatory practices. This opening of the medical profession caused a fundamental revision: it introduced standardised training rather than maintaining the old splits between 'pure' physicians and apprenticeship-trained surgeons, creating an interchangeable set of skills between the two branches. Male midwifery was one of the significant turning-points. The second was the Scottish practice of gaining two qualifications, in which Glasgow was exemplary, as James Adair Lawrie showed, most gaining a diploma in surgery from the FPSG and an MD from a university. Mobility and careers pushed training beyond the boundaries of local jurisdiction. This last issue, the national registration of medical qualifications, heralded a new era.

APPENDIX 1

Charter by King James VI to the Faculty of Physicians and Surgeons of Glasgow

JAMES, by the Grace of GOD, King of Scots, to all Provosts, bailies and burghs, sheriffs, stewards, bailies of regality, and other ministers of justice within the bounds following and their deputes, and all and sundry others [of] our lieges and subjects, whom it effects, [to] whose knowledge these our letters shall come, greeting, KNOW YOU WE, with advice of our council, understanding the great abuses which have been committed in time bygone, and yet daily continue by ignorant, unskilled and unlearned persons, who, under the colour of Surgeons, abuse the people to their pleasure, passing away but trial or punishment, and thereby destroy infinite number of our subjects, wherewith no order has been taken in time bygone, especially within our burgh and barony of Glasgow, Renfrew, Dunbarton, and our Sheriffdoms of Clydesdale, Renfrew, Lanark, Kyle, Carrick, Ayr and Cunningham; FOR avoiding of such inconveniencies, and for good order to be taken in time coming, to have made, constitute and ordained, and by the tenor of these our letters, makes, constitutes, and ordains Master Peter Low, our Surgeon and chief surgeon to our dearest son of the Prince, with the assistance of Mr. Robert Hamilton, professor of medicine, and their successors, indwellers of our City of Glasgow, GIVING and GRANTING to them and their successors, full power to call, summon, and convene before them, within the said burgh of Glasgow, or any others of our said burghs, or public places of the foresaid bounds, all persons professing or using the said art of Surgery, to examine them upon their literature, knowledge and practice; if they be found worthy, to admit, allow, and approve them, give them testimonial according to the art and knowledge that they shall be found worthy to exercise thereafter, receive their oaths, and authorise them as accords, and to discharge them from using any further than they have knowledge passing their capacity, lest our subjects be abused; and that every one cited report testimonial of the minister and elders, or magistrates of the parish where they dwell, of their life and conversation; and in case they be contumacious, being lawfully cited, every one to be fined in the sum of forty pounds, *toties quoties*, half to the judges, [the] other half to be disponed at the visitor's pleasure; and for payment thereof the said Mr Peter and Mr Robert, or visitors, to have our other letters of horning, on the party or magistrates where the contemptuous persons dwell, charging them to poind therefore, within twenty four hours, under the pain of horning; and the party not having gear poindable, the magistrate, under the same pain, to incarcerate them, until caution responsible be found, that the contumacious person shall compear at such day and place as the said visitors shall appoint, giving trial of their qualifications; *Next*, that the said visitors shall visit every hurt, murdered, poisoned, or any other person taken away extraordinarily, and to report to the Magistrate of the fact as it is: *Thirdly*, That it shall be lawful to the said visitors, with the advice of their brethren, to make statutes for the common good of our subjects, relating to the saids arts, and using thereof faithfully, and the breakers thereof to be punished and fined by the visitors according to their fault: *Forthly*, It shall not be lawful to any manner of persons within the foresaid bounds to exercise medicine without a testimonial of a famous university where medicine be taught, or at the leave of our and our dearest spouse's chief physician; and in case they fail, it shall be lawful to the said visitors to challenge, pursue and inhibit them through using and exercising of the said art of medicine, under the pain of forty pounds, to be distributed, half to the Judges, half to the poor, *toties quoties* they be found in using and exercising the same, *ay and until* they bring sufficient testimonial as said is: *Fifthly*, That no manner of person sell any drugs within the City of Glasgow, except the same be sighted by the said visitors, and by William Spang, apothecary, under the pain of confiscation of the drugs: *Sixthly*, That none sell rat poison, arsenic, or sublimate, under the pain of a hundred merks, except only the apothecaries who shall be bound to take caution of the buyers, for cost, harm, and damage: *Seventhly*, That the said visitors, with their brethren and successors, shall convene every first Monday of each month at some convenient place, to visit and give counsel to poor diseased folks gratis: and, *last of all*, Giving and granting to the said visitors indwellers of Glasgow, professors of the said arts, and their brethren, present and to come, immunity and exemption from all weapon showings, raids, hosts, bearing of armour, watching, warding, stenting taxations, passing on assizes, inquests, justice courts, sheriff or burgh courts, in actions criminal or civil, notwithstanding of our acts, laws, and constitutions thereof, except in giving their counsel in matters pertaining to the said arts: ORDAINING you, all the foresaid provosts bailies of burghs, sheriffs, stewards, bailies of regality, and other ministers of justice, within the said bounds, and your deputes, to assist, fortify, concur and defend the saids visitors, and their posterior, professors of the foresaid arts, and put the said acts made and to be made to execution; and that our other letters of our session be granted thereupon to charge them to that effect within twenty four hours next after they be charged thereto. GIVEN under our privy seal, at Holyrood house, the penultimate day of November, the year of God jmvc. And fourscore nineteen years, and of our reign the thirty third year.

(Written on the Tag thus) Per signaturam manu S. D. N. Regis,
Litera Mag"ri Petri Low, Chirurgi nec non manibus Dominorum Ducis
Et Mag"ri Roberti Hamiltone Lennocae Thesaurarii ac Scaccarii Dicti
Professoris Medicinae Domini Regis Subscriptam

(Written on the back thus)
Written to the Privy Seal, Penultimate [day of] November 1599

APPENDIX 2

Seal of Cause in Favour of the Surgeons and Barbers

To ALL and SUNDRY Whom it effects, TO whose knowledge these present letters shall come, We, John Anderson, Provost; John Anderson, John Walkinshaw, and William Neilson, Bailies of the Burgh of GLASGOW, senators and counsellors of the Same, Greeting in God everlasting: KNOW YOU, universities and all others whom it may concern, That there compeared before us sitting in our council house, JOHN HALL, present headman or deacon of SURGEONS AND BARBERS, within the same Burgh, for himself and in name and behalf of the said Surgeons and Barbers, did often diverse and sundry times, present to us and our council gathered together, the bill and supplication underwritten, of which the tenor follows:

UNTO the Right Honorable the Provost, Bailies, and Council of Glasgow, The humble petition of your servants and fellow burgesses, The Surgeons and Barbers, residents within the said city, humbly shows, -That where these Fifty seven years past, since the patent granted to us of the date the penultimate day of November, one thousand five hundred [and] Fifty-nine years, by the deceased King James, to your own and your predecessors' knowledge, we have been in use, yearly, to elect a deacon as visitor and overseer of the rest of the members of our calling as other callings have been in use, By virtue of any patent letter of deaconhead and seal of cause conferred upon them on this account, by any authority: And that it is incumbent to us to have a letter of deaconry of your honours, as others of these incorporations have granted to them by your predecessors, for a joint and harmonious correspondence of brotherhood, as brother citizens willing to sympathise with the rest of the body of the city, whereuntil we shall be concerned to the extent of our power, With the like privileges and liberties as that your authority may be interponed thereto, and we authorised thereby to use such power, observe such courses and customs as other callings have granted to them by their letter of deaconhead or seal of cause. THAT we convene at the ordinary time as other callings do yearly, before Michaelmas, in our ordinary place of meeting, in all time coming, and there, as use is, by plurality of votes, elect and make choice of one of our number to be visitor or deacon for a year thereafter to come, who shall be one of the most fit, qualified, and worthiest of the said calling, a Surgeon and burgess of the burgh, and he being sworn *de fideli administratione*, may appoint meetings for convening the calling, cause quarter masters be elected – the one – half of his own nomination and the other half by the calling itself, Who shall be authorized to impede any person whatsomever with concourse of their honours, To presume to exercise any point of the art of Surgery or Barbery, or set out any signs for either of them, till he be tried and admitted by the said calling, In manner of trial as shall be prescribed, being first admitted burgess of the town. NEXT, that a burgess's son serving his apprenticeship Five years as an apprentice, and two years for meat and fee, pay Forty merks Scots at his admission for his upset; And a stranger entering with the said calling, First being burgess, to pay for his admission, Four score merks, for the use of the poor of the calling. THIRDLY, That no freeman usurp the having of any more apprentices than one during the said seven years, without express warrant from the visitor and quarter masters. FOURTHLY, The said calling may fine any usurper that exercises the said arts, without their admission, tolerance, and licence, In the sum of Ten pounds Scots *totis quoties*, appropriating the one half to the bailies of this city, and the other to the box of the calling. FIFTHLY, That the visitor for the time appoint diets of four head courts or meetings of the calling, and oftener *pro re nata*, and cause poind the absentees in half a merk each time, to be employed for the use of the poor. SIXTHLY, That no freeman make use of an unfreeman under his tolerance, under the pain of a new upset, Neither take any other freeman's apprentices without his former master's leave, asked and granted, under the like pain. SEVENTHLY, That no freeman presume to take another freeman's cure off his hand until he be honestly paid for his bygone pains, and that at the sight of the bailies, with the advice of their visitor, Incase the patient find himself grieved by the surgeon, under the pain of a new upset. Excepting always, liberty to the visitor and quarter masters to take patients from a freeman not found qualified for the curing of them, and to put them to a more qualified person, as shall be thought expedient after exact trial. EIGHTHLY, That any member of the calling, of whatsomever quality, despiser of the visitor and his quarter masters, In any of the points aforesaid, or of their officer in execution of his office, (who is to be last entered freeman of the calling, and is to remain till another enter), pay a new upset, according to that he paid at his entry, To be qualified by the records of the calling. NINTHLY, That no brother within the said calling presume to meddle with any more points of surgery than those they are found qualified in at their admission and conform as they are booked, under the pains of the sums above-written respective as a new upset. AND, LASTLY, That the said visitor or deacon may judge between master and apprentices, at the bailies' sight, Incase any difference of importance arise; and between brother and brother of the calling in particular relating thereto; and give orders to poind absentees from courts and burials, Being warned for that effect, and for non-payment of quarter accounts. MAY it therefore please your honours, the premise being considerate, To grant a letter of deaconry or seal of cause to the said calling under the seal of the Burgh; And that in regard of our so long being a standing part of the crafts of this city, and contributing yearly in a constant proportion for the supplying of the poor of their hospital. To cause extend the same conform to the laudable custom observed to us and our successors Surgeons and Barbers burgesses of this city. AND to grant to us the liberties and privileges aforesaid granted to other callings as is above expressed in all points for removing of the disorders that may arise, AND your Lordships answer:

WHICH articles and statutes above-written Being often times Read, heard, understood, and maturely advised, By us, the said Provost, Baillies and Council of this Burgh of Glasgow, and we finding the same To tend to the good of the people, as well within as without the Burgh, and to the benefit of the said art and craft of Surgeons and Barbers: WE therefore by these presents, GRANT, Ratify, approve, and confirm the same, for us and our successors, IN the whole heads, articles, and clauses contained in the supplication above-written, TO the said JOHN HALL, present Deacon of the said Surgeons and Barbers, and all present brethren of that art and craft, and to their successors Surgeons and Barbers, Burgesses of this Burgh, in perpetual memory in all time coming, PROMISING faithfully to fortify and defend them concerning the matter abovementioned, By us and our successors and office-bearers for the time; AND the premisses, TO all and SUNDRY whom it effects, We make manifest and known, IN WITNESS of the WHICH, and for the greater verification of the same, WE have subscribed these presents, Together with our clerk-depute of court. Our common seal is hereto appended. AT GLASGOW, the sixteenth day of August, One thousand six hundred [and] fifty-six years.

(Signed)
    JO. ANDERSON, *Provost*
    JOHN ANDERSON, *Bailie*
    JOHN WALKINSHAW, *Bailie*
    WILLIAM NEILSON, *Bailie*
    JOHN BELL, *Dean of Guild*
    WALTER NEILSON, *Deacon Convenor*

W. ZAIR, clerk-depute of the said Burgh at command of the said Provost, baillies, and council, as witness my sign and subscription manual.

APPENDIX 3

Bonds of Desistance (from Practice) Given to the FPSG between 1657 and 1701

Date	Name, Designation and Location
1657	James Dougall, gardener
1657	John Logan, in the Gorbals
1659	Agnes Quantane, widow of H. Morae in Ayr
1659	James Harper in Govan
1661	James Faivie, hammerman in Rutherglen
1661	William Cuthbert, weaver in Hamilton
1661	John Simpson in Kilwinning
1670	Thomas Jackson, surgeon in Greenock
1670	Registered bond surgeons gratia Jackson
1670	Patrick Crawford of Dalgliesh
1671	James Loudon, gardener in Ayr
1671	John Loudon, gardener at St Quivox
1672	James Kerr in Hamilton
1673	John Paterson, surgeon in Paisley
1673	John Fenton in Hamilton
1673	Mr. Andrew Hamilton in Kilbride
1673	John Brown in Paisley
1673	John Couper in Lesmahagoe
1673	Alexander Wilson, gardener in Glasgow
1677	John Crawford in Holmhead of Kilmarnock
1678	William Wallace in Paisley
1678	Andrew Brown in Slipperfield
1679	John Adam at Inglistoune Bridge
1680	Mark Clifford in Lanark
1680	John Hutchison in Govan

1690	Margaret Anderson, wife of David Hutchison in Glasgow
1691	James Johnston in Hamilton
1692	Cornelius Jackson in Culraine [two bonds]
1692	John Ferguson in West Ferrie of Erskine
1692	James David in Possil
1693	Janet Hall, wife of John Wilson, skipper in Ayr
1693	Mr. Samuel Henderson in Kilmorres
1693	Mr. James Thomson in Dundonald
1693	Betha Hamilton, wife of William Smith in Kilmarnock
1693	Margaret Cromy in Ayr
1693	Thomas Gammell in Nether Blockwood of Granger
1693	Thomas Bisset, gardener in Kilmarnock
1693	John Small of Crocketshiell, for his wife
1693	John Paterson, gardener to Capringtoune
1693	Registered bond John Stewart in Glasgow
1693	James Richard in Ayr
1693	John Park of Dubbs
1693	Agnes Garven, wife of Andrew Cochran of Ayr
1693	Agnes Carmichael, wife of James Loudon in Ayr
1693	David Steven, weaver in Auchinreoch
1697	James Cuthbertson in Pollok
1697	Elizabeth Govan in Kilbride
1697	David Strachan in Dumbarton
1697	Isobel Crawford in Gorbals
1697	Hugh Pitrone in the parish of Dalry
1697	Robert Drummond, indweller in Glasgow
1698	Robert Young, hillman in Carntyne
1698	Francis Davidson, gardener in Castlemilk
1698	James Murdoch, indweller in Glasgow
1698	Janet Wilson, wife of John Thomson, glover there
1698	George Swan in Gorbals
1698	Bryce Macome in Blackstone
1698	George Hay in Drumm
1698	John Reid, gardener to Calderwood
1698	Mr John Semple

1698	Andrew Veitch
1699	James Steven, tailor in Glasgow
1699	William Jaffray, gardener in Broomhill
1699	James Brownlie, servant to Raploch
1699	James Stevenson in burgh of Lenzie
1699	Robert Forrest in Braehead of Dalserf
1699	William Kirkwood at the Castle of Crookston
1699	John Gray, tailor in Glasgow
1699	Hugh Mudie, gardener in Eglinton
1699	David Marshall, late baker in Hamilton
1699	John Naismith, apothecary in Hamilton
1699	James David in Possil
1699	James Hamilton, landlabourer in Hamilton
1700	John Smith
1700	William Howie
1700	William Hunter
1701	James Dunlop, writer in Renfrew, and Elizabeth Orr
1701	Hugh Gibson, cupper in Paisley, and Margaret Parkhill
1701	John Alexander, dyer in Glasgow
1701	Robert White in Inchinnan, and Genat Fulton, wife
1701	William Rowand of [blank]
1701	David Fleming of St Grinock
1701	John Davies in Hamilton
1701	Isobel Foster in Guivock
1701	John Mephan in Balfron

Source: SRO, CS233/97/10, appendix to case for the FPSG against Thomas Menzies and Others, 1827, pp. 9–11; RCPSG, Minutes of the Faculty of Physicians and Surgeons of Glasgow, 1599 to 1688.

APPENDIX 4

Presidents of the Faculty of Physicians and Surgeons of Glasgow

1602–1603	Robert Hamilton
1603–1604	Robert Hamilton
1604–1605	Robert Hamilton
1605–1606	Robert Hamilton
1606–1607	William Spang
1607–1608	Robert Hamilton
1608–1609	Robert Hamilton
1609–1610	Robert Hamilton
1610–1611	Robert Hamilton
1611–1612	Robert Allason
1612–1613	Robert Allason
1613–1614	John Hall
1614–1615	John Hall
1615–1616	Andrew Mill
1616–1617	No minutes for this year
1617–1618	Andrew Mill
1618–1619	John Hall
1619–1620	John Hall
1620–1621	Robert Hamilton
1621–1622	Robert Hamilton
1622–1623	Andrew Mill
1623–1624	Andrew Mill
1624–1625	John Hall
1625–1626	No minutes for this year
1626–1627	Andrew Mill
1627–1628	Andrew Mill
1628–1629	Andrew Mill
1629–1630	John Hall
1630–1631	John Hall
1631–1632	Andrew Mill
1632–1633	Andrew Mill
1633–1634	James Hamilton

APPENDIX 4

1634–1635	James Hamilton
1635–1636	Robert Archibald
1636–1637	Robert Archibald
1637–1638	John Hall
1638–1639	John Hall
1639–1640	John Hall
1640–1641	Daniel Brown
1641–1642	Andrew Muir
1642–1643	James Hamilton
1643–1644	George Michaelson
1644–1645	George Michaelson
1645–1646	Robert Maine
1646–1647	James Hamilton
1647–1648	Daniel Brown
1648–1649	John Hall
1649–1650	Andrew Muir
1650–1651	Andrew Muir
1651–1652	John Hall
1652–1653	John Hall
1653–1654	Daniel Brown
1654–1655	John Hall
1655–1656	John Hall
1656–1657	James Thomson
1657–1658	James Thomson
1658–1659	Thomas Lockhart
1659–1660	Thomas Lockhart
1660–1661	James Frank
1661–1662	James Frank
1662–1663	James Thomson
1663–1664	James Frank, elder
1664–1665	James Thomson
1665–1666	William Clydesdale
1666–1667	Archibald Bogle
1667–1668	Archibald Bogle
1668–1669	Archibald Graham
1669–1670	Archibald Bogle
1670–1671	Andrew Elphinstone
1671–1672	Archibald Bogle
1672–1673	Archibald Bogle
1673–1674	David Sharp

Surgeon Visitor

1674–1675	Archibald Bogle
1675–1676	Charles Mowat
1676–1677	Charles Mowat
1677–1678	Robert Houstoun
1678–1679	Robert Houstoun
1679–1680	Robert Houstoun
1680–1681	Charles Mowat
1681–1682	Lodovik Lindsay

There is no record of Presidents between 1682 and 1733 as the minute books were destroyed in a fire.

Second Minute Book 1733 Onwards

1733–1734	George Thomson
1734–1737	Thomas Brisbane
1737–1739	John Johnstoune
1739–1741	John Wodrow
1741–1743	David Paton
1743–1745	George Montgomery
1745–1747	Robert Hamilton
1747–1749	William Cullen
1749–1751	John Wodrow
1751–1753	Robert Dick
1753–1755	John Wodrow
1755–1757	John Gordon
1757–1759	Alexander Stevenson
1759–1761	Joseph Black
1761–1763	John Gibson
1763–1765	John Gordon
1765–1766	Joseph Black
1766–1769	Colin Douglas
1769–1771	Robert Marshall
1771–1773	Peter Wright
1773–1775	Alexander Stevenson
1775–1777	William Irvine
1777–1779	Peter Wright
1779–1781	Robert Marshall
1781–1783	Alexander Stevenson
1783–1785	William Irvine
1785–1787	Peter Wright
1787–1788	Robert Marshall

APPENDIX 4

1789–1791	Robert Cleghorn
1791–1793	Thomas Charles Hope
1793–1795	James Jeffray
1795–1797	Peter Wright
1797–1800	Robert Freer
1800–1802	Richard Millar
1802–1804	John Balmanno
1804–1806	Peter Wright
1806–1808	Richard Millar
1808–1810	John Nimmo
1810–1812	Robert Freer
1812–1814	John Balmanno
1814–1816	Robert Watt
1816–1818	Robert Graham
1818–1820	Richard Millar
1820–1822	James Monteith
1822–1824	William Coupar
1824–1826	John Robertson
1826–1828	Richard Millar
1828–1830	John McNish
1830–1832	John Gibson
1832–1834	John Macarthur
1834–1836	James Corkindale
1836–1838	Alexander Panton
1838–1839	Duncan Blair
1839–1841	James Watson
1841–1843	Francis Steel
1843–1845	Robert Perry
1845–1847	George Watson
1847–1849	William Weir
1849–1852	James Watson
1852–1855	Alexander D. Anderson
1855–1857	Robert Hunter

APPENDIX 5

Chronology

1598	Peter Lowe returned to Glasgow from France
1599	Royal charter granted on 29 November by James VI under the privy seal to Peter Lowe and Robert Hamilton, giving them and their successors power to examine all persons practising surgery in an extensive area of the west of Scotland
1600	Ratification of the charter on 9 February by Glasgow town council
1601	First meeting held in June by members of the new association. Adoption of the barbers as an appendage of surgery
1629	First use of the name 'Facultie' in Minutes
1635	General Letters Signet given by King Charles I
1654	First use of the title 'Facultie of Chyrurgeons and Physitians'
1656	Letter of deaconry or seal of cause erecting the Incorporation of Surgeons and Barbers
1672	Ratification of charter by Scottish Parliament
1673	Rapprochement leading to admission of physicians
1679	Faculty act restricting the admission of strangers to practise surgery and pharmacy
1697	Acquisition of first Faculty Hall beside Tron Church
1704	Physic Garden established. John Marshall, surgeon, became Keeper of the Physic Garden
1714	Founding of Chair of Medicine at Glasgow University. John Gordon, surgeon, taught anatomy at Glasgow University
1720	Founding of Chair of Botany and Anatomy at Glasgow University
1722	Formal separation of surgeons from barbers
1733	Second Minute Book accidentally destroyed by fire. Glasgow Town's Hospital founded
1740	FPSG first examined midwives. William Smellie in London
1741	William Hunter in London. Clinical teaching began in the infirmary of the Glasgow Town's Hospital
1744	William Cullen taught chemistry at Glasgow University
1751	William Cullen made Professor of Medicine

1757	James Muir gave 'private' lectures in midwifery
1763	Decision by the Court of Session against James Calder, a gardener who practised irregularly, and was the first person to challenge the Faculty's existence as a corporate body
1768	Thomas Hamilton lectured on midwifery at Glasgow University
1785	Faculty licentiateship created
1791	Acquisition of second Faculty Hall in St Enoch's Square
1792	Inauguration of Widows' Fund
1792	James Tower's dispensary and lying-in ward founded
1794	Opening of the Glasgow Royal Infirmary
1796	Foundation of Anderson's College
1797	College Street School founded (John and Allan Burns)
1802	FPSG inaugurated curriculum requirements (including midwifery)
1803	Glasgow University and Edinburgh University agreed mutual minimum curriculum requirements
1812	Glasgow University included midwifery and surgery as part of MD curriculum requirements. Act of Adjournal (prosecution of MDs and 'irregular' practitioners by FPSG). Society of Regular Physicians and Surgeons in the City of Glasgow founded. Prosecution by the FPSG of MDs for practising surgery
1815	*Chirurgiae Magister*, first university degree in surgery inaugurated by Glasgow University
1816	Bill to regulate surgeons' qualifications came before Parliament. Membership of Widows' Fund charged at £150
1817	Petition for Royal Charter
1819	Memorial of the town licentiates against elitism of the FPSG
1818	Botanic Gardens moved to Sandyford
1826	Portland Street School founded
1826	Royal Commission on the Universities of Scotland
1828	Anderson's College taught the full complement of medical courses
1830	Royal Commission on the Universities of Scotland
1832	Anatomy Act
1834	Glasgow Lying-In Hospitals founded
1840	House of Lords affirmed the Faculty's exclusive right to grant licences to practise surgery within their boundaries
1850	Act for Regulating the Privileges of the FPSG
1856	Scottish Branch of the General Conference of the Corporations of the UK founded

1857 FPSG gained representation on the proposed General Medical Council
1858 The Medical Act passed

Bibliography

MANUSCRIPT SOURCES

Edinburgh University Library, Special Collections

GEN874/v/13–14, Joseph Black to his father, 4 October 1763.
La. II 126, Letters of Mathew Millar of Glenlee, apothecary in Kilmarnock.
LHB, 1/1/2, Edinburgh Royal Infirmary Minute Book, 1742–49, 2, series 1.

Glasgow City Archives, Mitchell Library, Glasgow

A2:25, Decreet absolvitor in part and for expenses, the Magistrates and Town Council of Glasgow against the FPSG 1794.
TD589/20/3, Family correspondence and papers of the Hamiltons of Barns.
T-TH1/60, Foundation Charter of the Trades' Hospital.
T-TH14 5.2, Minutes of the Incorporation of Barbers, 1707–65.

Glasgow University Archives and Business Record Centre

26631, Faculty Meeting Minutes, 1702–20.
26639, University Meeting Minutes (Senate), 1730–49.
26643, University Meeting Minutes (Senate), 1763–68.
26645, Dean of Faculty's Minutes, 1732–68.
SEN 1/1/1–1/1/5, Senate Minutes, 1771–1845.

Glasgow University Library, Special Collections

Bh. 14-X5, N. 46, 'Horrible Seizure of Dead Bodies'.
Eph G/70, 'Trial and Execution of James Glen'.
Eph K/11G, 'Glasgow Lying-In Hospital and Dispensary'.
Eph K/92, 'The University Dispensary'.
Eph K/111, 'Prospectus of a Plan for the Extension and Improvement of the Lying-In Hospital in the City of Glasgow'.
Eph K/119, 'Prospectus of the Glasgow General Lying-In Hospital (1843)'.
MS Gen. 1356/1–78, Hamilton Manuscripts.
MSS 2255, William Cullen Papers.
MS Hamilton 82–85, William Hamilton: Medical Observations and Notes on Cases, 1778–86.

MS Hamilton 88–89, Heads of Lectures on Midwifery, (not dated).
MS Hamilton 113, Dr Hunter and Mr Cruickshank's Lectures, (not dated).
MS Hamilton 120/1 and 120/2, Dr Osborn's Lectures on Midwifery, London 1780.
MS Hamilton 120/2 and 120/3, Dr Denman's Lectures on Midwifery, London 1780.

Greater Glasgow Health Board

HB 14/1/1–5, Glasgow Royal Infirmary Managers' Minutes, 1787 to 1829.

Mitchell Library, Glasgow Rare Books and MSS Room

641982, Glasgow Town's Hospital, Minutes of the Directors' Meetings, 1732–64.

National Library of Scotland

MS 2524, holograph letters, No. 3, John Anderson to Baron Mure, 8 January 1763.

Public Records Office of Northern Ireland, Belfast

D. 1918/2, Papers of Dr John C. Ferguson (1818–63).

Royal College of Surgeons of Edinburgh

Minute book of the RCSEd, xiv, 1854 to 1859.

Royal College of Physicians and Surgeons of Glasgow

1/1/1/1b–1/1/1/8, Transcript minutes and minutes of the FPSG, 1599 to 1859.
1/5/3, Receipts, Bills and Papers relating to Lawyers' Fees and other business, 1844 to 1867.
1/5/4a, Bundle 11, Memorial and Queries for the Faculty anent Quartering of soldiers with the answers of Ferguson and Pringle advocates 1757.
1/5/5, Miscellaneous manuscript material.
1/10/2, Adversaria Cleghorn. Private case book of Dr Robert Cleghorn (first of three volumes, running from 1782 to 1816).
1/13/2/2B, (formerly 1/5/1), Weir's Faculty Memoranda.
1/13/33, a late seventeenth or early eighteenth-century MS herbal.
2/1/13, 20, Glasgow Medical Essays.
20/1/2, MS of John Rutherford, Clinical Lectures (no date).
20/1/3, MS of R. Whytt Clinical Lectures of 1760.
20/1/6/1, Lectures by Dr Young, Professor of Midwifery in the College of Edinburgh, November 22 1768.
24/2, Dr William Mackenzie's Letters.
40/1, Minutes of the Meetings of the Glasgow Faculty of Medicine, 1824–1907.

Royal Medical Society Library

MS 302, The Lectures of Thomas Young.
MS 96, Thomas Dale, MD, vii, Clinical Lectures by Dr Cullen, 1772–73.
MSS Cullen 101–104, 1772–73, vi, Pathology, Lectures on the Institutions of Medicine by W. Cullen MD.

Scottish Records Office, Edinburgh

CC9/7/7, Commissariot of Glasgow, Testaments, Last Will and Testament of Peter Lowe.
CS 232, box 170, Unextracted Processes, Court of Session.
CS 232, box 172, Court of Session Processes.
CS 233/97/10, Appendix to Case for the FPSG, against Thomas Menzies and Others, 1827.
RD3/93, Register of Deeds.

Strathclyde University Archives

B112, Minute Book of Anderson's Institution, 1799–1810.

ARTICLES AND PAMPHLETS

An Account of the Rise and Establishment of the Infirmary ... Erected at Edinburgh (Edinburgh, c. 1730).
Beekman, F., 'William Hunter's Early Medical Education', *Journal of the History of Medicine*, 5 (1950), pp. 72–84.
Brock, H.C., 'Dr William Hunter's Museum, Glasgow University', *Journal of the Society for the Bibliography of National History*, 9 (1980), pp. 403–12.
Buchanan, A., *Of Monopolies in Learning with Remarks on the Present State of Medical Education* (Glasgow, 1834).
Buchanan, G., 'On the Effects of Mr Warburton's Anatomy Bill, and the Facilities for the Study of Practical Anatomy on Glasgow', *Glasgow Medical Journal*, 2 (1855), pp. 431–42.
Buchanan, M., *Lecture Introductory to a Course of Anatomy* (Glasgow, 1842).
Buchanan, M., *Lecture Introductory to a Course of Clinical Surgery, Delivered to the Students of the Glasgow Royal Infirmary* (Glasgow, 1831).
Buchanan, M.S., *Remarks on Medical Reform Being the Substance of a Lecture Introductory to a Course of Anatomy Delivered to the Students of Anderson's University* (Glasgow, 1846).
Burns, J., *Observations on Abortion: Containing an Account of the Manner in Which it is Accomplished, the Causes Which Produced it and the Method of Preventing or Treating it* (London, 1806).
Butterton, J. R., 'The Education, Naval Service and Early Career of William Smellie', *Bulletin of the History of Medicine*, 60 (1986), pp. 1–18.

Chitnis, A., 'Provost Drummond and the Origins of Edinburgh Medicine', in idem, *The Scottish Enlightenment: A Social History* (London, 1976), pp. 86–97.

Cleghorn, R., 'A Biographical Account of Mr William Hamilton, Late Professor of Anatomy and Botany in the University of Glasgow', *Transactions of the Royal Society of Edinburgh*, 4 (1798), pp. 35–63.

Easton, J. A., *Lecture Introductory to a Course of Clinical Medicine Delivered in the Glasgow Royal Infirmary* (Glasgow, 1849).

Edington, G. H., 'The "Discourse" of Maister Peter Lowe: Extracts and Comments', *Glasgow Medical Journal*, 6th series, 98 (1922), pp. 43–50.

Finlayson, J., 'The Last Will and Testament, with the Inventory of the Estate, of Maister Peter Lowe, Founder of the Faculty of Physicians and Surgeons, Glasgow', reprinted from the *Glasgow Medical Journal*, new series, 50 (1898), pp. 241–44.

Finlayson, J., *Dr Sylvester Rattray, Author of the Treatise on Sympathy and Antipathy* (Glasgow, 1900).

Fulton, H. J., 'John Moore, the Medical Profession and the Glasgow Enlightenment', in A. Hook and R. B. Sher (eds), *The Glasgow Enlightenment* (East Linton, 1995), pp. 176–89.

Geyer-Kordesch, J., 'Comparative Difficulties: Scottish Medical Education in the European Context (c. 1690–1830)', in V. Nutton and R. Porter (eds), *The History of Medical Education*, (Amsterdam and Atlanta, 1995), pp. 94–115.

Geyer-Kordesch, J., 'Die medizinische Aufklärung in Schottland: nationale und internationale Aspekte', in H. Holzey, D. Brühlmeier, V. Murdoch (eds), *Die Schottische Aufklärung: 'A Hotbed of Genius'* (Frankfurt, 1996), pp. 91–106.

Goodall, A. L., 'The Royal Faculty of Physicians and Surgeons', *Journal of the History of Medicine and Allied Sciences*, 10 (1955), pp. 207–25.

Goodall, A. and Gibson, T., 'Robert Watt: Physician and Bibliography', *Journal of the History of Medicine*, 18 (1963), pp. 36–50.

Guthrie, D., 'The Achievement of Peter Lowe and the Unity of Physician and Surgeon', *Scottish Medical Journal*, 10 (1965), pp. 261–68.

Hannay, A. J., *Remarks on the Origin, Nature, and Importance of Medical Science: An Introductory Lecture to the Study of Medical Science* (Glasgow, 1837).

Herrick, J., 'Allan Burns, 1781–1813: Anatomist, Surgeon and Cardiologist', *Bulletin of the Society of Medical History of Chicago*, 4 (1928–35), pp. 457–83.

Hunter, J., 'Bicentenary Celebration: The Hunters and the Hamiltons. Some Unpublished Letters', *Lancet* (1928), pp. 354–60.

Jardine, R., 'The Glasgow Maternity Hospital Yesterday and Today', in Mrs Robert Jardine, *The Chapbook of the Rottenrow* (Glasgow 1913), pp. 13–17.

Jardine, R., 'The Glasgow Maternity Hospitals: Past and Present', *Glasgow Medical Journal* (1901), pp. 28–42.

Kendall, J., 'Thomas Charles Hope, MD', in A. Kent (ed.), *An Eighteenth Century Lectureship in Chemistry* (Glasgow, 1950), pp. 157–63.

Kent, A., 'William Irvine, MD', *An Eighteenth-Century Lectureship in Chemistry* (Glasgow, 1950), pp. 140–50.

Lowe, P., *An Easy Certaine and Perfect Method, to Cure and Prevent the Spanish*

Sickness: Whereby the Learned and Skilfull Chirurgian May Heale a Great Many Other Diseases (London, 1611).

Lumsden, H., 'Bibliography of the Guilds of Glasgow', *Records of the Glasgow Bibliographical Society*, 8 (1930), pp. 1–43.

Macdonald, F. A., 'The Infirmary of the Glasgow Town's Hospital, 1733 to 1800: A Case for Voluntarism?', *Bulletin of the History of Medicine*, 73 (1999), pp. 64–105.

McDonald, S. W., 'The Life and Times of James Jeffray, Regius Professor of Anatomy, University of Glasgow, 1790–1858', *Scottish Medical Journal*, 40 (1995), pp. 119–22.

McGrath, J., 'The Medieval and Early Modern Burgh', in T. M. Devine and G. Jackson (eds.). *Glasgow Volume*, i, *Beginnings to 1830* (Manchester and New York, 1995), pp. 17–62.

MacKinlay, C. J., 'Who is Houston? A Biography of Robert Houston, MD, FRS, 1678–1734', *Journal of Obstetrics and Gynaecology of the British Commonwealth*, 80 (1973), pp. 193–200.

Markus, T. A., 'Domes of Enlightenment: Two Scottish University Museums', *Art History*, 8 (1985), pp. 158–77.

Mathew, M. V., 'James Sutherland (1638(?)–1719): Botanist, Numismatist and Bibliophile', in H. M. Wright (ed.), *The Bibliotheck* (Glasgow, 1987), pp. 1–29.

Monro, A., 'An Essay on the Art of Injecting the Vessels of Animals', *Medical Essays and Observations* (Edinburgh, 1732–33), i, pp. 94–111.

Pattison, G. S., 'Life of Allan Burns', introduction, in A. Burns, *Surgical Anatomy of the Head and Neck* (Baltimore, 1823), pp. vii–xxiv.

Patton, D., 'The British Herbarium of the Botanical Department of Glasgow University', *Journal of the Glasgow and Andersonian Natural History and Microscopical Society*, 17 (1954), pp. 105–26.

Peachey, G. C., 'The Homes of the Hunters', *Lancet* (1928), pp. 360–67.

Robertson, H., *A Letter Communicated to the Monthly Gazette of Health Addressed to the Graduates of the Scotch Universities and General Practitioners on the Illegal and Unwholesome By-Laws of the College of Physicians of London, Establishing a Monopoly in Favour of the Graduates of Oxford and Cambridge, in Opposition to the Articles of Union between England and Wales* (London, 1827).

Ross, R. M., 'Peter Lowe: Founder of the Faculty, Man of Mystery', *Dental Historian*, 28 (1995), pp. 3–11.

Thornton, J. L., 'William Hunter (1718–1783) and his Contributions to Obstetrics', *British Journal of Obstetrics and Gynaecology*, 90 (1983), pp. 787–94.

Watt, R., 'An Address to Medical Students', in *A Catalogue of Medical Books* (Glasgow, 1812), pp. 1–14.

Wilson, J., 'Report of the Glasgow Lying-in Hospital and Dispensary for the Year 1851–52', *Glasgow Medical Journal*, 2nd series, 1, no. 1 (April, 1853), pp. 1–10.

Young, J., 'Dr Smellie and Dr W. Hunter: An Autobiographic Fragment', *British Medical Journal* (1896), pp. 514–16.

'Zeno', 'The Medical Schools of Glasgow', *Glasgow Medical Examiner*, 1 (1831–32), pp. 184–86.

BOOKS

Addison, W. I., *Roll of Graduates of the University of Glasgow, 1727–1897* (Glasgow, 1898).
Aird, A., *Glimpses of Old Glasgow* (Glasgow, 1894).
Alison, R., *The Anecdotage of Glasgow* (London, 1892).
Anderson, J. R. (ed.), *The Burgess and Guild Brethren of Glasgow, 1573–1750*, Scottish Record Society (Edinburgh, 1925).
Anderson, R., *The Life of John Moore, MD* (Edinburgh, 1820).
Anderson, W., *The Popular Scottish Biography* (Edinburgh, 1842).
Baillie, M., *The Morbid Anatomy of Some of the Most Important Parts of the Human Body* (London, 1793).
Bell, R., *A Dictionary of the Law of Scotland*, 2 vols (3rd edn, Edinburgh, 1826).
Brisbane, Thomas, *The Anatomy of Painting: or A Short and Easy Introduction to Anatomy. Being a New Edition, on a Smaller Scale, of Six Tables of Albinus, with their Linear Figures* (London, 1769).
Brock, H. C., *Dr James Douglas's Papers and Drawings in the Hunterian Collection, Glasgow University Library: A Hand List* (Glasgow, 1994).
Bruce, D., *Radical Doctor Smollett* (London, 1964).
Buchanan, M. S., *History of the Glasgow Royal Infirmary* (Glasgow, 1832).
Burns, A., *Observations on the Surgical Anatomy of the Head and Neck: Illustrated by Cases and Engravings, with a Life of the Author and Additional Cases and Observations by Granville Sharp Pattison* (2nd edn, Glasgow and London, 1824).
Burns, J., *Handbuch der Geburtshülfe mit Inbegriff der Weiber und Kinderkrankheiten nach der achten, vollständig und gleichsam ein neues Werk bildenden Ausgabe*, ed. H. F. Killian, i (Bonn, 1834).
Burns, J., *Practical Observations on the Uterine Hemorrhage: With Remarks on the Management of the Placenta* (London, 1807).
Burns, J., *The Anatomy of the Gravid Uterus with Practical References Relative to Pregnancy and Labour* (Glasgow, 1799).
Butt, J., *John Anderson's Legacy: The University of Strathclyde and its Antecedents, 1796–1996* (East Linton, 1996).
Christie, J., *The Medical Institutions of Glasgow* (Glasgow, 1888).
Companion to the Glasgow Botanic Garden or Popular Notices of Some of the More Remarkable Plants Contained in it (Glasgow, 1818).
Comrie, J. D., *History of Scottish Medicine*, 2 vols (London, 1932).
Cooper, S., *A Dictionary of Practical Surgery: Containing a Complete Exhibition of the Present State of the Principles and Practice of Surgery, Collected from the Best and Most Original Sources of Information, and Illustrated with Critical Remarks* (London, 1809).
Cordasco, F., *A Bibliography of Robert Watt, MD: Author of the Bibliotheca Britannica* (New York, 1980).
Couper, W. J., *Robert Wodrow* (n.p., 1828).
Coutts, J., *A History of the University of Glasgow from its Foundation in 1451 to 1909* (Glasgow, 1909).

Cowan, J. M., *Some Yesterdays: With a Note upon the Development of Hospitals by Joshua Ferguson* (Glasgow, 1949).
Crowther, A. M. and White, B., *On Soul and Conscience: The Medical Expert and Crime* (Aberdeen, 1988).
Daiches, D., *Glasgow* (London, 1977).
Davies, J., *Douglas of the Forests: The North American Journals of David Douglas* (Edinburgh, 1980).
Davies, T., *Some Instructions for Collecting* (n.p., 1790).
Denholm, J., *The History of the City of Glasgow and Suburbs* (Glasgow, 1804).
Devine, T., *The Tobacco Lords: A Study of the Tobacco Merchants of Glasgow and their Trading Activities, c. 1740–90* (Edinburgh, 1975).
Dow, D. A., *The Rottenrow: History of the Glasgow Royal Maternity Hospital, 1834–1984* (Carnforth, 1984).
Duncan, A., *Memorials of the Faculty of Physicians and Surgeons of Glasgow, 1599–1850: With a Sketch of the Rise of Progress of the Glasgow Medical School and of the Medical Profession in the West of Scotland* (Glasgow, 1896).
Eyre-Todd, G., *History of Glasgow: From the Revolution to the Passing of the Reform Acts, 1832–33* (Glasgow, 1934).
FPSG, *The Royal Charter and Laws of the Faculty of Physicians and Surgeons of Glasgow* (Glasgow, 1821).
Gibson, T., *The Royal College of Physicians and Surgeons of Glasgow: A Short History Based on the Portraits and Other Memorabilia* (Edinburgh, 1983).
Glaister, J., *Dr William Smellie and his Contemporaries* (Glasgow, 1894).
Hamilton, D., *The Healers: A History of Medicine in Scotland* (Edinburgh, 1981).
Hannay, A. J., MD, *On the Pathology of Puerperal Fever* (Glasgow, 1825).
Hill, Ninian, *Notes upon the Insufficiency of the Aortic or Semilunar Valves of the Heart* (Glasgow, 1836).
Hill, W. H., (ed.), *View of the Merchants' House of Glasgow: Containing Historical Notices of its Origin, Constitution and Property, and of the Charitable Foundations Which it Administers* (Glasgow, 1866).
Houstoun, J., 'The Memoirs of his Life', as contained in *The Works of James Houstoun, MD* (London, 1753).
Hutchison, J., *The Associations between the Faculty of the Physicians and Surgeons of Glasgow and the Herbal and the Botanic Gardens*, RCPSG, 1/13/7/9.
Innes, C., (ed.), *Munimenta Alme Universitatis Glasguensis: Records of the University of Glasgow from its Foundation till 1727*, 4 vols, Maitland Club, 72 (Glasgow, 1854).
Innes, J., *A Description of the Human Muscles with their Several Uses, and the Synonyma of the Best Authors: A New Edition with Notes, Practical and Explanatory by Robert Hunter* (Glasgow, 1822).
Jackson, A. M., *Glasgow Dean of Guild Court: A History* (Glasgow, 1983).
Jenkinson, J., Moss, M., Russell, I., *The Royal: The History of the Glasgow Royal Infirmary, 1794–1994* (Glasgow, 1994).
Kilpatrick, R., 'Nature's Schools: The Hunterian Revolution in London Hospital Medicine, 1780–1825' (Unpublished Ph.D. thesis, University of Cambridge, 1988).

Laskey, J., *A General Account of the Hunterian Museum, Glasgow* (Glasgow, 1813).
Lawrie, J., *Essay on Cholera* (Glasgow, 1832).
Lawrie, J. A., *Letters on the Charters of the Scotch Universities and Medical Corporations and on Medical Reform in Scotland* (Glasgow, 1856).
Lonsdale, H., *A Sketch of the Life and Writings of Robert Knox the Anatomist* (London, 1871).
Lowe, P. *The Whole Course of Chirurgerie*, Classics of Medicine Library (Birmingham, facsimile reprint of 1597 edn, 1981).
Lowe, P., *A Discourse of the Whole Art of Chyrurgerie* (London, 1612).
Lumsden, H., (ed.), *Records of the Trades House of Glasgow*, 2 vols, i, *1605–78* (Glasgow, 1910); ii, *1713–77* (Glasgow, 1934).
Lumsden, H., and Henderson, P., *History of the Hammermen of Glasgow: A Study typical of Scottish Craft Life and Organisation* (Paisley, 1912).
MacGregor, G., *The History of Burke and Hare and of the Resurrectionist Times: A Fragment from the Criminal Annals of Scotland* (Glasgow, 1884).
Mackie, J., *The University of Glasgow, 1451–1951: A Short History* (Glasgow, 1954).
Marwick, J. D. (ed.), *Extracts from the Records of the Burgh of Glasgow, AD 1663–1690*, iii, Scottish Burgh Records Society (Glasgow, 1905).
Marwick, J. D. (ed.), *Extracts from the Records of the Burgh of Glasgow, 1691–1717*, Scottish Burgh Records Society, iv (Glasgow, 1908).
Marwick, J. D. (ed.), *Records of the Convention of the Royal Burghs of Scotland, with Extracts from Other Records Relating to the Affairs of the Burghs of Scotland, 1295–1738*, 5 vols (Edinburgh, 1866–90).
McClintock, A. H. (ed.), *Smellie's Treatise on the Theory and Practice of Midwifery: Edited with Annotations* (London, 1876).
McGrath, J. S., 'The Administration of the Burgh of Glasgow, 1574–1586', 2 vols (unpublished Ph.D. dissertation, University of Glasgow, 1986).
McLellan, D., *Glasgow Public Parks* (Glasgow, 1894).
McUre, J., *History of Glasgow* (2nd edn, Glasgow, 1831).
Millar, R., *Medical and Surgical Establishments of the Infirmary* (Glasgow, 1828).
Moore, J. C., *The Life of Lieutenant General Sir John Moore, KB*, 2 vols (London, 1834).
Moore, J. N., *The Maps of Glasgow: A History and Cartobibliography to 1865* (Glasgow, 1996).
Murray, D., *Glasgow and Helensburgh: As Recalled by Sir J. D. Hooker* (Helensburgh, 1918).
Pattison, F. L. M., *Granville Sharp Pattison: Anatomist and Antagonist, 1791–1851* (Edinburgh, 1987).
Perry, R., *Facts and Observations on the Sanitary State of Glasgow* (Glasgow, 1844).
Renwick, R. (ed.), *Extracts from the Records of the Burgh of Glasgow, AD 1718–1738* (Glasgow, 1909).
Renwick, R. and Marwick, J. D. (eds.), *Extracts from the Records of the Burgh of Glasgow, 1573–1642*, Scottish Burgh Records Society (Glasgow, 1874).
Robertson, E., *Glasgow's Doctor James Burn Russell, 1837–1904* (East Linton, 1998).

Robinson, E. and McKie, D., (eds), *Partners in Science: James Watt and Joseph Black* (London, 1970).
Senex, (Robert Reid), *Glasgow, Past and Present: Illustrated in Dean of Guild Court Reports and in the Reminiscences and Communications of Senex, Aliquis, J. B*, 3 vols (Glasgow, 1884).
Senex (Robert Reid), *Old Glasgow and its Environs* (Glasgow, 1864).
Spittal, J., MD, *A Summary View of the Practical Utility of the Stethoscope in Diseases of the Chest* (Glasgow, 1827).
Strang, J., *Glasgow and its Clubs* (London, 1856).
Tennent, J., *Records of the Incorporation of Barbers, Glasgow, Formerly the Incorporation of Chirurgeons and Barbers* (Glasgow, 1930).
The Old Country Houses of the Old Glasgow Gentry (2nd edn, Glasgow, 1878).
Thomson, J., *An Account of the Life, Lectures and Writings of William Cullen MD* (Edinburgh and London, 1854).
Utz, H., *Schotten and Schweizer: Brother Mountaineers. Europa entdeckt die beiden Völker im 18. Jahrhundert* (Frankfurt, Main, 1995).
Veitch, J., *Memoir of Sir William Hamilton, Bart* (Edinburgh, 1869).

Index

Plate illustrations are shown in bold

1745 Rebellion 307

Abercorn, Hamilton Earls of 38
Abercorn, Marion Boyd, Countess of 38–39
Aberdeen: 'mediciner' 38; medical school 125; King's College 18, 232; Marischal College 392–93; curriculum requirements 343; MD degrees 348, 363, 375; midwifery training 254
abnormal births, *see* midwifery
'abusers' 44, 49; *see also* irregular practitioners
accidents 9, 314, 416
accoucheurs 254, 262, 269, 282, 287, 291; Royal Accoucheur 262; *see also* midwifery
accreditation, *see* curriculum requirements; licensing to practise; qualifications
Acland, Dr, MP 390n., 391
act of adjournal 363, 369; *see also* prosecutions and litigation
act for compearance 101–2; *see also* prosecutions and litigation
Act for Regulating the Privileges of the Faculty of Physicians and Surgeons of Glasgow (1850) 379; *see also* legislation
Adam, James 186
Adam, Robert 186
Adams, Dr Andrew 401
Aditus novus ad occultas sympathiae et antipathiae causas inveriendas (Rattray) 127
Albermarle, Earl of 235
Albinus, Bernhard Siegfried 176, 218
Albinus, C. B. 219
Alderton, Dr (RCPE delegate) 389n.
Alexis of Pimunth 59
Algeo, Thomas (1628) 39

alkali industry 245
Allason, Robert 12, 85n., 86
almshouse, *see* Crafts' Hospital
Alston, John 291
amputations pl. 8, 28, 54–56, 79, 124, 215, 343
Amsterdam 123n., 163, 168, 172, 294; *see also* Holland
anaesthetics 315
Anatomy Act (1832) 210, 211, 343, 399, 408, 409, 410; *see also* corpses; grave-robbing; legislation
anatomy and surgery: the terms 43, 97 and n.; academic importance of 40, 204, 341, 343, 344, 397; Hippocrates' maxim 69n.; humanist surgery 45, 47; importation of ideas and techniques 155; international agenda 172–78; teaching of 210–29; lecture system 172–78, 216, 347, 350, 354; dissection, *see* dissection; demonstrations 71, 95, 176, 178, 214–15; clinical teaching 219, 233; animal use in 174, 178, 216; specimens 233; anatomical preparations, *see* preparations; artistry 218–19; engravings 215, 219; Glasgow University 160, 194, 196, 241, 398, 403, 411–12; Anderson's College 186, 240–42, 412; Edinburgh 95–96, 172, 403; England 175, 216, 372; Paris 173–77 passim, 216, 218, 312; physicians' interest in 312; irregular practitioners and 233–34; popularity among working men 241–42; Inspector of Anatomy 411; *see also* blood-letting; corpses; grave-robbing; Lowe, Peter; military surgery; naval surgeons; preparations; surgeons; wound surgery

Anatomy of the Gravid Uterus, The (John Burns) 272, 340–41, 405
Anatomy of the Head and Neck (Allan Burns) 404–5
Anatomy of Human Bodies, The (Cowper) 219
Anderson family 160
Anderson, Dr Alexander Dunlop 353, 382, 384, 386, 387, 388
Anderson, David 94
Anderson, Isobel (née Neilson) 94
Anderson, Sir James 381
Anderson, Janet 147
Anderson, John, apprentice 94
Anderson, John, bailie 420, 422
Anderson, John, Provost 17, 420, 422
Anderson, Prof. John 239, 245, 246, 267
Anderson, Marion 188
Anderson, William, bailie 84
Anderson, William, surgeon 323
Anderson's College (from 1828 University): early days of 239–40; John Anderson 239, 245, 246, 267; extra-mural colleges 155, 287, 357; development as a medical school 240–42, 390; 'private' institution 347, 372; Glasgow University's rival 239, 246; library 240; museum 186, 240, 241; 'dome of Enlightenment' 186, 240; Faculty accreditation 240, 287, 357, 362, 372, 411, 413–14; curricula 240, 343, 412, 415; anatomy and surgery 186, 240–42, 412; chemistry 242, 246–47, 412; midwifery 213, 271, 278, 280, 281, 282–83, 412; botany 240, 412; materia medica 209; teaching aids 186, 241, 412; qualifications 372, 413–14; Mechanics' Institute 241, 247; George Birkbeck 246, 361; Birkbeck College 241, 246; student numbers 240, 362; English and Irish students 411; working tradesmen 241–42, 361; *see also* extra-mural colleges; Portland Street School of Medicine; universities
Andreas Perforatus (Andrew Boorde) 46–47, 46n.
Angers 126
animal use in anatomy 174, 178, 216
antimony 124–25

Aphorisms (Mauriceau) 258–59
Aphorisms, Institutes and (Boerhaave) 163, 165, 259
Apothecaries Act (1815) 346, 352, 368; *see also* legislation
Apothecaries Bill (1827) 375, 376
Apothecaries Company, London 349, 376; *see also* Society of Apothecaries
Apothecaries' Hall, Glasgow 203, 248, 287, 333
Apothecaries, Society of, *see* Society of Apothecaries
apothecaries and surgeon-apothecaries: the term 'Chyrurgion' 43; art of the surgeon-apothecary 16n., 128–29, 132–34; pharmacy 26, 111–12, 341, 343, 344, 351; 'surgeon-pharmacists' 193–94, 203; ruling on healing arts 368; status 31, 34–35; encroachment on physicians' preserve 16, 128–31, 350, 399; surgeons' right to practise pharmacy (Edinburgh) 16; general practitioners 12, 129, 349–50, 352, 399; pharmacopoeia 128, 197, 201; corporate association 2, 11, 12, 80; for Glasgow, *see* Faculty of Physicians and Surgeons; guilds outside Scotland 11–12, 27–28; accreditation 9, 14, 80, 128–36 passim, 193–94, 346, 374; apprentices, *see* apprenticeship system; curriculum requirements 337, 343, 344; midwifery 128, 351; Apothecary Hall, Glasgow 203; irregular practitioners 233–34; John Moore 234; *see also* barbers; botany; general practitioners; licensing to practise; materia medica; physicians; Society of Apothecaries; surgeons
Apothecaries and Surgeon-apothecaries, Fraternity of, Edinburgh 16
Apothecary, The History of Medicine ... (Good) 41 and n., 43
Apothecary's Company, Glasgow 244
apparatus, *see* teaching aids
apprenticeship system 91–101, 94n., 99n., 113, 137, 138, 215; indentured apprentices 26, 92–95, 113, 294; servants 91; women 144, 145; Holland 92, 294; Norwich 92; William Cullen 181; Faculty

library 230; university alternative 181, 296; *see also* craft guilds; Faculty of Physicians and Surgeons
Arbuckle, James 195–96
Archeologica Britannica (Lhwyd) 182
Archibald, Robert 122
Argyll, Duke of 203
Aristotle 51; *Ethics* 158
Armour, Prof. James 281, 282, 356–57, 372
armour-bearing exemption 10
Army and Navy boards: accreditation by 32, 347, 363; others' accreditation 240, 365, 397; clinical training 333; midwifery 254, 287; *see also* military surgery; naval surgeons; surgeons
Arouet, François Marie (Voltaire) 186
Arthur, Patrick 94–95
artists (anatomy) 218–19
Ashmolean Museum 160, 182
assize, passing on (exemption) 10
astrologers 115; *see also* irregular practitioners
Astruc, Jean 235
Asylum for the Insane, Glasgow 202
Atholl, Dukes of 201
Auchincloss, John 302
Auchmowtie, Robert 48
Auld Alliance 40
Austen, Jane 187, 190
Austria 273–74
autopsies 178, 219, 221, 276, 312, 316
Avicenna, Persian physician 72
Ayr 7, 136, 417

Bachelor of Medicine degree 413
Badham, Prof. Charles 373
Baillie, Alexander 6n.
Baillie, Matthew 190, 219; *The Morbid Anatomy ... of the Human Body* 219, 237
Baker, George, surgeon 47
Balfour, Sir Andrew 160, 168, 197–98; Balfour collection 160
Balmanno, Mrs 200, 202–3
Balmanno, Dr John 202–3
banana tree 205
Banks, Sir Joseph 168
Bannatyne, Lord 367
Barber-Surgeons Company 11–12, 21, 37n., 80, 95–96, 181; *see also* barbers and barber-surgeons; London
barbers and barber-surgeons: status 7–8, 17, 25, 35, 79, 105, 113; Surgeons of the Short Robe 116–17; Peter Lowe's opinion of 45, 48, 50, 62, 64; early barber – surgeon associations 79–80, 147; separation of 81 and n., 103; for Glasgow, *see* Faculty of Physicians and Surgeons; Edinburgh 2n., 81n., 103, 104; England 103, 147; France 81n., 103, 176, 295; services offered 79 and n., 117–18; beard-dressing 50, 79n., 93, 118; haircutting 79n., 93; periwig-making 93, 118 and n.; blood-letting 79 and n., 113, 117; licensing to practise 8, 117–19, 120, 124; practice signs 74n., 118; *see also* apothecaries; apprenticeship system; Barber-Surgeons Company; craft guilds; licensing to practise; surgeons
Barclay, Dr John 403
bardic medical schools 123n.
Barns, Laird of 132–34
Barony Parish Fever Hospital 278
Barr, J. pl. 12
Barrowfield, Lady 297
basin hanging 118; *see also* practice signs
Basle 174
Bathurst, Dr John 37n.
Baxter, Daniel 230
Bean, Sawney, gibe 185
beard-dressing 50, 79n., 93, 118; *see also* barbers; haircutting; periwig-making
Beaton family 123n.
bedside teaching 125n., 166, 274, 305, 310, 326, 335–36, 343; *see also* Boerhaave; clinical teaching; lying-in hospitals; teaching hospitals
Belfast 236; *see also* Ireland
Bell, Dr 138
Bell, Henry 248
Bell, James 77n.
Bell, John 422
Berlin 154
Berrell, George 87, 118
Bibliotheca Britannica 231
Bill for Regulating the Practice of Surgery (1816) 369, 370

Binns, Joseph 37n., 59n.
Birkbeck, Dr George 246, 361; Birkbeck College, London 241
births, *see* midwifery
Bishop, Archibald 28
Black, Adam, MP 382n.
Black, Prof. Joseph pl. 13; chemistry 193, 242, 243–44, 244, 249; *Lectures on the Elements of Chemistry* 244; on lecturing practice 195; MD degrees 196; Faculty library 230; other mentions 165, 245, 318, 321, 398
Blackburn, John 6n.
Blackwood, William 190
bladder stones, *see* lithotomy
Blair, Duncan 366
Blair, Elizabeth 265
Blair, Rev. Hugh 297
bleaching 194, 203, 245, 248, 295n., 298
Blisard, Sir William 374
blood-letting: Peter Lowe's *Chirurgerie* 59, 60, 74–76, 74n.; barber-surgeons and 79 and n., 113, 117; gardeners 34, 203; leeches 74, 76, 117
Blythswood Place, St Vincent Street (Fleming) pl. 16
BMA 351, 352 and n., 380, 394 and n.
boards (medical), *see* Army and Navy boards
Bobart family 169
bodies for dissection, *see* corpses; grave-robbing
Boerhaave, Herman: Leiden 166, 171, 218; teaching methods 125n., 164–65, 306; *Institutes and Aphorisms* 163, 165, 259; plant classification 167; anatomy 218; Cullen's view of 165, 166, 180; other mentions 230, 236
Bogle, Archibald 18, 22, 24n., 95 and n., 102–3, 118
Bogle, John 298
Bogle, Dr Robert 304
Bogle, William 95 and n., 118
bonds of desistance 423–25; *see also* prosecutions and litigation
bondsmen 81; *see also* freemen
bone-setting 115, 120, 145; *see also* fractures; irregular practitioners; skeletons

Bonham, Dr Thomas 11 and n.
Bonhill, Laird of 204
Bonn 273
Booke of the Infantment (Lowe) 49n., 64
Booke of Women's Diseases (Lowe) 49n., 53
booking (apprenticeship) 93–94, 95, 113
books, medical, *see* library resources; medical books
Booly, William Henry 340
Boorde, Andrew, *Breviarie of Health* 46–47, 46n.
botany: botanical networks 155, 166–71; gardens and plants 166–71, 197–210, 240; physic gardens, *see* physic gardens; exotica 167, 168, 169, 170, 197, 201; Jamaica 181–82; catalogues 167–71; compendia 198; teaching curricula 194, 196, 240, 341, 343, 412; *see also* apothecaries; Chelsea; Hooker family; Kew; Kirklee; Langside; materia medica; natural history; Royal Botanic Institution; Sandyford
'box' common (Faculty) 88–89, 96, 106; boxmasters 88–89, 106, 131
Boyd, Adam 313–14
Boyd, John 30n., 31n., 110n., 302
Boyd, Marion, Countess of Abercorn 38–39
Boyd, Robert 111, 113, 119
Boyl, George 136n.
Boyle, Robert 162–64, 164, 166
Brady, William, *see* Brodie, William
Braidwood, James, barber 117–18
Braidwood, James, younger 118n.
Breviarie of Health (Boorde) 46–47
Bridewell prison 321
Brindley, James 247
Brisbane, Dr John 195, 304
Brisbane, Dr Mathew/Matthew 21, 194–95
Brisbane, Prof. Thomas 194–96, 205, 307
British Medical Association (BMA) 351, 352 and n., 380, 394 and n.
British Medical Journal 352
British Museum 186
Brodie, William 18, 19
Broomielaw, Shipping (Fleming) pl. 2
Brown, Mr (Glasgow) 357
Brown, Mr (London) 301
Brown, Daniel 19, 25n., 90, 91, 94, 144n.

Brown, Prof. James 281, 282, 283, 290, 291
Brown, James 359, 360–61
Brown, John 204
Brown, Marion (née Jeffrey) 204
Brown, Robert 94
Brown, Dr Thomas 204, 205, 232
Brown, Thomas, surgeon 204
Brunt, Mr (Society of Apothecaries) 390n.
Buchan, William, *Domestic Medicine* 70
Buchanan, Prof. Andrew 315, 316, 347, 357–58, 376, 409n.
Buchanan, George 409 and n., 411
Buchanan, John 203
Buchanan, Prof. Moses Steven: Portland Street 240, 347; Anderson's University 240, 241, 399, 409n., 412; teaching aids 241; discussing hospitals 318, 412–13; health statistics 314; *Remarks on Medical Reform* 410; his son George 409 and n.
Buchanan, Thomas 110n., 293, 297, 304
burgesses 81 and n., 83; burgess fee 99; *see also* freemen
Burke, William 401, 402
Burmaster, Sarah 253
Burnell, George, *see* Berrell, George
Burns, Allan 212, 213, 233, 289, 356, 403–5, 406–8; *Observations on ... Diseases of the Heart* 212; *Observations on the Surgical Anatomy of the Head and Neck* 404–5
Burns, Sir George 211, 407
Burns, James 211, 407
Burns, Rev. John 407
Burns, Prof. John: pl. 14; surgeon 212; his portrait 211–12; family background 211, 406–8; Episcopalian 209, 210; CM degree 210, 211, 270, 364–65; MD degree 213; Royal Society 209; Glasgow Royal Infirmary 213, 267, 327–28; city practice 253, 267; death aboard the *Orion* 210, 211; his contribution 209–10, 211–12, 254; anatomy teacher 209–10, 211–12, 271–72, 412; private teaching 212, 271, 335, 372; College Street anatomical school 403–4, 408; grave-robbing 210, 212, 404, 408; Anderson's College 213, 239, 271–72, 281, 404, 406; midwifery 272–73, 275, 276, 281, 286, 289, 342; *The Anatomy of the Gravid Uterus* 272, 340–41, 405; *Popular Directions for the Treatment of the Diseases of Women and Children* 272, 405; *The Principles of Midwifery: Including the Diseases of Women and Children* 272–73, 405–6; clinical teaching 326, 327–28, 332, 333; Faculty and 210, 364–65; political lobbying 211; university curriculum 373; Glasgow Apothecary's Company 244
Burns and Laird (later Cunard Line) 211, 407
Burrows, Dr (RCPE delegate) 389n.
Bushell, Captain 300
business signs, *see* practice signs
Bute, Earl of 201
Bute Hall 212
Bylebyl, Jerome 47

cadavers, *see* corpses
Calcutta 315
Calder, James, quack 34
Calder, James junior, surgeon 253, 298, 304
Calder, James senior, surgeon 110n., 297, 304
Calder, John 220
Calendar of Scottish Papers 43–44
Calvinism, *see* Protestantism
Cambridge University 343, 346–47, 349
Cameron, Walter 203–4
Cameronians 158
Campbell family, Glasgow 160
Campbell, Daniel, MP 300
Campbell, Duncan 123n.
Campbell, Dr John 363
Campbell, John (of Loudon) 130–31
Canada 169
cancer 53, 57–58, 220
Cape of Good Hope 198
cardiac disease 405
Carmichael, Gershom 157, 162
Carrick, John 222–23
Carrick, Robert 204, 320
Carrick sheriffdom 7, 138, 417
Carruthers, Robert 90
Carssmuir, Laird of 148
Carstares, William 157

Cartesius 158
case histories 164–65, 309, 310, 315, 416; see also empirical observation
Catalogue of Medical Books (Watt) 231, 233
Catalogue of the Plants in the Physic Garden at Edinburgh (Sutherland) 168
catalogues (botany) 167–71
cataracts (eye) 60–61, 79n., 280, 343
Cathedral, View of the ... (Swan): pl. 15
Catholicism 38–39 and n., 158, 186, 258; Catholic League 41; see also Protestantism; religious strife
cauterisation 54, 55–56, 79
Cavenagh, Francis 363
Cecil Street Medical Society, London 276
Celsus, Aurelius Cornelius 54, 68, 122
Celts 182
certificates, see course certificates; qualifications; testimonials
cess rates 32
character references, see testimonials
Charité Hospital, La, Paris 177, 178, 256, 264
charity, see poor, provision for; sick poor
charlatans 49; see also irregular practitioners
Charles II: 19, 37n.
Charles Tennant & Co. 244
charmers 49; see also irregular practitioners
charter of 1599: James VI's reasons for 44, 46; facsimile of 417–19; loss of original 6n., 19; regulating scope of 1–2, 34; powers vested 6–10, 11, 34, 84–85; rights and privileges 10–11, 10n., 13, 32, 33, 358; practitioners constituted under 2, 11, 23, 80, 110; the craft guild and 84–85; ratification moves 11, 19–20; reiteration of powers 12–13; barbers incompatibility with 296; CM degree conflict 334–35, 366, 367–68; see also Faculty of Physicians and Surgeons; licensing to practise
charter petition (1817) 369–71, 374, 375
charter proposal (1844–49) 377, 379
Chartered Rights and Privileges regulations (Faculty) 343–44
Chelsea Physic Garden 168, 169
chemistry 242–49; free trade lecturing in 155; experiments 161, 162, 165, 193; graduate lecturers 170; Glasgow University 194, 223, 244; Anderson's College 242, 246–47, 412; curriculum requirements 341, 343; industrial uses 194, 203, 242, 244, 245, 247–48; laboratories 193, 244, 247, 248
chemists, see apothecaries; materia medica; physicians
Cheselden, William 172–73
children 82, 172, 268–69, 279, 284, 287
Children ... Woman and, Diseases of (John Burns) 272–73, 405–6
Children, Inquiry into ... Principal Diseases of (Watt) 231–32
Children, Treatment of ... Women and (John Burns) 272, 405
China 198
Chirurgerie: The Whole Course of and *A Discourse of the Whole Art of:* Peter Lowe's treatises 50–78; biographical content 41; dedications 38, 44n., 46; publication 49; 1599 charter petition 44; Hippocrates' *Prognostics* 47; internal medicine 45; *Portrait of the bones* 73; *Portraiture of a cloven lip* 63; *Portraiture of a drie suture* 67; see also Guillemeau; Lowe, Peter; Paré
Chirurgiae Magister, see CM degree
Chirurgians, Worshipfull companie of 44n.
chlorine 245, 248; see also bleaching
chloroform anaesthetic 315
cholera 314, 315, 316
Christie, David 227
Christie, Prof. James 284
Churchill, Dr (Dublin) 276
'Chyrurgion' (the term) 43
civic exemptions 10 and n., 32, 33, 358
Civil Service 395
Clark, L., *The City of Glasgow* pl. 1
classification (science) 167–68, 170–71, 182, 198, 201, 206, 215, 315
cleft lip 62, **63**, 66
Cleghorn, Margaret (née Thomson) 320
Cleghorn, Dr Robert: background 320; lecturing 232, 245, 308, 312, 320, 322, 326; professorship denied 321; Glasgow Royal Infirmary 308, 312, 319, 320, 322, 324, 330, 331; clinical case notes 312, 313–14; Langside Botanic Garden 205;

Glasgow Asylum 202; William Hamilton 228–29, 267–68, 308; Allan Burns 404; the Millars 321–22
Cleland, James 289
clerks (Faculty) 87; *see also* physicians' clerks; surgeons' clerks; ward clerks
Clinical Lectures on the Contagious Typhus (Millar) 317
clinical teaching 219, 233, 305–18, 323–37; Thomas Sydenham 164–65; Glasgow in the 1840s 410–11; comparisons 154, 413; becomes standard 414; *see also* bedside teaching; lying-in hospitals; teaching hospitals
clinical traditions, *see* medical schools
clinics, *see* dispensaries
cloven lip 62, 63, 66
Clowes, William 56 and n., 66n.
Clydesdale, Catherine (née Muir) 124–25
Clydesdale, William 102–3, 124–25, 125n.
Clydesdale sheriffdom 7, 417
clysters 46, 59, 61
CM degree: initiation of (Glasgow University) 364, 369; John Burns and 210, 211, 270, 364–65; Faculty opposition to 334–35, 364–66, 374, 375–76; Passenger Vessels Acts 365–66, 371; Edinburgh University 365; curricula 343, 347; graduates 345–46, 413; James Towers 210, 270, 364–65; W. J. Hooker 208–9; *see also* qualifications
coffee 167
Cointret, John 50
Collection of Cases in Midwifery (Smellie) 264–65
collections, *see* anatomy and surgery; botany; library resources; museums; natural history; preparations; teaching aids
College of the Faculty of Medicine of London 7n.; *see also* London: *College of Physicians*
College garden, Glasgow 169, 205
College of Physicians, Dublin 7, 342, 344, 363, 373, 390n.; *see also* Dublin; Ireland
College of Physicians, Edinburgh, *see* Edinburgh: *College of Physicians*
College of Physicians of England 389n.
College of Physicians, London, *see* London: *College of Physicians*
College of Physicians and Surgeons of Ireland 379; *see also* Dublin; Ireland
College of Surgeons, Dublin 342, 344, 347, 363, 368, 369, 373; *see also* Dublin; Ireland
College of Surgeons, Edinburgh, *see* Edinburgh: *College of Surgeons*
College of Surgeons of England 333, 384–85, 385n., 386, 387, 389–90n.; *see also* Company of Surgeons; London: *College of Surgeons*
College of Surgeons of Ireland 333, 385 and n., 386, 390n.; *see also* Dublin; Ireland
College of Surgeons, London, *see* College of Surgeons of England; London: *College of Surgeons*
College of Surgeons of Scotland, proposals for 378, 379, 388
College Street anatomical school, Glasgow 212, 271, 403–4, 408, 410; *see also* Burns, Allan; Burns, Prof. John
colleges of further education, *see* extra-mural colleges
Collegium Medicum, Prussia 197
Collins, Robert 277
Colquhoun, Dr John 21, 22–25, 132 and n.
Colquhoun, John, Faculty clerk 298
comfrey, roots of 140–41, 140n.
Commelin, Jan, *Horti medici Amstelodamensis rariorum plantarum descriptio et icones* 168, 198
common property (Faculty) 88–89, 96, 107, 108–9, 112
common register, *see* Medical Register
Company of Apothecaries, London 349, 376; *see also* Society of Apothecaries
Company of Surgeons (England) 103
Company of Surgeons, Paris 81n.
compearance, act for 101–2
compendia (botany) 198
Compendium of Voyages (Smollett) 187
Compleat Treatise of the Stone and Gravel (Greenfield) **121**
conferences, *see* General Conference of Corporations

conjoint examining boards 388–89; *see also* curriculum requirements; licensing to practise
Connell, William 209
consultants 348, 349, 399
contagion 277–78, 284–86
Contagious Typhus, Clinical Lectures on (Millar) 317
Coplande, Roger 47
copper engravings 215, 219; *see also* anatomy and surgery
Corbett, Dr David 304
Corkindale, James ('Corky') 246, 290, 362, 366, 373–74
coronary system 176, 177, 212
corporate associations, *see* craft guilds; Faculty of Physicians and Surgeons; General Conference of Corporations; Incorporation of Surgeons and Barbers; and Colleges indexed separately
corpses: Glasgow 96, 343, 409–10, 411, 415; Edinburgh 17, 95–96, 218; London 95–96, 403, 409; Paris 173, 176, 408; body-snatching, *see* grave-robbing; foundlings 172; shipping of bodies 222; condemned persons 222; Anatomy Act (1832) 210, 211, 343, 399, 408, 409, 410; embalming 50, 71, 72–74, 132, 133; 'like Yarmouth herrings' 402, 403; *see also* anatomy and surgery; dissection; preparations
cosoners 49; *see also* irregular practitioners
couching 60–61, 280; *see also* cataracts
country practices 6–7, 89–90, 123 and n., 136–38, 203, 254, 342, 346
Couper, John 289
Couper, Prof. William 203, 205, 244–45, 248, 324, 406
course certificates 212, 264, 342, 344, 397, 413–14; *see also* licensing to practise; qualifications
Covenanters 153, 161
Coventry, Captain 308
Coventry, Alexander 174, 308–9
Cowan family 199
Cowan, Prof. John Black 26n.
Cowan, Prof. Robert 26n., 199, 324
Cowan, Robert, merchant 26n.

Cowper, William F. 380, 382, 393–94, 395; *The Anatomy of Human Bodies* 219
craft guilds: function of 81–82, 81n.; early barber – surgeon associations 79–80, 147; 16th century Glasgow 4; seal of cause 81–82, 420–22; deaconry 16, 25, 80, 81–82, 81n., 85 and n.; Dean of Guild 82, 422; guild box 88–89; apprentices, *see* apprenticeship system; for Glasgow, *see* Faculty of Physicians and Surgeons; Edinburgh associations, *see* Edinburgh; associations outside Scotland 11–12, 27–28, 79–80, 81n., 103, 176; *see also* apothecaries; barbers; freemen; licensing to practise; surgeons
Crafts' Hospital 15, 83 and n., 84, 88
Craig, Andrew 188
Craig, Hendrie 136n.
Crawford, Mrs Anna 148
Crawford, Isabell 149–50, 424
Crawford, John 221, 223, 310
Crichton, Dr 18, 99, 135, 144n., 148
Cromwell, Oliver 14, 17, 19
Crosbie, John 141
Cross family, Glasgow 160
Cross, Dr John 175, 364, 366–67
Cruikshanks, William 225, 226, 227, 268
Cullen, Prof. William: pl. 9; background 178, 181, 223, 307; character and appearance 308–9; biography 181, 196; Jamaican experience 181–82, 183; John Paisley and 181, 229, 260; John Gordon 156, 181, 406; MD degree 155, 181, 307; Hamilton practice 181, 308, 310; Faculty role 181, 307–8; Glasgow Professor of Medicine 179; Latin tradition and 184, 309; *Dr Puff* 184; materia medica 181–83, 201, 205; *Materia Medica* 183; John Coakley Lettsom and 167, 183; chemistry 162, 182, 193, 242, 244, 398; Edinburgh Professor of Chemistry 174, 179; free trade lecturing and 156, 184; influence on teaching methods 179–81, 308–9; empirical observation 162, 163, 165, 179, 182, 309, 310; on Boerhaave 163, 165, 180; clinical teaching 307–9, 309n., 310; nosology 315; William Hunter and 156, 184, 260, 310; John

INDEX 451

Moore and 188, 234–35, 235–36, 259, 406; Prof. Thomas Hamilton and 310; Glasgow Literary Society 195
Cultivated Fruits of Britain (Roach) 168–69
Cumin, Prof. William 277, 284, 289, 290, 332
Cuming, Dr William 312n.
Cunard Line (formerly Burns and Laird) 211, 407
'cunning men of that art' 4, 6
Cunningham, Gabriel 103
Cunningham, Sir John 27
Cunningham sheriffdom 7, 138, 417
Cunninghame family 256
Cunninghame, James 166–67
cupping pl. 7, 74, 75–76, 79, 117; *see also* ventousing
curriculum requirements: Faculty 342–44, 370, 373, 374–75, 389, 399; Glasgow University 342–43, 347, 370, 373, 374–75, 399, 415; Anderson's College 240, 343, 412, 415; Edinburgh University 342–43; Aberdeen 343; MD degrees 212–13, 341, 344, 347; CM degrees 343, 347; botany and materia medica 193, 194, 196, 240, 341, 343, 412; chemistry 341, 343; apothecary experience 337, 343, 344; general practitioners 389, 398, 399; development of curricula 414–16; *see also* licensing to practise; qualifications
Currie, William, complainant 102–3
Currie, William, herbalist 139–40
cutting for the stone, *see* lithotomy
Cyril Thornton (Hamilton) 190

Dale, David 320–21
Dalrymple, Sir Hugh 27n.
Dalton, John 242
Daniel, John 142
Davidson, Marion 227
Davidson, Dr Thomas 159
Dawson, John 246
De Grave, Dr (Society of Apothecaries) 390n.
de la Faye (instrument maker) 280
de la Mettrie, Julien O., *L'Homme machine* 186

deaconry: Deacon 85 and n.; deacon convenor 81n., 82; letters of 16, 25, 80, 81–82, 85; *see also* craft guilds; Faculty of Physicians and Surgeons; Visitors
Dean of Guild 82, 422
deaths, *see* corpses; forensic medicine; lying-in hospitals; midwifery; murders
degrees (university), *see* CM degree; MD degree; qualifications
Deloraine's foot regiment 300–301
demonstrations, *see* anatomy and surgery; teaching aids
Denham, Thomas 227, 268 and n., 270, 273, 279–80
Descartes, René 158
Description of the Human Muscles (Innes) 214–15
Desnoues (surgeon) 177
destitution, *see* poor, provision for; sick poor
Deventer (author) 259
diagnostics 219, 312, 313, 405
Dick, Dr Robert 307
'Diet and Lodging' 195
Digby, Sir Kenelm 127
Dillenius, Johann Jacob 169
dinner money 99
Diploma in Surgery 333, 347, 366, 413; *see also* licensing to practise; qualifications
'discharge' of apprenticeship 94–95, 94n.
Discourse of the Whole Art of Chyrurgerie (Lowe), *see*: Chirurgerie
dislocations 71, 79, 227
dispensaries 278–79, 281, 284, 286, 287, 344, 414; *see also* apothecaries; materia medica; medicines, dispensing of; physicians
dissection: for practical training 172, 174, 237–38, 343; Paris manner of 173, 174, 175, 176; Glasgow 221, 411–12; condemned persons 222; dissecting-tables 403; public dissections 216; 'trial for qualification' 98, 217; *see also* anatomy and surgery; corpses; grave-robbing; preparations
Doctor of Medicine, *see* MD degree
'domes of Enlightenment' 186, 239, 240; *see also* Enlightenment; museums
Domestic Medicine (William Buchan) 70

Don Quixote (Miguel de Cervantes) 164, 187
Don Quixote translation (Smollett) 187
Dort 294
Double Qualification 347–49, 389, 413–14, 416; see also qualifications
Dougal, Thomas 293
Dougall, James 139, 423
Douglas, David 202
Douglas, Dr James 184, 219, 258, 258–59
Douglas, John 258
Dovehill estate 188
dressers 326, 336; see also anatomy and surgery
drugs and prescription, see apothecaries; botany; licensing to practise; materia medica
Drumakill, Laird of 203
drummers (Faculty) 88
Drummond, George 157
dry suturing 67, 68; see also suturing; wound surgery
Dublin: College of Physicians 7, 342, 344, 363, 373, 390n.; College of Surgeons 342, 344, 347, 363, 368, 369, 373; Dublin University 390n.; lying-in hospitals 275, 276, 277, 287–88; medical education 340, 347, 370; medical reform 370; see also Ireland
dummies 261–64, 267, 269, 271, 275, 280; see also midwifery; Smellie; teaching aids
Dunbarton 7, 417
Duncan, Dr (Edinburgh) 370
Duncan, Dr (RCPD delegate) 390n.
Duncan, Andrew 373
Dundonald, Lord 143
Dunlop, Alexander 267, 319, 323, 324, 328 and n., 355n.
Dunlop, Elizabeth 150
Dunlop, William (Alexander's son) 267, 328n.
Dunlop, William, University Principal 157, 159
Dureau, Dr 42n.
'Dutch connection' 163, 165, 294; see also Holland
'Dutch Edition of the French Memoirs' 230
Duvernay anatomical museum 177

East India Company (British) 166, 198, 333
East India Company (Dutch) 198
East Indian plants 168
Easton, J. A. 309–10
Edinburgh: 17th century status 3; training of physicians 125; University medical faculty 155; part of 'grand tour' 173; female practitioners 144, 149; attitudes to Glaswegians 184–86; corporate associations 10, 16, 81n., 89, 103, 104, 266; civic duty exemptions 33; prosecutions 142–43n., 350; midwifery 252 and n., 259, 262, 265–66; Park Place Lying-In Hospital 276n. Royal Medical Society 312;
College of Physicians: Visitors 9; regulation 17; bid for influence 17–19; certificates for licences 342, 344, 413–14; medical degrees 342, 348, 373, 375, 389; liaison with others 377–78, 379, 386, 390n.; Faculty and 363, 369–70, 370–71, 375; conjoint examining boards 388–89; Edinburgh Royal Infirmary 304–5; pharmacopoeia 197; Westminster medical reform 370
College of Surgeons: establishment of 82n.; Incorporation of Surgeons and Barbers 2n., 19 and n., 80, 81n., 96–97, 105n., 137, 149; College 'box' 88n.; Worshipfull companie of Chirurgians 44n.; medical education 215, 358, 373; literacy in Latin 145; qualifications 333, 342, 344, 348, 389; midwifery 252 and n.; Faculty and 137, 363, 369–70, 388–89; legislation 347, 368, 369, 375, 382–83, 382n.; liaison with others 377–79, 384–90 passim; Westminster medical reform 370; university and hospital ward education 266; Town's College 215; botany 168, 197; anatomy 95–96, 172, 403; corpses 17, 95–96, 218; clinical teaching 215, 327–28, 329, 413; student fees 413
University: medical faculty 155, 281, 373; CM degrees 365; MD degrees

340, 346, 347, 375; curriculum requirements 342–43; graduate numbers 346–47; the Medical Bills 392–93; midwifery 254, 266; standards 370
Edinburgh Medical Journal 316
Edinburgh Review 183, 204
Edinburgh Royal Infirmary 157, 164, 266, 275–76, 304–5, 306–7, 309
Edington (*British Medical Journal* contributor) 78
Education (Mill) 249
Edward: Various Views of Human Nature ... Chiefly in England (Moore) 189
Ehret, G. D. 171
Elcho, Lord (Earl of Wemyss) 379, 380, 381, 382, 392, 393, 395
Elphinstone, Andrew 25n., 98–99
Elphinstone, Sir George 84
Elphinstone, James 77n., 98
embalming 50, 71, 72–74, 132, 133
empirics 17, 65, 78, 120–22, 139, 148, 156; *see also* irregular practitioners
empirical observation 161–65, 179, 182, 309, 310; *see also* case histories
Encyclopaedia Britannica 242
enemas (clysters) 46, 59, 61
England: Henry VIII's 1512 Act 6n., 349; city colleges and guilds 27; surgeons and barbers 103, 128, 147, 185; College of Surgeons of England 333, 384–85, 385n., 386, 387, 389–90n.; female practitioners 146 and n., 147; general practitioners 349, 351–52; testimonials 143n.; medical degrees 340, 346–48; botany 168, 169, 194, 206, 207, 208; anatomy and surgery 175, 216, 372; midwifery 144, 254, 272; students 411, 413; English and Scottish comparisons 154, 347–48, 413; *see also* Cambridge; Hunter; individual Royal Colleges; London; Oxford; Smellie; Society of Apothecaries
engravings 215, 219; *see also* anatomy and surgery
Enlightenment, the 153, 158, 161, 191, 351; 'domes of Enlightenment' 186, 239, 240
Enniskillen 245
entry fees 26, 89, 93–94, 99–101, 108–9,
355, 358; *see also* freedom fines; quarter accounts; strangers
epidemiology 314–18
Episcopalians 158, 209, 210
Essay on Cholera (Lawrie) 315
Essay on Education (Mill) 249
ether anaesthetic 315
European medicine 179–86; *see also* medical schools
Ewing, John 99
examination procedure, *see* conjoint examining boards; curriculum requirements; licensing to practise
exemptions from civic duty 10 and n., 32, 33, 358
exhumed bodies, *see* grave-robbing
exotica 167, 168, 169, 170, 197, 201; *see also* botany
extra-mural colleges 155, 239–41, 274, 281, 287, 357, 361–62, 372; *see also* Anderson's College; Birkbeck, Dr George; mechanics' institutes; Portland Street School; universities
extractions (teeth) 62–64, 79 and n., 115
eye hospitals 239, 321, 344, 414
eye diseases 58–61, 59n.

Facts and Observations on the Sanitary State of Glasgow (Perry) 317
Faculté de Médecine, Paris 12, 42, 176
Faculty Act (1850) 379
Faculty Hall, The Second pl. 6
Faculty of Physicians and Surgeons of Glasgow:
  founding of: James VI's charter, *see* charter of 1599; the term 'Faculty' 12, 13 and n.; founding members 4, 12, 37–40, 77; majority barber-surgeons 80; first meeting (1602) 12, 85n.; second meeting 86; text book imprimatur 357; *see also* Hamilton, Robert; Lowe, Peter
  medical regulation: Faculty monopoly 1–2, 13, 27, 325, 334–35, 339, 342, 345, 357–62; resistance to 358–59; *see also* irregular practitioners; licensing to practise; prosecutions and litigation

craft guild: Incorporation of Surgeons and Barbers 2 and n., 11–12, 14–17, 80, 82–85, 112–13, 295; civic accreditation 14, 20, 80, 100; exclusion of physicians 106–7; letter of deaconry 16, 25, 80, 81–82, 85; freemen/burgesses 25–28, 81 and n., 83, 99–100, 101, 102, 109; seal of cause 81–82, 420–22; *see also* apprenticeship system; craft guilds; deaconry; Trades' House

rights, privileges and obligations: immunities & civic protection 10 and n., 32, 33, 80, 358–59; Faculty v. magistrates 12–13, 25–35 passim; *Chartered Rights and Privileges* regulations 343–44; 1850 Faculty Act 379; provision for the poor 10, 15, 26, 82, 83–84, 100; 'Act against Violating the Graves of the Dead' 221; Sabbath observance 103, 158

administration: Faculty officers 86–88, 90–91, 106, 131; meeting places 88; Faculty minutes 13 and n., 298; common property 88–89, 96, 107, 108–9, 112; Faculty 'box' 88–89, 96, 106, 131; common hall 104, 105; financial affairs 83–84, 86–91, 89n., 90n.; Widow's Fund 315, 353–55, 379, 414; teaching aids 96, 107, 220–21; gardens 198–200, 206–7; libraries 229–30, 297, 354; loss of documents 6n., 19, 298; *see also* freedom fines; quarter accounts; strangers

Faculty admission and membership: Henry Marshall test-case 25–29, 35, 198, 294–95, 294n.; William Stirling 98, 107–8, 110n., 113, 295, 296, 298; nepotism 296–97; membership restrictions 353–56, 358–59; Faculty Fellows 355; entry fees 26, 89, 93–94, 99–101, 108–9, 355, 358; entry standards 233–34; internal discipline 101–3

surgeons, apothecaries and barbers: surgeon domination 7, 25, 35, 105, 150; surgeons v. barbers 2, 35, 102–13, 110n., 294, 296, 297, 298; apothecaries 9, 11, 14, 80, 131–32, 193–94, 374; Visitors 8–10, 24, 82, 85–86, 85n., 88–89, 106–7, 135–36; Passenger Vessels Acts 365, 366, 368, 371, 375; Faculty and College of Surgeons, Edinburgh 137, 363, 369–70, 388–89; *see also* anatomy and surgery; apothecaries and surgeon-apothecaries; barbers and barber-surgeons; surgeons

physicians: status 2, 12, 14, 106–7, 127–29; surgeons v. physicians (1671–72) 21–25; Praeses (Physician-Visitor) 86, 107; proposed Edinburgh college of physicians 17–19; Faculty and College of Physicians, Edinburgh 363, 369–70, 370–71, 375; *see also* botany; materia medica; physicians

medical education: Faculty teaching role 96–97, 194, 355–56; Anderson's College 240, 287, 357, 362, 372, 411, 413–14; Portland Street School of Medicine 287, 357, 362; Faculty opposition to CM degree 334–35, 364–66, 374, 375–76; John Burns and 210, 364–65; MD degree holders 348–49, 363, 364, 365, 366–67, 375; curriculum requirements 342–44, 370, 373, 374–75, 389, 399; double qualification 347–48; *see also* apprenticeship system; clinical teaching; extra-mural colleges; Glasgow University

hospital care: hospitals, *see* Crafts' Hospital; Glasgow Lying-In Hospital; Glasgow Royal Infirmary; Town's Hospital; Faculty controlling role 304, 306; Faculty v. GRI managers 323–26, 332

medical reform: elitist self-interest 345, 359; 1817 charter petition 369–71, 374, 375; charter proposal (1844–49) 377, 379; liaison with other bodies 377–79, 380–95 passim; 'Treaty of London' 385 and n., 386, 390; General Medical Council

INDEX

representation 379; general practitioners 350, 366–68, 374; *see also* General Medical Council; legislation; Medical Act (1858); medical reform
Faculty of Surgery, Paris 50
Fairbairn, T., *The Old Town's Hospital* pl. 4
Fairie, James 149
Farr, Walter 232
fees, *see* booking; burgesses; entry fees; freedom fines; licensing to practise; quarter accounts; strangers; Widow's Fund
Felix, Charles François 176
Fellowship of Surgeons, London 80
female practitioners 17, 115, 141, 144–50, 144n., 146n.; midwives 144, 145, 253–54, 265, 266, 280, 281, 290; *see also* irregular practitioners; midwifery
fertilisers 194, 247
fevers 57, 278, 314, 317, 322; fever hospitals 278, 316; *see also* puerperal fever; rotten fever; typhoid; typhus
fines 8, 89, 137; *see also* freedom fines; prosecutions and litigation
Finlay, Kirkman, MP 369, 374
Finlayson, Dr James 42n., 309n.
Fleming, Adam 12, 85n., 86
Fleming, Mrs 200
Fleming, J.: *Broomielaw, Shipping* pl. 2; *Hunterian Museum, View of the* pl. 3; *Blythswood Place, St Vincent Street* pl. 16
Fleming, James 119
Fleming, John 25n., 94, 103
Fleming, Susanna (née Morrison) 94
Fleming, William 140
Forbes, Duncan, Lord Advocate 301
forceps 258, 259, 261–62, 265n., 270, 273, 280; *see also* midwifery
forensic medicine 8–9, 356, 415; *see also* murders
Forrester, James 8n.
Forret, James 6n., 77n., 84
Forster, John 135
Fothergill, John 312
Foulis, John 94
foundlings 172; *see also* children
FPSG, *see* Faculty of Physicians and Surgeons of Glasgow

fractures 71–72, 79, 115, 120, 145, 227
France: barbers and surgeons 81n., 103, 176, 295; corporate associations 10, 27–28; botany 168; forensic tradition 8–9; midwifery 173, 235, 258, 262, 273–74; religious wars 41, 50; *see also* Lowe, Peter; medical schools; Paris
Frank, James 86–87, 138, 141
Frankenstein, or the Modern Prometheus (Shelley) 401
Fraternity of Apothecaries and Surgeon-apothecaries, Edinburgh 16
Frederick Wilhelm I: 173
free trade lecture system 155–56, 176, 212, 236, 340, 373; *see also* private teaching
freedom fines 26, 87, 90, 100–101, 108, 120, 355; *see also* entry fees; fines; poor, provision for; quarter accounts; sick poor; strangers
freemen 25–28, 81 and n., 83, 90, 99–100, 101, 102, 109; *see also* craft guilds
Freer, Dr Robert 315, 321, 331
French, Mr, MP 377
French, John 31n.
French language proficiency 174, 344, 389; *see also* France
'French Memoirs, Dutch Edition of the' 230
French pox (syphilis), *see* venereal disease

Gaelic physicians 123n.
Gairner, John 139
Gale, Thomas 47 and n.
Galen (Claudius Galenus) 45n., 46, 47–48, 51, 58, 76, 117, 127; *De temperamentis et libris de symptomatum causis* 53; Golden Galen's Head 202, 203
Galloway 277–78
Gamble, Josias Christopher 245
gangrene 55, 79, 124
Gardener, Moses 363
gardener practitioners 17, 139–40, 203–4; *see also* irregular practitioners
gardens, *see* botany; materia medica; physic gardens
Gardiner, John 135–36
Garnett, John 361
Garnett, Dr Thomas 246
Garrow, Attorney General 371

Garvan, George 130 and n.
Gateshead 315
Gavan, Mr (Greenock) 143
General Conference of Corporations 380–95 passim
General Lying-In Hospital, Glasgow 282, 287; *see also* lying-in hospitals
General Medical Council 339, 379–98 passim
general practitioners: physicians' role undermined 12, 399; surgeon-apothecaries 12, 129, 349–50, 352, 374, 399; no 'pure' physicians left 348–49; Faculty action against 350, 366–68, 374; in England 349, 351–52; Society of Apothecaries 351, 387; Provincial Medical and Surgical Assocation (from 1855 British Medical Association) 351, 352 and n., 380, 394 and n.; medical reform 349, 351–53; recognition 397, 398, 399; national register 380, 399; General Medical Council 392; Scottish accreditation 340, 388, 389; MD degree entitlement 377; curriculum requirements 389, 398, 399; midwifery 273, 351; obstetrics and gynaecology 269; *see also* apothecaries; irregular practitioners; physicians
General Treatise of Midwifery (La Motte) 265n.
George IV: 248; as Prince Regent 371
Germany 27–28, 154, 164, 273–74, 406; *History of the German Empire* (Smollett) 187; proficiency in German 344, 389
Gesner, Johannes 173, 174, 175, 176, 177
Gibson family 160
Gibson, Andrew 136n.
Gibson, John, examiner 363
Gibson, John, library committee 230
Gil Blas (Smollett) 187
Glasgow: plans of 5, 250, 338; *The City of Glasgow* (Clark) pl. **1**; *Broomielaw, Shipping* (Fleming) pl. **2**; *The Old Town's Hospital* (Fairbairn) pl. **4**; *University of Glasgow*, pl. **5**; *View of the Cathedral, Infirmary* (Swan) pl. **15**; *Blythswood Place, St Vincent Street* (Fleming) pl. **16**; royal burgh (1611) 3; development of 3–4, 14 and n., 20; town council 81n.; trade links 4, 153, 169, 294, 295–96; centre of learning 173–74, 410–16; chemical industry 203, 244, 247–48; linen manufacturing 295, 299
Glasgow Apothecaries' Hall 203, 248
Glasgow Apothecary's Company 244
Glasgow Asylum for the Insane 202
Glasgow Courier 279, 281
Glasgow Eye Infirmary 239, 321, 414
Glasgow Herald 212, 328n.
Glasgow Journal 266, 306
Glasgow Literary and Commercial Society 357
Glasgow Literary Society 195
Glasgow Lying-In Hospital 246, 274, 285, 286, 287, 289, 290–92, 414; *see also* lying-in hospitals
Glasgow Mechanics' Institute, *see* mechanics' institutes
Glasgow Medical Examiner 359, 360, 364, 366, 368, 373
Glasgow Medical Journal 316
Glasgow Medical School consortium 325
Glasgow Medical Society 232, 283, 316, 339, 344, 359, 360
Glasgow Royal Infirmary (GRI): launching of 318–21; David Dale 320–21; philanthropy and clinical teaching 283, 286, 305–6; clinical lectures 308, 314, 326–27, 328–34, 343, 412–13, 414; controversial rotation system 322–25; Faculty objections 325–26, 332; rules regulating attendance 348; treatment statistics 314, 317; 'Medical School' 333; maternity policy 264, 274; hospital garden 200; staff salaries 329–30, 329n.; student fees 333, 413; Alexander Stevenson 245, 318, 319–20; Robert Cleghorn 308, 312, 319, 320, 322, 324, 330, 331; Richard Millar 317, 320, 322, 331; *Medical and Surgical Establishments of the Infirmary* (Millar) 322; John Burns 213, 267, 327–28; *see also* Town's Hospital, Glasgow
Glasgow University: Faculty lecturers 194; botany and anatomy chairs 194; anatomy and surgery 160, 194, 196, 241,

398, 403, 411–12; anatomical collections 222–23; Old Humanities Classroom 217; Hamilton Building 403; library 158, 229, 230; clinical teaching 307, 326–35, 410–11; midwifery 155, 194, 223, 254, 267–90 passim, 343, 414; chemistry 194, 223, 244; Cullen's lectures in English 184; John Burns 209–10, 211–12, 213, 271; medical degrees, *see* CM degree; MD degree; London-Scottish connection 223–24, 259–60; rival Anderson's College 239, 246; curricula 342–43, 347, 370, 373, 374–75, 399, 415; graduate numbers 345–46; body acquisition riots 221–22; 1745 Rebellion 307; Dr Thomas Wakley's view of 370; English and Irish students 411; medical reform 392–93; *see also* extra-mural colleges; universities
Glen, James 222
GMC, *see* General Medical Council
Golden Galen's Head 202, 203; *see also* Galen (Claudius Galenus); practice signs
Good, John Mason, *The History of Medicine ... the Apothecary* 41 and n., 43
Gordon, John: William Stirling partnership 188, 296, 297, 299, 310; Robert Wallace and 297, 299, 302; John Moore 185, 188, 234, 267, 310, 406; Gordon and Stirling's linen manufactory 299; Faculty role 194, 210, 217 and n., 230, 267, 296, 297; Incorporation of Surgeons and Barbers 110n., 297, 298, 302; anatomy lectures 172, 194, 214, 215, 217, 220, 259–60, 415; midwifery 251, 253, 254, 259–60, 267; MD degree 267, 310; provision for sick poor 298, 299, 302–4; Town's Hospital 303, 304; Edinburgh Royal Infirmary 304; Shawfield riots 293, 297, 301; civic affairs 301–2; William Hunter 156, 260; William Cullen 181, 260, 308, 406; Tobias Smollett 185, 217n.; William Smellie 217n., 260; Prof. Thomas Hamilton 310
Göttingen university 171
GPs, *see* general practitioners
Graaf, Regnier de 163
Graham, Archibald: Visitor's court 24n.;

proposed college of physicians 17, 18; pharmacy examinations 99, 134; Dr Sylvester Rattray 126–27; irregular practitioners 135, 142, 144n., 148; William Clydesdale 103
Graham, Sir James 370
Graham, John 108–9
Graham, Prof. Robert 207, 331
Graham, Dr Thomas 246–47
Graham-Gilbert, John: pl. 14
Grahame family 160
Grahame, Alexander 381 and n., 391
Grahame, W. 377
Grahame, Weems & Grahame 381n.
'grand tour' 173–75; *see also* medical schools
Granfield, Margaret 147–48
grave-robbing 210, 212, 221–22, 226, 272, 399–402, 404, 408–10; *see also* anatomy and surgery; corpses; dissection
gravel, cutting for, *see* lithotomy
gravid uterus 237, 260, 268; *The Anatomy of the Gravid Uterus* (John Burns) 272, 340–41, 405; *see also* midwifery
Gray, Adam 25n., 93, 95 and n.
Gray, Roderick 363, 364, 367
Greek and Graeco-Roman medicine 38, 46, 47–48, 51, 52, 62n., 74; Galen, *see* Galen (Claudius Galenus); *see also* Roman medicine
Greek literacy 38, 185, 344, 389
'Greek Temples to Arts and Science' 238, 239, 240–41
Green, Mr (RCSE delegate) 389n.
Greenfield, John, *A Compleat Treatise of the Stone and Gravel* 121
Greenock 153
Grégoire (Parisian teacher) 262
Grey, Nathan 122
GRI, *see* Glasgow Royal Infirmary
Grotius, Hugo (Huig de Groot) 162
Guido d'Arezzo (Guy of Arezzo) 72
guilds, *see* craft guilds
Guillemeau, Jacques 55 and n., 75; *Oeuvres de chirurgie* 62
Gundelsheimer, Andreas von 182, 183
gunpowder wounds 66 and n., 68–69; *see also* wound surgery
Guthrie, Douglas 78

Guy's Hospital, London 173, 413
gynaecology 268–69, 406; *see also* midwifery; obstetrics

Haarlem 295n., 299; *see also* Holland
haircutting 79n., 93; *see also* barbers; bearddressing; periwig-making
Haliday, Alexander Henry 179, 236
Hall, David 105, 107
Hall, John (early member) 101
Hall, John senior: Deacon 15, 16, 420, 422; meeting venues 88; Faculty documents 19; proposed college of physicians 17, 18; conflict with Faculty physicians 22, 24n.; Henry Marshall test case 27; regulation of practice 141
Hall, John junior 25n.
Hall, John, barber (Edinburgh) 16, 100
Hall, John, surgeon (Paisley) 143–44
Halle university 164, 405
Haller, (Viktor) Albrecht von 171, 173, 175, 176, 177, 230
Hallthorne (soldier) 141
Hamilton family (of Abercorn) 38
Hamilton, Dr, Physician-Visitor 119 and n.
Hamilton, Laird of Barns 132–34
Hamilton, Prof. Alexander 266, 273, 276 and n., 279
Hamilton, Andrew 127–28, 423
Hamilton, Rev. Andrew 39
Hamilton, Sir Claud 38, 39n.
Hamilton, Douglas, 8th Duke of 188, 189
Hamilton, Elizabeth (née Stirling) 268
Hamilton, Sir George 38–39n.
Hamilton, Grace 280
Hamilton, James, Earl of Abercorn 38
Hamilton, James, physician 18, 126–27
Hamilton, James, surgeon 293, 302, 304
Hamilton, Prof. James, Visitor 13
Hamilton, Prof. James (Edinburgh) 266, 276 and n., 279
Hamilton, John, Jesuit theologian 38n.
Hamilton, John, patient 131
Hamilton, Robert: background 37–39, 37–38n.; 1599 royal charter 1, 6 and n., 7, 11, 44, 417, 418, 419; Faculty founding member 12, 37, 80; craft guild associations 82, 84–85, 85n.; first Visitor 8, 85, 87; Catholic faith 39; death (1629) 39
Hamilton, Robert (of Torrence) 77n.
Hamilton, Prof. Robert: background 183, 307, 310; Faculty role 110n., 229, 307–8; Faculty and University libraries 229–30; Glasgow Literary Society 195; anatomy and botany lecturer 188, 221, 222, 234–35, 307; anatomical preparations 222; body acquisition 221
Hamilton, Thomas, surgeon 110n., 293, 297, 304
Hamilton, Captain Thomas 190–91; *Cyril Thornton* 190; *Men and Manners in America* 190
Hamilton, Dr Thomas 21, 22, 24, 111, 125 and n.
Hamilton, Prof. Thomas: background 183–84, 190, 224, 253, 268, 270, 280–81, 310–11; his son, *see* Hamilton, Prof. William; anatomy and botany lecturer 205, 222, 234, 239, 257, 308; midwifery lectures 257, 263–64, 267; Faculty role 223, 230; practice partner William Cullen 190, 223, 308; William Hunter 223–24, 225; William Stark 236, 237
Hamilton, William (Countess of Abercorn's son) 39
Hamilton, William, advocate 27
Hamilton, Rev. William 183–84
Hamilton, Sir William 190
Hamilton, Prof. William: background 190, 224, 281, 406; his father, *see* Hamilton, Prof. Thomas; sons, *see* Hamilton, Captain Thomas; Hamilton, Sir William; William Hunter's London school 223–27; Hunter's testimonials 224–25, 311–12; surgical lectures 227–29, 239; *System of Surgery* (unpublished) 228–29; anatomical preparations 226–27; midwifery 239, 253, 254, 267–70, 273, 275, 279–80, 281; puerperal fever 276; botany 205, 206; clinical science 308, 312; Glasgow Royal Infirmary 319; Sawney gibe 185–86
hanging basins 118; *see also* practice signs
Hannay, Prof. Alexander 277–78, 285, 343

INDEX

Hanoverians 301; *see also* Gordon, John; Shawfield riots
Hare, William 401, 402
harelip 62, **63**, 66
Harper, Thomas 120
Harris, Robert 18, 90–91, 118–19
Harrison, Dr (Dublin University) 390n.
Harrison, Dr Edward 352
Hart, Barbara 149
Harvey (or Harvie), James 44n., 49, 57n.
Harvey, William 87
Hastings, Dr Charles 352, 380
Hawkins, C. (RCSE delegate) 389n.
Hawkins, Dr Francis 389n.
Hay, Rev. John 39
Headlam, Thomas Emerson, MP 380, 381, 382, 383, 391, 392, 393, 394
healing practice 115–16; *see also* irregular practitioners
Heathcote, Sir William, MP 391
Heister, Lorenz 173
Henderson, Abram 301
Henderson, James 71
Henry IV of Navarre 41
Henry VIII: 6n., 7n., 46, 349
herbalists 115, 139–41; *see also* irregular practitioners; materia medica
Herbertson, Robert 87
herbs, *see* materia medica
hereditary medical families 123n.
Hermann, Paul 167
hernias 45, 64–66, 65n., 78, 343; herniotomy 120–22
Herron, William 363
Hewson, William 219
Highland medicine 123 and n.
Hill, James 242, 243
Hill, John, *The Vegetable System* 201
Hill, Rev. Laurence 243, 407
Hill, Laurence, lawyer 243
Hill, Ninian 193, 230, 242–44, 249, 406, 407
Hippocrates 46, 51, 52, 69n., 259, 309, 341; *Prognostics* 47
History of the German Empire (Smollett) 187
History of Medicine ... the Apothecary, The (Good) 41 and n., 43
History of Physick (Le Clerc) 230
Hoffmann, Friedrich 180, 198, 230, 236, 309; *Medicina rationalis* 166; *Observations* 164–65
Hohenheim, Philippus A. T. B. von (Paracelsus) 127 and n.
Holland: city colleges and guilds 27–28; apprenticeships 92, 294; anatomical preparations 172, 176; botany 167, 168, 198; East India Company 198; linen manufacturing 295, 299; Scottish 'Dutch connection' 163, 165, 294; Stirling family 295; *see also* Amsterdam; Leiden
Home, Mr (Hunter school of anatomy) 226
Home Civil Service 395
home deliveries 275, 277, 284–85, 287; *see also* lying-in hospitals; midwifery
Homme machine, L' (de la Mettrie) 186
Hood, Janet 149
Hood, William 359, 363
Hooker, Sir Joseph Dalton 208, 209
Hooker, Dr William 208, 209
Hooker, Sir William Jackson 170, 194, 202, 206, 207–10; *British Flora* 208; *Flora Scotica* 208
hooks 260, 269; *see also* midwifery
Hope, Mr (Edinburgh lecturer) 370
Hope, Prof John 201
Hope, Prof. Thomas Charles 245–46, 320, 321, 323, 324
Hopkirk, Thomas 205, 206
horning, letters of 8, 19, 137, 142–43n., 325; *see also* prosecutions and litigation
Hors(e)burgh family 199
Hors(e)burgh, Alexander 26n., 199, 251, 293, 304
Hors(e)burgh, Lillian (née Marshall) 26n.
Horti medici Amstelodamensis rariorum plantarum descriptio et icones (Commelin) 168, 198
Hortus Medicus Edinburgensi (Sutherland) 168, 198
hospitals, *see* lying-in hospitals; teaching hospitals; and specific institutions indexed separately
hosts exemption 10
Hôtel Dieu Operations, Paris 176, 256, 264
House of Commons select committees 211, 352, 374, 376, 380, 381–82, 384; *see also* legislation

Houston, Laird of 50
Houston, Robert 25n., 90n., 138
Houstoun family 256
Houstoun, Lady Anne 255
Houstoun, James 253, 255, 256–57, 257–59; *Aphorisms* translation 258–59
Houstoun, Sir John 255
Houstoun, Dr Robert 253, 254, 255–56, 257
Houstoun, Robert senior 255
Hughes, Dr James Stannus 385n., 390n.
Humane Society 320
Humani corporis fabrica, De (Vesalius) 71, 175
humanism, *see* medical humanism
Humphrey Clinker (Smollett) 186
Hunter, Hew 132
Hunter, John: background 183, 310, 311, 406; London 173, 184; Prof. Thomas Hamilton 224, 311; Prof. William Hamilton 225–26, 227, 268, 311; Matthew Baillie 219; scientific research 405; teaching aids 398, 405; on Latin literacy 184–85; *see also* Hunter, William
Hunter, Prof. Robert 215, 239, 241, 381
Hunter, Samuel 328n.
Hunter, William: pl. 10; background 183, 260, 310, 311, 406; Prof. William Cullen and 156, 184, 260, 310; free trade lecturing 156, 184, 236; anatomy 156, 221, 237, 312, 398; his London school 223–27, 238–39, 241, 268–70, 281; *Introductory Lectures* 233; specimen collections 186, 237, 405; William Smellie and 260, 261; midwifery: his interest in 253, 255, 260; Royal Accoucheur 262; midwifery teaching 156, 254, 263, 273, 274, 275; 'Hunter's Anatomy of the pregnant uterus' 230; gravid uterus 237, 260; Hunterian Museum pl. 3, 238–39, 240, 412; Prof. Thomas Hamilton 224, 311–12; Prof. William Hamilton 224–26, 227, 311–12; Matthew Baillie and 190, 219; Alexander Monro *secundus* 219–20; *see also* Hunter, John
Hunterian Museum, View of the (Fleming) pl. 3
'Hunter's Anatomy of the pregnant uterus' 230

Hutcheson, Elizabeth (née Dunlop) 150
Hutcheson, George 150
Hutcheson's Hospital, Glasgow 200, 243
hydropsy 45 and n., 49–50, 227

ice research 245
Incorporation of Surgeons and Barbers, Edinburgh 2n., 19 and n., 80, 81n., 96–97, 105n.; *see also* craft guilds; Edinburgh
Incorporation of Surgeons and Barbers, Glasgow, *see* Faculty of Physicians and Surgeons
indentured apprentices 26, 92–95, 113, 294; *see also* apprenticeship system
India 198, 315
industrial accidents 314
industrial chemistry 194, 203, 242, 244, 245, 247–48; *see also* chemistry
Industrial Revolution 244, 247–48, 275, 314
infanticide 285; *see also* midwifery
Infantment, Booke of the (Lowe) 49n., 64
infection 277–78, 284–86
Infirmary, *see* Glasgow Royal Infirmary
Inglis (Smellie's partner) 260, 261
Innes, John, *A Description of the Human Muscles* 214–15
inquests exemption 10
Inquiry into ... Principal Diseases of Children (Watt) 231–32
inspections, *see* materia medica; Visitor
Institute of Medicine and Surgery 378
'institutes', *see* physiology
Institutes and Aphorisms (Boerhaave) 163, 165, 259
instrument makers 280
internal medicine 45, 231
Introductory Lectures (Hunter) 233
Ireland: College of Physicians, Dublin 7, 342, 344, 363, 373, 390n.; King's and Queen's College of Physicians 379, 386; College of Physicians and Surgeons 379; College of Surgeons, Dublin 342, 344, 347, 363, 368, 369, 373; College of Surgeons of Ireland 333, 385 and n., 386, 390n.; shipping of bodies 222; students 411, 413; Irish Medical Charities Bill 377; bleaching 245; Belfast

236; Ulster Plantation 38–39 and n.; see also Dublin
irregular practitioners: Peter Lowe's view of 4–7, 45; Faculty measures to regulate 1–2, 22, 34–35, 138–51, 362; offering a service 115–16, 150–51; anatomy lectures for 233–34; English general practitioners 351; 'abusers' 44, 49; astrologers 115; bone-setters 115, 120, 145; charlatans 49; charmers 49; cosoners 49; drug peddlers 135; empirics 17, 65, 78, 120–22, 139, 148, 156; female practitioners 17, 49, 115, 141, 144–50; gardeners 17, 139–40, 203–4; herbalists 115, 139–41; itinerants (land loupers) 50, 65, 115; mountebanks 28, 46n.; quacks 17, 22, 34, 48, 115; quack-salvers 49; surgeon-apothecaries 233–34; tooth-pullers 115; traditional healers 1, 9, 34, 150; 'wise women' 145; witches 49; see also licensing to practise; prosecutions and litigation
Irvine, Grace (née Hamilton) 280
Irvine, James 130
Irvine, William 193, 244, 249, 280
Italy 27–28, 140, 168; proficiency in Italian 344; see also Latin literacy; Roman medicine
itinerants 50, 65, 115; see also irregular practitioners

Jacob (female patient) 278
Jacquin, Nicolas von 171
Jamaica 181–82, 183, 187, 209
James I: 85 and n.
James IV: 40, 80, 82n., 85n.
James VI: 6, 13, 44, 46, 417–19; see also charter of 1599
Jamieson, John 88
Jardine, Mr (University dissector) 402
Jardine, Prof. George 319, 330
Jeffray, Prof. James: anatomy lecturer 211, 241, 282, 356, 403, 411–12; botany 204; University hospital 289; appointments disputes 321, 325
Jeffrey, Lord Francis 183, 185, 204
Jeffrey, Marion 204
Jeffrey, Mary 185

Jena university 164
Jenkins case (England) 128
Johnston, Margaret (née Thomson) 320
Johnstoun, Dr 130 and n.
Johnstoun(e), Prof. John 130n., 194–96, 214, 257, 304, 307
Johnstoune, Joseph 196
Johnstoune, Thomas 196
Joseph II, Emperor 177–78
jury duty exemption 10

keeper of the keys (Faculty box) 88–89, 106, 131
Kelly, Mr, medical student 340
Kennedy, Dr (Wodrow physician) 199
Kennedy, David 137
Kennedy, Gilbert 137
Kennedy, Dr Thomas 159
Kennedy, William 203
Kew Gardens 194, 206, 207
Killian, Prof. H. F. 273, 406
'Killing Times' 158; see also religious strife
Killmakell, Laird of 141
Kilmarnock 138
King's College 18, 232; see also Aberdeen
King's and Queen's College of Physicians of Ireland 379, 386; see also Dublin; Ireland
kirk session 4–6, 38–39, 39n., 252; see also Protestantism
Kirklee Botanic Gardens 206
Kirkwood, Mr (Prof. Pagan's assistant) 285
Kirkwood, Allan 120
Knaut, Christian 167
Knox, Robert 401–2
Kyle 7, 138, 417

La Mettrie, Julien O. de, *L'Homme machine* 186
La Motte, Guillaume Mauquest de 235, 259; *General Treatise of Midwifery* 265n.
laboratories, see chemistry
Lamb, George 136n.
Lamb, John, elder 136n.
Lamb, John, younger 136n.
Lanark 7, 320, 417
Lancet 352, 403
land loupers (itinerants) 50, 65, 115; see also irregular practitioners

Lander, George 31n.
Lanfranc of Milan 42–43
Lang, William 204–5
Langside Cottage botanic garden 205
Langside House 204
latent heat discovery 244, 249
Latin literacy 38, 145 and n., 184–85, 309, 341, 344, 389; *see also* Greek and Graeco-Roman medicine; Italy; Roman medicine
Lawrence, Mr (RCSE President) 385n., 390n.
Lawrie, Prof. James Adair 315–16, 347–48, 375, 377, 398, 416; *Essay on Cholera* 315
Le Clerc, D., *History of Physick* 230
Le Dran, Henri-François 177
Lebot, Johanne 58
leeches, medical use of 74, 76, 117
legislation 332, 347, 349, 368–96; *see also* Anatomy Act; Apothecaries Act; Faculty Act; Medical Act; medical reform; Passenger Vessels Acts
Leiden (Leyden): medical 'grand tour' 126, 173; private teaching 218; St Caecilia Hospital 166; anatomical preparations 172, 176; barbers and surgeons 295; French language lessons 174; Bernhard Albinus 176, 218; Herman Boerhaave 166, 171, 218; Carolus Linnaeus 171, 183; Alexander Monro 166, 217–18; Archibald Pitcairne 125, 166; Frederick Ruysch 163, 176, 177, 218; *see also* Holland
Leith 142–43n.
Lennox, John 131–32
Leoniceno, Nicolao 47
lepers 4, 58
Leslie, Bishop John 3
letters of deaconry, *see* deaconry
letters of horning, *see* horning, letters of
Lettsom, Dr John Coakley 167, 183, 312n.; *The Natural History of the Tea Tree: ... Medical Qualities of Tea* 167
Levant 168, 182
Levret, André 273
Leyden, *see* Leiden
Lhwyd, Eduard 160, 182; *Archeologica Britannica* 182
library resources 229–35, 240, 265, 297, 299; *see also* apprenticeship system; Faculty of Physicians and Surgeons; Glasgow University; museums; teaching aids; Watt, Robert
Library Society 320
licensing to practise: Faculty licences 1–2, 8, 143, 334, 353, 357–62; monopoly powers 1–2, 334, 357–62; surgeon domination 150; Faculty and local magistracy 11, 25–35 passim; Henry Marshall test-case 25–29, 35, 198, 294–95, 294n.; licentiates 83, 193, 342, 346, 355–64 passim, 365, 379; outwith Glasgow 136–38; country licentiates 342, 346; Faculty licence v. University CM degree 334–35, 364–66, 374, 375–76; resistance to Faculty 358–59; 'trial for qualification' 94, 96–99, 99n., 117, 156, 217, 251, 341; action against offenders, *see* prosecutions and litigation; 'papers' (licence petitions) 119–20, 119n.; fees 89, 252, 346, 358, 359, 360, 413–14; testimonials 8, 117, 123, 136–37, 143–44, 143n., 196, 266; practice signs 74n., 118, 202, 203; Faculty curriculum requirements 342–44, 370, 373, 374–75, 389, 399; course certificates 212, 342, 344, 413–14; 'private' lecturers 347; surgeons 1, 8, 9, 117–24, 145, 210, 345, 346; Diploma in Surgery 333, 347, 366, 413; military surgeons 123–24, 141; physicians 9, 126–28; barbers 8, 117–19, 120, 124; surgeon-apothecaries 9, 128–32, 134, 135–36, 346; 'surgeon-pharmacists' 193–94, 203; drug selling 134–35, 135–36; female practitioners 144–50, 144n., 146n.; midwives 144, 145, 251–53, 346; apprentices 137, 138; 'licenced charlatans' 339; unqualified and irregular practitioners 1–2, 138–51; *see also* apprenticeship system; charter of 1599; conjoint examining boards; Faculty of Physicians and Surgeons; irregular practitioners; malpractice; prosecutions and litigation; qualifications
Liddell, John 25n., 118
Lies, John 117
ligatures 54–55, 56, 60, 75, 215

Linacre, Thomas 37n., 88
Lincolnshire Benevolent Medical Society 352
linen manufacturing 295, 299
Linlithgow 33
Linnaeus, Carolus (Carl von Linné) 171, 183, 201, 206; *Species plantarum* 167–68; *Systema naturae* 167–68
liquorice roots ('liquory sticks') 200
Literary and Commercial Society, Glasgow 357
Literary Society, Glasgow 195
literati 186–91
lithotomy 58, 120, 122–23, 123n., 173, 215, 343; *A Compleat Treatise of the Stone and Gravel* (Greenfield) 121
litigation, *see* prosecutions and litigation
Lock Hospital 315, 414
Locke, John 164
Lockhart, George 25n., 97
Lockhart, James 19
Lockhart, Thomas 18, 19, 135, 136n., 144n., 148
Logan, John 140
London: medical training 154, 156, 340; part of 'grand tour' 173, 174, 175; hospitals 173, 236, 237, 275, 413; Scottish connection 156, 173, 223–29; attitudes to the Scots 184–86; clinical teaching 154, 329, 343, 413; private lectures 172–73, 218, 223, 264, 347; anatomy 216; corpses 95–96, 403, 409; autopsies 312; midwifery 254, 259, 261, 275; student costs 413; Birkbeck College 241; medical reform 370, 380, 385–95; 'Treaty of London' 385 and n., 386, 390; *College of Physicians*: founders 7n., 37n.; Thomas Linacre 37n., 88; Henry VIII's charter 7n., 349; 1663 charter 21; early records of 14n.; monopoly and privilege 1n., 10; status of physicians 17; jurisdiction 7n., 9–10, 9n., 349, 363; Bonham's Case 11 and n.; certificates 342, 344, 373; Oxford and Cambridge 346–47; the poor 10; liaison with others 377–78, 379, 380, 386, 387; attitude to female practitioners 144; midwifery 144, 254; general practitioners 351–52 *College of Surgeons*: humanist surgery 47;

Barber-Surgeons Company 11–12, 21, 37n., 80, 95–96, 181; jurisdiction 349; certificates for licences 342, 344, 363, 369, 373; Oxford and Cambridge 346–47; Passenger Vessels Acts 368; liaison with others 377–78, 379, 380; Fellowship of Surgeons 80; Worshipfull companie of Chirurgians 44n. *see also* England; Hunter; physic gardens; Royal College of Physicians of England; Royal College of Surgeons of England; Royal Society; Smellie; Society of Apothecaries
Long Calderwood farm 183
Lonsdale, Henry 401, 402
Louis XIV: 176
Love, John 217, 220–21
Love, Peter 44
Lowe, Christian 77 and n.
Lowe, Helen (née Wemyss) 77, 295
Lowe, John (early member) 12, 85n.
Lowe, John (Peter's brother) 42
Lowe, John (Peter's son) 50, 53, 77 and n., 126–27
Lowe, Peter: 36; life of 37, 40–44; surgeon, ambassador and spy in France 6, 41–42; Catholic? 39 and n.; Glasgow town surgeon (1599) 4; death of (1610) 76–77; his contribution to medicine 77–78; 1599 royal charter 1, 6, 7, 11, 44, 417, 418, 419; Faculty founding member 12, 37, 80; craft guild associations 82, 84–85, 85n.; power of Visitor 8; Faculty quartermaster 86; master of Crafts' Hospital (1605) 84; his opinion of barbers 48, 50, 62, 64; on surgeon's essential qualities 50; medical humanism 47–48; embalming skills 50, 71, 72–74; amputations 54–56, 124; forensic experience 8–9; publications: *Chirurgerie* books, *see Chirurgerie: The Whole Course of* and *A Discourse of the Whole Art of*; *Booke of the Infantment/The Sicknes of Women* 49n., 64; *Booke of Women's Diseases* 53; *De partu mulierum* 49n.; *Spanish Sickness* 45–46, 46n., 49n.; *The Poore Man's Guide* 49, 50, 51
Lower, Richard 52
Lowery, Blais 6n.

'Lucius', *Remarks on Medical Reform and on Sir James Graham's Bill* 370
Ludwig, Christian Gottlieb 167
Luke, Janet 160, 199
Luke, John 160
lunatics 202, 303
Lyes, Alexander 117
Lyes, John, *see* Lies, John
lying-in hospitals 264, 274–92, 344; deaths 283–86, 287; Glasgow 281–82, 288, 359, 414; Glasgow Lying-In Hospital 246, 274, 285, 286, 287, 289, 290–92, 414; Glasgow University facilities 274, 277, 278–79, 281, 284, 286, 288–90, 414; General Lying-In Hospital, Glasgow 282, 287; Dublin 275, 276, 277, 287–88; Edinburgh 276n., 287; London 275; *see also* bedside teaching; clinical teaching; midwifery; sick poor; teaching hospitals
Lyle, Thomas 402–3

MA degree 38, 39
MacAllister, Mrs, body of 410
McArthur, Mr (Infirmary house surgeon) 327
MacBrayne, David 211
McClae, John 139
McCormick, Samuel 196
McDougal, Dr Peter 364, 367
McDougall, Dr (Infirmary appointee) 325
McDougall, Mrs, midwife 279
MacDougall, J. 241
McDowell, Ephraim 256
McFarland, John 217
Macfarlane, Laird of 203
Macfarlane, Dr John 316
MacGregor, Mr, MP 381
machines (midwifery), *see* teaching aids
Macintosh, Charles 209
McKechny, Alexander 265
McKechny, Elizabeth (née Blair) 265
Mackenzie, James 356
Mackenzie, Dr William 239, 313–14, 356–57
MacKindo, Duncan 118
Mackinnis, Alexander 136n.
MacLachlan family 123n.
Maclagan, Dr (Edinburgh) 389
McLauchlan, Donald 203–4

McLean, Dr Hector 221, 304
McLean, Prof. John 321
Maclehose, James 271
McNeil, Elizabeth 265
McNeil(l), Evir/Iver, lithotomist 25n., 123 and n.
McNeilage, Elizabeth (née McNeil) 265
McNeilage, James 265
McNeill, Duncan bane 123n.
McNeill, Evir, apprentice 123n.
McNeill, Lachlan 123n.
maize 198
male midwifery, *see* midwifery
M'Allay, Thomas 105
malpractice 124–25, 140, 339, 341; *see also* irregular practitioners; licensing to practise; prosecutions and litigation
man-midwifery, *see* midwifery
manikins, *see* teaching aids
Marischal College 392–93; *see also* Aberdeen
Marshall, Henry: background 26 and n., 29n., 32n., 198–99, 294; Faculty membership test case 25–29, 35, 198, 294–95, 294n.; nepotism 296–97; surgeons v. barbers 105–6, 110n., 294, 296, 298; library books 297; *see also* Marshall, John
Marshall, Humphrey 77n.
Marshall, John: background 29n., 32n., 198–99, 204, 294; training 32n., 294–95; Andrew Reid and 29, 30–31, 30n., 93; Keeper of the Physic Garden 194, 196, 198, 296–97, 298; *see also* Marshall, Henry
Marshall, Lillian 26n.
Marshall, Patrick 26, 29n., 294
Marshall, Robert, maltman 140
Marshall, Robert, medical practitioner 319
Martine, Andreu 301
Mary, Queen of Scots 10, 85n.
Master of Surgery, *see* CM degree
materia medica: curricula requirements 193, 194, 341, 343; drugs from plants 167, 169, 182, 198, 201, 203; apothecaries 200, 202–3, 206, 344; apothecary gardens 197, 200–201; herbal drugs 200–201, 205, 206; herbaria 198, 205, 208; regulation of 9–10, 34, 39, 134,

135–36; pharmacopoeia 128, 197, 201; drug peddlers 135; *see also* apothecaries; botany; licensing to practise; natural history; physic gardens; physicians
Materia Medica (Cullen) 183
maternity care, *see* lying-in hospitals; midwifery
Mathies, John 120
Maunsell, H. 385n., 390n.
Mauriceau, François 235; *Aphorisms* 258–59
Maxwell, John, surgeon 363
Maxwell, John, wright 28
Mayne, Prof. Robert 21
Mayo, Dr (RCPE President) 389n.
Maze (patient) 183
MB degree 413
MD degree: Glasgow University 155, 195–96, 255, 273, 345–46, 347, 366, 375; Edinburgh 340, 346, 347, 375; Aberdeen 348, 363, 375; St Andrews 348, 363, 375; university comparisons 346–47; free trade education towards 212–13, 223, 340; testimonials and certificates 196, 212; curriculum requirements 212–13, 341, 344, 347; residency 212, 347; graduates 195–96, 255, 345–46, 347–48, 366, 413; Faculty and 348–49, 363, 364, 365, 366–67, 375; surgeons 255, 273, 343, 367; midwifery and 255, 273, 343; double qualification 347–48; general practitioner entitlement to 377; *see also* qualifications
mechanics' institutes 241, 247, 283, 361; *see also* extra-mural colleges
Meckel, Prof. Johann Friedrich 405
Medical Act (1858) 1, 343, 353, 365, 379, 395, 415; bills preparatory to 347, 379, 380–95; *see also* legislation; medical reform
medical and surgical degrees, *see* CM degree; MD degree; qualifications
medical books, acquisition of 174; *see also* library resources
Medical Books, Catalogue of (Watt) 231
Medical Charities Bill 377
Medical Council, *see* General Medical Council
Medical Essays and Observations 220
Medical Examiner, see Glasgow Medical Examiner

medical 'grand tour' 173–75; *see also* medical schools
medical humanism 38 and n., 45, 47–48, 123n., 126
medical jurisprudence 343
medical literati 186–91
Medical Observations (Sydenham) 164, 170
medical philanthropy, *see* poor, provision for; sick poor
medical reform 339, 345, 349–50, 351–53, 368–96, 416; *see also* legislation; Medical Act
Medical Reform, Remarks on (Buchanan) 410
Medical Reform, Remarks on, and on Sir James Graham's Bill ('Lucius') 370
Medical Register 380, 388, 399, 416
medical regulation, *see* legislation; licensing to practise; medical reform; prosecutions and litigation
medical schools: continental schools 38, 125–26, 154; Scottish model 154, 155, 156; medical 'grand tour' 173–75; Angers 126; Basle 174; Berlin 154; Edinburgh 125, 155, 173; Glasgow 34, 155, 194, 325, 371; Leiden (Leyden), *see* Leiden; London 154, 156, 173, 174, 175; Montpellier 125; Padua 47, 125; Paris 42, 125, 154, 173; Rheims 125, 126, 256; Vienna 154, 171, 177–78; Zurich 174
Medical Society, *see* Glasgow Medical Society
medical standards 339–48
medical statistics 231–32, 314, 315, 411, 413
Medical and Surgical Establishments of the Infirmary (Millar) 322
Medical Times and Gazette 284, 337, 413
Medicina rationalis (Hoffmann) 166
medicines, dispensing of 128–34; *see also* apothecaries; botany; dispensaries; materia medica; physicians
Medico-Chirurgical Society 315
medieval medicine 38
Meikleham, Mr (Glasgow Infirmary) 331
Melvill(e), John 110n., 132–34, 132n.
Men and Manners in America (Hamilton) 190
Menzies, Archibald 202
Merchant Hospital 83 and n.
Merchants' House 82

Merry Andrew (clown) 46
Metcalf, John 247
M'Gow(a)n (pregnant woman) 278
Michaelson, George 94
microscope 161, 163
midwifery 251–92; female practitioners 144, 145, 253–54, 265, 266, 280, 281, 290; male midwifery 150, 254–65 passim, 273, 274, 278, 369, 416; surgeons 128, 254, 255, 273, 351; surgeon-apothecaries 128, 254, 351; physicians 254, 273; general practitioners 273, 351; obstetrics 256, 258–59, 260, 268–69, 280, 406, 414; Faculty and 155, 251–54, 265; licensing to practise 144, 145, 251–53, 266, 346, 387; London-Scottish connection 156, 223–34, 259–60, 268–70; Glasgow University 155, 194, 223, 254, 267–90 passim, 343, 414; Anderson's College 213, 271, 278, 280, 281, 282–83, 412; Portland Street School of Medicine 281, 282, 291; Edinburgh 252 and n., 254, 259, 262, 265–66, 276n.; Aberdeen 254; England 144, 254, 272; 'British school' of 273–74; France 173, 235, 258, 262, 273–74; German and Austrian practice 273–74; diplomas 254, 346; degree courses 255, 273, 343; teaching aids 261–64, 267, 269, 271, 275, 280; legislation 369; moral issues 263, 278–79, 283–84; unmarried mothers 283–84; gravid uterus 237, 260, 268, 272, 340–41, 405; pelvic deformities 279, 280; extra-uterine pregnancy 256; abnormal births 79, 254, 260, 262, 269, 273, 279–80; deaths 270, 275–78, 283–86, 287; abortion 149, 272, 405; sterility 268; hospital care, *see* lying-in hospitals; puerperal fever 264, 266, 275–78, 284, 285, 287, 291; 'midwives tea' 285; home births v. hospital deliveries 275, 277, 284–85, 287; natural approach to childbirth 270, 273–74; forceps use 258, 259, 261–62, 265n., 270, 273, 280; hooks 260, 269; chloroform 315; *see also* Burns, Prof. John; Hunter, William; La Motte; Mauriceau; Smellie

Midwifery, A Collection of Cases in (Smellie) 264–65
Midwifery, General Treatise of (La Motte) 265n.
Midwifery, Praeternatural Cases in (Smellie) 264–65
Midwifery, The Principles of: Including the Diseases of Women and Children (John Burns) 272–73
Midwifery, Treatise on the Theory and Practice of (Smellie) 261, 264–65
Miligan, Dr (RCPD delegate) 390n.
military balloting 358
military surgery 66 and n., 68–69, 69n., 294, 416; army surgeons 32, 34, 40, 123–24, 141, 208, 369; Prussian and Russian surgeons 173; Vienna school 177–78; *see also* Army and Navy Boards; Lowe, Peter; naval surgeons
Mill, James, *Education* 249
Millar, Prof. John 354
Millar, Margaret, complainant 124–25
Millar, Margaret, patient 255
Millar, Prof. Richard 283, 315, 317–18, 320, 321–22, 331, 369; *Clinical Lectures on the Contagious Typhus* 317; *Medical and Surgical Establishments of the Infirmary* 322
Miller, Provost 298–99, 300–301
Miller, Hugh 363, 369
Miller, John, apothecary 129
Miller, John (of Skeinstown) 130
Miller, Mathew 129–31, 129n.
Milligan, Dr (RCPD delegate) 390n.
Mitchell, Joseph 253
Mitchell, Margaret 253
Moffat, Alexander 110n.
Mondeville, Henri de 71
monks 170; Guido d'Arezzo 72
Monro, Alexander *primus*: his training 217–18; anatomy lectures 172, 214, 215, 218, 223, 236; injected preparations 214n., 218; anatomical amphitheatre 312; bedside teaching 166; *Medical Essays and Observations* journal 220
Monro, Alexander *secundus* 174, 219–20, 308–9, 309n., 405
Monro, Alexander *tertius* 370

INDEX 467

Monro, Donald 218
Monro, John 215, 217
Montague House, London 186
Monteath, James 203, 248, 280, 319, 323, 406
Montgomerie, Robert 234
Montgomery, 3rd Viscount 87n.
Montgomery, Dr George: MD degree 196, 307; Faculty membership 307–8; examinations 251, 252, 253, 257; libraries 229, 230; lectures by 188, 235; clinical teaching 196, 251, 306–7; midwifery 251, 252, 253; sick poor provision 304
Montgomery, Hugh 87 and n.
Montpellier 125
Montrose, Duke of 224, 311
Moore, Jean (née Simson) 189
Moore, Sir John 191
Moore, John: pl. 12; personality and background 187–89, 223, 234–35, 253; apprenticeship 185, 188, 234; travels 189, 235;army service 234, 235; Duke of Hamilton 188, 189; Earl of Albermarle 235; Paris training 235, 259; Faculty involvement 188, 223, 234, 236; partnerships 267, 310, 406; John Gordon 185, 188, 234, 267, 310, 406; Tobias Smollett 185, 186, 187, 190; literary works 189–90; *Edward: Various Views of Human Nature ... Chiefly in England* 189; *Mordaunt: Sketches of Life, Character and Manners* 189; *Zeluco: Various Views of Human Nature* 189; on Alexander Stevenson 318; his garden 200
Moore, Marion (née Anderson) 188
Morbid Anatomy ... of the Human Body, The (Baillie) 219, 237
Mordaunt: Sketches of Life, Character and Manners (Moore) 189
Morris, Dr Andrew 188, 307
Morrison, Prof. Robert 167, 168, 169
Morrison, Susanna 94
Morton, James 94–95
mountebanks 28, 46n.; *see also* irregular practitioners
Mowat, Charles 25n., 92
Mowat, James 117, 136n.
Muir, Archibald 77n.

Muir, Catherine 124–25
Muir, James, lecturer 263, 265, 266–67
Muir, James, merchant 149
Muir, James, surgeon 232
Muir, Quentin 129
Muir, Robert, barber 118
Muir, Robert, burgess 77n.
murders 8–9, 17, 48, 285, 401–2; *see also* forensic medicine
Murdoch, James 111 and n.
Mure, Baron 267
Murray, Adam 175–76
Murray, Elspeth 144, 146
Murray, Patrick 197
Murray, William 175–76
Muscles, Human, Description of the (Innes) 214–15
museums: 'domes of Enlightenment' 186, 239, 240; complementary learning tools 219, 232, 233; Ashmolean Museum 160, 182; William Hunter's museum: pl. 3, 238–39, 240, 412; Anderson's College 186, 240, 241; British Museum 186; Duvernay museum 177; *see also* library resources; teaching aids
Muspratt, James 245
Mylne, Andrew 93

Nairne, Dr (RCPLond delegate) 389n.
Naper, Robert 141
Napier, Dr Andrew 341
Napier, Joseph, MP 391
Napoleonic wars 400
narcotica 167; *see also* materia medica
National Institute of Medicine and Surgery 378
national register, *see* Medical Register
natural history 160, 166–67, 186, 344; *see also* botany; materia medica
Natural History of Jamaica, The (Sloane) 181–82
Natural History of the Tea Tree: ... Medical Qualities of Tea, The (Lettsom) 167
natural law theory 162
natural philosophy 246, 389
naval surgeons 32, 34, 57, 70n., 369, 411, 412; *see also* Army and Navy boards; military surgery; surgeons

Neill, John 87
Neilson, Gilbert 134
Neilson, Isobel 94
Neilson, Walter 422
Neilson, William 420, 422
neo-Platonism 186–87
Nessmith, John 70
Netherlands, *see* Holland
New Kirk Steeple House 88
New Lanark 320; *see also* Lanark
Newcastle upon Tyne 315
Nihell, Mrs (London) 263
Nimmo, Alexander 325
Nimmo, John 363
Niven, John 119 and n.
nomenclature 167–68, 198, 215; *see also* science
non-freemen 81, 90; *see also* craft guilds; freemen
Norris, Mr 374
North America 169–70
Norwich 92
Nostradamus (Michel de Notredame) 58
notary (Faculty officer) 87
Nuremberg 127
nutrition 198, 201

observational methodology, *see* case histories; empirical observation
Observations on ... Diseases of the Heart (Allan Burns) 212
Observations (Hoffmann & Stahl) 164–65
Observations on the Surgical Anatomy of the Head and Neck (Allan Burns) 404–5
obstetrics 256, 258–59, 260, 268–69, 280, 406, 414; *see also* midwifery
occupational risks 314
Oeuvres ... de chirurgerie (Paré) 62, 65n.
Oeuvres ... de chirurgie (Guillemeau) 62
officer role (Faculty) 87–88
ointments 134; *see also* materia medica
'Old Independents' 321; *see also* Protestantism
Old Town's Hospital, The (Fairbairn): pl. 4
onions 167, 201
ophthalmy 58–59; *see also* eye diseases
Orion shipwreck 210, 211
orphans 82; *see also* children

Osborn, William 227, 268 and n., 269, 273, 279–80
Oswald, John 313
outpatient facilities, *see* dispensaries
ovariotomy 255–56
Owen, Robert 321
Oxford University 169, 343, 346–47, 349, 390n.

Padua 47, 125
Pagan, Prof. John 277, 282, 284–85, 316
Paisley, John 181, 196, 217, 220, 221, 229, 259–60, 304
Paisley Medical Society 232
Paisley parish 138
Palgrave, Francis Turner 209
Palmerston, Henry John Temple, Viscount 391–92, 393
Panton, Dr Alexander 377
Panton, John 123–24
'papers' (licence petitions) 119–20, 119n.; *see also* licensing to practise
Paracelsus (Philippus A. T. B. von Hohenheim) 127 and n.
Paré, Ambroise 55, 56 and n., 68, 71; *Les oeuvres ... de chirurgerie* 62, 65n.; *Traité des rapports* 8–9
Paris: recognised medical school 42, 125, 154, 173; St-Côme community 7, 42, 80, 103, 116, 126, 176; Charité Hospital 177, 178, 256, 264; Faculté de Médecine 12, 42, 176; Faculty of Surgery 50; anatomy and surgery 173, 174, 175, 176, 216, 218; 'Paris manner of dissection' 173, 408; autopsies 312; midwifery 173, 258, 262, 273–74; private teaching 177, 218; gardens 171; barbers and surgeons 81n., 176, 295; surgeons of the long and short robes 116–17; instrument makers 280; John Moore 235, 259; James Houstoun 258; Siege of Paris 41, 50; *see also* France; medical schools
Park Place Lying-In Hospital, Edinburgh 276n.
parliamentary measures, *see* House of Commons select committees; legislation
Partu mulierum, De (Lowe) 49n.
Passenger Vessels Acts 365, 366, 368, 371, 375; *see also* legislation

passing on assize (exemption) 10
Pate, Lawrence 204
Paterson, Prof. James 160, 241, 278, 280, 281, 282
Paterson, John 119
pathology 179, 219, 310; *see also* clinical teaching
patients: poaching of 101; patient-power 120
Patoun, Dr David 304
Patoun, Dr Peter 105, 220, 304
Pattison, Granville Sharp 213, 239, 331, 356, 404–5, 407, 408, 410
'pensionaires' 173
Peregrine Pickle (Smollett) 186–87
periwig-making 93, 118 and n.; *see also* barbers; beard-dressing; haircutting
Perry, Robert 315, 316–17; *Facts and Observations on the Sanitary State of Glasgow* 317
Perth 33
Peter I, the Great 173
petitions to practise 119–20, 119n.; *see also* licensing to practise
Pettegrew, John 111 and n.
Pettigrew, Thomas 6n.
phantoms, *see* teaching aids
pharmacopoeia 128, 197, 201
pharmacy, *see* apothecaries; materia medica; physicians; Society of Apothecaries
philanthropy, *see* poor, provision for; sick poor
Phillips, Prof. 265n.
Philosophical Transactions (Ray) 182, 220
phlebotomy, *see* blood-letting
phosphates 194, 247
Physic Garden at Edinburgh, Catalogue of the Plants in the (Sutherland) 168
physic gardens: value to medicine 170, 197–98; Glasgow 158, 159, 160, 168, 169, 194–205 passim, 296–97; Chelsea 168, 169; Edinburgh 168; Oxford 169; Vienna 171; *see also* botany; materia medica
Physician-Visitor (Praeses) 86, 107; *see also* Deacon; Visitor
physicians: status 2, 14, 17–18, 43, 45, 126; realm of healing 79, 214, 368, 399; formation of colleges 12, 27–28, 126; for Glasgow, *see* Faculty of Physicians and Surgeons; proposed separate Glasgow college 17–19; encroachment by apothecaries 16, 128–31, 350, 399; Society of Apothecaries 21, 128, 135; licensing to practise 9, 126–28; medical prescription rights 128–29; emergence of the general practitioner 12, 348–49, 399; consultants 348, 349, 399; university training 14, 125–28, 155; Latin literacy 38, 341, 389; interest in anatomy 312; midwifery 254, 273; Faculty pioneer, *see* Hamilton, Robert; Dr Archibald Pitcairne 125, 166; Gaelic physicians 123n.; *see also* apothecaries; botany; general practitioners; licensing to practise; materia medica; qualifications
physicians' clerks 322, 327
Physicians and Surgeons, Society of Regular, Glasgow 358
physiology 180, 181, 182
Pitcairne, Dr Archibald 125, 166, 265
plague epidemics 4
plants, *see* botany
plates (anatomical art) 218–19
Plato 48, 186–87
PMSA, *see* Provincial Medical and Surgical Association
poison 10, 136; 'poisoners' 8, 149
polling, *see* beard-dressing; haircutting
poor, provision for 10, 15, 26, 32, 82, 91, 100; medical care, *see* lying-in hospitals; midwifery; public health; sick poor; *see also* freedom fines
Poore Man's Guide, The (Lowe) 49, 50, 51
Port Glasgow 153
Porterfield, Alexander 110n., 293, 297, 298, 304
Portland Street School of Medicine: function of 239–40; 'private' institution 347, 399; Faculty accreditation 287, 357, 362; midwifery 281, 282, 291; *see also* Anderson's College
Portrait of a cleft lip (Lowe) 63
Portrait of a dry suture (Lowe) 67
post-mortem examinations, *see* autopsies
potatoes 198, 201
poverty, *see* poor, provision for; sick poor
practice signs: barber's pole 74n.; Golden

Galen's Head 202, 203; hanging basins 118
Praeses (Physician-Visitor) 86, 107; *see also* Deacon; Visitor
Praeternatural Cases in Midwifery (Smellie) 264–65
pregnancies, *see* midwifery
preparations: product of Dutch science 172; pioneers: Albinus, Rau and Ruysch 176, 218; made locally or imported 174–75; skeletons 175; preserved in spirits 172, 238; injected preparations: 'Ruyschian art' 218; secret formulas 214n.; wax-based solutions 177–78, 239; Paris 177; William Hunter 226, 237, 238, 239; William Hamilton 226–27; Allan Burns 405; in diagnostic precision 233, 405; Faculty collection 96, 107, 220–21; *see also* anatomy and surgery; corpses; dissection
Presbyterianism, *see* Protestantism
prescription of medicines 128–34; *see also* apothecaries; botany; dispensaries; materia medica; physicians
Primrose, Gilbert 44n., 49, 70
Prince Regent 371; as George IV: 248
Principles of Midwifery: Including the Diseases of Women and Children, The (John Burns) 272–73, 405–6
'Principles and Practise of Medicine' (Watt) 231
private gardens 168, 205, 206–7; *see also* botany; materia medica; physic gardens
private teaching: London 172–73, 218, 223, 264, 347; Paris 177, 218; Leiden 218; Glasgow 212, 231, 271, 403–4, 408, 410; Dublin 347; William Cheselden 172–73; Hunter brothers 173, 223; William Smellie 264; Burns brothers 212, 271, 403–4, 408; Robert Watt 231; extra-mural colleges 347, 372; 'private' lecturers 347; Royal Commission's query 335, 372; formula for success 356–57; course certificates 264, 397; teaching certificates 374; modernising 398–99; *see also* anatomy and surgery; extra-mural colleges; free trade lecture system
Prognostics (Hippocrates) 47

prosecutions and litigation: procedure 142–43, 142–43n.; magisterial support 1–2, 11; Bonham case 11 and n.; letters of horning 8, 19, 137, 142–43n., 325; act for compearance 101–2; act of adjournal 363, 369; bonds of desistance 423–25; fines 8, 89, 137; first recorded case (1669) 124–25; irregular practitioners 35, 139; female practitioners 145; general practitioners 350, 366–68, 374; to enforce Faculty jurisdiction 362–68, 374; MD degree holders 363–64, 367; CM degree holders 365–66, 375; Edinburgh practice 350; immunity from prosecution 358–59; *see also* irregular practitioners; licensing to practise
Protestantism 153–54, 157–64, 185–86, 258, 321; *see also* Catholicism; kirk session; religious strife
Provincial Medical and Surgical Association (PMSA) 351, 352 and n., 380, 394n.
Prussia 173, 197
public disorder: over grave-robbing 221–22, 408; Shawfield riots 293, 296, 297, 298–99, 300–301; 1745 Rebellion 307
public health 4, 275, 277–78, 286, 314–18
puerperal fever 264, 266, 275–78, 284, 285, 287, 291; *see also* fevers; lying-in hospitals; midwifery; teaching hospitals
Puff, Dr, see Cullen, Prof. William
Puffendorf, Samuel, Freiherr von 162
Purden, John 302
'pure surgeons' 7, 42
Puritans 164
Purtabgurh Malwa 315
Purves, Dr George 17

quacks 17, 22, 34, 48, 115; quack-salvers 49; *see also* irregular practitioners
qualifications 333–35, 339–49, 350, 372, 413–14; university degrees 137, 211, 340, 345–47, 373; double qualification 347–49, 389, 413–14, 416; MA degree 38, 39; Doctor of Medicine, *see* MD degree; Chirurgiae Magister, *see* CM degree; Diploma in Surgery 333, 347, 366, 413; Edinburgh medical degrees 342, 348, 365, 373, 375; English degrees

INDEX   471

340, 346–48; MB degree 413; midwifery diplomas 254, 346; teaching certificates 374; *see also* curriculum requirements; irregular practitioners; licensing to practise; 'trial for qualification'
quarter accounts (Faculty) 86–87, 89–90, 90n.; *see also* entry fees; freedom fines; strangers
quartermasters (Faculty) 86, 106, 131

Rae, Thomas 302
Raeburn, Sir Henry pl. 13
raids exemption 10
Rainy, Prof. Harry 356–57
Ralstoun, Andrew 25n., 140
Rampoorah Local Battalion 315
Ramsay, Allan: pl. 10
Ramsay, John 203–4
rat poison 10, 136
Rattray, Dr Sylvester 117, 126–27; *Aditus novus ad occultas sympathiae et antipathiae causas inveriendas* 127
Rau, Johannes 176
Ray, John 167, 168, 170, 182, 206; *Philosophical Transactions* 182; *The Wisdom of God Manifested in the Works of Creation* 161, 163–64
RCPE, *see* Royal College of Physicians of England
RCPEd (Royal College of Physicians of Edinburgh), *see* Edinburgh: *College of Physicians*
RCPL (Royal College of Physicians of London), *see* London: *College of Physicians*
RCSEng, *see* Royal College of Surgeons of England
RCSEd (Royal College of Surgeons of Edinburgh), *see* Edinburgh: *College of Surgeons*
RCSI, *see* Royal College of Surgeons of Ireland
Read, Alexander 68n.
Rebellion of 1745: 307
reference libraries, *see* library resources
references (character), *see* testimonials
Reformation 153
Regality Court 88

register of qualified practitioners, *see* Medical Register
regius professors 334
regulation of practice, *see* legislation; licensing to practise; medical reform; prosecutions and litigation
Reid, Dr Andrew 363, 364, 367
Reid, Andrew, surgeon 29–32, 30n., 31n., 32n., 93
Reid, G. 225, 226
Reid, John, irregular practitioner 139
Reid, John, *Scots Gardener* contributor 200
Reid, Robert 200
religious strife 38–39 and n., 41, 158–59, 162–63, 209, 335; *see also* Catholicism; Protestantism
Remarks on Medical Reform (Buchanan) 410
Remarks on Medical Reform and on Sir James Graham's Bill ('Lucius') 370
remedies (drug), *see* materia medica
Renfrew 7, 417
Renger (Halle printer) 405
'resurrectionists' 226, 399, 401–2, 408, 409–10; *see also* anatomy and surgery; corpses; dissection; grave-robbing
Rheims 125, 126, 256
rhubarb 201
Riddell, John 323
riots, *see* public disorder
Risk, John 124–25
Risk, Margaret (née Millar) 124–25
Ritchie, Adam 136n.
Roach, F. A., *Cultivated Fruits of Britain* 168–69
Robertson, Lord 367
Robertson, Dr Henry 340
Robertson, Walter 109
robes, surgeons of the long and short 116–17
Robieson, Walter 105
Robison, Catherine 144
Robison, James 137
Robison, John, chemist 193, 244, 249
Robison, John, surgeon-apothecary 25n., 97, 141
Rodger, James 147–48
Rodger, Janet (née Anderson) 147
Rollo, John 146
Roman Catholicism, *see* Catholicism

Roman medicine 38, 54, 62n., 68, 74, 122; see also Greek and Graeco-Roman medicine; Italy; Latin literacy
roots of comfrey 140–41, 140n.
Rosse, George 229
rotten fever 58; see also fevers
Rottenrow Lying-In Hospital, see Glasgow Lying-In Hospital
Rotterdam 294
Rotunda Lying-In Hospital, Dublin 275, 277; see also lying-in hospitals
Rowland, Robert 89
Royal Accoucheur 262
Royal Botanic Institution 315
royal charters (Faculty), see charter of 1599; charter petition (1817); charter proposal (1844–49)
Royal College of Physicians, Dublin (RCPD) 7, 342, 344, 363, 373, 390n.; see also Dublin; Ireland
Royal College of Physicians, Edinburgh (RCPEd), see Edinburgh: *College of Physicians*
Royal College of Physicians of England (RCPEng) 389n.
Royal College of Physicians of London (RCPL), see London: *College of Physicians*
Royal College of Physicians and Surgeons of Ireland 379; see also Dublin; Ireland
Royal College of Surgeons, Dublin 342, 344, 347, 363, 368, 369, 373
Royal College of Surgeons, Edinburgh (RCSEd), see Edinburgh: *College of Surgeons*
Royal College of Surgeons of England (RCSEng) 333, 384–85, 385n., 386, 387, 389–90n.; see also Company of Surgeons; London: *College of Surgeons*
Royal College of Surgeons of Ireland (RCSI) 333, 385 and n., 386, 390n.; see also Dublin; Ireland
Royal College of Surgeons, London, see London: *College of Surgeons*; Royal College of Surgeons of England (RCSEng)
Royal College of Surgeons of Scotland (RCSS), proposals for 378, 379, 388
Royal commissions 332, 335, 372

Royal Infirmary, Edinburgh, see Edinburgh Royal Infirmary
Royal Infirmary, Glasgow, see Glasgow Royal Infirmary (GRI)
Royal Medical Society, Edinburgh 312
Royal Society, London 127, 209, 255, 256
Russell, Andrew 404
Russell, James Burns 317
Russia 173, 174, 403
Rutherford, Andrew 355n.
Rutherford, John 164, 236, 305, 306–7
Ruysch, Frederick 163, 176, 177, 218; 'Ruyschian art' 218; see also preparations

Sabbath observance 103, 158
St Andrews University 127, 343, 347, 348, 363, 375, 398
St Bartholomew's Hospital, London 173, 413
St Caecilia Hospital, Leiden 166
St-Côme community, Paris 7, 42, 80, 103, 116, 126, 176
St Enoch Square: *The Second Faculty Hall* pl. 6
St George's Hospital, London 173, 236, 237
St Helens, Lancashire 245
St Ninian's leper hospital 4
St Rollox Chemical Works 244, 247, 248
St Thomas' Hospital, London 173
St Vincent Street: *Blythswood Place* (Fleming) pl. 16
Sandyford Botanic Gardens 205, 206–7, 208
sanitary precautions 277–78; see also public health
Sanitary State of Glasgow (Perry) 317
Sawney Bean gibe 185
scarification 74, 76; see also cupping; ventousing
Scharp, James 136n.
Scheele, Carl Wilhelm 248
schools of medicine, see extra-mural colleges; medical schools; universities
science: in medical education 154, 155, 161–62; classification 167–68, 170–71, 182, 198, 201, 206, 215, 315
Scot, Andrew 61–62
Scot, James 139
Scots' character 153–54, 202
Scots Gardener 200

Scott, Sir Walter 154, 183
Scott and Watt, architects 240
seal of cause 81–82, 420–22; *see also* craft guilds; Faculty of Physicians and Surgeons
Second Faculty Hall, The: pl. 6
select committees, *see* House of Commons
Seller, Dr William 390n.
Semmelweis, Ignaz Philipp 278
Sempill (of Mylnebank) 77n.
Sempill, Sir James 77n.
Sempill, Lord Robert 77n.
Sempill, William 138
'Senex' (Robert Reid) 200
servants (pre-apprentice) 91
Sharp, David 24n., 90, 94, 97, 103, 134, 136n., 142
shaving, *see* beard-dressing
Shawfield riots 293, 296, 297, 298–99, 300–301; *see also* public disorder
Sheap, Robert 140
Shelley, Mary W., *Frankenstein, or the Modern Prometheus* 401
Sherard, William 169
Shiells, John 196
Ship Bank, Glasgow 204, 223
ship surgeons, *see* naval surgeons
Sibbald, Sir Robert 160, 168, 197; Sibbald collection 160
sick poor: guild responsibilities to 10, 15, 26, 82, 83–84, 100; Crafts' Hospital 83 and n.; gentlewomen and 141, 145; John Balmano 203; increased provision for 302–6, 314–15, 318; Gordon, Stirling and Wallace 298, 299, 302–4; effects of poverty 275, 315, 317; lying-in hospitals 275, 279, 284, 286; serving the needs of the medical profession 264, 278–79, 283–84, 314, 318, 327, 416; disadvantages of the system 283; moral issues 279, 283–84; David Dale 320–21; *see also* lying-in hospitals; midwifery; poor, provision for; public health
Sicknes of Women, The (Lowe) 49n.; *see also*: *Booke of Women's Diseases*
Siege of Paris 41, 50
signs (business), *see* practice signs
Simeons, Mr (Society of Apothecaries) 390n.

Simon, Sir John 393, 395
'simples' 168, 169, 170, 201; *see also* botany
Simpson, Rev. Patrick 8n.
Simson, Jean 189
Simson, Prof. John 160, 188–89
Simson, Robert 188
Simson, Thomas 217
Sinclair, Alexander 203–4
Sinclair, Walter 174
Sinclair, William 295n.
single mothers 283–84
skeletons 71, 73, 96, 107, 175
sketches in anatomy 218
skin diseases 414
slander 102–3
Sloane, Sir Hans 160, 167, 168, 169, 181–82, 183, 186; *The Natural History of Jamaica* 181–82
small pox 232; *see also* vaccinations
Smellie, William: Lanark background 260, 261, 265; Faculty membership 260; contribution to midwifery 253, 254, 255, 260, 261; London practice 259, 260, 268; hospital for the poor 264; teaching methods 261–65, 265n., 274, 275, 279, 281; his demonstration dummies 261–64, 267, 269, 271, 275, 280; forceps 261, 265n.; Tobias Smollett 186, 187; publication 187, 190; 'Smellie's Cutts' 230; *Treatise on the Theory and Practice of Midwifery* 261, 264–65; *A Collection of Cases in Midwifery* 264–65; *Praeternatural Cases in Midwifery* 264–65; John Gordon 217n.; John Moore 235
Smith, Adam 156
Smith, Archibald 340
Smith, Thomas 134 and n.
Smollett, Tobias George 185, 186–87, 189–90, 217n.; *Compendium of Voyages* 187; *Don Quixote* translation 187; *Gil Blas* 187; *History of the German Empire* 187; *Humphrey Clinker* 186; *Peregrine Pickle* 186–87; *Universal History* 187
Snodgrass, James 146
Society of Apothecaries: physicians and 21, 128, 135; general practitioners 351, 387; college certificates 240; midwifery

diplomas 254; Scottish reformers and 349–50, 376; legislation 376, 377–78, 379, 386, 390n.; *see also* apothecaries
Society of Barbers, Edinburgh 81n., 103, 104
Society of Regular Physicians and Surgeons, Glasgow 358
soldiers, billeting of 32, 33, 358
Soutar, John 128
South Africa 169, 198
Souttar, William 123
Spain 168, 193; Spanish regiments 41, 46n., 50
Spang, William, apothecary 9, 11, 12, 39–40, 85n., 86, 418
Spang, William (the younger) 40
Spanish Sickness (Lowe) 45–46, 46n., 49n.; *see also* venereal disease
Species plantarum (Linnaeus) 167–68
specimen collections, *see* anatomy and surgery; botany; museums; natural history; preparations; teaching aids
Stahl, Georg Ernst, *Observations* 164–65, 166, 180, 182, 309, 310
standards (medical) 339–48
Stanley, Mr (RCSEng delegate) 390n.
Stark, William, architect 238
Stark, William, surgeon 236–37
statistics 231–32, 314, 315, 411, 413
Steel, Dr John 363, 367, 364
stent 10 and n., 29, 31n., 32
Stephenson, George 248
Stevenson, Prof. Alexander 245, 318 and n., 319–20
Stewart, Sir Archibald 77n.
Stewart, Sir James 27n.
stillborn births, *see* midwifery
Stirling, Charles 289
Stirling, Elizabeth 268
Stirling, Helen (née Wemyss) 77, 295
Stirling, John (Glasgow University) 157, 160, 198
Stirling, John (Walter's son) 295
Stirling, John, bailie (John's son) 296, 298
Stirling, Walter (the first) 295
Stirling, Walter (William's son, hunchback) 295n.
Stirling, Walter, magistrate (John's son) 296
Stirling, William, industrialist 298–99
Stirling, William, surgeon: background 295–96, 295n.; Faculty and 98, 107–8, 110n., 113, 295, 296, 298; business partnerships 188, 297, 299, 302, 310; sick poor provision 298, 299, 302, 304; John Moore 188; Shawfield riots 293
Stirling & Sons 298
Stirling's Public Library 299
stitching, *see* suturing
stone-cutters, *see* lithotomy
Strabane 38n.
Strang, Dr 285
strangers (outside practitioners) 90 and n., 94n., 100–101, 104; *see also* freedom fines; quarter accounts
strong 'box' (Faculty) 88–89, 96, 106, 131
sudden deaths, *see* forensic medicine; murders
Sunderland 315
surgeon-apothecaries/pharmacists, *see* apothecaries and surgeon-apothecaries; surgeons
Surgeon-Visitor 86; *see also* Deacon; Visitor
surgeons: the term 'Chyrurgion' 43; status 17, 34–35, 40, 43, 45; 'pure surgeons' 7, 42; Surgeons of the Long Robe 116–17; military surgeons, *see* Army and Navy boards; military surgery; naval surgeons; early barber – surgeon associations 79–80, 147; separation of 81 and n., 103; for Glasgow, *see* barbers and barber-surgeons; Faculty of Physicians and Surgeons; Society of Regular Physicians and Surgeons 358; Edinburgh 2n., 81n., 103, 104; England 103, 128, 147; France 81n., 103, 176, 295; guild surgeon's work 79; master-surgeons 113; ruling on healing arts 368; Jenkins case (England) 128; licensing to practise 1, 8, 9, 117–24, 145, 210, 345, 346; apothecaries, *see* apothecaries and surgeon-apothecaries; midwifery 128, 254, 255, 273, 351; university training 155, 215; Latin literacy 145, 344, 389; Chirurgiae Magister, *see* CM degree; MD degrees 255, 273, 343, 367; *see also* anatomy and surgery; apprenticeship system; craft guilds; licensing to practise; Lowe, Peter

surgeons' clerks 322, 326, 336
Surgeon's Hall, Glasgow 297
Surgeon's Hall, London 368
Surgeons, Regular Society of Physician and, Glasgow 358
surgery, *see* anatomy and surgery; Lowe, Peter; military surgery; wound surgery
Surgery, Bill for Regulating the Practice of (1816) 369, 370
Surgical Anatomy of the Head and Neck, Observations on the (Allan Burns) 404–5
Surgions Mate, The (Woodall) 70n.
Sutherland, Prof. James 169, 170, 197, 198; *A Catalogue of the Plants in the Physic Garden at Edinburgh* 168, 198
Sutton, Hugh 110n.
suturing 55, 56, 62, 63, 64 and n., 66–68, 67; *see also* wound surgery
Swammerdam, Jan 163
Swan, Joseph, *View of the Cathedral, Infirmary* pl. 15
Swan, William 93
Swiss 154
Sydenham, Thomas 164–65, 166; *Medical Observations* 164, 170
syphilis, *see* venereal disease
System of Surgery (Hamilton, unpublished) 228–29
Systema naturae (Linnaeus) 167–68

Taggart, Mr (Society of Apothecaries) 390n.
taxation 10 and n., 32, 33
tea 167
teaching aids: in anatomy 216, 398; collections 237–39, 241; artistry 218–19, 241; engravings 215, 219; anatomical preparations, *see* preparations; Smellie's midwifery 'machines' 261–64, 267, 269, 271, 275, 280; museums 219, 232, 233, 238–39, 240, 241, 412; Anderson's College 186, 241, 412; *see also* anatomy and surgery; botany; clinical teaching; library resources; museums; natural history; skeletons
teaching gardens 197, 199–200; *see also* botany; materia medica; natural history
teaching hospitals 154, 274, 304, 410, 416; *see also* bedside teaching; clinical teaching; Edinburgh Royal Infirmary; Glasgow Royal Infirmary; lying-in hospitals; Town's Hospital
teeth extraction 62–64, 79 and n., 115
Telford, Thomas 247–48
Temperamentis et libris de symptomatum causis, De (Galen) 53
Tennant, Charles 203, 247–48, 289; Charles Tennant & Co. 244
Tennant, Captain David 248
Tennoch, W. 301
testimonials: petitioning to practise 8, 9, 136–37, 143–44, 143n.; from patients 123; as licences to practise 117; MD degrees 196; midwifery 266; William Hunter's 224–25, 311–12; *see also* licensing to practise; qualifications
Thoburn, John 203–4
Thomson, Andrew 320
Thomson, Dr George 293, 304
Thomson, James, editor 242
Thomson, James, surgeon 25n., 135
Thomson, Margaret 320
Thomson, Thomas 12, 85n., 101
Thomson, Prof. Thomas 209–10, 242, 246, 289
Thomson, William 105, 110n.
tickets for lectures 179
Tobias, James 136
Tobias, John 136
Tod, John 137
Todd, John 363
Todd, Sweeney 185
Todd, William 203–4
tolls exemption 358
tomatoes 198
tonsils 61–62, 62n.
tooth-pullers 62–64, 79 and n., 115
Torrell, David 147, 148
Torrell, Margaret (née Granfield) 147–48
Tournefort, Joseph Pitton de 167, 183; *Voyage into the Levant* 182
Towers, Prof James: CM degree 210, 270 and n., 364–65; midwifery 253, 254, 267, 270–71, 272, 275, 281, 286; Glasgow University lying-in ward 276–77, 278–79, 288; Glasgow Royal Infirmary 271, 323, 324

Towers, James junior 270n.
Towers, Prof. John 270n., 271, 277, 279, 288–89
Town's Hospital, Glasgow: *The Old Town's Hospital* (Fairbairn): pl. 4; philanthropy & clinical teaching 302–6, 308, 314, 326, 415; George Montgomerie 196, 251; Robert Cleghorn 308, 312, 326; William Cullen 308; William Hamilton 312; David Dale 320–21; no lying-in facilities 274; *see also* Glasgow Royal Infirmary
tracheotomy 61–62
Trades' Hall 246
Trades' House 20–21, 28, 81n., 82, 89n., 107, 288, 295; Crafts' Hospital 15, 83 and n., 84, 88; *see also* craft guilds; Faculty of Physicians and Surgeons
traditional healers 1, 9, 34, 150; *see also* irregular practitioners
Traité des rapports (Treatise on Reports)(Paré) 8–9
Travers, Benjamin 386, 389n.
Treatise on the Theory and Practice of Midwifery (Smellie) 261, 264–65
Treatment of the Diseases of Women and Children (John Burns) 272, 405
'Treaty of London' 385 and n., 386, 390; see also General Conference of Corporations
tree importation 201–2; *see also* botany
trepanation 70 and n., 79
Trew, Dr Johann Jacob 171
'trial for qualification' 94, 96–99, 99n., 156, 217, 251, 341; *see also* licensing to practise
tuberculosis 237, 344
Tübingen 127
Tucker, Thomas 14
tulips 167
tumours 53, 57–58, 215, 227
Turnbull, A., empiric 156n.
Turnbull, Andrew 234, 239
Turner, Dawson 209
turnips 201
typhoid 316; *see also* fevers
typhus 314, 316–17, 322; *see also* fevers

ulcers 70–71, 71n., 79, 119n., 227

Ulster Plantation 38–39 and n.; *see also* Ireland
unfreemen 81, 90; *see also* craft guilds; freemen
Union of England and Scotland (1707) 153, 158
Universal History (Smollett) 187
universities 125–28, 154–55, 181, 332, 340, 346–47, 399, 410–16; *see also* Aberdeen; Anderson's; Cambridge; CM degree; Edinburgh; Glasgow; MD degree; Oxford; qualifications; St Andrews
unmarried mothers 283–84
unqualified practice, *see* irregular practitioners; licensing to practise
Ure, Dr Andrew 239, 246, 247

vaccinations 232, 284
van Swieten, Gerhard 171, 309
Vancouver, Captain George 202
vascular preparations 405; *see also* preparations
vegetables 167, 198, 201; *Vegetable System, The* (Hill) 201
venereal disease 79, 146, 227; syphilis 45–46, 46n., 47n., 49n., 56, 58, 79n., 99n., 141, 344; Lock Hospital 315, 414
venesection, *see* blood-letting
ventousing 59, 61, 74, 75–76, 117; *see also* cupping
Vesalius, Andreas 40, 47; *De humani corporis fabrica* 71, 175
Viador, General Dondego de varro 50
Vienna 154, 171, 177–78
View of the Cathedral, Infirmary (Swan): pl. 15
View of the Hunterian Museum (Fleming): pl. 3
Vinniell, Andrew 297
Virginia 169
Visitors 8–10, 24, 82, 85–86, 85n., 88–89, 106–7, 135–36; *see also* craft guilds; Faculty of Physicians and Surgeons
visual aids, *see* teaching aids
Voltaire (François Marie Arouet) 186
voluntary charities, *see* poor, provision for; sick poor
Voyage into the Levant (de Tournefort) 182

Wade, General George 300
Wakely, Mr (Commons committee) 378
Wakley, Dr Thomas 352, 370
Walkinshaw, John 420, 422
Wallace, Elspeth (née Murray) 144, 146
Wallace, John 146
Wallace, Dr Michael 24, 138
Wallace, Robert: and Faculty 110n., 299; practice partnership 297, 299; sick poor provision 299, 302, 304; Glasgow Royal Infirmary 323, 324; on William Cullen 181
Walpole, Sir Robert, 1st Earl of Orford 300
Walpole, Spencer 394, 395
'wappen-shaw' (weapon-showing) 10 and n.
'war is the only real school for the surgeon' (Hippocrates) 69n.
Warburton, Mr, MP 211, 373
ward clerks 156
warding exemption 10, 33
warrants, *see* horning, letters of
watching exemption 10, 33, 358
watercolour sketches (anatomy) 218
Waterhough estate 204
Watson, Dr James 379–92 passim
Watt, Scott and, architects 240
Watt, Prof. George 316, 339, 344, 381, 400
Watt, James 243–44, 248
Watt, Dr James 364, 367
Watt, Robert: pl. 11, 230–33, 241–42, 281, 315, 331; *Catalogue of Medical Books* 231, 233; *Inquiry into ... Principal Diseases of Children* 231–32; 'Principles and Practise of Medicine' 231
wax preparations 177–78, 239
weapon-showing 10 and n., 33
Webster (Edinburgh lecturer) 174
Weems, Helen 77, 295
Weems: Grahame, Weems & Grahame 381n.
Weir, Dr Daniel 378
Weir, James 25n., 30–31 and n., 105, 132
Weir, John, apothecary 134–35
Weir, John, messenger 89–90
Weir, Dr William 316, 402
Wemyss, Earl of, *see* Elcho, Lord
Wemyss, Rev. David 295
Wemyss, Helen 77, 295
Wengal (instrument maker) 280

West Indies plants 168
Western Medical Club 316
Whole Course of Chirurgerie, The (Lowe), *see*: Chirurgerie
Whyt, John 129
Whyte, Robert 309 and n.
widows 82, 146
Widow's Fund (Faculty) 315, 353–55, 379, 414
wigs, *see* periwig-making
Wigton, Earl of 198, 297
Williams, Dr R. C. 385n., 390n.
Wilson, Alexander 140, 141, 423
Wilson, Charles 324
Wilson, Gilbert 120
Wilson, James, barber 87, 118
Wilson, James, medical superintendent 281, 282–83, 285–86, 290, 291
Wilson, James George 282–83
Wilson, Nathan 217
Winslow, Jacques-Bénigne 176, 178
Wisdom of God Manifested in the Works of Creation, The (Ray) 161, 163–64
'wise women' 145; *see also* female practitioners; irregular practitioners
Wiseman, Richard 37n.
witches 49; *see also* irregular practitioners
Wodrow, Elizabeth 159
Wodrow, Prof. James 158–59, 160, 199
Wodrow, James, librarian 160
Wodrow, Janet (née Luke) 160, 199
Wodrow, Dr John 158, 159–60, 199–200, 206, 251, 304
Wodrow, Robert 157–59, 160–61, 199
Wolsey, Cardinal Thomas 7n.
women, diseases of 279, 284, 287
Women and Children ..., Diseases of (John Burns) 272–73, 405–6
Women and Children, Treatment of the Diseases of (John Burns) 272, 405
women and medicine, *see* female practitioners; irregular practitioners; midwifery
Women's Diseases, Booke of (Lowe) 49n., 53
Wood, Dr Andrew 382n., 385n., 389, 390n., 391, 395
Wood, John 241
Woodall, John 70; *The Surgions Mate* 70n.

workhouses 303, 304; *see also* poor, provision for; sick poor
working-class education 241–42, 361; *see also* Birkbeck; mechanics' institutes
Worshipfull companie of Chirurgians 44n.
wound surgery 66–70, 78, 79, 227; *see also* anatomy and surgery; suturing
Wright, James 253
Wright, Janet 253
Wright, Dr Peter 319, 342
Wrightman, General 297

York 144n.

Young, Dr 316
Young, Archibald 319, 324
Young, Isabell (née Crawford) 149–50
Young, J., student 225, 226
Young, James, apprentice 94
Young, Prof. Thomas 236, 252n., 259–67 passim, 275–76
Young, William 149
Younger, Thomas 135

Zair, W. 422
Zeluco: Various Views of Human Nature (Moore) 189
Zurich 174